nbranes

tic

l

HANDBOOK OF BIOLOGICAL PHYSICS

Series Editor: A.J. Hoff

Volume 1:	Structure and Dynamics of Me
A:	*From Cells to Visicles*
B:	*Generic and Specific Interaction*
Volume 2:	Transport Processes in Eukary and Prokaryotic Organisms
Volume 3:	Molecular Mechanisms in Visu Transduction

HANDBOOK OF BIOLOGICAL PHYSICS

VOLUME 3

Molecular Mechanisms in Visual Transduction

Editors:

D.G. Stavenga

Department of Neurobiophysics, University of Groningen
The Netherlands

W.J. DeGrip

Department of Biochemistry, University of Nijmegen and
Department of Biophysical Organic Chemistry
University of Leiden
The Netherlands

E.N. Pugh Jr

Department of Ophthalmology and Institute of Neurological Sciences
University of Pennsylvania, Philadelphia
USA

2000

ELSEVIER

Amsterdam - London - New York - Oxford - Paris - Shannon - Tokyo

ELSEVIER SCIENCE B.V.
Sara Burgerhartstraat 25
P.O. Box 211, 1000 AE Amsterdam, The Netherlands

First edition 2000

Library of Congress Cataloging in Publication Data
A catalog record from the Library of Congress has been applied for.

ISSN: 1383-8121
ISBN: 0 444 50102 9

∞ The paper used in this publication meets the requirements of ANSI/NISO Z39.48-1992 (Permanence of Paper).
Printed in The Netherlands.

General Preface

Biological Physics encompasses the study of the processes of life with physical concepts and methods based on the laws of nature, which are assumed to be equally valid for living and dead matter. A multidisciplinary approach brings together elements from biology – knowledge of the problem that is attacked – and from the physical sciences – the techniques and the methodology for solving the problem. In principle, Biological Physics covers the physics of all of biology, including medicine, and therefore its range is extremely broad. There is a need to bring some order to the growing complexity of research in Biological Physics, to present the experimental results obtained in the manifold of its (sub)fields, and their interpretation, in a clear and concise manner. The Handbook of Biological Physics answers this need with a series of interconnected monographs, each devoted to a certain subfield that is covered in depth and with great attention to the clarity of presentation. The Handbook is structured so that interrelations between fields and subfields are made transparent. Areas, in which a concentrated effort might solve a long-standing problem, are identified. Evaluations are presented of the extent to which the application of physical concepts and methodologies (often with considerable effort in terms of personal and material input) have advanced our understanding of the biological process under examination.

Individual volumes of the Handbook are devoted to an entire "system" unless the field is very active or extended (as e.g. for membranes or vision research), in which case the system is broken down into two or more subsystems. The guiding principle in planning the individual volumes is that of going from simple, well-defined concepts and model systems on a molecular and (supra)cellular level, to the highly complex structures and functional mechanisms of living matter. Each volume contains an introduction defining the (sub)field and the contribution of each of the following chapters. Chapters generally end with an overview of areas that need further attention, and provide an outlook into future developments.

The first volume of the Handbook – Structure and Dynamics of Membranes – deals with the morphology of biomembranes and with different aspects of lipid and lipid–protein model membranes (Part A), and with membrane adhesion, membrane fusion and the interaction of biomembranes with polymer networks such as the cytoskeleton (Part B). The second volume – Transport Processes in Eukaryotic and

Prokaryotic Organisms – continues the discussion of biomembranes as barriers between the inside of the cell and the outside world, or between distinct compartments of the cellular inner space, across which a multitude of transport processes occur.

The present volume extends the scope of the previous volumes to the molecular mechanisms of phototransduction in vertebrates and invertebrates. The molecular properties and the primary photoreactions in rhodopsin are treated in depth. The structure and comparative molecular biology of numerous visual pigments are discussed, and the microvillar and ciliary photoreceptors of invertebrates are examined and compared. Furthermore, modeling approaches of the visual processes in vertebrate and insect photoreceptors are extensively reviewed.

Rhodopsin, as one of the best studied G-protein coupled receptors (GPCR), has become a touchstone for research in other areas of signal transduction. We expect, therefore, that the present volume will provide a valuable reference source for workers in these related fields.

Planned volumes

The "bottom-up" approach adopted for individual volumes of the Handbook, is also the guideline for the entire series. Having started with two volumes treating the molecular and supramolecular structure of the cell, Volume 3 is the first of several volumes on cellular and supracellular systems. Volume 4, to be published early in 2001, is on Neuro-informatics – neural modelling and information processing. The next two planned volumes are on Molecular motors as chemo-mechanical transduction devices, and on Biological electron transport processes. Further planned volumes are:

- Vision – perception, pattern recognition, imaging
- The vestibular system
- Hearing
- The cardio-vascular system, fluid dynamics and chaos
- Electro-reception and magnetic field effects

Further volumes will be added as the need arises.

We hope that the present volume of the Handbook will find an equally warm welcome in the Biological Physics community as the first two volumes, and that those who read these volumes will communicate their criticisms and suggestions for the future development of the project.

Leiden, Fall 2000
Arnold J. Hoff
Editor

Preface to Volume 3 Molecular Mechanisms in Visual Transduction

Visual transduction is presently one of the most intensely studied areas in the field of signal transduction research in biological cells. Because the sense of vision plays a primary role in animal biology, and thus has been subject to long evolutionary development, the molecular and cellular mechanisms underlying vision have a high degree of sensitivity and versatility. The aims of visual transduction research are first to determine which molecules participate, and then to understand how they act in concert to produce the exquisite electrical responses of the photoreceptor cells.

Since the 1940s [1] we have known that rod vision begins with the capture of a quantum of energy, a photon, by a visual pigment molecule, rhodopsin. As the function of photon absorption is to convert the visual pigment molecule into a G-protein activating state, the structural details of the visual pigments must be explained from the perspective of their role in activating their specific G-proteins. Thus, Chapters 1–3 of this Handbook extensively cover the physico-chemical molecular characteristics of the vertebrate rhodopsins. Following photoconversion and G-protein activation, the phototransduction cascade leads to modifications of the population of closed and open ion channels in the photoreceptor plasma membrane, and thereby to the electrical response. The nature of the channels of vertebrate photoreceptors is examined in Chapter 4, and Chapter 5 integrates the present body of knowledge of the activation steps in the cascade into a quantitative framework.

Once the phototransduction cascade is activated, it must be subsequently silenced. The various molecular mechanisms participating in inactivation are treated in Chapters 1–4 and especially Chapter 5.

Molecular biology is now an indispensable tool in signal transduction studies. Numerous vertebrate (Chapter 6) and invertebrate (Chapter 7) visual pigments have been characterized and cloned. The genetics and evolutionary aspects of this great subfamily of G-protein activating receptors are intriguing as they present a natural probe for the intimate relationship between structure and function of the visual pigments. Understanding the spectral characteristics from the molecular composition can be expected to progress soon from the stage of ordering the visual pigments into spectral classes based on their known primary structure, towards quantitative predictions, with quantum chemical theory.

Although the rhodopsins of vertebrates and invertebrates appear to have much in common (Chapters 6 and 7), their phototransduction processes have diverged. A survey of visual transduction in invertebrates, together with the morphological

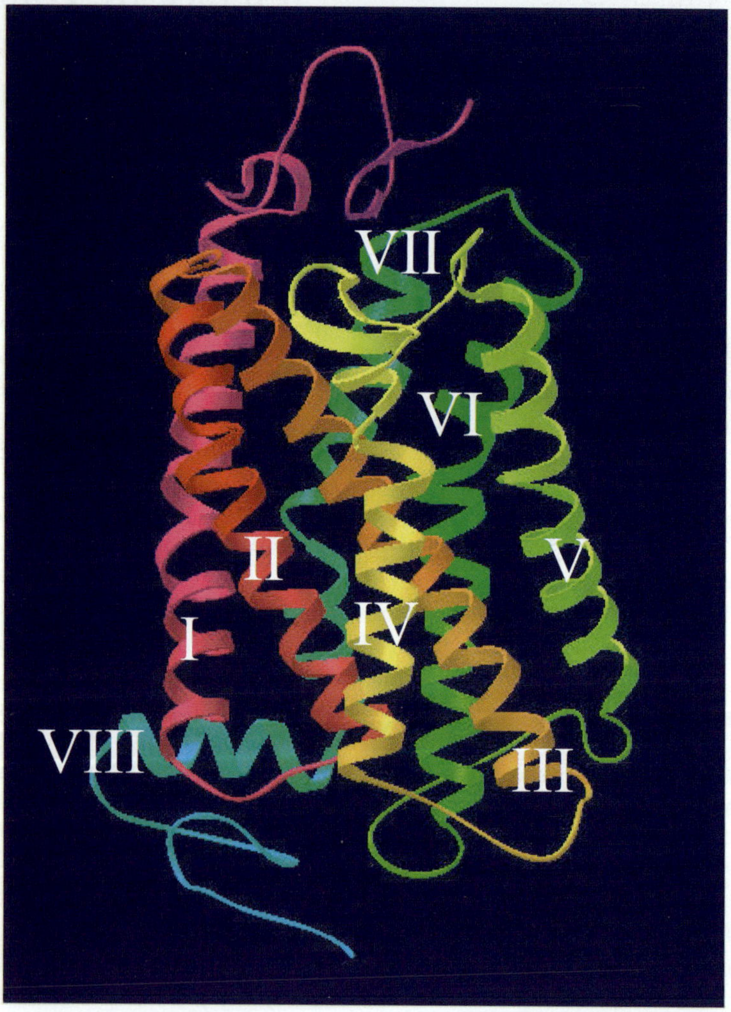

Ribbon-structure of bovine rhodopsin based upon the recently published crystal structure [2]. This ribbon-structure was generated at the CMBI, Nijmegen (www.cmbi.kun.nl). The N-terminus is at the top, the intracellular surface at the bottom. Note the amphipathic helix conformation of the "8th" helix (bottom) and the remarkable slant of the third transmembrane helix.

differences of the photoreceptor cells, is provided in Chapter 8. Chapter 9 focuses on the fruitfly *Drosophila*, which is not only a classic preparation for the investigation of neural development, but also a rich source for molecular and physiological studies of visual transduction. Finally, Chapter 10 reviews a number of cases where quantitative, modeling approaches have been performed on insect photoreceptors. As is shown in particular in Chapter 5, such an analysis demonstrates that crucial insights into visual transduction can be gained this way.

The relevance of visual pigments and visual signal transduction extends beyond the field of vision. As one of the best studied G-protein coupled receptors (GPCR), rhodopsin and its signaling mechanism have become a prototype for the entire GPCR field. The relative high levels of transduction components in the photoreceptor outer segment has advanced identification and purification of key elements in GPCR transduction, including the G-protein, arrestin, G-P kinase, recoverin, phosducin and RGS-family (Chapter 5). The low-resolution structure of rhodopsin has inspired many protein modelers, seeking a better molecular grip on their favorite GPCR. In this context the very recent publication of the first crystal structure of bovine rhodopsin at better than 3 Å resolution again is a major step forward [2]. The transmembrane α-helices and the intradiscal (extracellular) connecting loops now have been quite well defined (see figure), allowing reliable modeling of the membrane domain of other GPCRs (Chapter 1).

Hopefully this volume will provide a valuable reference source for a wide audience, not only students and investigators interested in vision, but also for those working in neighbouring areas. Our understanding of visual transduction has now reached a stage where most of the instruments playing in the orchestra are known. We are getting sufficiently familiar with this elegant music, so that its basic themes are becoming well recognizable. But much of its depth and beauty remains to be savored.

References

1. Hecht, S., Shlaer, S. and Pirenne, M.H. (1942) J. Gen. Physiol. **25**, 819–840.
2. Palczewski, K., Kumasaka, T., Hori, T., Behnke, C.A., Motoshima, H., Fox, B.A., LeTrong, I., Teller, D.C., Okada, T., Stenkamp, R.E., Yamamoto, M. and Miyano, M. (2000) Science **289**, 739–745.

Doekele G. Stavenga
Willem J. DeGrip
Edward N. Pugh Jr

Contents of Volume 3

Contributors to Volume 3

W.J. DeGrip, Department of Biochemistry UMC-160, Institute of Cellular Signalling, University of Nijmegen, P.O. Box 9101, 6500 HB Nijmegen and Department of Biophysical Organic Chemistry, Gorlaeus Lab, University of Leiden, P.O. Box 9502, 2300 RA Leiden, The Netherlands

M. del Pilar Gomez, Department of Physiology, Boston University School of Medicine, Boston, MA 02118, USA

W. Gärtner, Max-Planck-Institut für Strahlenchemie, Stiftstrasse 34–36, D-45470 Mülheim an der Ruhr, Germany

R.C. Hardie, Department of Anatomy, University of Cambridge, Cambridge, CB2 3DY, UK

K.P. Hofmann, Institute for Medical Physics and Biophysics, Charité Medical School, Humboldt University, Berlin, Germany

U.B. Kaupp, Forschungszentrum Jülich, Institut für Biologische Informationsverarbeitung, 52425 Jülich, Germany

T.D. Lamb, Department of Physiology, University of Cambridge, Downing Street, Cambridge CB23EG, UK

J. Lugtenburg, Leiden Institute of Chemistry, 2300 RA Leiden, The Netherlands

R.A. Mathies, Chemistry Department, University of California, Berkeley, CA 94720, USA

B. Minke, Department of Physiology, The Kühne Minerva Center for Studies of Visual Transduction, Hadassah Medical School, The Hebrew University, Jerusalem 91120, Israel

R.S. Molday, The University of British Columbia, Department of Biochemistry, 2146 Health Sciences Mall, Vancouver, BC Canada V6T 1Z3

E. Nasi, Department of Physiology, Boston University School of Medicine, Boston, MA 02118, USA

J. Oberwinkler, Department of Neurobiophysics, University of Groningen, Nijenborgh 4, NL-9747 AG Groningen, The Netherlands

R. Payne, Department of Biology, University of Maryland, College Park, MD 20742, USA

M. Postma, Department of Neurobiophysics, University of Groningen, Nijenborgh 4, NL-9747 AG Groningen, The Netherlands

E.N. Pugh Jr, F.M. Kirby Center for Molecular Ophthalmology, Department of Ophthalmology and Institute of Neurological Sciences, Stellar-Chance Laboratories, University of Pennsylvania, 422 Curie Boulevard, Philadelphia, PA 19104-6069, USA

K.J. Rothschild, Department of Physics and Molecular Biophysics Laboratory, 590 Commonwealth Avenue, Boston University, Boston, MA 02215, USA

D.G. Stavenga, Department of Neurobiophysics, University of Groningen, Nijenborgh 4, NL-9747 AG Groningen, The Netherlands

R. Yokoyama, Department of Physiology, State University of New York Health Science Center at Syracuse, 750 East Adams St., Syracuse, NY 13210, USA

S. Yokoyama, Department of Biology, Biological Research Laboratories, Syracuse University, 130 College Place, Syracuse, NY 13244, USA

CHAPTER 1

Structure and Mechanism of Vertebrate Visual Pigments

W.J. DEGRIP

Department of Biochemistry,
University of Nijmegen and
Department of Biophysical Organic Chemistry,
University of Leiden

K.J. ROTHSCHILD

Department of Physics and
Molecular Biophysics Laboratory,
Boston University

Handbook of Biological Physics
Volume 3, edited by D.G. Stavenga, W.J. DeGrip and E.N. Pugh Jr

Contents

1. Introduction

1.1. General features of visual pigments

1.1.1. Common structural elements

All visual pigments thus far discovered belong to the super-class of hepta-helical membrane proteins. The bovine rod pigment rhodopsin was the first representative of this superfamily to be recognized. Its amino acid sequence was first elucidated by direct protein sequencing [1,2] and later from its cDNA sequence [3]. This hepta helical membrane protein family exhibits a broad variety of functional properties, ranging from ion translocation to signal transduction. However, they all share the common structural motif predicted by their amino acid sequences of seven, largely α-helical, transmembrane segments (TM-domain), which are connected by peptide sequences of variable structure, located outside of the membrane (loops; Fig. 1). The TM-domain folds to present a binding site for a large variety of ligands. In the case of vertebrate visual pigments, the ligand consists of either 11-*cis* retinal (an A_1 retinoid) or 11-*cis* 3,4-dehydroretinal (an A_2 retinoid), which are derived from all-*trans* retinol (vitamin A) or the provitamin β-carotene.

Visual pigments all belong to the G-protein coupled receptors (GPCRs), the largest family of hepta-helical proteins. All GPCRs share common sequence motifs, which are structurally or functionally relevant [4–6]. In addition, there exists a variety of common post-translational modifications including N-glycosylation at Asn residues in the extracellularly located N-terminal region; disulfide bridge formation, linking TM-helix III and the loop between IV and V; and thiopalmitoylation of Cys residues in the intracellularly located C-terminal region [7,8]. The number of thiopalmitoylation sites varies within sub-families of visual pigments, with cone pigments exhibiting zero to one and rod pigments up to two sites [9–11]. Among the G-protein coupled receptors the visual pigments exhibit a unique common structural element: their ligand is covalently attached in the binding site. The linkage consists of a protonated Schiff base with a lysine residue. This configuration is stabilized by a nearby carboxylate anion. This capacity to bind the ligand covalently through a retinylidene Schiff base enables a variety of elegant studies with ligand analogs, that can also be accommodated in the retinal binding pocket [12–14]. The molecular weight of the vertebrate visual pigments varies between 39 and 42 kDa, including ligand and post-translational modifications.

1.1.2. G-protein coupled receptor family (GPCR)

Studies of rhodopsin, a member of the GPCR family, present a unique opportunity both at a basic research level and for pharmacological applications in the treatment of human diseases. The GPCR family comprises a large number of receptors with

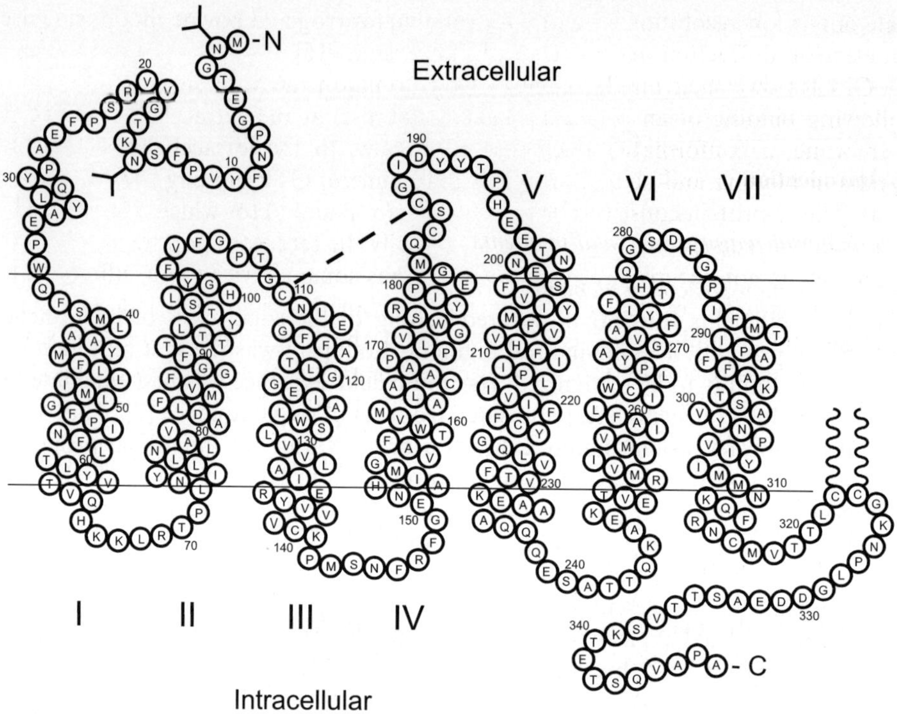

Fig. 1. Amino acid sequence and putative topography of bovine rhodopsin. The arrangement of loops and transmembrane (TM) segments is based on the current state of evidence discussed in the text. Horizontal lines represent the membrane interface. The numbers I–VII indicate the seven TM segments. The corresponding intracellular helical extensions outside the membrane are designated helix A–G, respectively (see text). The following post-translational modifications are indicated: glycan moieties at Asn-2 and Asn-15 in the N-terminal region, disulfide bridge between Cys-110 and Cys-187, and thiopalmitoyl esters at Cys-322 and Cys-323. Numbers indicate protein residue number. Amino acid residues are indicated according to the one-letter convention (A = Ala, etc.). Extracellular corresponds to intradiscal.

hormones, neurotransmitters, peptides, glycoproteins and even small ions as their ligand. Because GPCRs mediate a broad array of important physiological processes they are frequently the target of pharmacological intervention [4,5,15]. The various GPCR sub-classes all contain a variety of isoforms with a different pharmacological profile. Isoform-specific drug design and protein engineering for this receptor family would thus benefit strongly from 3D information on receptor structure and its mode of interaction with its ligands at atomic resolution. Due to their high concentration and specific location in the characteristic ciliary outgrowth of the photoreceptor cell (the outer segment), visual pigments virtually are the only GPCRs, that have thus far been isolated and purified from native sources in sufficient amounts to allow detailed structural studies by biophysical techniques, like crystallography, Fourier transform infrared (FTIR) spectroscopy and NMR spectroscopy. Nevertheless, to date only a low resolution structure is available for frog and bovine rhodopsin based on electron diffraction and imaging of 2D crystals [16].

GPCRs owe their family name to their common mode of signal transduction; following binding of an activating ligand (agonist) at the extracellular side of the membrane, a conformational change propagates to the intracellular side, which results in binding and activation of a heterotrimeric GTP-binding protein (G-protein). The G-protein consists of three subunits (α, β and γ) for which a large number of partially tissue-specific isoforms exist. Usually the receptor apoprotein, where the ligand site is not occupied (R state), already has some activity ("low-affinity state" or "basal activity" or "constitutive activity"), but this is strongly enhanced upon agonist-binding and subsequent formation of the "high-affinity" or "activated" state (R-A or R*). In addition, the activity of the receptor or the effect of an agonist can be modulated by other ligands. For instance, partial agonists only activate some of the receptor functions; antagonists counteract the effect of agonists, but do not affect the basal activity, while inverse agonists counteract the effect of agonists and reduce the basal activity as well.

The evolution of GPCRs towards a function in vision required additional specialization [6,8,17,18]. Proper functioning under dim light conditions ("scotopic vision") requires very low noise levels, i.e., a very low basal activity of the receptor. Rapid signal turnover, essential for the necessary fast image processing, requires rapid activation and inactivation kinetics. These two requirements were elegantly realized by occupying the binding site of the receptor in the "dark state" with a photoactive molecule, sensitive to visible light, which acts as an inverse agonist: 11-*cis* retinal. This combination strongly enhances the thermal stability of the protein [19] and very effectively reduces basal receptor activity to such an extent that the rod photoreceptor cell can detect a single photon [20]. Upon absorption of light the 11-*cis*, inverse agonist configuration of the ligand is converted to the all-*trans* configuration, which acts as a full agonist, triggering receptor activation by at least six orders of magnitude (cf. Chapters 3–5 and 8 in this volume). Owing to the inherent high rate of photochemical processes and the fact that ligand-conversion occurs *in situ*, receptor activation does not depend on diffusional processes and is both very efficient and complete within several milliseconds at physiological temperature. In addition, the binding mode of the ligand, a protonated, hence positively charged, Schiff base conjugated with the polyene skeleton, allows modulation of its ground and excited state electronic levels by the electronic charge distribution in the binding site. Hence, through proper selection of relevant residues, the absorbance band of the visual pigment can be shifted over the entire visible range ("spectral tuning") while maintaining the required receptor function (cf. Chapters 6 and 7 in this volume). This is a very important feature, since combining pigments with different spectral sensitivity allows color discrimination (photopic vision). Since the ligand in visual pigments is essential for their spectral properties in the visible region of the spectrum, it often is referred to as the "chromophore".

A major experimental advantage in the study of photosensitive receptors is the ability to synchrously activate an entire population by a light pulse, thereby allowing time-dependent studies. In addition, light activation can be performed at low temperatures, allowing intermediate states to be trapped and their structure characterized. In this way it should be possible to map the structural steps involved in receptor activation at atomic resolution. Thus far, structural details about the pathway for receptor activation are only available for visual pigments and in particular the bovine rod visual pigment rhodopsin. However, it is likely that similar structural studies will become possible for other GPCRs. Recently larger quantities of these proteins can be expressed in heterologous systems, permitting more detailed analysis through mutagenesis and non-radioactive isotope labeling [21–23]. Hence, it is expected that the paradigms evolving for the visual pigment family can soon be tested in other receptors to assess their general value.

1.1.3. Vertebrate versus invertebrate visual pigments

As outlined in Chapters 6 and 7 of this volume, vertebrate and invertebrate visual pigments share a common ancestor and thus show a fair degree of sequence and structural homology [24]. On the other hand, the invertebrate pigments use a signal transduction pathway, as discussed in detail in Chapters 8–10 of this volume, which is more closely related to that of other "non-visual" GPCR's than to that of the vertebrate pigments. In addition, the invertebrate pigments do not release their ligand subsequent to photoactivation, but use instead a reverse photochemical process, which is activated in a different spectral region, in order to return to their "dark state". These elements are discussed in Chapters 7 and 10 of this volume. Furthermore, their chromophore selection also includes the 3- and 4-hydroxyderivatives of 11-*cis* retinal. Finally, invertebrate pigments become very unstable upon solubilization in detergent solution, which strongly hampers analysis or purification. So far no successful strategies have been reported allowing *heterologous* expression in a functional state. As a consequence, knowledge of structure and structure–function relationships of invertebrate visual pigments is quite limited at present in comparison to their vertebrate counterparts.

1.2. Aim of this review

This review will focus on biophysical studies of the structure of vertebrate visual pigments and the conformational changes accompanying photoactivation. Most of these studies have been performed on the bovine rod visual pigment (rhodopsin). While the resulting description probably is applicable to other receptors in a more global sense, details are likely to diverge already between visual pigments (e.g., cone pigments versus rod pigments). It should also be recognized that despite current progress, there are still significant gaps in our knowledge, especially at the level of atomic resolution. Nevertheless, we will attempt to forge the available evidence into a comprehensive and coherent picture of the receptor activation process. As we fill in disputed or blank elements, this will unavoidably include some subjective or provocative interpretations.

The following sections will discuss relevant data obtained by means of UV/visible spectroscopy, solid-state magic-angle spinning NMR spectroscopy, ESR-spectroscopy and FTIR spectroscopy, respectively. Where appropriate, data obtained in mutagenesis studies will also be included. The results will be discussed in the context of our present knowledge about the rhodopsin structure. Structural data on the invertebrate pigments can be found in Chapter 7 of this volume.

2. Light-triggered steps in rhodopsin activation (photointermediates)

2.1. General aspects

The first step in the photoactivation of visual pigments, the photo-induced isomerization of the ligand, presents some fascinating photochemistry, that triggers the cascade of structural transitions producing the active receptor conformation. The combination of protein and bound chromophore has evolved in remarkable symbiosis to fine-tune the photoisomerization reaction to near perfection. In model compounds, even in solutions approaching the alleged dielectric conditions in the protein interior, 11-*cis* retinal and derivatives exhibit quantum yields from only 0.05 to 0.15, producing a mixture of stereoisomers with kinetics of several pico-seconds [25,26]. In contrast, photoisomerization in rhodopsin proceeds with an unprecedented quantum yield of 0.67, i.e., every two out of three photons absorbed are effective in activating the ligand and thereby triggering receptor activation [27]. Equally surprising, the reaction is fully stereospecific, producing only the all-*trans* configuration, and is complete within 200 fs [28], making it the fastest photochemical reaction observed to date. Clearly, a special interreactive mechanism must underlie this process. This is discussed in detail in Chapter 2 of this volume.

It was recognized by visual inspection over a century ago that upon light exposure the hue of an isolated frog retina rapidly discolors from red to light-yellow [29,30]. This process has therefore often been referred to as "bleaching", and more recently and more appropriately as "photolysis". It has now been established that the light-yellow state represents not only an intermediate step in a complex photolytic sequel but the active receptor state for the visual transduction process, as well. Several decades ago a number of intermediate steps in this photocascade was already identified by their characteristic spectral absorbance bands. This was possible because the photocascade can be fully or partially blocked at characteristic transition temperatures allowing the UV/visible absorption spectrum of intermediates to be obtained by cryospectroscopy [31,32]. The sequence of intermediates discovered in this way is shown in Fig. 2A and represents the classic photocascade. Subsequent rapid progress in kinetic spectroscopy down to femtosecond resolution made it possible to study photoreceptor activation under more physiological conditions. This work confirmed the basics of the classic scheme, but increased its complexity by adding new intermediates (Fig. 2B) (see Section 2.2). As it stands now, the transition steps in the photocascade are well established up to the Meta I intermediate. However, kinetic analyses have not yet yielded a fully consistent picture for the subsequent steps.

2.2. Static versus kinetic spectroscopy

The classic picture for the photocascade presented in Fig. 2A was constructed by means of cryospectroscopy, analyzing spectral properties and single transitions

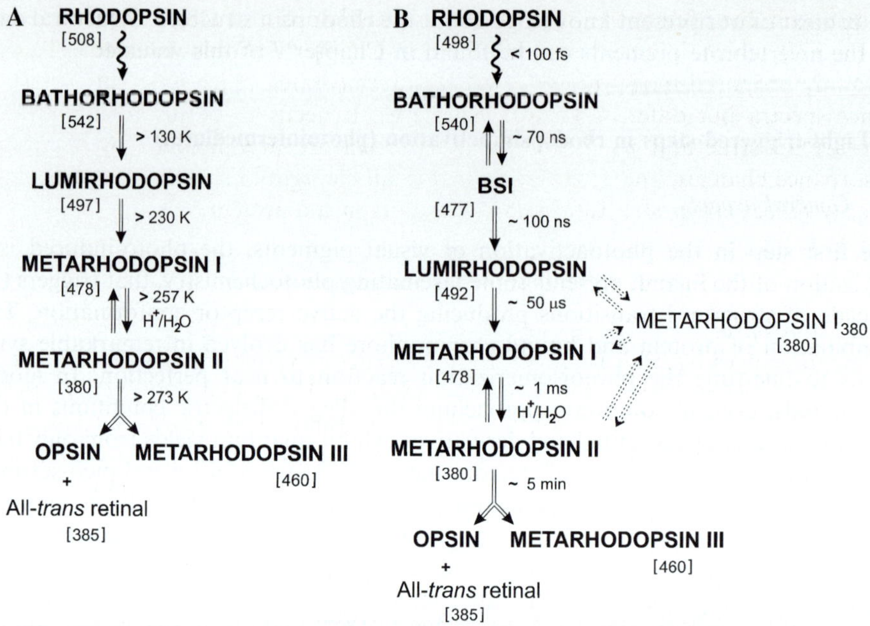

Fig. 2. Intermediates in the rhodopsin photocascade. (A) The photointermediate sequence according to cryospectroscopic studies ("Classic scheme"; see text). Numbers between parenthesis indicate the λ_{max} of that intermediate below the transition temperature. The transition becomes allowed at the indicated temperature. (B) The photointermediate sequence according to kinetic spectroscopy ("Kinetic scheme", see text). BSI stands for blue-shifted intermediate (see text). Intermediates and transitions, which also have been structurally resolved, are shown in bold and with continuous arrows, respectively. The pathways involving metarhodopsin I_{380} need to be further elucidated. Numbers between parenthesis indicate the λ_{max} of that intermediate at 20°C. The other numbers give approximate half lives of the corresponding intermediate at 20°C.

at specific temperatures [17,31,33–36]. In the scheme obtained [rhodopsin ⇝ batho(rhodopsin) → lumi(rhodopsin) → meta(rhodopsin) I ↔ meta(rhodopsin) II → meta(rhodopsin) III and opsin] only the first step is light dependent; the subsequent ones are thermodynamically driven. The transition from Meta I → Meta II is accompanied by a large absorbance change from orange to light-yellow. Meta II represents the active receptor, which binds and activates the G-protein (cf. Chapter 2 of this volume). This thermal trapping approach has some disadvantages, however. It basically is restricted to intermediates having increasing enthalpy barriers, thereby obscuring transitions with a significant entropic component, and may be complicated by phase transitions in the lipid matrix or thermal effects on the protein structure [32,33]. Indeed, kinetic spectral studies, performed in the range of room temperature to physiological temperatures with microsecond to picosecond time resolution, have revealed that the photolytic cascade probably is more complicated than suggested by cryospectroscopy, in particular during the Batho → Lumi transition and in the Meta transitions [33,37]. A putative scheme [33,38–40] is presented in Fig. 2B which is based on global analysis [39]. It should be noted that the currently proposed model may not reflect intermediates which have similar absorbance spectra but different protein structures. It needs to be further established, whether all structural transitions in the protein are accompanied by significant absorbance changes, and vice versa whether all changes in spectral properties necessarily reflect larger structural rearrangements in the protein.

2.3. *Transitions and spectral properties of photointermediates*

2.3.1. *Rhodopsin*

As mentioned above, the visible spectral properties of vertebrate visual pigments are due to the presence of a ligand or chromophore, 11-*cis* (3,4-dehydro)retinal [17] covalently bound in a protein binding pocket. Binding of dehydroretinal (fairly common in fish and amphibians) results in a red-shifted absorbance spectrum relative to retinal itself (cf. Chapter 6 of this volume). Protein residues in the binding site interact with the chromophore to modulate its electronic properties. In this way a visual pigment can "select" the position of its major absorbance band in the visual spectrum of light (spectral tuning). Taking only 11-*cis* retinal based pigments into account, the following spectral distributions are found: the absorbance band of most rod pigments is centered close to 500 nm [17,18] with some scattered outliers in the range between 480–490 and 505–510 nm (cf. Chapter 6 in this volume). In cone pigments several distinct sub-families can be distinguished, that show relatively high sequence homology and cover specific regions of the visible spectrum [41–47]. The UV/blue sensitive short-wavelength (SW) subgroup covers the range 360–430 nm. The blue-sensitive medium-wavelength (MW1) sub-group is spectrally quite homogeneous: 440–460 nm. The blue/green sensitive medium-wavelength (MW2) subgroup spans a range of 465–510 nm. Finally, the green/red sensitive long-wavelength (LW) sub-group covers the range 508–570 nm. The spectral range of the LW group is quite broad, because the more red-sensitive pigments contain a specific anion-binding site, that under physiological conditions binds chloride, and red-shifts the absorbance spectrum. This red-shift can amount to ca. 30 nm.

At least seven protein residue positions have already been identified which contribute to spectral tuning of visual pigments. Their individual effects are to a large extent additive, in particular within a sub-group. Molecular descriptions of wavelength tuning have been proposed [48–52], though an accurate theoretical explanation will most likely depend on the availability of high resolution structures of the binding site in the various pigment sub-groups. Details and references are given in Chapters 2, 6 and 7 of this volume.

2.3.2. *Batho(rhodopsin)*

As mentioned before, the formation of this first ground-state product of photoexcitation of visual pigments involves fascinating photochemistry with unprecedented kinetics, as extensively discussed in Chapter 2. This intermediate stores 32–35 kcal/mole of the photon energy [53] in a highly twisted all-*trans* chromophore. Vibrational spectra are comparable at ambient and cryotemperatures [54], suggesting that at this stage only small local changes in protein backbone structure occur, which are most likely largely restricted to protein residues lining the chromophore binding site [55,56]. Evidence has already been presented for "perturbation" of a cysteine, a tryptophan and a tyrosine residue, presumably all located in TM segment VI [57–59] and for participation of Glu113 and Gly121 in TM segment III [60,61]. This is consistent with findings, that the C-3 region of the cyclohexene ring is in close contact with TM VI [62] and upon photoactivation moves closer to TM III [62,63]. In addition, rearrangement of water molecules in the Schiff base region [57,64–67] probably facilitates changes in the protein–chromophore interaction.

The chromophore photoisomerization path is likely to be trapped in Batho in a high-energy intermediate state due to steric interaction with the protein. The contact sites serve to translate the conformational energy, transiently stored in the chromophore, into protein structural changes which eventually uncover latent interaction sites for the G-protein transducin at the intracellular surface. The 9- and 13-methyl groups of the chromophore seem to participate in such energy-transfer. The 9-desmethyl analog of bovine rhodopsin exhibits a significantly altered photocascade, which lacks a well-defined Batho intermediate [36,68] and shows alterations in the subsequent intermediates, as well as perturbed signaling activity [68–70]. The 13-desmethyl analog adopts a chromophore configuration which exhibits a twist different from the native chromophore for the dark-adapted state as well as the Batho intermediate [71–73], but then proceeds to a fairly normal active state [71,74]. Adding a methyl group at the 10-position increases the thermal barrier for decay of Batho [75]. Evidently, chromophore and opsin binding site have been remarkably fine-tuned for optimal energy-transfer [76–79].

2.3.3. Blue-shifted-intermediate (BSI)

Athough low temperature data do not provide a clear indication for the existence of an intermediate between Batho and Lumi in native rhodopsin, evidence for such an intermediate first was found from studies of the rhodopsin analog, 5,6-dihydro-isorhodopsin [55,80] and for cyclohexene-ring-modified analogs [81,82]. Subsequent time-resolved studies on native rhodopsin could be interpreted by assuming a Batho ↔ BSI equilibrium, with BSI decaying to Lumi [82,83]. This equilibrium is strongly temperature dependent, and at the low temperatures where Batho is trapped in cryostudies, the BSI component would be undetectably small. Hence, structural information on BSI is very limited.

The concept that formation of BSI involves the first stage of relaxation of a steric interaction between the twisted chromophore and the protein, is supported by several observations. First, formation of BSI is much more rapid in the 13-desmethyl rhodopsin analog [84] and BSI may be stabilized at low temperatures in this analog [85]. In addition, the formation rate of BSI is affected by increasing substituent volume at the 9-position of retinal, or at position Gly-121 of rhodopsin [61,86,87], proposed to be located close to the 9-methyl-group [70]. The available evidence suggests that formation of BSI is accompanied by small structural transitions in the protein, which allow repositioning of the C13–C15 section of the polyene chain [56,84,88–90] and re-orientation of the polyene chain with respect to the ring segment. This latter aspect agrees with a report, that a 6-S-*cis*-locked chromophore which strongly restricts flexibility of the chain relative to the ring segment, does not proceed beyond Batho and thermally reverts to the parent pigment [36,91]. Interestingly, a similar behavior is observed in long-wavelength-sensitive (LWS) cone pigments like the chicken red cone pigment iodopsin [92–94] and enhanced by chloride binding [95]. Binding of a chloride ion to LWS pigments effectuates appreciable red shifts in the absorbance spectrum [96–99], and also induces subtle changes near the C14 position of the chromophore in the Batho state [100]. This could indicate, that binding of chloride to a specific binding site in the second extracellular loop [98,101] increases the thermal barrier for Batho decay by rendering the protein structure near the end of the polyene chain of the chromophore more rigid and/or more compact. Time-resolved spectroscopy, on the other hand, reveals a normal Batho → BSI → Lumi sequence for all cone pigments studied at ambient temperatures [35,96,102–105], except that the lifetime of the Batho intermediate is much shorter, while that of BSI is similar to the corresponding intermediate of bovine rhodopsin. Apparently, Batho decay in cone pigments has a much larger temperature coefficient than in rod pigments. The present evidence suggests that the first structural rearrangements in the protein, together with the initial relaxation of the twisted all-*trans* chromophore occurs during formation of BSI, and that this change relieves a constraint for subsequent formation of Lumi.

2.3.4. Lumi(rhodopsin)

Evidence indicates that formation of Lumi involves a major relaxation of the chromophore strain (see [56] and Section 4.4.2) with a small increase in reaction volume (ca. 30 ml/mol) relative to rhodopsin [56,106,107]. The reversal of the sign of the predominant CD band from strongly negative in Batho to slightly negative in Lumi [108–110] along with the re-orientation of the major absorption dipole closer to its orientation in rhodopsin [89,111–113] also reflects this relaxation. Nevertheless, there is no evidence for larger conformational changes at the protein level [58,89,108,114,115], suggesting that these have not yet propagated far from the binding site and are still largely confined to the environment of the chromophore [116]. This agrees with the fact that Lumi can form at low temperature (130 K) [31,117] as well as in extensively dehydrated photoreceptor membrane [66,118–120], conditions which do not allow much conformational flexibility of the protein. This picture of a Lumi intermediate consisting of a largely relaxed chromophore and structural rearrangement of the binding-site which has absorbed most of the stored photon energy also agrees with several other observations [33]. For instance, in contrast to Batho, Lumi is not easily photoconverted back to rhodopsin. Instead, illumination produces a mixture of several *cis*-isomers of retinal [32,115]. Furthermore, in the case of the 10-methyl rhodopsin analog, the Batho intermediate is stable up to a much higher temperature (180 K) and retains all the characteristics of chromophore strain until it decays into Lumi [75].

2.3.5. *Meta(rhodopsin) I, II and III*

While a one-step transition from Lumi to Meta I_{480} was suggested by low temperature studies on bovine rhodopsin, similar analyses of chicken rhodopsin indicated that the situation is more complex [121]. Time-resolved analyses at ambient up to physiological temperatures also can only be interpreted by assuming a more complex situation [33,38,40,122–124]. Although several schemes have been presented, most require an intermediate absorbing at 380 nm, that is in equilibrium with Lumi and precedes Meta I_{480}; it therefore has been dubbed Meta I_{380}. Global analyses indicate that Meta I_{480} only would be a minor intermediate at temperatures above ambient. Nevertheless, bulkier retinal analogs (10-methyl-, 12-methyl-) stabilize a 480 nm-absorbing intermediate with all characteristics of Meta I_{480} up to at least 30°C [75]. This does not really comply with parallel pathways originating from Lumi. Further studies will be needed to clarify the interrelationship between Meta I_{480} and Meta I_{380} and to determine if these states represent structurally distinct intermediates in the photocycle or rather spectral isoforms of the same global protein conformation, only differing in very local structural elements controlling Schiff base protonation [125].

The Meta I → Meta II transition is the first one that does not proceed under dehydrated conditions [66,118,120,126], and depends on a variety of micro-environmental conditions like pH, surface potential, lipid unsaturation, pressure and lipid-to-protein ratio [114,127–151]. This transition is accompanied by a positive reaction volume of 108 ml/mol, deprotonation of the Schiff base and uptake of a proton from the cytosol [124] (cf. Chapter 3 in this volume), which largely explains the dependence on its micro-environment. Deprotonation and proton uptake can be uncoupled by solubilization in detergents [152]. This slows down the latter process, but in the membrane-bound state deprotonation is the rate-determining step. Hence, the Meta II_a → Meta II_b transition has been introduced, representing the deprotonated and the deprotonated, acidified isospectral forms, respectively. Schiff base deprotonation explains the large bathochromic shift (480 → 380 nm) between Meta I_{480} and Meta II. All available evidence indicates that Schiff base deprotonation is accompanied by protonation of its counterion Glu-113 [126,153].

Mutation of an essential glutamate residue (E134 in bovine rhodopsin) abolishes external proton uptake [154], but also increases the basal signal transduction activity of the apoprotein (constitutive activity) and the activity induced by free all-*trans* retinal [155–160]. Mutation of several histidine residues, in particular H211, also affects the Meta I ↔ Meta II equilibrium [161,162]. Cone pigments do not clearly exhibit a pH-dependent Meta I ↔ Meta II equilibrium, possibly because of a strong upshift in pK_a [35,99,163]. They do not have the equivalent of H211, but do contain the equivalent of E134. However, cone pigments have a much higher isoelectric point, and consequently a quite different charge distribution, and, in contrast to rod pigments, a net positive charge at physiological pH [164]. In addition, cone pigments exhibit a much shorter half-life of Meta II [35,99,165–167]. This has been connected to the absence of the glutamate residue at position 122 ([167], however see [99,168]), which is fully conserved only in rod pigments. There is increasing evidence that the surface charge strongly affects the pK_a of the Meta I ↔ Meta II equilibrium as well as the rate of Meta II formation [138,147,151,162,169,170]. One attractive mechanism to explain the above-mentioned observations is the existence of (an) H-bonded network(s) in rhodopsin, that act(s) as a structural element as well as helps convey the receptor activation signal from the ligand site to the intracellular membrane surface [171]. Such a network can easily undergo rearrangements through alteration in the internal hydrogen bonding pattern or through proton uptake and re-orientation of bound water molecules and would be expected to be sensitive to surface charge and/or membrane voltage. Several lines of evidence support the presence of such a voltage-sensitive hydrogen-bonding network in bacteriorhodopsin [172–178].

The Meta transitions and their relation to signal transduction are extensively discussed in Chapter 2. Here we will restrict ourselves to structural data. On the basis of volume changes, the formation of Meta II appears to involve the largest conformational change in the photocascade of rhodopsin. This is also suggested by FTIR data (see Section 4.4.4) and surface plasmon resonance studies [179,180]. Ample evidence has been provided for participation of aromatic residues in these conformational rearrangements in the protein [165,181–184]; some of them have been mapped to TM's III and VI [185]. Such changes should serve to expose binding sites for the G-protein transducin at the cytoplasmic segments of the receptor [144,186–188]. In addition to protein conformational changes, several lines of evidence indicate that this structural activity includes the chromophore, even as late as the Meta II step of receptor activation. For instance, chromophore homologs (10-methyl- or 12-methyl-) strongly retard or completely inhibit formation of Meta II [75], while deletions in the cyclohexene ring may abolish Meta II formation [189,190]. In addition, studies of oriented systems provide evidence for a re-orientation of the chromophore during the Meta I $\rightarrow$ Meta II transition [191,192].

Decay of Meta II into Meta III represents the first slow transition in the photocascade (half life of ~5 min at ambient temperature) and is accompanied by a nearly complete loss of signaling activity. The nature of the conformational changes accompanying this transition are not well understood. Although no direct evidence exists, the position of the absorbance band of Meta III (450–460 nm) suggests that it contains an unperturbed protonated Schiff base linkage with retinal. For instance, acid denaturation of rhodopsin generates a product absorbing at ca 445 nm [193,194]. Sulfhydryl activity [195], accessibility of a specific epitope [188] and FTIR analyses [196,197] indicate, that the protein largely refolds upon decay of Meta II into a conformation resembling that of rhodopsin. Regeneration studies demonstrate that Meta III binds 11-*cis* retinal and forms a pigment with spectral properties identical to rhodopsin [194,198]. Such data strongly indicate that the decay of Meta II involves the transfer of all-*trans* retinal from its original binding site to another, possibly non-specific site, along with the partial formation of a protonated Schiff base. Indeed, chemical analysis reveals phospholipids as well as protein amino groups as potential acceptor sites [194], while in Meta II retinal is not in close contact with lipids [199].

This interpretation is difficult to reconcile, however, with reports that Meta III is in equilibrium with Meta II [191,200], and might represent a temporary storage form to replenish Meta II inactivated by phosphorylation and arrestin binding. First of all, all-*trans* retinal cannot easily re-enter the opsin binding site [201,202]. In addition, it has only recently been established that the addition of all-*trans* retinal enhances the basal signaling activity of opsin [202–206]. This may have previously led to erroneous conclusions as to reformation of an active state (Meta II; 380 absorbance band very similar to all-*trans* retinal [128]) from a Meta III-like state. Upon addition to membranes, retinal easily and randomly forms Schiff bases with available amino groups, that under physiological pH are partially protonated and have absorbance bands peaking in the range 420–470 nm [193,207–209]. Finally, it should be realized that, while Meta III is quite stable *in vitro*, this is not the case *in vivo* where the retinal is transferred to a retinol oxidoreductase and reduced to all-*trans* retinol or vitamin A_1 [210–214]. This would not support a temporary storage function for Meta III. Since FTIR evidence also agrees with a very similar structure for opsin and Meta III (Section 4.4.5), we favor the interpretation that the decay of Meta II into Meta III or opsin releases the retinal from its binding pocket.

3. Probing structure and mechanism of rhodopsin through electron diffraction, and polarized light-, NMR- and ESR-spectroscopy

3.1. Electron diffraction and molecular modeling: Rhodopsin's global shape

Almost all high-resolution structural data of proteins have been obtained via X-ray diffraction of suitable 3D-crystals. While over a 1000 structures of soluble proteins have been solved in this way (cf. Protein database at http://www.rcsb.org/pdb/ or Swiss-Prot database at http://www.expasy.ch/sprot/sprot-top.html), very few integral membrane proteins are amenable to such analysis due to difficulties of forming 3D-crystals. Instead, most structures of membrane proteins solved at medium-to-high resolution have been obtained through electron diffraction or imaging of 2D crystals [215–218]. This technique generates a 2D-projection structure, from which, a 3D-structure can be constructed by repeating the analysis at various angles of the crystal relative to the electron beam.

In the case of visual pigments this approach has thus far succeeded in producing a medium resolution (5–10 Å) structure of bovine, frog and squid rhodopsin [16,24,219,220]. Significantly, this work is the first medium resolution structure available for any GPCR and provides the first experimental confirmation for the presence of seven transmembrane α-helices in this receptor family [16].

Most helices transverse the membrane at an angle which significantly deviates from the membrane normal (Fig. 3), in agreement with the average angle of 35–40° predicted form FTIR analysis of oriented membranes [192,221–223]. It can be estimated that about 50% of the protein mass resides in the membrane, which is in good agreement with earlier neutron diffraction studies [7,224,225]. Combining these projection data with sequence and residue conservation and mutation effects in a large number of GPCR's, Baldwin [226,227] has proposed a general model for the arrangement and orientation of the transmembrane α-helices in the GPCR 7TM-complex. This model has laid the basis for several refined molecular modeling studies, employing a variety of programs to describe molecular interactions and

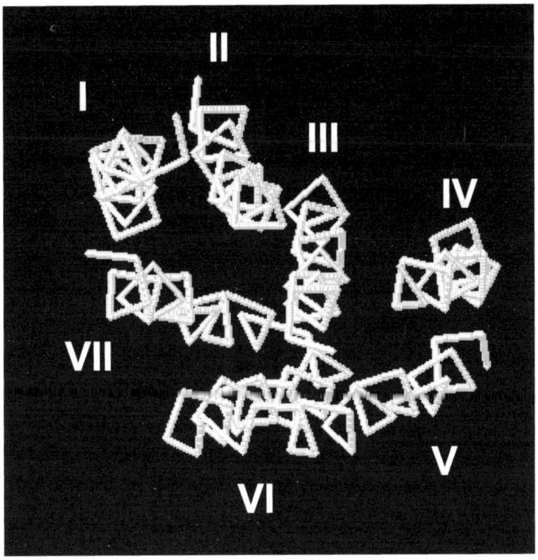

Fig. 3. Orientation of the transmembrane segments of bovine rhodopsin. The data is based on 2D projection structures and adapted from Baldwin [227]. Rhodopsin is viewed from the intracellular side. The numbering of the transmembrane segments corresponds to Fig. 1.

perform energy minimization [13,65,228–231]. However, a more detailed model which reveals atomic resolution information on the position of chromophore and protein residues is still difficult to achieve on the basis of the present structural information. For the time being, data from non-diffraction techniques will be essential to improve current models. For example, all models comply with a negative helicity in the C11–C13 segment of the chromophore, proposed on basis of earlier exciton coupled CD studies [232]. However, recent *ab initio* calculations of CD-spectra support the presence of a positive helicity in this segment of the chromophore ([233], Buss, V. unpublished). These first *ab initio* calculations of CD-spectra seem to be quite reliable, as they also agree with the highly twisted structure proposed by *ab initio* Car-Parinello calculations for the chromophore in Batho [234,235].

3.2. Polarized light: Relative orientation of structural elements

Intact rod outer segments can be oriented in a magnetic field due to the stacking of disc photoreceptor membranes in the rod, the intrinsic orientation of α-helices in rhodopsin and the diamagnetic anisotropy of peptide groups [236]. This has been put to elegant use in early studies with polarized light (Linear Dichroism, LD) [237,238]. First it was determined that the major absorption dipole of the chromophore was oriented nearly parallel to the membrane plane (angle of ca. 18° [239–241]). This is nearly optimal for light absorption in the intact eye, where the incoming light beam runs more or less parallel to the long axis of the photoreceptor outer segments. At the same time it was shown that vertebrate rhodopsin has a high translational diffusion constant and rotational relaxation time (ca. 5.10^{-9} cm^2/s^{-1} and 20 μs, respectively [242 245]). This is in contrast to arthropods and insects, where visual pigments are immobilized in the rhabdomeric membrane, which in principle allows detection of changes in light polarization (cf. Chapters 7 and 8 of this volume). LD-studies also provided the first evidence for conformational re-orientation in protein and chromophore upon formation of Meta II [191,246]. In another interesting application using magnetic orientation of rod outer segments, disk and plasma membrane lipids of the outer segment as well as the individual lipid species could be resolved by ^{31}P-NMR spectroscopy [247]. This also provided evidence that the rim of the disk membranes acts as a barrier for translational diffusion of lipids.

In addition to the use of magnetic fields to orient rod outer segments, an iso-potential spin-dry technique was developed in 1980, that allowed a film of highly oriented membranes to be deposited on any suitable substrate [248]. This approach has been used to study the orientation of rhodopsin structural elements, using a variety of polarization sensitive spectroscopic methods [192,223,249–251]. Such oriented films confirmed the specific orientation of the chromophore [223,249], and provided the first convincing evidence for an oriented α-helix bundle in the trans-membrane domain of rhodopsin [222,223] (see Section 4.2). Furthermore, this approach allowed the first orientational analysis of structural elements, including the chromophore, that participate in the conformational changes accompanying formation of Meta II ([192] and Section 4.5.6). More detailed analyses using orientation-dependent NMR and FTIR should be possible on rhodopsin and other GPCR members once a more detailed 3D structure will be available.

3.3. NMR-spectroscopy: High-resolution details

Although the molecular weight range for the 3D-structure determination of soluble proteins by narrow-bore high-resolution NMR spectroscopy has steadily increased over the last decade to the 30–40 kDa range [252–254], most membrane proteins such as rhodopsin can still not be analyzed using this technique. Upon solubilization by detergents, mixed micelles with molecular weights over 100 kDa are formed, and their slow rotational diffusion results in broad line-widths. In contrast, components of membrane proteins such as terminal domains and hydrophilic linkage regions are still accessible to this approach. For example, using a pars-pro-toto approach, Yeagle and coworkers [255–257] analyzed the structural preference of peptides corresponding to the C-terminal and intracellular loops of bovine rhodopsin. Remarkably, all peptides exhibit a high degree of organization, possibly due to the presence of stable β-turns which restrict their conformational freedom and show a high proportion of β-structure. A high content of β-structure in the domains outside the membrane also agrees with FTIR studies on the distribution of secondary structure in rhodopsin [258,259]. In an attempt to obtain a representation of the structure of the intracellular receptor domain of rhodopsin, the derived peptide structures were docked together and fitted onto the transmembrane segments

[260,261], the position of which was derived from 2D crystals [220]. Distances in this structure were compared to those obtained by other approaches (see Section 3.4) and showed global agreement [261], though the average accuracy is probably not better than 3–5 Å, still far from a reliable high-resolution structure. These distance data actually suggest that the structure, produced with this combinatorial solution NMR approach, resembles a photoproduct of rhodopsin rather than the dark state.

High-resolution ^{19}F-NMR studies have been performed on intact detergent-solubilized rhodopsin, using either fluorine-labeled retinals [78,262–264] or modification of single-cysteine mutants with fluoro-reagents [265]. The latter study provides a means of studying structural perturbations upon light activation at specific positions in rhodopsin, which can complement ESR studies (see Section 3.4). Single-cysteine substitutions at positions 67, 140, 245, 248, 311 and 316 in the cytoplasmic face in rhodopsin enabled targeted modification with a trifluoroethylthio (TET), CF_3–CH_2–S, group. All mutants showed chemical shifts downfield of free TET (6.5 ppm). Upon illumination a metarhodopsin II-like intermediate was formed, and upfield changes in chemical shift were observed at positions 67, 140 and downfield changes for positions 248 and 316. Upon decay of metarhodopsin II, the chemical shifts reverted largely to those originally observed in the dark. These results fully confirm earlier ESR studies (see Section 3.4). However, direct information about local structure and mobility will require a more detailed theoretical interpretation of fluorine chemical shifts.

In contrast to solution NMR studies, solid-state NMR spectroscopy is not restricted by molecular size and offers the ability to obtain high-resolution structural data of membrane proteins like the members of the GPCR family [266–270]. The broad lines normally present in solid-sate NMR due to dipolar coupling and chemical shift anisotropy can be dramatically narrowed by applying rotation around the magical angle (magic angle spinning) at high speed (5–15 kHz). In addition, the use of higher field strength, wide-bore magnets (750 MHz is the latest achievement), in combination with the development of highly sophisticated pulse sequences and multi-dimensional analysis, will soon allow the application of methods for macromolecular structure analysis comparable to those previously developed for solution NMR.

In early 1D measurements, solid-state NMR was used to probe chromophore properties *in situ* after incorporation of ^{13}C-labeled retinal [266]. This confirmed the 11-*cis* configuration of the chromophore and provided evidence for an S-*cis* configuration around the C6–C7 bond linking the cyclohexene ring and polyene chain [271,272], as was later corroborated by analog pigment studies [91]. The chemical shift of the carbon atoms at the end of the polyene chain (C11–C15) deviates significantly from model retinylidene Schiff bases [271,273]. The observed pattern can be simulated by positioning a negative charge in the vicinity of the C12 atom at a distance of about 3 Å [274,275]. This result nicely agrees with a model based on two-photon spectroscopy [276] and with the identification of a glutamate residue (E113) as the counterion for the protonated Schiff base by means of site-directed mutagenesis [277–279]. The recent ^{15}N solid-state NMR data obtained of ^{15}N-lysine labeled rhodopsin, the first NMR studies of a stable-isotope labeled eukaryotic membrane protein, directly confirms the presence of a protonated Schiff base [22,23]. In addition, since the ^{15}N chemical shift of a protonated Schiff base is strongly dependent on the distance to its counterion [280–282], this value could be derived quite accurately (4.3 ± 0.2 Å) for rhodopsin (Fig. 4), by comparison with spectral properties and chemical shifts of model compounds [22,23]. 1D solid-state NMR measurements have also been performed on photointermediates of rhodopsin, providing direct evidence for an unprotonated Schiff base in Meta II [283], in agreement with earlier resonance Raman studies (Chapter 2 of this volume). Remarkably, the chemical shift differences with model compounds are largely preserved in the chromophore of Batho, suggesting a similar electron density profile and perturbation at C12/13 as in rhodopsin [284].

State-of-the-art solid-state NMR techniques such as 1D or 2D rotational resonance (distance measurements [285–288]) and double-quantum heteronuclear field spectroscopy (bond torsional angle [289]), can generate high resolution structural information. Such studies, using multiple ^{13}C labeled chromophores in the case of rhodopsin, have generated the first detailed structural data on a GPCR ligand. Measurements of distances between C10–C20 and C10–C13 [79] and of the C10–C11 torsional angle [290,291] clearly demonstrate, that the C10–C13 region of the chromophore has a highly twisted conformation (angle of 40–50°). This twist is most likely due to sterical interaction between the 10-H proton and 20-methyl group, since it is reduced in the 13-desmethyl pigment [72] and increases to 60–70° in the 10-methyl pigment analog [79]. Such a twisted conformation probably has an important function in predisposing the excited chromophore towards a highly efficient isomerization pathway (cf. Chapter 2). Both distance and torsional angle data are in full agreement with a fully relaxed all-*trans* chromophore at the Meta I stage [79,291].

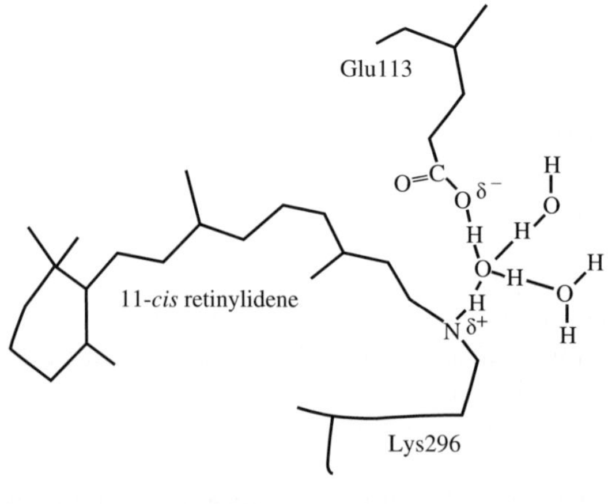

Fig. 4. Model of the protonated Schiff base region in the chromophore binding site. For simplicity, double bonds in the polyene chain have been omitted. This model is based on evidence obtained through ^{15}N-NMR analysis [22]. The 20-methyl group of the chromophore is twisted out of the plane of the paper towards the reader.

Another promising development in NMR spectroscopy is the use of oriented membranes. Two-dimensional orientation results in a significant reduction in chemical-shift anisotropy and hence in line-width. In order to implement this approach for the case of narrow-bore solution NMR spectroscopy, lipid–detergent combinations are utilized that form large micellar structures (bicelles), which adopt a preferential orientation in a magnetic field [267,292–295]. While this approach can orient "trapped" soluble proteins or small incorporated transmembrane peptides, the use with larger membrane proteins has not yet been reported. Another approach (magic angle oriented sample spinning, MAOSS) combines solid-state spectroscopy with membranes oriented on a suitable substrate [250,268,296,297]. This approach has already been applied to integral membrane proteins. If one can utilize the high intrinsic rotational diffusion rate of an uniaxially oriented membrane protein, low rotor spinning speeds already suffice to achieve a very high resolution, even allowing proton spectroscopy [250,298,299].

In the case of rhodopsin, oriented membranes combined with ^{2}H-labeling of the chromophore have been used to obtain structural information about the orientation of the chromophore [251,300,301]. This work supports earlier data showing that the chromophore exists in a twisted conformation, and provides also information on the orientation of the chromophore with respect to the membrane. Although the error still is substantial (± 5 Å), detailed conclusions based upon modeling were presented. The model suggests a 6-S-*trans* configuration of the chromophore [301], in contrast to earlier results, mentioned above. This feature clearly needs more detailed investigation. The results further indicate substantial rotation of the retinal plane during the photocascade ($\sim 55°$), and agree with a fully relaxed all-*trans* chromophore at the Meta I state.

3.4. Mutagenesis and ESR studies: Local structure

Site-directed ESR probes (nitroxide spin labels attached to a single cysteine residue introduced by site-directed mutagenesis) can report on mobility (line-widths in the ESR spectrum), accessibility (collision with hydrophilic or hydrophobic paramagnetic probes like oxygen) and secondary structure (periodicity of first two parameters) at specific positions in the protein structure [302–304]. This approach can thereby provide very useful structural information and can directly report on local changes in backbone conformation or dynamics.

Site-directed ESR labeling, for instance, confirmed the extended α-helical structure in the second and third intracellular loops [305,306], indicated a very compact structure for the fourth loop [307], and detected a high mobility in the C-terminal sequence [308,309]. This approach was also used to probe the interface between membrane-embedded and water-exposed structural elements [305,307,310,311].

Site-directed ESR probes can also be used to study structural changes which occur in rhodopsin upon photoactivation. For example, generation of the Meta II state induces selective changes in ESR-spectra, indicative of a movement of the α-helical segments in the second and third cytoplasmic loops and a rearrangement in the fourth loop [306,312,313]. These changes largely reverse upon decay of Meta II, in agreement with FTIR data (see Section 4.4.5). Furthermore, the interaction of two site-directed spin labels allows distance measurements [303,304] and permits more refined analysis of conformational changes. Such studies suggest a "rigid-body" movement of helix F with respect to C upon formation of Meta II [305,314] with a concomittant movement of the first and fourth cytoplasmic loops away from each other [315]. The reactivity pattern of single-cysteine mutants supports such a "rigid-body" movement [316], and rearrangement of α-helical elements is also indicated by a recent LD–FTIR study [192]. Photoactivation seems to involve a highly concerted cooperative effect of defined structural elements. Stochastic coupling of these events may explain the low basal activity of the apoprotein, and partial coupling may underly the activity of constitutively active mutants [144,156,159].

In an interesting approach, Albert et al. [317] used MAS ^{31}P-NMR to estimate the distance between a spin label attached to Cys-140 in bovine rhodopsin (second intracellular loop) and phosphate residues incorporated by means of rhodopsin kinase onto Ser338 and Ser343 (C-terminal region) by virtue of the relaxation enhancement induced by the spin label on the phosphate resonances. The estimated distance (15–20 Å) is not very accurate but provides very useful limits for modeling of the cytoplasmic surface of rhodopsin.

It should be noted, however, that ESR scanning mutagenesis has certain limitations. For example, since the approach requires modification of the protein with a fairly bulky reagent, tightly packed regions in transmembrane domains are usually not accessible. In addition, the spin label may not exactly reflect the position or properties of the residue it is attached to, due to its size and potential to disturb the microenvironment of the position probed. For this reason, the design of more compact ESR probes, which are incorporated as site-specific non-native amino acids via tRNA mediated protein engineering (TRAMPE), is highly desirable [318,319].

Global folding of the polypeptide backbone, the proximity of structural elements and structural changes during photoactivation can be probed by altering or introducing linkage sites (e.g., disulfides groups) via site-directed mutagenesis. For example, mutation of the cysteine residues participating in the disulfide linkage (C-110 and C-187 in rhodopsin) demonstrated that this link is not absolutely essential in opsin folding, but markedly contributes to the stability of the photoactive state (Meta II) and of Meta III [320,321]. In the reverse case, global topographical information can be obtained by probing the proximity relationship between two cysteine residues, introduced at selected positions, through their ability to form a disulfide bridge [322]. For example, this approach provided evidence for the proximity of the C-terminal to the helical extension of the sixth TM segment in the third cytoplasmic loop, and of the first and fourth cytoplasmic loops in the dark state of rhodopsin. In both cases the ability to produce a disulfide group diminished upon photoactivation [309,315]. The same approach also suggests close proximity of the second cytoplasmic loop to the cytoplasmic end of TM V, which, however, moved towards each other upon photoactivation [323,324]. These disulfide links usually inhibit structural rearrangements, required for activation of the G-protein [324]. A comparable, reversible cross-link approach utilizing site-directed mutation to incorporate histidine residues allowing cross-linking with bivalent ions like Zn^{2+} has also been reported. Cross-links could be generated between cytoplasmic ends of TMs III and VI (residues 138–251 and 141–251), which also blocked activation of transducin [325]. This again supports proximity of these loop segments and their structural involvement in exposing binding sites for transducin. No cross-links could be established between the cytoplasmic ends of TMs III and VII, however [325].

Proper rhodopsin folding is very sensitive to mutations and deletions in the intradiscal domain of rhodopsin [326,327]. Insertion of a c-myc epitope confirmed this sensitivity of the intradiscal face to structural perturbation [328]. On the other hand, insertional or deletional mutagenesis had much less effect on folding or stability of the resulting opsins when directed towards the cytoplasmic face, even when insertion into the second or third cytoplasmic loop affected or even abolished activation of transducin [328–330]. This suggests that the intradiscal domain has a very compact structure with quite stringent spatial conditions and might, for instance, serve as an anchor site allowing the light-induced structural changes to propagate to the cytoplasmic surface.

Points of close contact between protein and chromophore have also been scanned using a combination of modeling, mutagenesis and functional studies. The functional deficiencies induced by removal of a methyl group on the chromophore in 9-desmethyl-rhodopsin could be partially rescued by increasing the size of the Gly residue at position 121 in TM III of bovine rhodopsin (Gly → Ile, Leu or Trp) [70]. In a series of elegant studies, bulky substitutions at position 121, that strongly perturb functional activity of rhodopsin (partial agonist activity of 11-*cis* retinal, increased activity with free all-*trans* retinal), could be partially rescued by a concomittant size-reduction of the Phe residue at position 261 in TM VI [70,331–333]. Such data indicate a close structural relation and strong functional interaction of the 9-methyl group in the chromophore and elements in TM III and VI. This would agree with the "rigid-body" movement of the corresponding helices mentioned above. Mutation of Met-257, one helix turn away from Phe-261, resulted in a significant increase in constitutive activity of opsin, as well as a strong increase in activity-enhancement by free all-*trans* retinal [160]. This was interpreted as evidence for a specific interaction of Met-257 with the highly conserved NPXXY motif in TM VII (residues 302–306), that would function to restrict movement of helix F in the absence of photo-activated chromophore [160].

Several elegant mutagenesis studies have addressed how critical Schiff base formation and protonation, and counter-ion location are for proper structure and function of bovine rhodopsin. Mutation of Lys-296 into a smaller neutral amino acid did not inhibit binding of 11-*cis* retinal, but induced constitutive activity in the dark state, unless an entire retinylidene Schiff base was incorporated [155,156,334]. On the other hand, an aspartate or glutamate residue introduced at position 117, one helix turn away from Glu-113 in TM III, or at position 90 in TM II, opposite Glu-113, rescues the wild-type spectral properties of Glu-113 mutants [88,335–338]. Functional properties are only partially preserved, however, in these counter-ion analogs. Meta II-like photointermediates retain a protonated Schiff base at physiological pH, decay only slowly and have impaired signaling activity

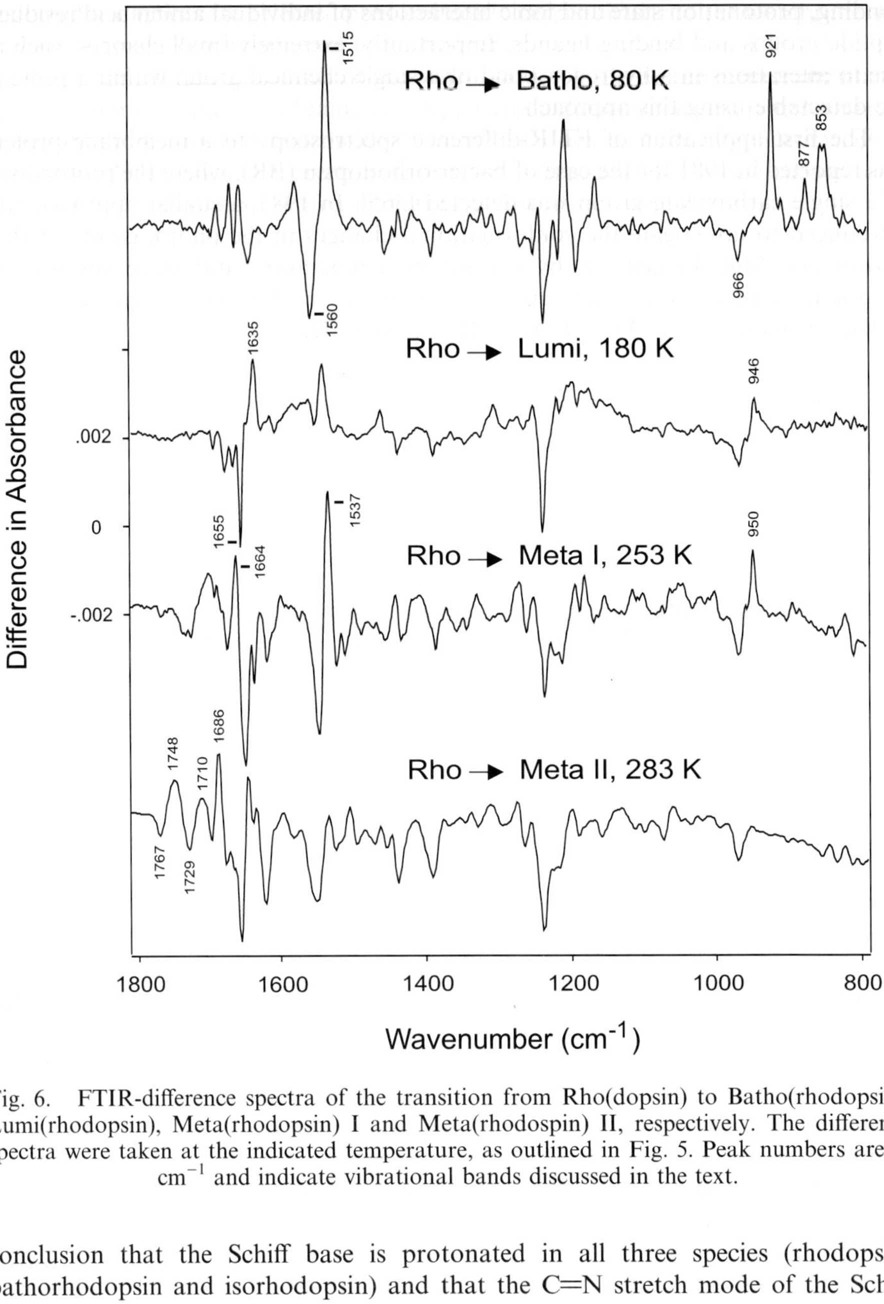

Fig. 6. FTIR-difference spectra of the transition from Rho(dopsin) to Batho(rhodopsin), Lumi(rhodopsin), Meta(rhodopsin) I and Meta(rhodospin) II, respectively. The difference spectra were taken at the indicated temperature, as outlined in Fig. 5. Peak numbers are in cm^{-1} and indicate vibrational bands discussed in the text.

conclusion that the Schiff base is protonated in all three species (rhodopsin, bathorhodopsin and isorhodopsin) and that the C=N stretch mode of the Schiff base appears at a similar frequency [346,367].

An important question relates to the response of the protein to retinal isomerization during the Rho → Batho transition. In order to address this question, rhodopsin was regenerated with isotopic labeled retinals [114] such as the 11,12-

dideuterio label in order to shift the major ethylenic bands (1560 and 1515 cm^{-1}). Residual bands still appear in the amide II region near 1560 and 1535 cm^{-1}, indicating that the rhodopsin peptide backbone undergoes structural changes at this early stage of photobleaching. Small bands in the 1730–1775 cm^{-1} frequency region are also detected [346,367] and attributed to either Asp/Glu carboxyl group stretch modes [346] or ester carbonyl groups present in photoreceptor membrane phospholipids [367]. More recently, these bands have been assigned to specific Asp/Glu residues by site-directed mutagenesis (see Section 4.5.1). Studies using rhodopsin reconstituted into an ether-phospholipid lacking the ester carbonyl groups indicate that contributions from phospholipids in this region of the spectrum are not strong [114].

Low temperature FTIR studies have also been reported recently for the Batho transition of iodopsin, the chicken red-sensitive cone visual pigment [100]. A pattern similar to the rod pigment was observed, with C—C and hydrogen-out-of-plane (HOOP) vibrations characteristic of a highly perturbed all-*trans* configuration of the chromophore. In this red cone pigment effects of anion binding can be studied. In the case of the dark state (iodopsin) changing the bound anion from chloride to nitrate has little effect on several chromophore vibrational modes, including the stretching frequency of the Schiff base (C=N stretch) and HOOP modes. This suggests that local chromophore interactions with the anion are not crucial for the spectral shift induced by the anion. In contrast, in the case of Batho-iodopsin, anion substitution did alter the C14–H wagging mode, suggesting that the anion does play a role in the thermal back-reaction of Batho-iodopsin to iodopsin, observed at low temperatures for the chloride bound form.

4.4.2. Lumi(rhodopsin)

Large changes in the configuration of the retinal chromophore upon formation of this photointermediate were deduced on the basis of FTIR differences [115], including a shift in the C=N Schiff base frequency and an upshift in the C_{15}—H rocking vibration compared to rhodopsin, bathorhodopsin and isorhodopsin. A structural rearrangement of the chromophore is strongly supported by the disappearance of the intense bathorhodopsin HOOP modes at 921, 874 and 851 cm^{-1} and the appearance of a new band at 946 cm^{-1} [114], close to the frequency of the HOOP mode measured in model all-*trans*-retinylidene protonated Schiff base by resonance Raman spectroscopy [370]. Based on stable-isotope labeling, the chromophore rearrangement at Batho and subsequent relaxation at Lumi occur near the C12–H group [371,372]. Recent studies of the Lumi chromophore configuration, based on picosecond time-resolved coherent anti-Stokes Raman spectroscopy (PTR/CARS) at ambient temperature [56], confirm that by Lumi the chromophore adopts an almost completely all-planar (non-twisted) all-*trans* configuration. This suggests that the structural changes in the chromophore between Batho and Lumi are associated with a transfer of energy between the chromophore and protein. Consistent with this picture, FTIR difference spectroscopy [114] detects the first major protein changes upon Lumi(rhodopsin) formation. This is indicated by the appearance of prominent new bands near 1655 and 1635 cm^{-1} in the amide I region, which were distinguished from the C=N stretch mode through H/D exchange and 14,15-D_2 labeling of the chromophore.

4.4.3. Meta(rhodopsin) I

The Rho → Meta I difference spectrum can be recorded at 250 K, where the decay of Meta I is blocked. Upon formation of the Meta I intermediate from Lumi, additional changes occur in the rhodopsin protein structure, as indicated by peaks appearing in the amide I region from 1600 to 1700 cm^{-1} [114]. Most prominent is a positive peak at 1664 cm^{-1}, consistent with alterations in α-helical structure or orientation. New bands also appear above 1700 cm^{-1}, representing alterations in the carboxyl groups of Asp and/or Glu residues [346,366,373]. Chromophore bands characteristic of Meta I also appear at 950 cm^{-1}, assigned to a HOOP mode, and at 1537 cm^{-1}, assigned to the ethylenic stretch of the Meta I chromophore. A stable Meta I-like intermediate is also formed at room temperature in air-dried films of rhodopsin which retains some features of the Lumi protein structure according to FTIR difference spectroscopy [120].

4.4.4. Meta(rhodopsin) II

FTIR differences of the Rho → Meta II transition reveal that major conformational changes occur in the protein upon decay of Meta I [114,373]. In addition to amide I and II bands present at the Lumi and Meta I stages, an intense positive band appears at 1686 cm^{-1}, characteristic of β-turns [352]. Several new D/H exchange sensitive bands appear in the region above 1700 cm^{-1}, most prominently at 1767/1748 (−/+) cm^{-1}, 1729 cm^{-1} (−) and near 1710 cm^{-1} (+), which have been assigned to Asp-83, Glu-122 and Glu-113, respectively, as discussed in Section 4.5.1.

4.4.5. Meta(rhodopsin) III and Opsin

In contrast to the Meta II intermediate, which appears to exhibit the most distorted protein conformation relative to rhodopsin, thermal decay of Meta II to Meta III was found to involve a slow return of the protein back to a rhodopsin-like conformation, as indicated by decay of bands assigned to protein backbone and carboxyl group C=O stretch [196]. In a subsequent study, FTIR differences were measured using time-resolved FTIR after photobleaching at three different pH values and fit globally to two exponential decay processes which correspond to Rho → Meta III and Rho → opsin. This work showed that Meta III and opsin share very similar protein structures distinct from both the Meta II and the rhodopsin state [197].

4.5. Recent progress

In this section, we review selected FTIR experiments which probe conformational changes in the protein portion of rhodopsin after photobleaching. The reader is also referred to two recent reviews [374,375].

4.5.1. Assignment of bands to specific carboxyl residues

Although the introduction of isotopically labeled amino acids can lead to assignment of specific bands to particular amino acids, it does not provide a more detailed assignment to individual residues. For this purpose several groups, following earlier work on bacteriorhodopsin [376–378], utilized site-directed mutagenesis (SDM). In this approach, vibrational bands are assigned to a specific residue on the basis of spectral changes induced by substitution of that residue by another.

In one study using this approach, bovine rod rhodopsin and membrane-carboxyl group mutants were expressed using a recombinant baculovirus expression system and investigated in the membrane-bound state [379]. The Rho → Batho FTIR difference spectra of the mutants Asp-83 → Asn (D83N) and Glu-134 → Asp (E134D) were found to be very similar to that of native rhodopsin in the photoreceptor membrane, demonstrating that the retinal chromophores of these mutants undergo a normal 11-*cis* to all-*trans* photoisomerization. However, two bands at 1767 (−) and 1748 (+) cm^{-1} in the Rho → Meta II FTIR difference spectrum are absent in the D83N mutant, and hence were assigned to the C=O stretching mode of Asp and/or Glu carboxyl groups. Corresponding changes are not observed in the carboxylate C=O stretching region, indicating that the 1767/1748 pair represents the carboxyl group of Asp-83, that remains protonated in rhodopsin and its bleaching intermediates but undergoes an increase in its hydrogen bonding during the Meta I → Meta II transition. In contrast, the mutant E134D produced a normal Rho → Batho and Rho → Meta II difference spectra, but a fraction of misfolded protein was observed, supporting earlier evidence that Glu-134 plays a role in proper protein insertion and/or folding in the membrane.

In a second study [380], wild-type rhodopsin and three mutants with amino acid replacements of membrane-embedded carboxylic acid groups Asp83 → Asn (D83N), Glu122 → Gln (E122Q), and the double mutant D83N/E122Q were expressed in COS cells and studied in detergent micelles. Each of the mutant opsins produced a Meta II-like species, with maximal absorption at 380 nm, which activated G_t. In addition to confirming the assignment of the 1767/1748 cm^{-1} difference bands to Asp-83, additional bands (1725/1734 cm^{-1}) were assigned to Glu-122. This indicates that both residues are protonated in the dark state (rhodopsin) as well as in Meta II, but undergo a perturbation during this transition. In the mutant E122I, which produces a much smaller shift in the visible absorption λ_{max} (498 → 495 nm) than substitution of Glu-122 to Gln, Asp or Ala (λ_{max} 475–480 nm) [278,381,382], perturbation of a 1725 cm^{-1} band was also detected, but suggested to involve backbone structure as well [168]. By studying the effects of substitutions of single histidine residues on the bands assigned to Glu-122 as well as of double mutants it was concluded that there exists an interaction between TM III and V, which involves Glu-122 and His-211 [383].

A positive band, observed near 1712 cm^{-1}, was not affected by any of the mutants discussed above and was assigned to the protonation of Glu-113 [380]. This assignment was further supported on the basis of a shift of this band in the mutant E113D to 1709 cm^{-1} and its disappearance in E113A. Importantly, this work supports the model that Glu-113 serves as the Schiff base counter-ion [277–279] and accepts the proton during formation of Meta II.

4.5.2. Probing structural changes in rhodopsin using stable-isotope labeling

Stable-isotope labeling of amino acids combined with FTIR-difference spectroscopy can be an effective probe of protein conformational changes. In particular, isotope-induced band shifts provide a means to assign bands in the FTIR-difference spectrum to specific amino acids. Normally, such studies involve introduction of the isotope label by growing bacteria on a stringent growth medium containing the labeled amino acid. For example, such an approach has been used previously to detect changes in tyrosines, tryptophans, prolines and threonines residues in bacteriorhodopsin grown in *Halobacterium salinarum* (see e.g., [384,385]).

Recently, FTIR difference spectroscopy and amino acid isotope labeling were also combined to analyze structural changes of tyrosines in rhodopsin [386]. Isotope labeling was accomplished by growing Sf9 insect cells on a stringent medium enriched in L-[*ring*-$^2H_4$]Tyr. The resulting labeled recombinant rhodopsin was isolated using a C-terminal His_6-tag [387]. Rho → Meta II difference spectra were obtained at 10°C for unlabeled and L-[*ring*-$^2H_4$]Tyr labeled recombinant Rho (Fig. 7). Tyrosine bands can be identified and assigned on the basis of the isotope-induced shifts and by comparison with a series of model tyrosine compounds at high and low pH. This work demonstrates that a set of bands in the Rho →

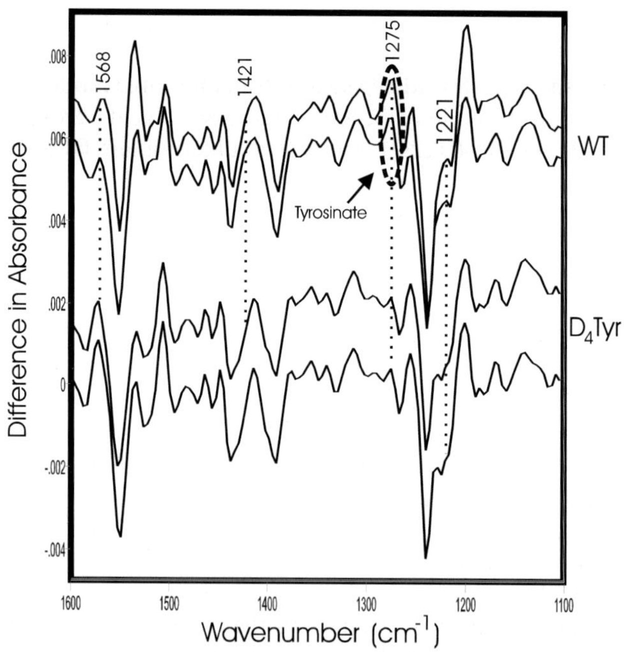

Fig. 7. FTIR-difference spectra of the Rho(dopsin) to Meta(rhodopsin) II transition of wild-type (WT) and of L-[ring-$^2H_4$]Tyr-labeled rhodopsin. Recombinant WT and labeled rhodopsin were expressed in Sf9 cells using recombinant baculovirus. Two independent measurements are shown for either conditions. The band intensity at 1275 cm^{-1} in the "labeled" difference spectrum drops due to the band shift induced by the tyrosine ring label. This vibration is typical of a tyrosinate C—O group and indicates the involvement of a tyrosinate group in the formation of the active state. This band shift is quite similar to one observed in the photocycle of bacteriorhodopsin (adapted from [386]).

Meta II difference spectra can be assigned to vibrational modes of tyrosine and tyrosinate. On this basis, it was concluded that one or more tyrosines participate in the rhodopsin photocascade up to formation of the Meta II intermediate, and that at least a partial protonation of a tyrosinate is involved. A possible involvement of tyrosine residues was previously proposed on the basis of UV-difference spectroscopy [183] and site-specific mutagenesis [382] and also suggested by a recent study using UV-resonance Raman spectroscopy [184]. A similar pattern of tyrosine/tyrosinate bands have also been observed in the BR $\rightarrow$ M transition in bacteriorhodopsin, though the sign of the bands is reversed. In bacteriorhodopsin, these bands were assigned to Tyr-185, which along with Pro-186 in the F-helix may form a hinge that facilitates α-helix movement.

4.5.3. Changes in water molecules during photoactivation

As first demonstrated in studies on bacteriorhodopsin [388,389], water molecules, which undergo a change in environment during individual steps during rhodopsin photoactivation, can be detected by FTIR-difference spectroscopy [126,390–392]. Such structurally active waters should exhibit bands in the 3400–3800 cm^{-1} region reflecting changes in the frequency and intensity of the OH stretching modes of the water molecule(s). In addition, bands should downshift out of this region upon substitution of D_2O for H_2O, provided that the water molecule is accessible for H/D exchange. However, the assignment of bands to structurally active water molecules is complicated by the appearance of bands in this region due to the OH- and NH-stretch modes, that arise from protein residues (e.g., threonine OH groups) and peptide groups (amide A NH-stretch mode) and also shift upon H/D exchange. For this purpose $H_2{}^{18}O$ is often used to identify water bands, since it causes a characteristic 8–12 cm^{-1} downshift in frequency upon substitution for H_2O and does not alter vibrations of protein OH and NH groups. In addition, site-directed mutagenesis can be used to provide clues to where the water molecule is located in the rhodopsin structure.

As an example Maeda and coworkers used site-directed mutagenesis to establish that one of the water molecules which is structurally active during the Rho $\rightarrow$ Batho transition forms hydrogen bonds with Glu-113 and the Schiff base, while another water molecule is linked to this structure through the peptide backbone [67]. This was accomplished by observing the effects of water assigned bands in the 3500–3650 cm^{-1} region for a variety of mutants, including E122Q, G121A, G120A, D83N, D83N/G120A, D83N/E113Q and E113Q/G120A.

In a different study, focused upon water changes during formation of the Meta I and Meta II intermediates [392], several negative bands in the region above 3500 cm^{-1} associated with water molecules in the dark state (rhodopsin) were detected, which shift to lower frequencies upon illumination (Fig. 8). These data indicate that at least one water molecule undergoes an increase in hydrogen bonding upon formation of the Meta I intermediate, while at least one other increases its hydrogen bonding during Meta II formation. Furthermore, these bands were not perturbed in the mutant D83N, and it was concluded that Asp-83, which undergoes a change in its hydrogen bonding during Meta II formation, does not appear to

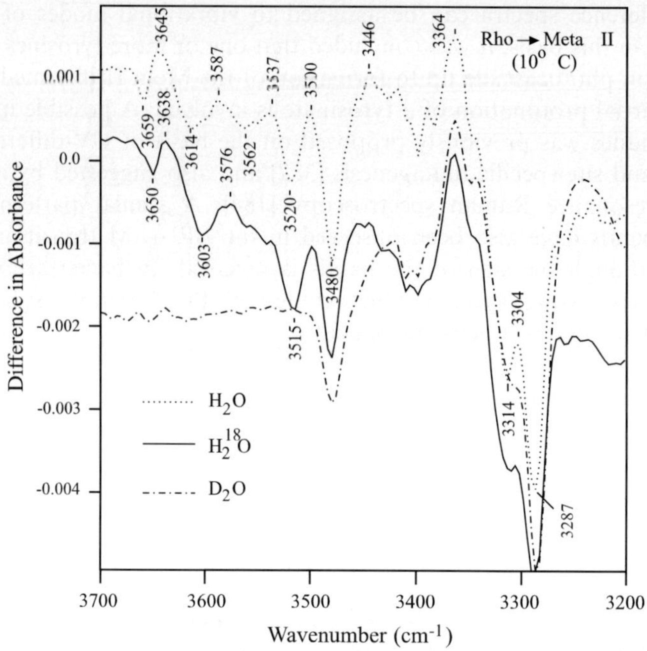

Fig. 8. Identification of H_2O vibrations in FTIR-difference spectra. FTIR difference spectra of the Rho → Meta II transition were recorded at 10°C for rhodopsin films hydrated with H_2O, D_2O or $H_2{}^{18}O$. The Y-scale shown is for the spectrum in H_2O. Negative/positive band pairs at 3659(−)/3645(+), 3614(−)/3587(+) and 3562(−)/3537(+) show band shifts of 5–10 cm^{-1} in $H_2{}^{18}O$ and originate in changes in water molecules involved in this transition (from [392]).

interact with any of the structurally active water molecules. Several NH and/or OH groups, which are inaccessible to hydrogen/deuterium exchange, also undergo alterations during Meta I and Meta II formation.

4.5.4. H/D exchange of core rhodopsin peptide groups upon photoactivation

As described previously, infrared spectroscopy can also be used to probe rhodopsin structure by monitoring the extent and kinetics of hydrogen/deuterium (H/D) exchange [344,356,357]. In a more recent study, time-resolved FTIR spectroscopy was used to measure the kinetics of hydrogen/deuterium exchange in rhodopsin upon photoexcitation [361]. The extent of hydrogen/deuterium exchange of backbone peptide groups can be monitored by measuring the integrated intensity of the amide II and amide II′ bands. In this study, both attenuated total reflection (ATR) and transmission techniques were used to monitor changes in the H/D exchange rate of rhodopsin upon photobleaching. In the case of ATR, membrane films can be measured in a bulk aqueous medium under near physiological conditions [347,393]. When rhodopsin films are exposed to D_2O in the dark for long periods, the amide II band retains at least 60% of its integrated intensity, reflecting a core of backbone

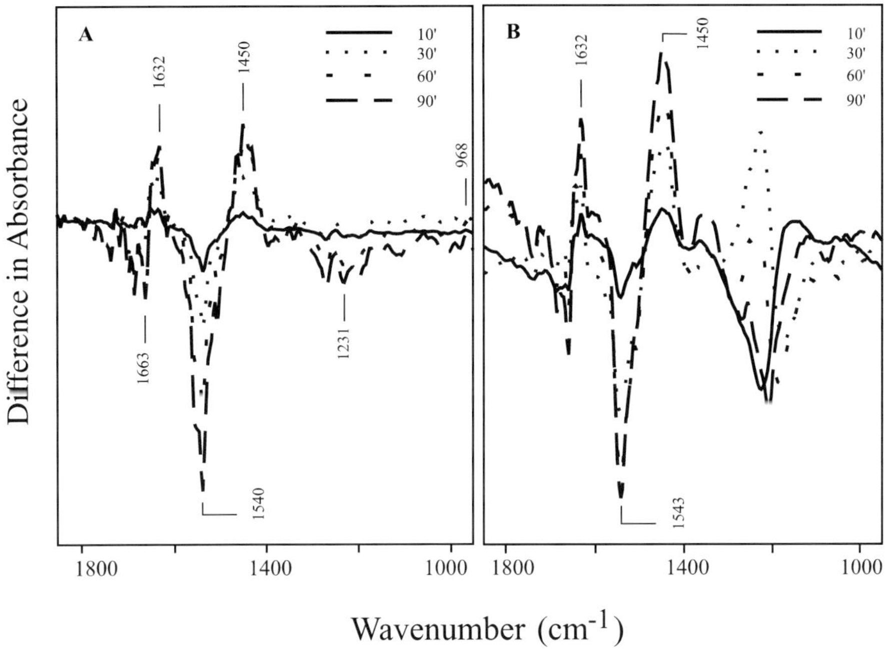

Fig. 9. H–D exchange of core peptide groups in rhodopsin triggered by illumination. FTIR-difference spectra of a rhodopsin film were recorded at 10°C using transmission (A) and ATR–FTIR (B) spectroscopy. Rhodopsin films were exposed in the dark to D_2O for more than 24 h prior to the measurements shown, and subsequently illuminated. The difference spectra represent the changes occurring after photobleaching and were obtained by subtracting the first spectrum of the photoreceptor membrane film recorded immediately after illumination from the successive spectra recorded thereafter at the times indicated. The major shift from 1540 to 1450 cm^{-1} is due to deuteration of additional peptide groups (adapted from [361]).

peptide groups which are resistant to H/D exchange. Upon photoactivation, rhodopsin in the presence of D_2O exhibits a new phase of H/D exchange which at 10°C consists of fast (time constant ~30 min) and slow (~11 h) components (Fig. 9). These results indicate that photoactivation causes buried portions of the rhodopsin backbone structure to become more accessible. This may correspond to the conformational changes, occurring during the Meta I → Meta II transition, that are partially reversed during Meta II decay. However, the region(s) of the rhodopsin structure exhibiting increased H/D exchange upon photoactivation remain(s) to be determined.

4.5.5. Changes in cysteine residues during photoactivation

In a study employing FTIR-difference spectroscopy, structural changes involving cysteine residues have been detected during the photoactivation of rhodopsin [394]. In particular, a positive band with a low frequency shoulder near 2550 cm^{-1} was detected during Meta II formation, which was assigned to the S–H stretching mode of one or more cysteine residues on the basis of its frequency and H/D induced isotope shift. Time-resolved studies show that the intensity of this band correlates with the formation and decay kinetics of the Meta II intermediate. Modification of rhodopsin with the reagent NEM, which selectively reacts with the SH groups of Cys-140 and Cys-316 on the cytoplasmic surface of rhodopsin, had no effect on the appearance of this band. Four other cysteine residues are also unlikely to contribute to this band because they are either thio-palmitylated (Cys-322 and Cys-323) or form a disulfide bond (Cys-110 and Cys-187). Thus, only four remaining cysteine residues are likely candidates to give rise to the SH bands observed. Difference bands in the Rho → Batho transition (positive band with shoulder at 2574 cm^{-1}) have also been assigned to a change in a cysteine SH stretch vibration, which along with the shifts in water and amide NH peptide backbone vibrations indicate an increase in hydrogen bonding around the chromophore upon Batho formation [57].

4.5.6. Probing intramolecular orientations by polarized ATR–FTIR difference spectroscopy

As described previously, polarized FTIR studies have been used to reveal information about the orientation of structural components of rhodopsin, e.g., α-helices, relative to the membrane plane [221,222]. In addition, attenuated total reflection infrared difference spectroscopy can be used to probe structural changes of rhodopsin [147,361,395] in an aqueous environment. These two methods have been combined in a recent study in order to investigate the structural changes of bovine rhodopsin as it undergoes the transition from the dark-state to the active (metarhodopsin II) conformation [192]. This method combines the sensitivity of infrared spectroscopy with the ability of polarized spectroscopy to probe the relative orientation of the transition dipole moment associated with specific bands. In the case of difference spectroscopy, however, information is obtained about the orientation of specific structurally active molecular groups relative to the membrane plane including previously assigned protein residues and retinylidene groups (or vibrational modes). This approach also complements other experimental approaches which probe orientation of structural components of rhodopsin including the visible dichroism studies [191,364] and, more recently, solid-state NMR studies [251].

The polarized difference spectra of the Rho → Meta II transition clearly reveal several new, previously unobserved bands throughout the entire spectral range (4000–800 cm^{-1}) [192]. Polarized bands in the amide I and amide II regions suggested that there occurs a small re-orientation of helical segments in the activation step of rhodopsin. A novel band at 939 cm^{-1}, assignable to a HOOP mode in metarhodopsin II, is believed to reflect a re-orientation of the retinal $HC_{11}{=}C_{12}H$ plane in the metarhodopsin I to metarhodopsin II transition. Previously unobserved bands at 1406, 1583 and 1736 cm^{-1} are tentatively assigned to the symmetric and asymmetric carboxylate and carboxyl stretching vibrations of Glu-134, which, on basis of evidence discussed before, was proposed to protonate upon formation of metarhodopsin II [154].

4.5.7. Lipid changes upon rhodopsin photoexcitation

Rhodopsin photoactivation is very sensitive to its environment. For example, increased lipid or detergent fluidity favors Meta II formation, possibly by accommodating a volume increase of the protein [134,138,141,396]. This may be one reason for the exceptionally high level of poly-unsaturated phospholipids in the photoreceptor disc membrane [397–400]. In spite of this high content of poly-unsaturated fatty acyl chains and the large proportion of the phospholipid phosphatidylethanolamine that tends to promote organization of lipids into non-bilayer structures [401–404], the rod outer segment membranes exhibit a normal bilayer organization [247,401,404,405]. Apparently, rhodopsin is able to stabilize the lipid matrix into a bilayer structure. In addition, it has recently been shown that there is a rapid flip/flop movement of phospholipids in the disc membranes, which might involve a low-order solubilization zone around rhodopsin [406]. However, the details of the interactions between rhodopsin and lipids in the photoreceptor membrane are poorly understood, and a molecular interpretation probably has to await a high-resolution 3D-structure.

Since FTIR-difference spectroscopy should be sensitive to alterations in lipid structure, this approach has been used in two different studies to probe for such changes during rhodopsin photoactivation. The basic approach consists of looking for contributions in the region above 1700 cm^{-1}, which normally reflect the C=O stretch mode of carboxyl groups such as found in aspartic and glutamic acid residues (see Section 4.5.1). However, vibrations in this region can also arise from the ester carbonyl group of phospholipid molecules such as present in the photoreceptor membrane. Normally, these groups produce a strong band centered near 1740 cm^{-1} in the absolute absorption spectrum. However, only if these groups are perturbed during photoactivation would bands be expected to appear in the difference spectrum.

An early work focusing on this question examined bands appearing in the difference spectrum between Rho and Meta I or II in the region from 1700–1770 cm^{-1}, which may be produced by the C=O stretch of carboxyl (COOH) or ester carbonyl (COOC) groups, the latter being found exclusively in membrane lipids [373]. In order to distinguish between these two types of groups, rhodopsin was reconstituted in membranes containing a synthetic phosphatidylcholine that lacks ester carbonyl groups. On this basis, it was concluded that the major changes in this region are due to rhodopsin carboxyls, which either undergo a change in local environment or a protonation/deprotonation reaction. However, additional small changes in this region were detected which also could reflect a direct involvement of phospholipids in the Meta I → Meta II transition [373]. Reconstitution of rhodopsin in ether-PC abolishes the band at 1735 cm^{-1}, appearing in the absolute absorption spectrum, that can be attributed to the ester carbonyl groups. In the Rho → Meta II difference spectrum reduced intensity was observed under the same conditions in the negative band at 1726 cm^{-1} compared to the native photoreceptor membrane, which was attributed to loss of a negative band near 1730 cm^{-1} and a positive band at a higher frequency. Both can be assigned to ester carbonyl groups of lipids which are perturbed during this transition.

lining the binding site, with only very minor involvement of the protein backbone. Participation of Tyr-268, Trp-265, Cys-264 (all in TM VI) and of Glu-113 and Gly-121 (both in TM III) is fairly well documented. This may reflect movement of the cyclohexene ring away from TM VI towards TM III. In addition, re-orientation occurs of at least one water molecule near Glu-113 and Gly-121. At this stage, the ligand is trapped in a highly perturbed all-*trans* configuration, twisted around several single C—C bonds, as indicated by the uncoupling and intensification of hydrogen-out-of-plane vibrations. However, direct structural information of the chromophore structure at an atomic level is not yet available. The twisted conformation may store a large part of the photon energy, which is used to drive subsequent structural changes in the protein. The transient trapping in this twisted state is probably due to steric interaction between specific sites in the protein and chromophore. This specific interaction will accomplish a selective vectorial transfer of energy to activate the required S–F modules in a cooperative way. In the case of the chromophore, possible interaction sites include the 19- and the 20-methyl groups. This interaction is finely tuned, since it is perturbed by most known modifications in either ring or polyene chain of the ligand.

Batho relaxes first into BSI, but structural information on this intermediate is very limited. The available evidence suggests that relaxation of the chromophore is initiated in this step, involving repositioning of the terminal section of the polyene chain and of the cyclohexene ring with respect to the chain. The 19- and 20-methyl groups are strong determinants of the rate of this transition, which may reflect their involvement in energy transfer between chromophore and protein interaction sites and the first conformational re-orientation in the receptor. Overall, these changes are likely to be very small since BSI is in dynamic equilibrium with Batho.

Larger changes are observed in the next intermediate, Lumi. Major relaxation of the ligand has occurred at this stage, with clear evidence for changes in protein secondary structure. Additional residues participating at this stage have not yet been identified. The structural changes will nevertheless be restricted to the vicinity of the binding site, since at low temperature Lumi can be efficiently photoconverted into rhodopsin.

Additional structural changes occur upon relaxation of Lumi → Meta I. Several lines of evidence support a relaxed all-*trans* configuration of the ligand in Meta I, along with additional changes in protein secondary structure. One or two water molecules are involved, and rearrangement in TM III may have reached Glu-122. Structural changes probably have not yet propagated to the surface regions of the protein, since under conditions that strongly restrict surface mobility (dehydrated membranes, extensive cross-linking, low lipid content, elevated pressure) a photo-intermediate with characteristics of Lumi or Meta I can still be generated.

The largest conformational changes in the protein are observed upon formation of Meta II, which constitutes the "active" state of the receptor. These conformational changes involve distinct proton movements, since a proton is transferred from the Schiff base to Glu-113 and proton uptake occurs at the cytosolic surface involving Glu-134. These are separate steps, defining a transition Meta $II_a \leftrightarrow$ Meta $II_b(H^+)$, which have not yet been structurally resolved. Some re-orientation of the ligand may occur, as indicated by the ability of additional methyl groups at the 10- or 12-position to strongly stabilize the Meta I intermediate. Participation of additional protein residues (Asp-83, Glu-122 and several Tyr, Cys and Trp residues) and at least two water molecules has already been established. This probably reflects movement of (parts of) TM II, III, IV and VI. This redistribution of membrane domain elements has a marked effect on the cytosolic domain, effectuating rearrangement of loop 4, retracting from loop 1, and of loop 2 moving away from loop 3 due to rigid-body rotation of helix F away from helix C. The latter changes will probably represent the largest changes at this stage, likely to be associated with exposure of sites for binding of the G-protein, and may also account for the increased exposure of buried peptide bonds to H–D exchange.

As noted above, many of the established changes in the Meta I → Meta II transition involve proton movements (proton transfer between Glu-113 and Schiff base, proton uptake via Glu-134, H-bonding redistribution involving Asp-83, Glu-122 and Cys and Tyr residues, H_2O redistribution). One interesting possibility is that the mechanism of transmitting conformational changes between the chromophore and the peripheral regions of the protein involves an extended H-bonded network. Proton-uptake by such a network will re-distribute the hydrogen bonds and associated water molecules and even may fragment the network in smaller units, and vice versa, at little energetic cost. This would constitute a highly efficient means to rearrange S–F modules, and would allow rapid long-range communication within the protein structure. An H-bonded network would also be able to shift the pK of participating residues (in fact may act as an acid/base micro-array) and to stabilize charged residues. Further, it would be sensitive to surface charge, membrane voltage and bulk water and ion concentration, and hence could accommodate the reversible transitions in the Meta I ↔ Meta II equilibrium.

Decay of Meta II into Meta III or the apoprotein opsin is accompanied by proton release and, at least partial, reversal of most of the light-triggered conformational changes. This suggests re-orientation of S–F modules towards their original position. Binding of G-protein to Meta II is likely to lock them in the active state, thereby explaining the observed blocking of Meta II decay. The conformation of Meta III appears to be quite similar to that of opsin, providing additional support to the concept that decay of Meta II involves release of the ligand from its original binding site. Rod pigments are characterized by a Glu residue at position 122 (TM III), that participates in Meta II formation. Presence of an apolar residue (I, V, M) at this site is characteristic for cone pigments and appears to accelerate the decay of Meta II. This suggests, that Glu-122 may be part of an extended H-bonded network that controls formation and decay of Meta II. His-211 at TM V has also been implicated in this process.

Even with an empty ligand binding site, opsin and Meta III still exhibit low constitutive activity, that we attribute to stochastic activity of S–F modules. This activity can be significantly enhanced by addition of excess all-*trans* retinal, but only if the binding site is not occupied and the protein is palmitoylated [202]. This suggests, that the receptor has a second low-affinity binding site for all-*trans* retinal, that can trigger receptor activation with lower efficiency, using at least part of the pathway involved in photoactivation. Studies of this second site using analog pigments should be quite informative.

5.2. Future perspectives

A major challenge for the future will be to elucidate the tertiary structure of rhodopsin as well as other GPCR's at high resolution. This will not only have a major impact on our insight in the principles of membrane protein structure and folding, it will also provide a reliable basis for homology modeling. In addition, it will be of enormous value to further enhance our understanding of the mechanism of receptor activation, for instance by means of NMR and FTIR studies in combination with site-directed mutagenesis.

The best prospects of obtaining detailed structural information within the next couple of years are presently offered by solid-state NMR spectroscopy and 3D-crystallography. The rapid progress in solid-state NMR (high field, 2D and 3D pulse sequences, oriented samples) already allows detailed structural analysis of the ligand *in situ*, and the major limitation to extend such studies to the entire receptor still lies in the difficulty of heterologous production of labeled functional recombinant protein in amounts of tens of milligrams. Major research efforts are underway to address this bottleneck and thus we expect significant progress in the next few years.

Recently, important progress has also been reported in the classical X-ray structural approach using 3D crystallization. The first 3D crystals of native bovine rhodopsin with a resolution better than 4 Å have been generated [409], a significant improvement over the resolution presently available through electron diffraction. A corresponding 3D structure is not yet available, and for mechanistic studies the resolution needs to be further improved, as for example demonstrated recently for bacteriorhodopsin [410,411]. Overall, these structural approaches are likely to have a major impact on this field in the next few years as well.

An essential issue yet to be addressed is whether the structure and mechanism of rhodopsin can be directly extrapolated to other GPCR's. Visual pigments are quite specialized in this family with regard to receptor activation (bound ligand, *in situ* activation, input of photon energy), that might have put special structural requirements and adaptations for proper function. For instance, mutation of an Asp residue in TM II results in impaired G-protein activation in a variety of aminergic receptors, but has very little effect on photocascade and transducin activation by rhodopsin. On the other hand, several sequence motifs are highly conserved within the GPCR family, and several mechanistic aspects of rhodopsin activation (ion–pair interactions in membrane domain, TM VI ↔ TM III movements) seem to play a role in other receptors as well [412,413] and may indicate a generic principle for receptor activation.

A first step in this direction would be a structural comparison of rod and cone pigments. While the mechanism of both classes is very similar in principle, they exhibit qualitative differences on a protein level. For instance compared to rod pigments, the thermal stability of cone pigments and cone opsins is much lower, the K_a for ligand binding is lower, while the on-rate of the binding reaction is much higher, and the decay rates of the Batho and Meta II intermediates also are significantly higher [414]. An explanation of such differences at the structural level would provide considerable insight into the principles governing protein folding and stability, and will also help to identify structural elements which contribute to functional properties.

High-resolution structures, once available, will provide a solid base for detailed mechanistic studies. These mechanistic studies of rhodopsin and related receptors will be facilitated by a variety of biophysical approaches. For example, solid-state NMR studies should be able to considerably refine details of receptor–ligand interactions. FTIR studies using low-temperature, time-resolved and ATR methods will be able to identify intermediate steps in receptor activation and corresponding protein residues and backbone structure. Cryo-electronmicroscopy and atomic force or atomic interaction microscopy will provide additional insights on G-protein binding and activation.

In conclusion, while significant progress has recently been made in understanding of the mechanism of visual pigments, we expect even more dramatic progress to occur in the next few years. This will in turn help clarify the mechanism of G-protein coupled receptors. Insight into whether there is a general common principle in the mechanism of G-protein coupled receptors, and whether this includes elements like distinct intermediate steps and conservation of S–F modules, is of crucial importance. Such progress will have a tremendous impact on GPCR related applications in physiology and pharmacology.

Note added in proof

A very recent study on the 9-desmethyl analog of rhodopsin indicates that its impaired signaling ability is due to ineffective proton transfer, affecting the Meta I → Meta II transition [415]. This observation presents additional evidence, that the ligand participates up to the last stage of receptor activation and can directly modulate S–F modules through interacting H-bonds.

Proofreading witnessed a major breakthrough in this field. The nearly complete 3D structure of native bovine rhodopsin was reported at 2.8 Å resolution [416]. About 93% of the amino acid side chains are resolved, and the chromophore is represented by a well-defined electron density profile. This will trigger a major (re)modeling wave through the entire GPCR family. Some major implications will be considered in the preface to this volume. Structural elements relevant for this chapter include the following: As predicted by NMR and analog studies, the chromophore is bound in the 6-s-*cis* configuration and exhibits a considerable twist in its C10–C13 segment. The magnitude of this twist cannot be accurately determined at this resolution. Furthermore, the CD-calculations prove to be correct, as this twist exhibits positive helicity. The distance of the centre of charge at Glu-113 to the Schiff base nitrogen is 3.5–4.0 Å, close to the value predicted by solid-state NMR studies. This fully agrees with a "soft" counterion. Any water molecules at this site could not yet be resolved, however. The intradiscal side of the Schiff base site is lined by residues of the second extracellular loop including C-187, probably correctly positioned by the C-110/C-187 disulfide bridge. This sequence includes residues that are involved in anion binding in LW cone pigments. This explains a direct effect of chloride binding on the electronic properties of the chromophore. The intradiscal domain is very compactly structured indeed, showing a high content of β-structure, as suggested by FTIR studies. Four well defined β-strands seem to put a firm lid on the binding site. The intracellular domain is less structured and more open. The second and third loops line the border of this domain. Remarkably, the fourth loop folds into an amphipathic α-helix, comprising residues 310–321, with its long axis about parallel to the membrane plane. The C-terminus folds back towards the first loop and then extends away from the protein surface, explaining the easy accessibility of the C-terminal part. Asp-83 as well as Glu-134 take part in extensive hydrogen-bonding, that provides interactions between several helices. These H-bonding networks may extend much further actually, but so far only very few structural water molecules have been resolved. It should be noted, that no interaction between Met-257 and the NPXXY motif in TM VII is apparent. Also, Gly-121 lines the polyene chain but is not close to the 9-methyl group. Rather, the 9-methyl group is located close to the side chains of Thr-118 and Tyr-268. This cautions again, that results from mutagenesis experiments should be interpreted very cautiously if a well-resolved 3D structure is not yet available. The 2D pattern of structural domains, as presented in Fig. 1, holds up quite well, except that the second extracellular loop as well as the third and fourth intracellular loops are longer. Thus, TM IV ends at residue Trp-175, and the subsequent loop spans through several turns and two well-defined β-strands all the way to residue Asn-200. TM V ends at residue Thr-229. The subsequent loop is not fully resolved, but ends at Thr-243 in the next helix. Position and structure of the fourth intracellular "loop" already were outlined above.

References

1. Ovchinnikov, Y.A. (1982) FEBS Lett. **148**, 179–191.
2. Hargrave, P.A., McDowell, J.H., Curtis, D.R., Wang, J.K., Juszcak, E., Fong, S.-L., Rao, J.K.M. and Argos, P. (1983) Biophys. Struct. Mech. **9**, 235–244.
3. Nathans, J. and Hogness, D.S. (1983) Cell **34**, 807–814.
4. Beck-Sickinger, A.G. (1996) Drug Discov. Today **1**, 502–513.
5. Bockaert, J. and Pin, J.-P. (1998) C. R. Acad. Sci. Ser. [III]-VIE. **321**, 529–551.
6. Sakmar, T.P. (1998) Prog. Nucl. Acid Res. Mol. Biol. **59**, 1–34.
7. Dratz, E.A. and Hargrave, P.A. (1983) Trends Biochem. Sci. **8**, 128–131.
8. Hargrave, P.A. and McDowell, J.H. (1992) FASEB J. **6**, 2323–2331.
9. Ovchinnikov, Y.A., Abdulaev, N.G. and Bogachuk, A.S. (1988) FEBS Lett. **230**, 1–5.
10. Papac, D.I., Thornburg, K.R., Büllesbach, E.E., Crouch, R.K. and Knapp, D.R. (1992) J. Biol. Chem. **267**, 16889–16894.
11. Moench, S.J., Moreland, J., Stewart, D.H. and Dewey, T.G. (1994) Biochemistry-USA **33**, 5791–5796.
12. Kropf, A., Whittenberger, B.P., Goff, S.P. and Waggoner, A.S. (1973) Exp. Eye Res. **17**, 591–606.
13. Mirzadegan, T. and Liu, R.S.H. (1991) Prog. Retin. Res. **11**, 57–74.
14. Nakanishi, K. and Crouch, R.K. (1995) Isr. J. Chem. **35**, 253–272.
15. Ji, T.H., Grossmann, M. and Ji, I.H. (1998) J. Biol. Chem. **273**, 17299–17302.
16. Unger, V.M., Hargrave, P.A., Baldwin, J.M. and Schertler, G.F.X. (1997) Nature **389**, 203–206.
17. Wald, G. (1968) Nature **219**, 800–807.
18. Crescitelli, F. (1991) Prog. Retin. Res. **11**, 1–32.
19. Khan, S.M.A., Bolen, W., Hargrave, P.A., Santoro, M.M. and McDowell, J.H. (1991) Eur. J. Biochem. **200**, 53–59.
20. Hecht, S., Shlaer, S. and Pirenne, M.H. (1942) J. Gen. Physiol. **25**, 819–840.
21. Klaassen, C.H.W., Bovee-Geurts, P.H.M., DeCaluwé, G.L.J. and DeGrip, W.J. (1999) Biochem. J. **342**, 293–300.
22. Creemers, A.F.L., Klaassen, C.H.W., Bovee-Geurts, P.H.M., Kelle, R., Kragl, U., Raap, J., DeGrip, W.J., Lugtenburg, J. and DeGroot, H.J.M. (1999) Biochemistry-USA **38**, 7195–7199.
23. Eilers, M., Reeves, P.J., Ying, W.W., Khorana, H.G. and Smith, S.O. (1999) Proc. Nat. Acad. Sci. USA **96**, 487–492.
24. Davies, A., Schertler, G.F.X., Gowen, B.E. and Saibil, H.R. (1996) J. Struct. Biol. **117**, 36–44.
25. Becker, R.S. (1988) Photochem. Photobiol. **48**, 369–399.
26. Kandori, H., Katsuta, Y., Ito, M. and Sasabe, H. (1995) J. Amer. Chem. Soc. **117**, 2669–2670.
27. Dartnall, H.J.A. (1972) in: Photochemistry of Vision, ed H.J.A. Dartnall. pp. 122–145, Springer, Berlin.
28. Peteanu, L.A., Schoenlein, R.W., Wang, Q., Mathies, R.A. and Shank, C.V. (1993) Proc. Nat. Acad. Sci. USA **90**, 11762–11766.

29. Kühne, W. (1877) Nat. Med. Verhandl. Heidelberg **1**, 484–492.
30. Boll, F. (1877) Arch. Anat. Physiol. 4–35.
31. Yoshizawa, T. (1972) in: Photochemistry of Vision, ed H.J.A. Dartnall. pp. 146–179, Springer, Berlin.
32. Yoshizawa, T. and Kandori, H. (1991) Prog. Retin. Res. **11**, 33–55.
33. Kliger, D.S. and Lewis, J.W. (1995) Isr. J. Chem. **35**, 309–323.
34. Imai, H., Imamoto, Y., Yoshizawa, T. and Shichida, Y. (1995) Biochemistry-USA **34**, 10525–10531.
35. Imai, H., Terakita, A., Tachibanaki, S., Imamoto, Y., Yoshizawa, T. and Shichida, Y. (1997) Biochemistry-USA **36**, 12773–12779.
36. Shichida, Y. (1986) Photobiochem. Photobiophys. **13**, 287–307.
37. Shichida, Y. (1990) Photochem. Photobiol. **52**, 1179–1185.
38. Straume, M., Mitchell, D.C., Miller, J.L. and Litman, B.J. (1990) Biochemistry-USA **29**, 9135–9142.
39. Szundi, I., Lewis, J.W. and Kliger, D.S. (1997) Biophys. J. **73**, 688–702.
40. Szundi, I., Mah, T.L., Lewis, J.W., Jäger, S., Ernst, O.P., Hofmann, K.P. and Kliger, D.S. (1998) Biochemistry-USA **37**, 14237–14244.
41. Nathans, J. (1987) Annu. Rev. Neurosci. **10**, 163–194.
42. Yoshizawa, T. (1994) Biophys. Chem. **50**, 17–24.
43. Okano, T., Fukada, Y. and Yoshizawa, T. (1995) Comp. Biochem. Physiol. Pt. B. **112**, 405–414.
44. Bowmaker, J.K., Heath, L.A., Wilkie, S.E. and Hunt, D.M. (1997) Vision Res. **37**, 2183–2194.
45. Jacobs, G.H. (1998) Invest. Ophthalmol. Visual Sci. **39**, 2205–2216.
46. Hunt, D.M., Dulai, K.S., Cowing, J.A., Julliot, C., Mollon, J.D., Bowmaker, J.K., Li, W.-H. and Hewett-Emmett, D. (1998) Vision Res. **38**, 3299–3306.
47. Yokoyama, S. (1999) Genes Genet. Syst. **74**, 189–199.
48. Honig, B., Dinur, U., Nakanishi, K., Balogh-Nair, V., Gawinowicz, M.A., Arnaboldi, M. and Motto, M.G. (1979) J. Amer. Chem. Soc. **101**, 7084–7086.
49. Lin, S.W., Imamoto, Y., Fukada, Y., Shichida, Y., Yoshizawa, T. and Mathies, R.A. (1994) Biochemistry-USA **33**, 2151–2160.
50. Kochendoerfer, G.G., Wang, Z.Y., Oprian, D.D. and Mathies, R.A. (1997) Biochemistry-USA **36**, 6577–6587.
51. Kochendoerfer, G.G., Lin, S.W., Sakmar, T.P. and Mathies, R.A. (1999) Trends Biochem. Sci. **24**, 300–305.
52. Vought, B.W., Dukkipatti, A., Max, M., Knox, B.E. and Birge, R.R. (1999) Biochemistry-USA **38**, 11287–11297.
53. Cooper, A. (1979) Nature **282**, 531–533.
54. Popp, A., Ujj, L. and Atkinson, G.H. (1996) Proc. Nat. Acad. Sci. USA **93**, 372–376.
55. Yoshizawa, T., Shichida, Y. and Matuoka, S. (1984) Vision Res. **24**, 1455–1463.
56. Ujj, L., Jäger, F. and Atkinson, G.H. (1998) Biophys. J. **74**, 1492–1501.
57. Kandori, H. and Maeda, A. (1995) Biochemistry-USA **34**, 14220–14229.
58. Lewis, J.W., Jäger, S. and Kliger, D.S. (1997) Photochem. Photobiol. **66**, 741–746.
59. DeGrip, W.J., Klaassen, C.H.W. and Bovee-Geurts, P.H.M. (1999) Biochem. Soc. Trans. **27**, 937–944.
60. Lin, S.W., Sakmar, T.P., Franke, R.R., Khorana, H.G. and Mathies, R.A. (1992) Biochemistry-USA **31**, 5105–5111.
61. Jäger, S., Han, M., Lewis, J.W., Szundi, I., Sakmar, T.P. and Kliger, D.S. (1997) Biochemistry-USA **36**, 11804–11810.
62. Zhang, H.Z., Lerro, K.A., Yamamoto, T., Lien, T.H., Sastry, L., Gawinowicz, M.A. and Nakanishi, K. (1994) J. Amer. Chem. Soc. **116**, 10165–10173.
63. Nakayama, T.A. and Khorana, H.G. (1990) J. Biol. Chem. **265**, 15762–15769.
64. Deng, H., Huang, L.W., Callender, R.H. and Ebrey, T.G. (1994) Biophys. J. **66**, 1129–1136.
65. Han, M. and Smith, S.O. (1995) Biochemistry-USA **34**, 1425–1432.
66. Ganter, U.M., Schmid, E.D. and Siebert, F. (1988) J. Photochem. Photobiol. B-Biol. **2**, 417–426.
67. Nagata, T., Terakita, A., Kandori, H., Shichida, Y. and Maeda, A. (1998) Biochemistry-USA **37**, 17216–17222.

228. Pogozheva, I.D., Lomize, A.L. and Mosberg, H.I. (1997) Biophys. J. **72**, 1963–1985.
229. Tang, L., Ebrey, T.G. and Subramaniam, S. (1995) Isr. J. Chem. **35**, 193–209.
230. Shieh, T., Han, M., Sakmar, T.P. and Smith, S.O. (1997) J. Mol. Biol. **269**, 373–384.
231. Herzyk, P. and Hubbard, R.E. (1998) J. Mol. Biol. **281**, 741–754.
232. Tan, Q., Lou, J.H., Borhan, B., Karnaukhova, E., Berova, N. and Nakanishi, K. (1997) Angew. Chem. Int. Ed. **36**, 2089–2093.
233. Buss, V., Kolster, K., Terstegen, F. and Vahrenhorst, R. (1998) Angew. Chem. Int. Ed. **37**, 1893–1895.
234. Buda, F., DeGroot, H.J..M. and Bifone, A. (1996) Phys. Rev. Lett. **77**, 4474–4477.
235. Bifone, A., DeGroot, H.J..M. and Buda, F. (1997) J. Phys. Chem. B **101**, 2954–2958.
236. Pauling, L. (1979) Proc. Nat. Acad. Sci. USA **76**, 2293–2294.
237. Chalazonitis, N., Chagneux, R. and Arvanitaki, A. (1970) C.R. Acad. Sci. [D]. **271**, 130–133.
238. Chabre, M. (1978) Proc. Nat. Acad. Sci. USA **75**, 5471–5474.
239. Liebman, P.A. (1962) Biophys. J. **2**, 161–177.
240. Hárosi, F. (1975) J. Gen. Physiol. **66**, 357–382.
241. Tokunaga, F., Kawamura, S. and Yoshizawa, T. (1976) Vision Res. **16**, 633–641.
242. Cone, R.A. (1972) Nature New Biol. **236**, 39–43.
243. Brown, P.K. (1972) Nature New Biol. **236**, 35–38.
244. Poo, M. and Cone, R.A. (1974) Nature **247**, 438–441.
245. Liebman, P.A. and Entine, G. (1974) Science **185**, 457–459.
246. Chabre, M. and Breton, J. (1979) Photochem. Photobiol. **30**, 295–299.
247. Mollevanger, L.C.P.J., Dratz, E.A., DeKruijff, B., Hilbers, C.W. and DeGrip, W.J. (1986) Eur. J. Biochem. **156**, 383–390.
248. Clark, N.A., Rothschild, K.J., Simon, B.A. and Luippold, D.A. (1980) Biophys. J. **31**, 65–96.
249. Hsiao, T.-L. and Rothschild, K.J. (1980) Biochem. Biophys. Res. Commun. **94**, 618–624.
250. Glaubitz, C. and Watts, A. (1998) J. Magn. Reson. **130**, 305–316.
251. Gröbner, G., Choi, G., Burnett, I.J., Glaubitz, C., Verdegem, P.J.E., Lugtenburg, J. and Watts, A. (1998) FEBS Lett. **422**, 201–204.
252. Clore, G.M. and Gronenborn, A.M. (1994) Protein Sci. **3**, 372–390.
253. VanNuland, N.A.J., Forge, V., Balbach, J. and Dobson, C.M. (1998) Acc. Chem. Res. **31**, 773–780.
254. Wider, G. and Wüthrich, K. (1999) Curr. Opin. Struct. Biol. **9**, 594–601.
255. Yeagle, P.L., Alderfer, J.L. and Albert, A.D. (1995) Nature Struct. Biology **2**, 832–834.
256. Yeagle, P.L., Alderfer, J.L. and Albert, A.D. (1995) Biochemistry-USA **34**, 14621–14625.
257. Yeagle, P.L., Alderfer, J.L., Salloum, A.C., Ali, L. and Albert, A.D. (1997) Biochemistry-USA **36**, 3864–3869.
258. Rothschild, K.J., DeGrip, W.J. and Sanches, R. (1980) Biochim. Biophys. Acta **596**, 338–351.
259. Pistorius, A.M.A. and DeGrip, W.J. (1994) Biochem. Biophys. Res. Commun. **198**, 1040–1045.
260. Yeagle, P.L., Alderfer, J.L. and Albert, A.D. (1997) Biochemistry-USA **36**, 9649–9654.
261. Yeagle, P.L. and Albert, A.D. (1998) Biochem. Soc. Trans. **26**, 520–531.
262. Colmenares, L.U., Asato, A.E., Denny, M., Mead, D., Zingoni, J.P. and Liu, R.S.H. (1991) Biochem. Biophys. Res. Commun. **179**, 1337–1343.
263. Colmenares, L.U. and Liu, R.S.H. (1996) Tetrahedron **52**, 109–118.
264. Colmenares, L.U., Zou, X.-L., Liu, J., Asato, A.E., Liu, R.S.H., DeLera, A.R. and Alvarez, R.A. (1999) J. Amer. Chem. Soc. **121**, 5803–5804.
265. Klein-Seetharaman, J., Getmanova, E.V., Loewen, M.C., Reeves, P.J. and Khorana, H.G. (1999) Proc. Nat. Acad. Sci. USA **96**, 13744–13749.
266. Lugtenburg, J., Mathies, R.A., Griffin, R.G. and Herzfeld, J. (1988) Trends Biochem. Sci. **13**, 388–393.
267. Cross, T.A. and Opella, S.J. (1994) Curr. Opin. Struct. Biol. **4**, 574–581.
268. Watts, A., Ulrich, A.S. and Middleton, D.A. (1995) Mol. Membrane Biol. **12**, 233–246.
269. Smith, S.O., Aschheim, K. and Groesbeek, M. (1996) Quart. Rev. Biophys. **29**, 395–449.
270. Watts, A. (1998) Biochim. Biophys. Acta **1376**, 297–318.
271. Mollevanger, L.C.P.J., Kentgens, A.P.M., Pardoen, J.A., Courtin, J.M. L., Veeman, W.S., Lugtenburg, J. and DeGrip, W.J. (1987) Eur. J. Biochem. **163**, 9–14.

272. Smith, S.O., Palings, I., Copié, V., Raleigh, D.P., Courtin, J.M.L., Pardoen, J.A., Lugtenburg, J., Mathies, R.A. and Griffin, R.G. (1987) Biochemistry-USA **26**, 1606–1611.
273. Smith, S.O., Palings, I., Miley, M.E., Courtin, J.M.L., DeGroot, H.J.M., Lugtenburg, J., Mathies, R.A. and Griffin, R.G. (1990) Biochemistry-USA **29**, 8158–8164.
274. Han, M., DeDecker, B.S. and Smith, S.O. (1993) Biophys. J. **65**, 899–906.
275. Han, M. and Smith, S.O. (1995) Biophys. Chem. **56**, 23–29.
276. Birge, R.R. (1990) Biochim. Biophys. Acta **1016**, 293–327.
277. Zhukovsky, E.A. and Oprian, D.D. (1989) Science **246**, 928–930.
278. Sakmar, T.P., Franke, R.R. and Khorana, H.G. (1989) Proc. Nat. Acad. Sci. USA **86**, 8309–8313.
279. Nathans, J. (1990) Biochemistry-USA **29**, 9746–9752.
280. Harbison, G.S., Herzfeld, J. and Griffin, R.G. (1983) Biochemistry-USA **22**, 1–4.
281. Hu, J.G., Griffin, R.G. and Herzfeld, J. (1994) Proc. Nat. Acad. Sci. USA **91**, 8880–8884.
282. Hu, J.G., Griffin, R.G. and Herzfeld, J. (1997) J. Amer. Chem. Soc. **119**, 9495–9498.
283. Smith, S.O., DeGroot, H.J.M., Gebhard, R. and Lugtenburg, J. (1992) Photochem. Photobiol. **56**, 1035–1039.
284. Smith, S.O., Courtin, J.M.L., DeGroot, H.J.M., Gebhard, R. and Lugtenburg, J. (1991) Biochemistry-USA **30**, 7409–7415.
285. Creuzet, F., McDermott, A.E., Gebhard, R., VanDerHoef, K., Spijker-Assink, M.B., Herzfeld, J., Lugtenburg, J., Levitt, M.H. and Griffin, R.G. (1991) Science **251**, 783–786.
286. Smith, S.O. (1993) Curr. Opin. Struct. Biol. **3**, 755–759.
287. Verdegem, P.J.E., Helmle, M., Lugtenburg, J. and DeGroot, H.J.M. (1997) J. Amer. Chem. Soc. **119**, 169–174.
288. Griffin, R.G. (1998) Nature Struct. Biology **5**, 508–512.
289. Feng, X., Lee, Y.K., Sandström, D., Edén, M., Maisel, H., Sebald, A. and Levitt, M.H. (1996) Chem. Phys. Lett. **257**, 314–320.
290. Feng, X., Verdegem, P.J.E., Lee, Y.K., Sandström, D., Edén, M., Bovee-Geurts, P.H.M., DeGrip, W.J., Lugtenburg, J., DeGroot, H.J.M. and Levitt, M.H. (1997) J. Amer. Chem. Soc. **119**, 6853–6857.
291. Feng, X., Verdegem, P.J.E., Edén, M., Sandström, D., Lee, Y.K., Bovee-Geurts, P.H.M., DeGrip, W.J., Lugtenburg, J., DeGroot, H.J.M. and Levitt, M.H. (2000) J. Biomol. NMR **16**, 1–8.
292. Sanders, C.R. II, and Prosser, R.S. (1998) Structure **6**, 1227–1234.
293. Ottiger, M. and Bax, A. (1998) J. Biomol. NMR **12**, 361–372.
294. Prosser, R.S., Hwang, J.S. and Vold, R.R. (1998) Biophys. J. **74**, 2405–2418.
295. Marassi, F.M. and Opella, S.J. (1998) Curr. Opin. Struct. Biol. **8**, 640–648.
296. Gröbner, G., Taylor, A., Williamson, P.T.F., Choi, G., Glaubitz, C., Watts, J.A., DeGrip, W.J. and Watts, A. (1997) Anal. Biochem. **254**, 132–138.
297. Moltke, S., Wallat, I., Sakai, N., Nakanishi, K., Brown, M.F. and Heyn, M.P. (1999) Biochemistry-USA **38**, 11762–11772.
298. Glaubitz, C., Burnett, I.J., Gröbner, G., Mason, A.J. and Watts, A. (1999) J. Amer. Chem. Soc. **121**, 5787–5794.
299. Grage, S.L. and Ulrich, A.S. (1999) J. Magn. Reson. **138**, 98–106.
300. Watts, A., Burnett, I.J., Glaubitz, C., Gröbner, G., Middleton, D.A., Spooner, P.J.R. and Williamson, P.T.F. (1998) Eur. Biophys. J. Biophys. Lett. **28**, 84–90.
301. Gröbner, G., Burnett, I.J., Glaubitz, C., Choi, G., Mason, A.J. and Watts, A. (2000) Nature **405**, 810–813.
302. Hubbell, W.L. and Altenbach, C.A. (1994) Curr. Opin. Struct. Biol. **4**, 566–573.
303. Hubbell, W.L., Mchaourab, H.S., Altenbach, C.A. and Lietzow, M.A. (1996) Structure **4**, 779–783.
304. Hustedt, E.J. and Beth, A.H. (1999) Annu. Rev. Biophys. Biomol. Struct. **28**, 129–153.
305. Farahbakhsh, Z.T., Ridge, K.D., Khorana, H.G. and Hubbell, W.L. (1995) Biochemistry-USA **34**, 8812–8819.
306. Altenbach, C.A., Yang, K., Farrens, D.L., Farahbakhsh, Z.T., Khorana, H.G. and Hubbell, W.L. (1996) Biochemistry-USA **35**, 12470–12478.
307. Altenbach, C.A., Cai, K.W., Khorana, H.G. and Hubbell, W.L. (1999) Biochemistry-USA **38**, 7931–7937.

308. Langen, R., Cai, K.W., Altenbach, C.A., Khorana, H.G. and Hubbell, W.L. (1999) Biochemistry-USA **38**, 7918–7924.
309. Cai, K.W., Langen, R., Hubbell, W.L. and Khorana, H.G. (1997) Proc. Nat. Acad. Sci. USA **94**, 14267–14272.
310. Altenbach, C.A., Greenhalgh, D.A., Khorana, H.G. and Hubbell, W.L. (1994) Proc. Nat. Acad. Sci. USA **91**, 1667–1671.
311. Altenbach, C.A., Klein-Seetharaman, J., Hwa, J., Khorana, H.G. and Hubbell, W.L. (1999) Biochemistry-USA **38**, 7945–7949.
312. Farahbakhsh, Z.T., Hideg, K. and Hubbell, W.L. (1993) Science **262**, 1416–1419.
313. Resek, J.F., Farahbakhsh, Z.T., Hubbell, W.L. and Khorana, H.G. (1993) Biochemistry-USA **32**, 12025–12032.
314. Farrens, D.L., Altenbach, C.A., Yang, K., Hubbell, W.L. and Khorana, H.G. (1996) Science **274**, 768–770.
315. Yang, K., Farrens, D.L., Altenbach, C.A., Farahbakhsh, Z.T., Hubbell, W.L. and Khorana, H.G. (1996) Biochemistry-USA **35**, 14040–14046.
316. Dunham, T.D. and Farrens, D.L. (1999) J. Biol. Chem. **274**, 1683–1690.
317. Albert, A.D., Watts, A., Spooner, P.J.R., Gröbner, G., Young, J.E. and Yeagle, P.L. (1997) BBA-Biomembranes **1328**, 74–82.
318. Cornish, V.W., Benson, D.R., Altenbach, C.A., Hideg, K., Hubbell, W.L. and Schultz, P.G. (1994) Proc. Nat. Acad. Sci. USA **91**, 2910–2914.
319. Rothschild, K.J. and Gite, S. (1999) Curr. Opin. Biotechnol. **10**, 64–70.
320. Karnik, S.S. and Khorana, H.G. (1990) J. Biol. Chem. **265**, 17520–17524.
321. Davidson, F.F., Loewen, P.C. and Khorana, H.G. (1994) Proc. Nat. Acad. Sci. USA **91**, 4029–4033.
322. Yu, H.B., Kono, M., McKee, T.D. and Oprian, D.D. (1995) Biochemistry-USA **34**, 14963–14969.
323. Yu, H.B., Kono, M. and Oprian, D.D. (1999) Biochemistry-USA **38**, 12028–12032.
324. Yu, H.B. and Oprian, D.D. (1999) Biochemistry-USA **38**, 12033–12040.
325. Sheikh, S.P., Zvyaga, T.A., Lichtarge, O., Sakmar, T.P. and Bourne, H.R. (1996) Nature **383**, 347–350.
326. Doi, T., Molday, R.S. and Khorana, H.G. (1990) Proc. Nat. Acad. Sci. USA **87**, 4991–4995.
327. Anukanth, A. and Khorana, H.G. (1994) J. Biol. Chem. **269**, 19738–19744.
328. Borjigin, J. and Nathans, J. (1994) J. Biol. Chem. **269**, 14715–14722.
329. Franke, R.R., König, B., Sakmar, T.P., Khorana, H.G. and Hofmann, K.P. (1990) Science **250**, 123–125.
330. Ernst, O.P., Hofmann, K.P. and Sakmar, T.P. (1995) J. Biol. Chem. **270**, 10580–10586.
331. Han, M., Lin, S.W., Smith, S.O. and Sakmar, T.P. (1996) J. Biol. Chem. **271**, 32330–32336.
332. Han, M., Lin, S.W., Minkova, M., Smith, S.O. and Sakmar, T.P. (1996) J. Biol. Chem. **271**, 32337–32342.
333. Han, M., Lou, J.H., Nakanishi, K., Sakmar, T.P. and Smith, S.O. (1997) J. Biol. Chem. **272**, 23081–23085.
334. Zhukovsky, E.A., Robinson, P.R. and Oprian, D.D. (1991) Science **251**, 558–560.
335. Zhukovsky, E.A., Robinson, P.R. and Oprian, D.D. (1992) Biochemistry-USA **31**, 10400–10405.
336. Zvyaga, T.A., Min, K.C., Beck, M. and Sakmar, T.P. (1993) J. Biol. Chem. **268**, 4661–4667.
337. Rao, V.R., Cohen, G.B. and Oprian, D.D. (1994) Nature **367**, 639–642.
338. Zvyaga, T.A., Fahmy, K., Siebert, F. and Sakmar, T.P. (1996) Biochemistry-USA **35**, 7536–7545.
339. Zvyaga, T.A., Fahmy, K. and Sakmar, T.P. (1994) Biochemistry-USA **33**, 9753–9761.
340. Fahmy, K., Siebert, F. and Sakmar, T.P. (1994) Biochemistry-USA **33**, 13700–13705.
341. Fahmy, K. and Sakmar, T.P. (1993) Biochemistry-USA **32**, 9165–9171.
342. Lewis, J.W., Szundi, I., Fu, W.Y., Sakmar, T.P. and Kliger, D.S. (2000) Biochemistry-USA **39**, 599–606.
343. Mathies, R.A., Lin, S.W., Ames, J.B. and Pollard, W.T. (1991) Annu. Rev. Biophys. Biophys. Chem. **20**, 491–518.
344. Osborne, H.B. (1977) FEBS Lett. **84**, 217–220.
345. Rothschild, K.J., Cantore, W.A. and Marrero, H. (1983) Science **219**, 1333–1335.

346. Siebert, F., Mäntele, W. and Gerwert, K. (1983) Eur. J. Biochem. **136**, 119–127.
347. Braiman, M.S. and Rothschild, K.J. (1988) Annu. Rev. Biophys. Biophys. Chem. **17**, 541–570.
348. Surewicz, W.K., Mantsch, H.H. and Chapman, D. (1993) Biochemistry-USA **32**, 389–394.
349. Rothschild, K.J. and Sonar, S. (1995) in: CRC Handbook of Organic Photochemistry and Photobiology, eds W.M. Horspool and P.-S. Song. pp. 1521–1544, CRC Press, London.
350. Pebay-Peyroula, E., Rummel, G., Rosenbusch, J.P. and Landau, E.M. (1997) Science **277**, 1676–1681.
351. Luecke, H., Schobert, B., Richter, H.-T., Cartailler, J.-P. and Lanyi, J.K. (1999) J. Mol. Biol. **291**, 899–911.
352. Parker, F.S. (1983) Applications of Infrared, Raman and Resonance Raman Spectroscopy in Biochemistry, Plenum Press, New York.
353. Rothschild, K.J., Andrew, J.R., DeGrip, W.J. and Stanley, H.E. (1976) Science **191**, 1176–1178.
354. Callender, R.H., Doukas, A.G., Crouch, R.K. and Nakanishi, K. (1976) Biochemistry-USA **15**, 1621–1629.
355. Rothschild, K.J. and Stanley, H.E. (1974) Science **185**, 616–618.
356. Englander, J.J., Downer, N.W. and Englander, S.W. (1982) J. Biol. Chem. **257**, 7982–7986.
357. Downer, N.W., Bruchman, T.J. and Hazzard, J.H. (1986) J. Biol. Chem. **261**, 3640–3647.
358. Blout, E.R., DeLozé, C. and Asadourian, A. (2000) J. Amer. Chem. Soc. **83**, 1895–1900.
359. Osborne, H.B., Sardet, C., Michel-Villaz, M. and Chabre, M. (1978) J. Mol. Biol. **123**, 177–206.
360. Haris, P.I., Coke, M. and Chapman, D. (1989) Biochim. Biophys. Acta **995**, 160–167.
361. Rath, P., DeGrip, W.J. and Rothschild, K.J. (1998) Biophys. J. **74**, 192–198.
362. Osborne, H.B. and Nabedryk-Viala, E. (1978) Eur. J. Biochem. **89**, 81–88.
363. Clark, N.A. and Rothschild, K.J. (1982) Meth. Enzymol. **88**, 326–333.
364. Rothschild, K.J., Rosen, K.M. and Clark, N.A. (1980) Biophys. J. **31**, 45–52.
365. Rothschild, K.J., Zagaeski, M. and Cantore, W.A. (1981) Biochem. Biophys. Res. Commun. **103**, 483–489.
366. Rothschild, K.J. and DeGrip, W.J. (1986) Photobiochem. Photobiophys. **13**, 245–258.
367. Bagley, K.A., Balogh-Nair, V., Croteau, A.A., Dollinger, G., Ebrey, T.G., Eisenstein, L., Hong, M.K., Nakanishi, K. and Vittitow, J. (1985) Biochemistry-USA **24**, 6055–6071.
368. Eyring, G. and Mathies, R.A. (1979) Proc. Nat. Acad. Sci. USA **76**, 33–37.
369. Eyring, G., Curry, B., Broek, A., Lugtenburg, J. and Mathies, R.A. (1982) Biochemistry-USA **21**, 384–393.
370. Smith, S.O., Myers, A.B., Mathies, R.A., Pardoen, J.A., Winkel, C., VanDenBerg, E.M.M. and Lugtenburg, J. (1985) Biophys. J. **47**, 653–664.
371. Palings, I., Pardoen, J.A., VanDenBerg, E.M.M., Winkel, C., Lugtenburg, J. and Mathies, R.A. (1987) Biochemistry-USA **26**, 2544–2556.
372. Ohkita, Y.J., Sasaki, J., Maeda, A., Yoshizawa, T., Groesbeek, M., Verdegem, P.J.E. and Lugtenburg, J. (1995) Biophys. Chem. **56**, 71–78.
373. DeGrip, W.J., Gillespie, J. and Rothschild, K.J. (1985) Biochim. Biophys. Acta **809**, 97–106.
374. Maeda, A., Kandori, H., Yamazaki, Y., Nishimura, S., Hatanaka, M., Chon, Y.-S., Sasaki, J., Needleman, R. and Lanyi, J.K. (1997) J. Biochem. Tokyo **121**, 399–406.
375. Fahmy, K., Sakmar, T.P. and Siebert, F. (2000) Meth. Enzymol. **315**, 178–196.
376. Braiman, M.S., Mogi, T., Marti, T., Stern, L.J., Khorana, H.G. and Rothschild, K.J. (1988) Biochemistry-USA **27**, 8516–8520.
377. Metz, G., Siebert, F. and Engelhard, M. (1992) FEBS Lett. **303**, 237–241.
378. Sasaki, J., Lanyi, J.K., Needleman, R., Yoshizawa, T. and Maeda, A. (1994) Biochemistry-USA **33**, 3178–3184.
379. Rath, P., DeCaluwé, G.L.J., Bovee-Geurts, P.H.M., DeGrip, W.J. and Rothschild, K.J. (1993) Biochemistry-USA **32**, 10277–10282.
380. Fahmy, K., Jäger, F., Beck, M., Zvyaga, T.A., Sakmar, T.P. and Siebert, F. (1993) Proc. Nat. Acad. Sci. USA **90**, 10206–10210.
381. Nathans, J. (1990) Biochemistry-USA **29**, 937–942.
382. Nakayama, T.A. and Khorana, H.G. (1991) J. Biol. Chem. **266**, 4269–4275.
383. Beck, M., Sakmar, T.P. and Siebert, F. (1998) Biochemistry-USA **37**, 7630–7639.

384. Liu, X.-M., Lee, M.J., Coleman, M., Rath, P., Nilsson, A., Fischer, W.B., Bizounok, M., Herzfeld, J., Karstens, W.F.J., Raap, J., Lugtenburg, J. and Rothschild, K.J. (1998) Biochim. Biophys. Acta **1365**, 363–372.
385. Kandori, H., Kinoshita, N., Yamazaki, Y., Maeda, A., Shichida, Y., Needleman, R., Lanyi, J.K., Bizounok, M., Herzfeld, J., Raap, J. and Lugtenburg, J. (1999) Biochemistry-USA **38**, 9676–9683.
386. DeLange, F., Klaassen, C.H.W., Wallace-Williams, S.E., Bovee-Geurts, P.H.M., Liu, X.-M., DeGrip, W.J. and Rothschild, K.J. (1998) J. Biol. Chem. **273**, 23735–23739.
387. Janssen, J.J.M., Bovee-Geurts, P.H.M., Merkx, M. and DeGrip, W.J. (1995) J. Biol. Chem. **270**, 11222–11229.
388. Maeda, A., Sasaki, J., Shichida, Y. and Yoshizawa, T. (1992) Biochemistry-USA **31**, 462–467.
389. Fischer, W.B., Sonar, S., Marti, T., Khorana, H.G. and Rothschild, K.J. (1994) Biochemistry-USA **33**, 12757–12762.
390. Nishimura, S., Kandori, H., Nakagawa, M., Tsuda, M. and Maeda, A. (1997) Biochemistry-USA **36**, 864–870.
391. Nagata, T., Terakita, A., Kandori, H., Kojima, D., Shichida, Y. and Maeda, A. (1997) Biochemistry-USA **36**, 6164–6170.
392. Rath, P., DeLange, F., DeGrip, W.J. and Rothschild, K.J. (1998) Biochem. J. **329**, 713–717.
393. Marrero, H. and Rothschild, K.J. (1987) Biophys. J. **52**, 629–635.
394. Rath, P., Bovee-Geurts, P.H.M., DeGrip, W.J. and Rothschild, K.J. (1994) Biophys. J. **66**, 2085–2091.
395. Fahmy, K. (1998) Biophys. J. **75**, 1306–1318.
396. Gibson, N.J. and Brown, M.F. (1991) Photochem. Photobiol. **54**, 985–992.
397. Anderson, R.E. and Maude, M.B. (1970) Biochemistry-USA **9**, 3624–3628.
398. Nielsen, N.C., Fleischer, S. and McConnell, D.G. (1970) Biochim. Biophys. Acta **211**, 10–19.
399. Borggreven, J.M.P.M., Rotmans, J.P., Daemen, F.J.M. and Bonting, S.L. (1971) Arch. Biochem. Biophys. **145**, 290–299.
400. Aveldaño, M.I. (1988) Biochemistry-USA **27**, 1229–1239.
401. DeGrip, W.J., Drenthe, E.H.S., VanEchteld, C.J.A., DeKruijff, B. and Verkleij, A.J. (1979) Biochim. Biophys. Acta **558**, 330–337.
402. Cullis, P.R. and DeKruijff, B. (1979) Biochim. Biophys. Acta **559**, 399–420.
403. Mollevanger, L.C.P.J. and DeGrip, W.J. (1984) FEBS Lett. **169**, 256–260.
404. Deese, A.J. and Dratz, E.A. (1986) in: Progress in Protein-Lipid Interactions, eds A. Watts and J.J.H.H.M. DePont. pp. 45–82, Elsevier, Amsterdam.
405. Deese, A.J., Dratz, E.A. and Brown, M.F. (1981) FEBS Lett. **124**, 93–99.
406. Hessel, E., Herrmann, A., Müller, P., Schnetkamp, P.P.M. and Hofmann, K.P. (2000) Eur. J. Biochem. **267**, 1473–1483.
407. Nishimura, S., Sasaki, J., Kandori, H., Matsuda, T., Fukada, Y. and Maeda, A. (1996) Biochemistry-USA **35**, 13267–13271.
408. Nishimura, S., Kandori, H. and Maeda, A. (1998) Biochemistry-USA **37**, 15816–15824.
409. Okada, T., LeTrong, I., Fox, B.A., Behnke, C.A., Stenkamp, R.E. and Palczewski, K. (2000) J. Struct. Biol. **130** , 73–80.
410. Luecke, H., Schobert, B., Richter, H.-T., Cartailler, J.-P. and Lanyi, J.K. (1999) Science **286**, 255–260.
411. Subramaniam, S., Lindahl, I., Bullough, P.A., Faruqi, A.R., Tittor, J., Oesterhelt, D., Brown, L.S., Lanyi, J.K. and Henderson, R.A. (1999) J. Mol. Biol. **287**, 145–161.
412. Sheikh, S.P., Vilardarga, J.P., Baranski, T.J., Lichtarge, O., Iiri, T., Meng, E.C., Nissenson, R.A. and Bourne, H.R. (1999) J. Biol. Chem. **274**, 17033–17041.
413. Lu, Z.-L. and Hulme, E.C. (2000) J. Biol. Chem. **275**, 5682–5686.
414. Rieke, F. and Baylor, D.A. (2000) Neuron **26**, 181–186.
415. Meyer, C.K., Bohme, M., Ockenfels, A., Gärtner, W., Hofmann, K.P. and Ernst, O.P. (2000) J. Biol. Chem. **275**, 19713–19718.
416. Palczewski, K., Kumasaka, T., Hori, T., Behnke, C.A., Motoshima, H., Fox, B.A., LeTrong, I., Teller, D.C., Okada, T., Stenkamp, R.E., Yamamoto, M. and Miyano, M. (2000) Science **289**, 739–745.

CHAPTER 2

The Primary Photoreaction of Rhodopsin

R.A. MATHIES

*Chemistry Department,
University of California,
Berkeley*

J. LUGTENBURG

*Institute of Chemistry,
University of Leiden*

*Handbook of Biological Physics
Volume 3, edited by D.G. Stavenga, W.J. DeGrip and E.N. Pugh Jr*

Contents

1. Introduction

Light is essential for living organisms because it provides the energy for life through photosynthesis as well as the means to effectively interact with the environment through visual sensation. It is estimated that 70% of the information in the human brain is introduced by visual processes. In the eyes of the three phyla of animals – vertebrates, mollusks and arthropods – light from the outside world is converted into neural information by pigmented membrane proteins called rhodopsins that start the transduction process. In the human eye the rod cells initiate dim light vision and the cones trigger the process of color vision. For the majority of visual pigments, the chromophore is 11-*cis* retinal that is linked to the surrounding opsin protein via a Schiff base linkage to a specific lysine residue. In other cases, visual pigments exploit 3,4-didehydro retinal, 3-hydroxy retinal or 4-hydroxy retinal as the chromophore (this volume Chapters 6, 7). Visual pigments can have absorptions that range from the UV out to the near IR region of the spectrum (Ref. [1], Chapters 6 and 7). For example, bovine rhodopsin has a broad unresolved visible absorption spectrum with a maximum absorbance (λ_{max}) at 498 nm, while the human cone pigments exhibit electronic absorptions with λ_{max} values ranging from 425 to 560 nm. These differences in absorbance maxima, called the opsin shift, are determined by interactions of the chromophore with the surrounding protein. Fig. 1 presents a model of the rhodopsin chromophore–protein complex indicating the trans-membrane α-helices, the bound retinal chromophore, as well as some of the chromophore–protein interactions that mediate the opsin shift.

Classic studies by Wald and coworkers first revealed the photochemical process that initiates visual excitation [2]. In particular, Wald first showed that the primary action of light is to drive the 11-*cis*-to-*trans* isomerization of the bound retinal chromophore. Bovine rhodopsin is completely stable in the dark, but when illuminated it bleaches efficiently to form the free protein, called opsin, and all-*trans* retinal (see Fig. 2). In the eye, the all-*trans* retinal is enzymatically converted back to 11-*cis* retinal [3], which reacts with bovine opsin to regenerate rhodopsin. This spontaneous regeneration is useful because it allows the facile preparation of rhodopsins with isotopically modified or chemically modified chromophores. Irradiation of rhodopsin at liquid nitrogen temperatures showed that light causes the formation of a primary photoproduct called bathorhodopsin that absorbs maximally at 543 nm. Above 140°C bathorhodopsin thermally relaxes to lumirhodopsin (λ_{max} = 498 nm), which relaxes to metarhodopsin I followed by metarhodopsin II initiating the visual transduction process.

The chemical mechanism that generates the visual nerve pulse is now well understood [4]. Metarhodopsin II interacts with transducin, which is a heterotrimeric G-protein containing T_α, T_β and T_γ subunits. After exchange of GDP for GTP, this

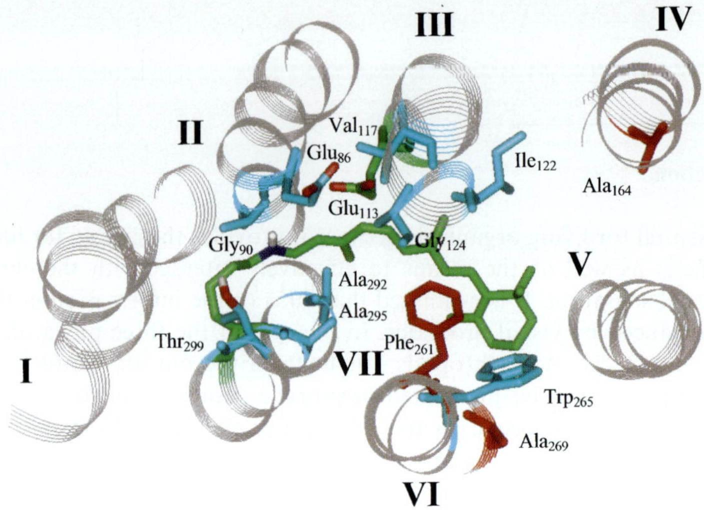

Fig. 1. Structural model of the human green-cone pigment viewed from the cytoplasmic side of the membrane. The 11-*cis*-retinal chromophore and its Glu113 counterion are shown in green. The green-pigment residues whose alteration is important for the color shift from the green to the blue pigment are shown in blue; green pigment residues whose alteration is important for the color shift from the green to the red pigment are shown in red. Transmembrane helices I–VII are indicated.

complex dissociates yielding T_{α}-GTP and the $T_{\beta\gamma}$ subunits. The T_{α}-GTP subunit activates a phosphodiesterase that catalyses the hydrolysis of 3′,5′-cyclic GMP thereby reducing the concentration of cGMP in the cell. In the dark, the cGMP effects the opening of the Na^+, K^+-channels in the plasma membrane of the rod cell, keeping the cell depolarized. The decrease in cGMP concentration after light absorption causes the closure of the Na^+, K^+-channels leading to the hyperpolarization of the plasma membrane and the generation of a nerve signal. Metarhodopsin II catalyses the activation of ~500 transducin molecules. In the other steps of the sequence, amplification of the signal molecules also takes place such that in a fully dark-adapted eye the absorption of one photon is sufficient to generate a nerve impulse [5]. Metarhodopsin II is deactivated through phosphorylation by rhodopsin kinase and binding to arrestin [6].

Rhodopsin is one of the most prominent members in the class of G-protein coupled receptors, to which many important receptor systems in the body belong [7]. G-protein coupled receptors, like rhodopsin, are seven α-helical *trans*-membrane proteins. One important difference between rhodopsin and the other G-protein coupled receptors is that activation of rhodopsin hyperpolarizes the cell by decreasing the concentration of the second messenger (cGMP) whereas most G-protein coupled receptors act by increasing the concentration of cGMP. In general the active site of such receptors interacts with small molecule ligands to effect the signal generation. In most cases no covalent bond is present between the receptor

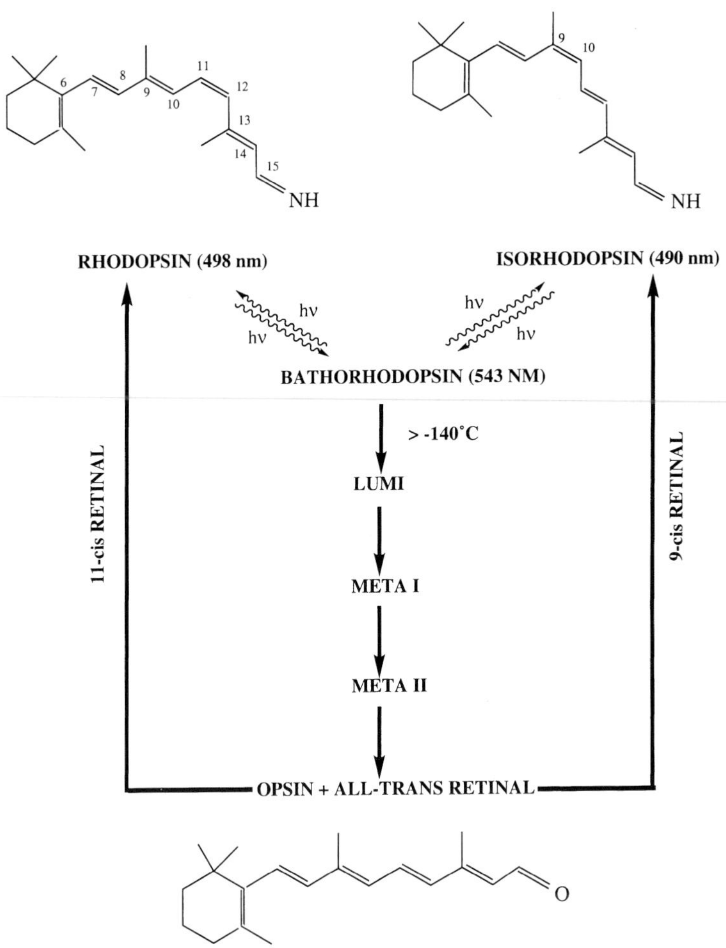

Fig. 2. Bleaching scheme of the visual pigment rhodopsin. Below −140°C the primary photoproduct bathorhodopsin can be trapped and studied in a photostationary steady-state mixture with 11-*cis* rhodopsin and 9-*cis* isorhodopsin. At physiological temperatures, bathorhodopsin decays through a series of intermediates to all-*trans* retinal plus opsin. Addition of exogenous 11-*cis* (or 9-*cis*) retinal spontaneously regenerates the pigment.

and antagonist or agonist. The rhodopsins are unique in that the agonist and antagonist are bound to the receptor and interconverted by light. Site-directed mutagenesis studies have produced rhodopsins where there is no covalent link between chromophore and protein and these systems exhibit all the properties of the native receptor system [8]; this work makes the relation between rhodopsin and other G-protein coupled receptors even closer. In visual pigments, 11-*cis* retinal thus acts as an antagonist while the agonist is the all-*trans* chromophore in metarhodopsin II. This unique trait allows us to study the molecular mechanism of receptor activation in rhodopsins with exquisite chemical detail from the primary photochemical process that switches 11-*cis* to all-*trans* retinal to the mechanism of the subsequent G-protein-triggered enzymatic cascade.

This chapter is focused on our understanding of the molecular mechanism of the primary events that initiate visual excitation. Our goals are to provide an understanding of the structure of the 11-*cis* retinal chromophore in rhodopsin, the interactions of the chromophore with the protein that determine its absorption maximum, the excited-state dynamics that lead to an efficient isomerization, and the structure of the primary photoproduct that is formed by *cis*-to-*trans* isomerization. Our understanding of these important parts of the visual excitation process has been advanced dramatically by the development of modern spectroscopic techniques, through the advancement of molecular biological methods for cloning and site-specific mutagenesis of receptor proteins, and by the marriage of these techniques through the unique capability of bio-organic chemists to make modified and isotopically labeled retinal agonists and antagonists. This work has led to a new understanding of not only the chemical interactions that determine the absorption maximum of visual pigments but also of the molecular mechanism of the *cis*-to-*trans* isomerization and the factors that determine the all-important quantum yield of the visual process. We are at last achieving a mechanistic understanding at the molecular level of the chemical reactions that are the basis of visual transduction.

2. Structure and interactions of the chromophore in rhodopsin

The definition of the structure of the retinal chromophore in rhodopsin is critical for understanding its chemical, spectroscopic and photochemical properties. The first key information was the determination that the chromophore is linked to the protein via a Schiff base linkage and that this linkage is protonated [9,10]. Subsequently the chromophore structure in rhodopsin has been detailed by vibrational structural measurements, solid-state NMR, CD measurements, chemical analog studies, and site-specific mutagenesis of the surrounding protein. These enhanced measurements now permit a more complete definition of the structure so that the functional spectroscopic and photochemical properties of the chromophore can be understood at the molecular level.

2.1. Raman vibrational studies of rhodopsin chromophore structure

Resonance Raman (RR) provided the first spectroscopic vehicle for obtaining structural information about the retinal chromophore in rhodopsin. Early studies with laser Raman spectroscopy were insightful but the spectra were complicated by the photochemical degradation of the photosensitive chromophore. The development of rapid flow techniques in 1975 [11] resolved this problem by flowing the rhodopsin sample through the laser beam sufficiently rapidly that there was insignificant build-up of photoproducts; by this means pure Raman spectra of rhodopsin could be obtained. The technology has continued to evolve with the development of modern spectrographs and highly sensitive CCD detectors to the point where it is now routine to obtain rhodopsin Raman spectra in minutes on a few ml of sample and low temperature microscope systems can generate high quality spectra on a few μl of pigment solution [12], facilitating studies of expressed rhodopsins [13].

The initial studies of rhodopsin vibrational structure were necessarily qualitative but very informative. For reference a high-quality RR spectrum of rhodopsin is reproduced in Fig. 3. First, a line was observed at 1655 cm^{-1} in a frequency range characteristic of a C=NH Schiff base stretch. When the vitamin A aldehyde binds with the primary aminogroup on the active site lysine side chain, water is released and a C=N Schiff base bond is formed. The frequency shifts observed in deuterated media demonstrated that this Schiff base linkage was protonated in the native protein [10,11,14]. Thus the chemical nature of the chromophore–protein linkage was identified and the major chemical factor (protonation) that produces a visible prosthetic group absorption was defined. Comparison of the RR spectra of the 11-*cis* protonated Schiff base (PSB) model compound in methanol with that of rhodopsin revealed that the C—C stretch and C—C—H rocking modes in the 1100–1400 cm^{-1} fingerprint region were virtually identical, indicating that the basic 11-*cis* chromophore structure was also preserved in the protein [15]. However, the ethylenic stretching mode at 1548 cm^{-1} was significantly lower in frequency than that of the 11-*cis* PSB (1556 cm^{-1}); this observation is diagnostic of the presence of additional interactions that red shift the absorption from 440 to 500 nm and produce a more delocalized vibrational structure. Furthermore, the line at 970 cm^{-1} was much more intense in the protein indicating altered excited-state dynamics as a result of protein binding. This important observation remained unexplained until much later (see Section 3.3). Finally, comparison with model compounds indicated that the modes between ~1000 and 1018 cm^{-1} in retinals were sensitive to the conformation

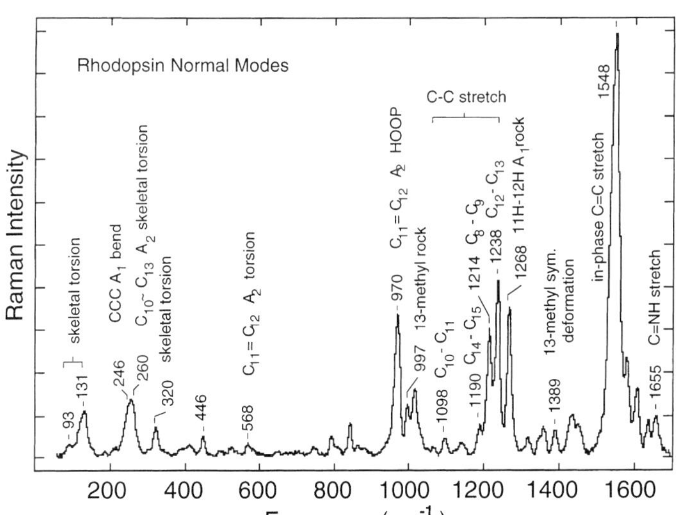

Fig. 3. Resonance Raman vibrational spectrum of the 11-*cis* retinal protonated Schiff base chromophore in rhodopsin with the vibrational assignments indicated [25].

about the C_{12}—C_{13} bond. These modes were initially thought to be due to isoprenoid methyl stretching modes. Their eventual assignment as methyl rocks [16] does not alter their significance. The steric interaction between the C_{10} H and C_{13}—CH_3 groups in the 12-s-*trans* conformer splits the degeneracy of the C_{13}—CH_3 and C_9—CH_3 rocks producing rhodopsin lines at 997 and 1018 cm^{-1}, respectively [14]. In the 11-*cis* retinal crystal, the 12-s-*cis* conformer with less steric interaction is preferred and the methyl rocks have similar frequencies. The protein, however, selects the 12-s-*trans* conformer giving rise to the characteristic splitting of these methyl modes. This selection of a single ligand conformer from an ensemble distribution in solution is very relevant for understanding the isomerization mechanism.

Since these early studies, a great deal of work has been performed to obtain vibrational assignments of retinals [17], of retinal PSB model compounds [18], of bacteriorhodopsin [19] and of rhodopsin and its intermediates [20–23]. This work was dependent on the development of methods for the synthesis of a wide variety of specifically ^{13}C- and ^{2}H-substituted retinals [24], their incorporation into rhodopsin, and the vibrational analysis of the isotopically induced shifts.

The assignments for the retinal chromophore in rhodopsin are summarized in Fig. 3, and a complete vibrational analysis will be found in Ref. [25]. The C_{10}—C_{11}, C_{14}—C_{15}, C_8—C_9 and C_{12}—C_{13} stretches of rhodopsin have been assigned at 1098, 1190, 1214, and 1238 cm^{-1}, respectively. The C_{10}—C_{11} stretch is characteristically low in wavenumber because it is adjacent to the C_{11}=C_{12} *cis* bond and, unlike the C_8—C_9 and C_{12}—C_{13} stretches, there is no methyl stretch interaction pushing the mode up [17]. The prominent band at 1268 cm^{-1} is the C_{11}=C_{12} A_1 in-plane hydrogen rocking mode that is characteristic of the 11-*cis* isomer. The significance of the methyl rocking modes at 997 and 1018 cm^{-1} for defining the 12-s-*trans* conformation was discussed above. The prominent $C_{11}H$=$C_{12}H$ A_2 hydrogen out-of-plane (HOOP) mode at 970 cm^{-1} also provides a nice probe of local molecular structure. An A_2 mode is Raman forbidden in a local C_{2v} point group; the observation of this mode in the RR spectrum demonstrates the reduction in symmetry to C_2 by out-of-plane distortion about the C_{10}—C_{11}, C_{11}=C_{12} and C_{12}—C_{13} bonds due to the C_{13}—CH_3 to 10-hydrogen interaction. By a similar argument, the torsional modes at 260 and 568 cm^{-1} are assigned as A_2. The 260 cm^{-1} mode is a delocalized skeletal torsion in the $C_{10}\cdots C_{13}$ region. The 568 cm^{-1} mode is assigned as a relatively localized C_{11}=C_{12} skeletal torsional mode.

While the frequencies in a RR spectrum are informative about the ground state chromophore structure, the intensities provide mode-specific information about the excited state structure and dynamics. When the RR spectrum of rhodopsin was first recorded, the vibronic sum methods used to analyze RR intensities were too difficult to apply to molecules with 25–30 coupled modes like rhodopsin and the theory made the relationship between intensities and excited-state dynamics distant. The development of wavepacket methods for analyzing RR intensities [26], improved computers, and the evolution of femtosecond studies of molecular dynamics dramatically altered this situation. The observation of more than 25 modes in rhodopsin with significant RR intensity demonstrates that there is a large reorganization of the excited state molecular structure about these degrees of freedom after excitation. Each, of course, reacts on its characteristic time scale ranging from ~10 fs for the highly displaced ethylenic stretch to ~500 fs for the lowest frequency skeletal torsional modes. The RR spectrum thus provides direct information about the complex multimode relaxation that occurs immediately after optical excitation. The availability of specific normal mode assignments allows us to focus on the subset of modes that are expected to contribute to the reactive portion of the overall chemical relaxation. These modes, the 970 cm^{-1} A_2 HOOP mode, the 568 cm^{-1} A_2 $C_{11}{=}C_{12}$ torsional mode and the ~250 cm^{-1} $C_{10} \cdots C_{13}$ skeletal torsion, are of most interest from a photochemical point of view because they are expected to contribute significantly to the isomerization reaction coordinate. A more quantitative discussion of the importance of this observation for the excited-state dynamics of rhodopsin will be found in Section 3.3.

2.2. Solid-state NMR studies of rhodopsin chromophore structure

Solid-state NMR techniques in combination with selective isotope enrichment provide another powerful method for investigating membrane protein receptors [27]. The labeling of the chromophore in vertebrate (bovine) rhodopsin is affected by regeneration of opsin with deuterium and/or ^{13}C-labeled 11-*cis* retinal. The specific isotopically labeled 11-*cis* retinals are prepared via total organic synthesis [24]. Biotechnological approaches have been developed to express eucaryotic membrane proteins such as bovine rhodopsin on media that allow the incorporation of isotopically labeled amino acids [28–30]. Site directed ^{13}C- and ^{15}N-isotopically labeled amino acids have been prepared for 10 of the 20 amino acids. Furthermore, all amino acids are available as uniformly ^{13}C- and ^{15}N-labeled isotopomers [31].

In a magic angle sample spinning (MASS) NMR experiment, the chemical shift anisotropy broadening of the NMR resonance is suppressed by macroscopic sample rotation around an axis at the magic angle (54°44′) with respect to the applied magnetic field. A detailed treatment of the MASS averaging of the NMR response can be found elsewhere [32]. The key point is that with specific isotopic labeling it is possible to perform NMR examinations of unique nuclei in receptor proteins like rhodopsin having an effective molecular weight of 40–50 kDa. In initial studies of rhodopsin, Smith and coworkers measured the isotropic chemical shift values of the carbon atoms at positions 5–15 using samples stabilized at −20°C [33]. Similarly, the isotropic chemical shift values of the model protonated 11-*cis* retinal Schiff bases have been obtained [34]. In Fig. 4 (top) the differences in isotropic chemical shift between rhodopsin and the model 11-*cis* retinal protonated Schiff bases have been plotted. For most of the atoms, the chemical shift differences are 2 ppm or less. In view of the fact that chromophore binding leads to an almost unperturbed molecule, the bond lengths should be similar to those in a recently determined X-ray structure of the all-*trans* protonated Schiff base [35]. The distances observed show that protonation leads to a delocalized structure at the Schiff base end and more pronounced bond alteration closer to the β-ionone ring. Looking again at Fig. 4 the differences range from 3 to 4 ppm at atoms 11 and 12 to 6 ppm at atom 13. The larger shift values at these positions correspond to about 0.1 higher positive electric charge at

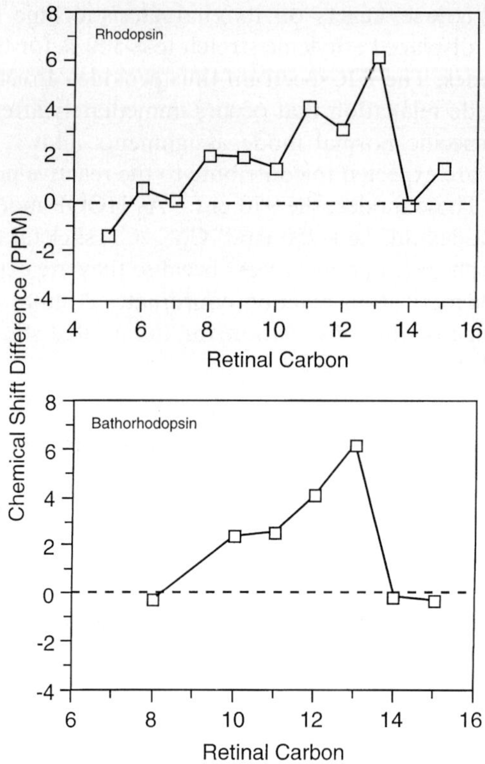

Fig. 4. Plots of the ^{13}C-chemical shift differences between carbon atoms in the rhodopsin chromophore vs. the solution spectra of the 11-*cis* retinal protonated Schiff base (top), and of the chemical shift differences between the carbon atoms in bathorhodopsin and the all-*trans* protonated Schiff base in solution (bottom). The perturbation of the chemical shift at the $C_{11}\cdots C_{13}$ positions is not significantly altered by photoconversion. Adapted from Refs. [33,108].

these atoms. This effect is depicted more clearly in Fig. 5 where the shift difference is color coded on the retinyl framework to indicate the orientation of the perturbation. This perturbation results from the polarization effects by the glutamate counterion (Glu113) near these positions.

MASS NMR spectra of rhodopsin with ^{15}N-labeling of the lysine residues have also recently been obtained [29,30]. The baculovirus Sfg expression system was used to incorporate 99% $^{15}N_2$-enriched L-lysine. The isotropic chemical shift of ^{15}N-lys 296 that links the retinylidene chromophore to trans-membrane helix 7 is 155.9 ppm. This value is appropriate for a protonated Schiff base structure stabilized by a counterion with an effective center-to-center distance between the counterion and the PSB of 4.3 ± 0.1 Å. In another study, the incorporation was effected by expression of the protein in cultures of HEK 293S cells [30]. The protein was isolated and introduced in 1.2-dialeoyl-sn-glycero-3-phosphochlorine. The ^{15}N-resonance corresponding to the protonated retinylidene Schiff base nitrogen was

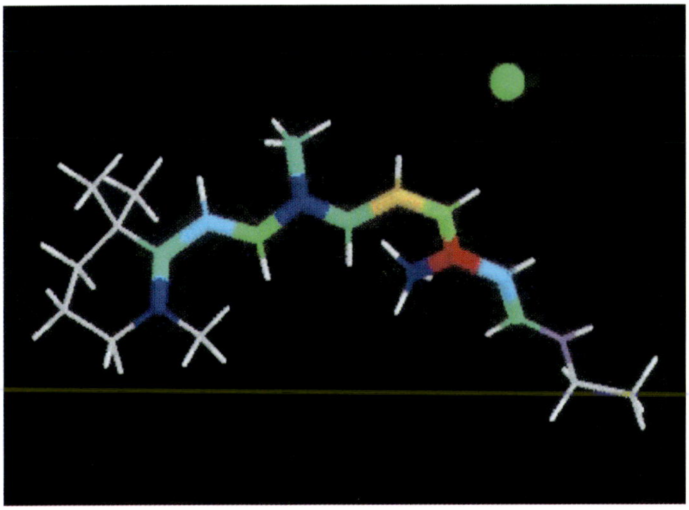

Fig. 5. Depiction of the NMR chemical shift for the chromophore in rhodopsin relative to the shifts for the 11-*cis* retinal protonated Schiff base in solution. The shift difference increases as the color goes from blue through green and yellow to red. The location of the counterion is schematically indicated by the green ball. Figure courtesy of Steven O. Smith.

observed at 156.8 ppm in the MASS spectrum suggesting an effective counterion distance of 4.4 Å. The difference between these two values (~1 ppm) is within the range expected as a result of the different lipid environments.

The principal values of ^{13}C-shift tensors can also be measured and used to obtain precise structural information. The ^{13}C-shift principal tensor elements can be determined by measuring the relative intensities of the rotational side bands and center band in MASS spectra as a function of the rotational speed and then fitting the intensities based on the calculations of Herzfeld and Berger [36]. These studies were carried out for $^{13}C_5$-rhodopsin to establish the conformation of the 6–7 single bond [33,37] and for $^{13}C_{14}$ to establish the C_{15}=N configuration [33]. In Fig. 6 the principal tensor element values for $^{13}C_5$-rhodopsin (A) are depicted together with those for $^{13}C_5$ 6-s-*cis* retinoic acid (B) and $^{15}C_5$ 6-s-*trans* retinoic acid (C). The tensor element σ_{33} is the diagnostic marker for conformation around the C_6—C_7 bond. The value of 210 ppm in rhodopsin is very close to that of the non-planar 6-s-*cis* model and very different from the planar 6-s-*trans* model. This demonstrates that in rhodopsin the conformation around this bond is non-planar 6-s-*cis*. The fact that the β-ionone moiety binds in a 6-s-*cis* conformation is consistent with chemically modified derivative studies on other rhodopsins [38]. Similarly, the anisotropy values of $^{13}C_{14}$-rhodopsin provide information on the configuration of the C_{15}=N double bond (Fig. 6). The σ_{22} and σ_{11} values for $^{13}C_{14}$-rhodopsin (D) are close to those of the C_{15} anti model (E) and different from the C_{15} syn model (F). Furthermore, the isotropic value of the $^{13}C_{14}$-rhodopsin resonance at 121.2 ppm is also more consistent with the C=N anti models (120–123 ppm) than the syn model (110.5 ppm).

angle for the C—C_{20} bond with respect to the membrane normal is 30 ± 5°, roughly opposite to that of C—C_{19}. These values define the orientation of the chromophore with respect to the membrane normal vector and are in agreement with the structure of the chromophore of rhodopsin presented in Fig. 7.

It is clear that with solid-state NMR techniques, very precise conformational and structural information on the chromophore or ligand within a membrane protein can be obtained even though these intrinsic membrane systems have no translation symmetry. This ability is particularly important when other structural techniques such as diffraction and solution NMR are not applicable or do not give sufficiently detailed information.

2.3. Retinal analog and CD studies of rhodopsin structure

Early work provided information on the importance of the shape of the chromophore for its binding and regeneration by opsin. In the active site of rhodopsin there is a binding pocket that recognizes the β-ionone group [49]. Molecules that have the structural elements of 11-*cis* retinal from the β-ionone ring up to the C_{11}-double bond, form complexes with opsin [50–52]. These analogs bind better as the conjugated tail is made longer, because these systems have more structural elements in common with 11-*cis* retinal. The rate of regeneration of these rhodopsin analogs is inversely related to their stability, meaning they are competitive inhibitors of rhodopsin formation. Larger systems such as the C_{18}-ketone, 13-*cis* retinal and all-*trans* retinal do not form complexes and they are not competitive inhibitors. Systems with an acetylenic bond at the 7,8 position have a more stick-like conformation in the β-ionone region which also prevents interaction with the β-ionone binding site of opsin [53].

The active site of the protein can accommodate significant structural changes in the chromophore, illustrating the induced-fit model for protein ligand interaction [54]. The stereochemistry of the modified 10-methyl rhodopsin chromophore has been studied with NMR [42]. The $r_{10\text{–}20}$ distance in this system is 3.5 ± 0.2 Å and that between $10CH_3$ and C_{13} groups is 3.1 ± 0.2 Å. The signals are as sharp as in the native case showing that again the protein selects only one chiral form. Due to the steric interaction between the two methyl groups, this system is more twisted (about 65°) than the native system. The conformation around the 12–13 bond is more affected in view of the larger $r_{10\text{–}20}$ distance (3.5 vs 3.1 Å). The much smaller ε_{max} (24,000 M^{-1} cm^{-1}) of 10-methyl rhodopsin vs 40,600 M^{-1} cm^{-1} for rhodopsin is a reflection of the larger twist in the central part of the modified structure. The ε_{max} values of conjugated systems do not change appreciably from a fully planar to a slightly twisted state; however, much more twist leads to a lowering of the ε_{max} value. 10–20 methanorhodopsin has also been studied. It has a more planar central structure than 10-methylrhodopsin. The additional covalent bond between 10-methyl and 13-methyl group in 10–20 methanorhodopsin locks the central part in an 11-*cis* structure preventing photoisomerization. Other than the photostability and 10-fold longer regeneration time, this chemically modified rhodopsin mimics rhodopsin reasonably well: it has approximately the same λ_{max} (~510 nm) and it has the same stability toward $NaBH_4$ and NH_2OH, but a lower ε_{max} value of 34,000 M^{-1} cm^{-1}.

Circular dichroism (CD) allows the establishment of the absolute chirality of the retinal chromophore in the $C_{10} \cdots C_{13}$ region. Rhodopsin has a strong positive CD effect in its long wavelength absorption band. Buss et al. [55] have been able to establish the absolute sense of twist of the C_{12}—C_{13} bond in the retinal chromophore based on calculations of the chiroptical data. The chromophore has the positive helicity as indicated in Fig. 7. This means that the interaction of the protein leads to a very strong chiral recognition. 11-*cis* retinal has 8 possible chiral forms due to the non-planarity of the 10-s-*trans*,12-s-*trans* and 10-s-*trans*,12-s-*cis* forms that are in dynamic equilibrium. The rhodopsin receptor protein only selects the 10-s-*trans*,12-s-*trans* form having positive helicity. This is the first system as far as we know where such strict chiral recognition in protein ligand interaction has been discovered. This chiral selectivity has profound implications for the photochemistry.

2.4. *Mechanism of the opsin shift*

While the protonated Schiff base of 11-*cis* retinal in organic solvents absorbs at 440 nm, this same 11-*cis* chromophore in human cone pigments absorbs either in the blue (~425 nm), green (~530 nm) or red (~560 nm), depending on its protein environment [56]. How does one chromophore detect light over such a wide spectral region? Chromophore–protein interactions that may cause this opsin shift include: (i) a weakening of the interaction of the positive charge of the retinal protonated Schiff base with its negative counterion or hydrogen bonding partner [9,57]; (ii) the placement of full or partial charges [58–62] or polarizable groups [63,64] near the polyene chain; (iii) planarization of the polyene chain caused by the protein environment [65,66]. The most commonly accepted explanation for the opsin shift has been the point charge model which regulates the absorption maximum by altering the placement of full or partial charged groups in the chromophore binding pocket [67]. The majority of the measurements in support of the point charge model were indirect in that they relied on comparisons with model compounds or on the introduction of non-isomorphous chemically modified retinals.

More recently, by combining site directed mutagenesis together with resonance Raman structural measurements, it has been possible to make a direct comparison of the structure of the retinal chromophore in different visual pigments and thereby elucidate the physical mechanism underlying the opsin shift. Figure 8 compares Raman spectra of the 11-*cis* retinal PSB in methanol (absorbing at 440 nm), a blue absorbing rhodopsin mutant (438 nm), the human green cone pigment (530 nm), and the human red cone pigment (560 nm) [68]. The blue-rhodopsin mutant protein was expressed using a rhodopsin gene that had been modified to replace nine key amino acid residues with the homologous residues present in the human blue pigment (M86L/G90S/A117G/E122L/A124T/W265Y/A292S/A295S/A299C). These residue replacements have recently been shown to account for the majority of the green-to-blue opsin shift [13]. The first striking observation is that the hydrogen

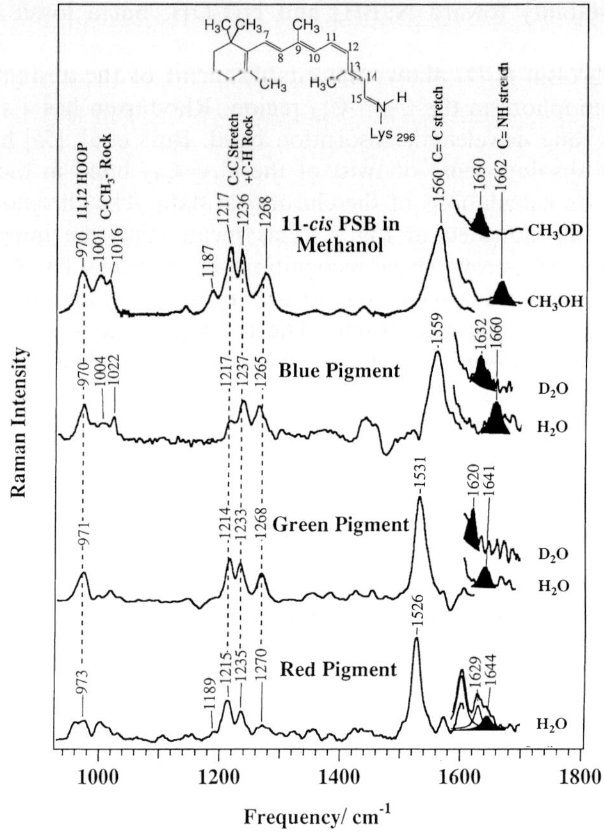

Fig. 8. Resonance Raman vibrational spectrum of the 11-*cis* retinal protonated Schiff base chromophore in methanol, of the blue-rhodopsin mutant pigment, of the human green cone pigment, and of the human red cone pigment. These data permit a direct visualization of the structural differences in the chromophore in blue, green and red visual pigments. The insets present portions of spectra recorded from D_2O buffers. Data adapted from Ref. [68].

out-of-plane, C—C, C=C, and Schiff base modes of the 11-*cis* PSB in the blue pigment are virtually identical to those of the isolated PSB in methanol. The blue pigment thus provides a chromophore environment that has no local perturbations of the chromophore vibrational structure and that solvates the Schiff base group like methanol.

Comparison of the blue pigment spectrum with that of the green pigment reveals a number of important differences associated with the amino acid alterations. There is a dramatic downshift of the C=C mode from 1559 to 1531 cm^{-1}, because the chromophore–protein interactions produce a more delocalized electronic structure. However, this enhanced delocalization is associated with only one specific local structural alteration. There is a significant downshift in the Schiff base mode by 19 cm^{-1} to 1641 cm^{-1}. Furthermore, the deuteration-induced shift of the Schiff base mode is reduced from ~28 cm^{-1} in the blue absorbing pigment to 21 cm^{-1}. These alterations reveal that formation of a green pigment involves a significant change in the Schiff base structure that is most likely due to a reduction of the dielectric interaction between the Schiff base group and its protein environment. These dielectric interactions keep the charge on the retinyl cation localized in the blue pigment; their removal in the green pigment permits enhanced charge delocalization and a green-shifted absorption. Consistent with these observations, the blue pigment residue alterations are dominated by the introduction of polar side chains (Ser90, Ser292, Ser295 and Cys299) in the chromophore binding pocket near the Schiff base group [13]. These dipolar residues differentially stabilize the ground state charge distribution of the chromophore and produce the blue shift of the pigment absorption.

Comparison of the green and the red pigment spectra reveals yet another pattern. The fingerprint, HOOP and other skeletal modes are identical in the green and red pigments, demonstrating that no new local perturbations are responsible for the green-to-red opsin shift. Consistent with the red-shifted absorption and delocalized electronic structure, the ethylenic mode has shifted down to 1526 cm^{-1}. After correcting for the fact that the red pigment spectrum was obtained at 77 K, its 1644 cm^{-1} Schiff base mode is structurally identical to that observed for the green pigment. Thus the shift in the absorption from the green to the red pigments must be caused by indirect electrostatic interactions. Site-specific mutagenesis shows that the three amino acid residues primarily responsible for the green-to-red opsin shift are Ser164, Tyr261, and Thr269 [13,59,60]. Each of these alterations involves the introduction of a dipolar residue in the chromophore binding pocket near the ionone ring end of the chromophore. The fact that no specific local perturbations of the vibrational structure are observed in the red pigment suggests that these dipolar residues interact electrostatically with the chromophore's excited state charge distribution to lower the transition energy. Retinyl polyenes experience a dramatic change in charge distribution upon excitation. The change in dipole moment is of the order of ~15 D [69] and oriented such as to shift positive charge toward the ionone ring in the excited state. The Ser164, Tyr261, and Thr269 residues in the red pigment are ideally positioned to stabilize this excited state charge distribution and lower the electronic transition energy.

The vibrational data together with sequence alignment and molecular modeling [68] allow us to advance our understanding of color pigment spectral tuning beyond previous explanations based on differential point-charge perturbations [67]. This new model is summarized for clarity in Fig. 9. We now understand that direct dipolar electrostatic interactions with the ground state chromophore charge distribution dominate the shift of the absorption maximum from the green to the blue pigments. Longer range, dipolar interactions between polar protein hydroxy dipoles and the change in electric dipole moment upon electronic excitation are responsible for the shift of the absorption maximum from the green to the red pigments. Nature thus exploits the dielectric interaction of polar protein residues with the asymmetrical and polarizable charge distribution of the retinal prosthetic group to give us the vivid sensation of color.

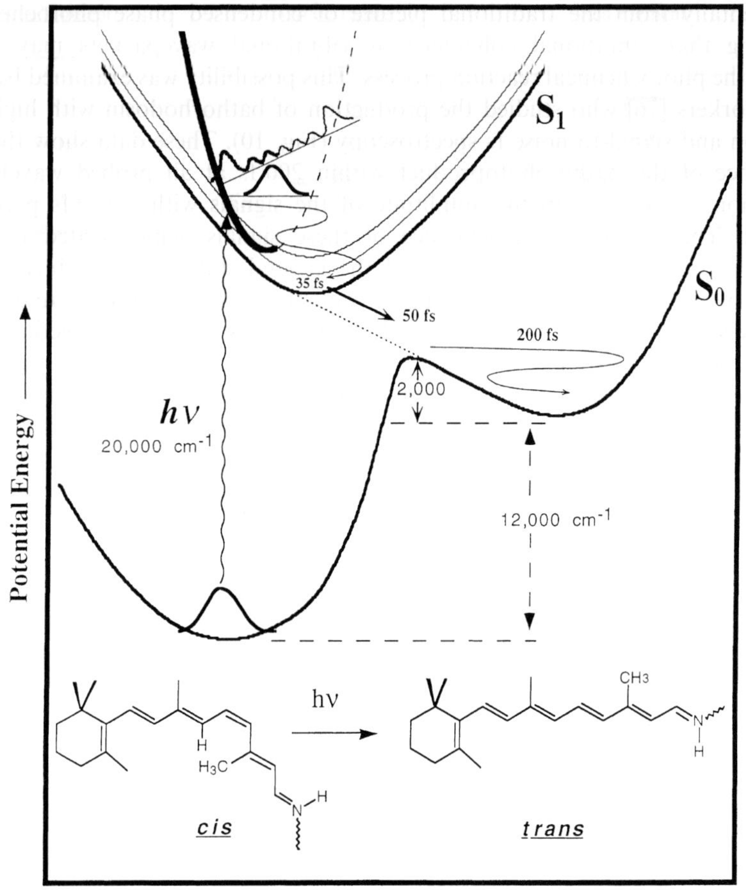

Fig. 11. Schematic potential energy surfaces and reaction dynamics for the femtosecond isomerization of rhodopsin. The dotted line indicates the diabatic pathway along which the reaction proceeds.

3.2. Spontaneous and stimulated emission measurements

Measurements of the spontaneous emission spectrum and quantum yield provide the most basic information on rhodopsin's excited state dynamics. The fluorescence quantum yield of rhodopsin has been determined to be $\sim 10^{-5}$ [77,78]. Together with the radiative lifetime of 1 ns, a Strickler–Berg analysis suggests that the excited-state lifetime is $\sim$100 fs. Furthermore the emission spectrum is found to be excitation wavelength dependent, shifting from 16,800 to 14,200 cm^{-1} when the excitation is shifted from 472 to 568 nm [78]. This observation is consistent with the idea that the emission arises from a non-stationary excited-state population. The spontaneous emission is difficult to time resolve because of its ultra-low yield. Nevertheless

Shichida and coworkers have recently succeeded in measuring the time-resolved spontaneous emission, observing decay times of 146 fs (80%) and 1.5 ps at 578 nm and 330 fs and 1.7 ps at 635 nm [79]. They concluded that the fast decay component was indicative of the coherently reacting excited state population, consistent with the 200 fs product appearance time measured by transient absorption. They further concluded that the slow ps decay component was due to vibrationally relaxed components of the excited state population that dephased and followed more conventional relaxation pathways primarily back to unreacted rhodopsin.

The emission from the excited state of rhodopsin can also be probed by stimulated emission in conventional pump-probe experiments. This approach has the disadvantage that the signals are combined with all transient excited state absorption, reactant bleaching and photoproduct formation signals. This overlap can make the emission signals hard to observe and even harder to unambiguously interpret. Such stimulated emission signals from octopus rhodopsin were first reported by Kobayashi and coworkers [80]. They observed a feeble stimulated emission signal near 630 nm with a subpicosecond appearance and decay time as well as a significant stimulated emission at 860–920 nm that decayed with a 2.2 ps time constant. These traces correspond well with the observations of Shichida and coworkers [79] and were interpreted as measurements from the fast coherently evolving population and a dephased and partially thermalized population that slowly relaxes back to the rhodopsin ground state accounting for the unsuccessful fraction of the isomerization quantum yield. In later measurements on rhodopsin, including transient anisotropy measurements, stimulated emission signals at 900 nm were observed that rose and decayed within ~100 fs [81]. The inconsistency of the stimulated emission decay time with the earlier measurements on octopus rhodopsin was not resolved. The observation of rapid red-shifted stimulated emission is best explained by the Stokes relaxation of the excited state along both high and low frequency vibrational modes consistent with multimode Raman intensity analyses of these excited states.

3.3. Resonance Raman intensity analysis of multimode isomerization dynamics

The measurement of absolute Raman scattering cross sections for each of rhodopsin's RR active modes permits a F–C vibrational analysis of the electronic excited state to determine the vibrational modes that are strongly coupled to the optical absorption and to quantitate the excited-state displacements of their potential minima [82]. These displacements reflect the initial motion of the molecule along the excited-state potential surface out of the F–C region. Since the intensities of each mode can be measured, this provides a direct multimode probe of the excited-state nuclear dynamics. The intensities can also be related to the mode-specific, excited-state reorganization energies since the displacement for a particular mode Δ is related to its contribution to the reorganization energy λ by $\lambda = 0.5\Delta^2\omega$, where ω is the vibrational frequency. Reference to the Raman spectrum in Fig. 3 reveals that more than 25 normal modes are displaced in the excited state including C=C, stretches, C—C stretches, HOOP modes and skeletal torsional modes. The higher the intensity, the greater is the displacement as well as the reorganization energy along that degree of freedom. The rhodopsin cross sections were first measured and quantitatively analyzed by Loppnow and Mathies [83]. The slopes of the torsional modes in the F–C region calculated from the RR intensities predicted a 100 fs 90° twist time and a 200 fs isomerization time, consistent with later experiments. These data also permit the calculation of the contribution of each of the RR active modes to the reorganization energy: for example the C=C mode ($\Delta = 0.9$) is predicted to contribute a dominant 627 cm^{-1} to the reorganization energy and all the RR active modes together contribute a total relaxation of 2400 cm^{-1}. The idea that C=C and other modes contribute to the reorganization of the excited state of retinals has reemerged in time resolved studies of bacteriorhodopsin and more recently in rhodopsin [81,84]. The previous RR intensity analysis of rhodopsin, bacteriorhodopsin and retinals provide a mode specific and quantitative examination of this effect [83,85,86]. Very recently the Raman intensities of rhodopsin have been successfully modeled using *ab initio* methods providing further support for this multimode relaxation about both in-plane and out-of-plane degrees of freedom [87].

It has been proposed that the ultrafast isomerization in rhodopsin is driven by the instantaneous relaxation of the excited polyene via torsional modes that project directly along the $C_{11}=C_{12}$ torsion. These dynamics may be realized by steep slopes in the fundamental vibrational transitions of $C_{11}=C_{12}$ torsional modes at 260 and 568 cm^{-1}. The displacement along the 568 cm^{-1} mode indicates that the retinal chromophore on the S_1 surface instantaneously experiences a torque that twists the molecule about the $C_{11}=C_{12}$ bond. This initial rotation is accompanied by skeletal distortion, represented by the displacement of the 260 cm^{-1} mode, resulting from concerted A_2 torsions about other C—C and C=C bonds. The slopes along the 260 and 568 cm^{-1} coordinates accelerate and help guide the nuclei toward isomerization about the $C_{11}=C_{12}$ bond. Consistent with this interpretation, the isomerization rates of isorhodopsin and 13-demethyl rhodopsin, which lack RR activity in the relevant torsional coordinates, are a factor of 3 and 2, respectively, slower than rhodopsin [88,89]. The importance of initial excited-state displacements to the overall photochemistry has also been suggested for the conrotatory ring-opening dynamics of cyclohexadiene [90] and the isomerization of *cis*-stilbene [91].

3.4. *Dynamical information from Fourier transformation of the optical absorption*

Time-dependent information about rhodopsin's excited-state dynamics can also be obtained through Fourier transformation of the optical absorption spectra. The frequency-dependent optical absorption spectra of native, 11 D and 11–12 D_2 rhodopsin have been measured; their difference spectra show specific changes in the electronic absorption induced by the isotopic substitution [92]. After Fourier transformation, the absolute values of the time correlation function reveal that the effects of deuterium substitution are not evident in the first 20 fs of the excited-state dynamics, weak changes appear in the 20–60 fs time range, more significant changes appear in the 70–120 fs time range, and the time range from 120–170 fs exhibits even more complicated changes. This indicates that the motions of hydrogen atoms at C_{11} and C_{12} are significantly coupled with the skeletal motion of the chromophore in the time range 70–100 fs. This observation agrees with the slopes of the torsional modes in the F–C region that were calculated from RR intensities leading to a prediction of a 100 fs 90° twist time. The hydrogen motions after 100 fs contribute to the excited-state dynamics in a more complicated way. Around this time the branching between the 2/3 of the excited species that will lead to ground state bathorhodopsin and the other 1/3 that remains in the excited state rhodopsin occurs. The torsional dynamics of molecules following these two different pathways may be effected differently by deuterium substitution. Overall these results show that the information obtained via Fourier transformation of optical absorption spectra of rhodopsin and the information obtained via RR intensity analysis are in agreement.

3.5. Excited state dynamics from chemically modified chromophores

Studies of the effect of chromophore chemical modification on the photochemistry also provide important information about the nature of the excited-state dynamics. In addition to rhodopsin, 10-methylrhodopsin (λ_{max} = 506 nm, 24,000 M^{-1} cm^{-1}), 13-demethyl rhodopsin (λ_{max} = 500 nm, 44,000 M^{-1} cm^{-1}) and 10-methyl, 13-demethyl rhodopsin have been studied. These four systems form a set that differs in the possible methyl substitution pattern at the 10- and 13-positions. Based on the structural information derived for rhodopsin and 10-methyl rhodopsin it is to be expected that the 13-demethyl and 10-methyl,13-demethyl chromophores bind in the same chiral conformation. This will lead to a more planar form in the case of 13-demethyl rhodopsin and a conformation that is similar to rhodopsin in the isomerization region for 10-methyl,13-demethyl rhodopsin (however, in the latter form additional twists should exist around the 8–9 bond due to the $10CH_3 \cdots 8H$ interaction). The quantum yields of the native ($\phi = 0.67$), the 10-methyl analog ($\phi = 0.55$), the 13-demethyl ($\phi = 0.47$), and the 10-methyl-13-demethyl ($\phi = 0.33$) forms have been determined [93]. Based on these data it is clear that the effect of the 10- and 13-methyl groups follows a simple addition rule. The presence of a $13CH_3$ increases the quantum yield by 0.20, whereas the presence of a $10CH_3$ group decreases the quantum yield by ~0.12. Methyl substitution may affect the quantum yield through both an electronic and a steric or conformational effect. The $13CH_3$ group is important because it influences the twist around the C_{12}—C_{13} bond whereas the $10CH_3$ group acts by influencing the electronic charge in the excited state. This suggests that alkyl substitution at various positions may exhibit different effects, because in the ground state the positive change is located on the odd numbered carbon atoms, while in the electronically excited state the positive change is more delocalized.

Furthermore, there is a relation between the ground state structures, the quantum yield and the formation kinetics. The native bathorhodopsin photoproduct forms within 200 fs while that of the 13-demethyl chromophore forms within 400 fs [89]. 13-demethyl rhodopsin has a more planar structure due to the absence of the $C_{10}H \cdots C_{13}H_3$ interaction, and this is reflected by the somewhat higher extinction coefficient (44,000 M^{-1} cm^{-1}) compared to that of the native pigment (40,600 M^{-1} cm^{-1}). Similarly, if the *cis* bond is moved to the unhindered C_9=C_{10} position in 9-*cis* isorhodopsin, the quantum yield drops to 0.20 and the isomerization time lengthens to 600 fs [88]. A plot of these available quantum yields vs. the isomerization time is presented in Fig. 12. The observed linear correlation shows that the speed and the efficiency of the photochemical reaction are directly related and that the ground state conformation plays a critical role. This important observation was anticipated in early studies of triene photochemistry, where it was discovered that there is a direct relation between the ground state conformation of alkyl substituted trienes and the products formed after photochemistry with UV irradiation [94].

When photochemistry is performed within a protein active site, these conformational questions become even more subtle and important. Since the barriers for conformational transitions in trienes are low, the various chiral conformers are in rapid dynamical equilibrium and cannot be easily isolated or frozen out. However, in the case of the chromophore of rhodopsin, only one chiral conformation with positive helicity is selected by the protein. This conformational selectivity provides a powerful tool for studying how the ground state conformation determines the photochemical properties. First, the conformation present in the protein determines the double bond that preferentially isomerizes, in this case the $C_{11}{=}C_{12}$. In Fig. 7 the C_{20}-methyl group projects out of the plane and the nonbonded interaction with the 10-hydrogen contributes to the high regioselectivity of the isomerization. The molecule is in effect distorted along the reaction coordinate and primed for 11-*cis*-to-*trans* photochemistry. Secondly, the conformation also determines the direction of the rotational motion of the $C_{12}H$ fragment that converts the rhodopsin structure into the batho structure. In conventional solution phase photochemistry the methyl group could be in front of, or behind, the plane in Fig. 7 in a 50/50 mixture. Isomerization through the methyl-10-hydrogen non-bonding barrier is

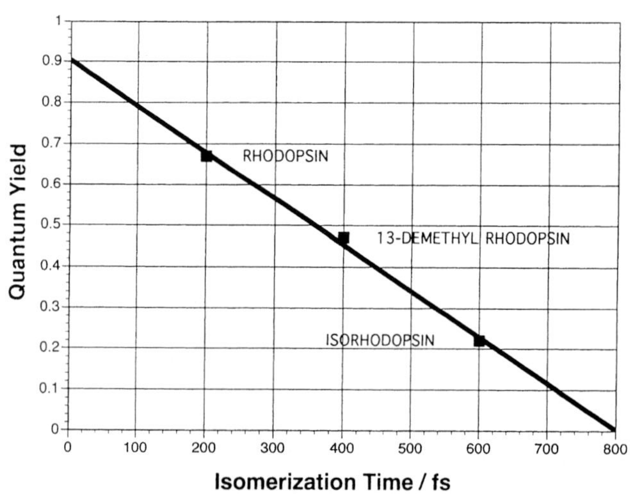

Fig. 12. Plot of the photochemical quantum yield for some rhodopsins vs their photoproduct formation times. It is seen that the quantum yield is directly related to the speed of the isomerization.

unlikely, so some molecules will isomerize in a clockwise sense and some will go counterclockwise. In the protein these two different rotational directions are inequivalent because of the protein environment. One direction is very efficient and the other might be very inefficient. Thus by selecting one and only one conformer, the protein opsin may enhance the quantum yield by as much as a factor of two. Rhodopsin thus provides a dramatic and important example of the role of ground state conformational control over photochemical reactions. By comparing the photochemistry of rhodopsin and carefully selected rhodopsin mutants (as well as chemical chromophore analogs) it should be possible to obtain a set of rhodopsins with positive helicity that differ in their precise conformational geometry and environmental constraints. Such a study would further our understanding of the ground state structural factors that determine the quantum yield as well as the femtosecond dynamics of the isomerization process. In this way studying the photochemistry of rhodopsin also serves as a paradigm for discovering the factors that govern the photochemistry of other light-sensing and energy-transducing systems.

4. Structure and interactions of the chromophore in the primary photoproduct

4.1. Vibrational studies of chromophore structure in bathorhodopsin

Since the existence of the primary bathorhodopsin photoproduct was first demonstrated by Yoshizawa and Wald [95], the structure of the chromophore in this intermediate state has remained an enigma. The bathochromically shifted absorption, the fact that this intermediate can only be trapped at very low temperatures (77 K), the high 35 kcal/mole enthalpy and $>60\%$ energy storage efficiency [96], as well as the rapid 200 fs formation time [72] are all very unique and unusual properties. The first structural data that reported on the bathorhodopsin chromophore was obtained by Oseroff and Callender [10] who trapped bathorhodopsin at 77 K and recorded the RR spectrum of the photoproduct. They observed that bathorhodopsin had a very delocalized electronic structure, indicated by its 1536 cm^{-1} ethylenic line and three vibrational lines at 850, 870 and 920 cm^{-1}, whose intensities and frequencies were unprecedented. It was clear that the primary photoproduct chromophore was distorted, but the configuration of the chromophore and the nature of this distortion was unknown. The most interesting speculation proposed proton transfer and the formation of a fully tautomerized chromophore in the primary step [97,98].

The definition of the structure of the chromophore in the primary photoproduct was first addressed through the combination of physical organic chemistry and RR spectroscopy. A very extensive series of isotopically substituted retinal derivatives were synthesized and incorporated into rhodopsin to assign the Raman vibrational spectra of rhodopsin and, in particular, of bathorhodopsin. The first studies showed that bathorhodopsin has a protonated Schiff base linkage that is nearly structurally identical to that of rhodopsin [99]. This observation disproved proton tautomerism models for the batho structure. Proposals for the energy storage that involved Schiff base charge separation from its counterion or translocation of the Schiff base into a new protein dielectric environment were also inconsistent with these observations [100]. Deuterium substitution of the chain methyl groups and the vinyl hydrogens showed that the unique bathorhodopsin "low wavenumber" lines were in fact due to the out-of-plane wagging of the 10, 11, 12 and 14 hydrogens [23,101,102]. The satisfactory comparison of the fingerprint vibrational frequencies of bathorhodopsin with those of the all-*trans* PSB argued that the chromophore had adopted a formal *trans* configuration about the $C_{11}{=}C_{12}$ bond [21]. Fig. 13 presents a summary of the vibrational structure of the bathorhodopsin chromophore with the vibrational assignments and a depiction of the chromophore structure from Palings et al. [21]. The placement of a negative charge perturbation, now recognized to be the Glu113 counterion, in closest proximity to the 10, 12 and 14 hydrogens was based on the fact that the vibrational analysis of the C—H wagging modes required a significant

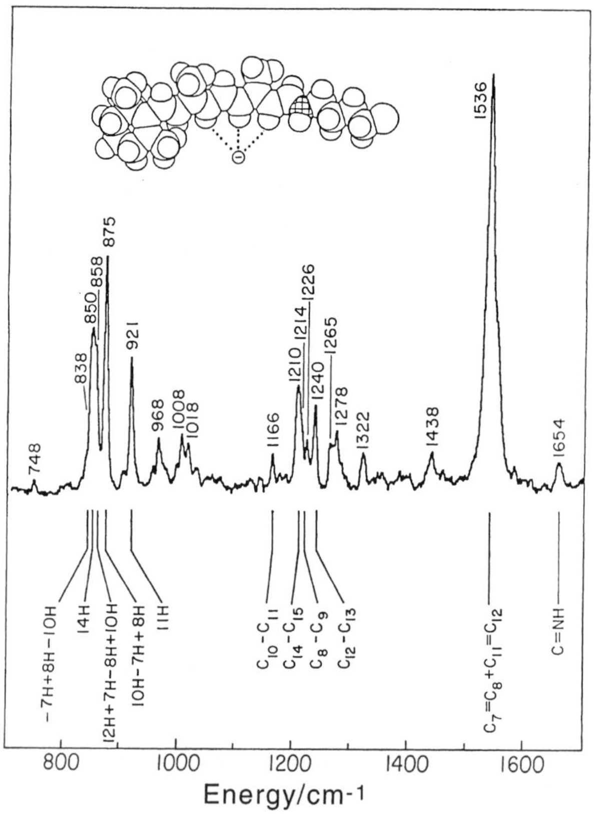

Fig. 13. Summary of the normal mode assignments for bathorhodopsin. Skeletal stretching assignments are taken from Palings et al. [21]. The inset presents a model for the chromophore in bathorhodopsin that is twisted 20° about the $C_8{-}C_9$, $C_{10}{-}C_{11}$, $C_{12}{-}C_{13}$ and $C_{14}{-}C_{15}$ bonds and perturbed by a negatively charged residue near C_{12}. While the twist magnitude was properly predicted in these earlier studies, the arbitrarily selected chiral sense was opposite to that eventually shown by CD analysis.

reduction of the diagonal force constants of only the 10, 12 and 14 hydrogens. The large reduction of the C_{12}—H wag force constant and consequently its low intrinsic frequency removed its degeneracy with the C_{11}—H wag, thereby causing the very unusual uncoupling of the C_{11}H and C_{12}H wagging modes in bathorhodopsin. This model was used to explain the unique bathorhodopsin vibrational lines and nicely anticipated the location of the Glu113 counterion. Furthermore, these observations suggest that the majority of the energy storage in the bathophotoproduct is due to the interaction of Glu113 with the chain portion of the chromophore in concert with the indicated structural distortions.

It is also pertinent to consider the role of the protein environment in the rhodopsin-to-bathorhodopsin transition. The magnitude of the protein structural changes can be qualitatively estimated by considering a few facts. First, the rhodopsin-to-batho transition occurs at temperatures as low as 4 K and the available measurements indicate that the low temperature trapped chromophore structure is identical to the room temperature time-resolved structure obtained with 1–2 ps time resolution [103,104]. These data suggest that only small protein structural changes are occurring in the primary transition. An additional probe of protein structural changes is provided by low-temperature FTIR studies of the rhodopsin-bathorhodopsin difference spectrum [105,106]. These data are dominated by chromophore modes of the rhodopsin reactant disappearing and of the bathorhodopsin photoproduct appearing. Only very weak protein difference features were observed at ~1760 cm^{-1}, indicative of a small number of minor carboxyl environmental changes. However, later studies over a wider spectral range by Maeda and coworkers [107] revealed significant difference signals in the water OH stretching band as well as indole N–H difference features. These results support the idea that while small protein structural changes occur in the primary step, there are significant alterations of water structure in the binding site as well as changes of the hydrogen bonding of a tryptophan residue upon batho formation. The counterion location in batho as well as the overall shape of the binding pocket are probably identical to that in rhodopsin. The lack of protein relaxation means that once the chromophore formally isomerizes to C_{11}=C_{12} *trans*, the isomerized chromophore–protein complex is held in a highly strained conformation that stores considerable energy despite the water structure and hydrogen bonding relaxations.

4.2. Solid-state ^{13}C-NMR studies of chromophore structure in bathorhodopsin

Magic-angle sample spinning NMR spectra have also been obtained of the bathorhodopsin photointermediate trapped at low temperature (< 130 K) with retinals specifically ^{13}C-labeled at positions 8, 10, 11, 12, 13, 14 and 15 [108]. Comparison of the chemical shifts of the bathorhodopsin resonances with those of an all-*trans* retinal PSB chloride salt show the largest difference (6.0 ppm) at position 13 of the protein-bound retinal. The differences in chemical shift between bathorhodopsin and the all-*trans* protonated Schiff base are summarized in the bottom panel of Fig. 4. The chemical shift value of C_{14} (120.0 ppm) is in agreement with a C=N anti configuration as found in rhodopsin [27]. Small differences in chemical shift between bathorhodopsin and the all-*trans* PSB model compound are also observed at positions 10, 11, and 12. The effects are almost equal in magnitude to those previously observed in rhodopsin. Similarly, as in the case of rhodopsin, glutamate 113 induces a change of 0.1 electronic charge on the central positions of the chromophore, especially C_{13}. This amazing similarity in the rhodopsin and bathorhodopsin structures indicates that the energy stored in the primary photoproduct bathorhodopsin is not associated with any substantial change in the average electron density at the labeled positions. These data indicate that the electronic and structural properties of the protein environment are similar to those in rhodopsin. In particular, a previously proposed perturbation near position 13 of the retinal (due to Glu113) appears not to change its position significantly with respect to the chromophore upon isomerization. This result excludes charge separation between the chromophore and a protein residue as the dominant mechanism for energy storage in the primary photoproduct and argues that the light energy is stored primarily in the form of distortions of the bathorhodopsin chromophore and steric chromophore–protein interactions.

4.3. Modeling bathorhodopsin chromophore structure and CD spectra

A variety of first semi-empirical [109,110] and more recently *ab initio* [111,112] methods have been used to model the excited-state structure and reaction dynamics of the retinal chromophore in rhodopsin. Modeling the structure of the chromophore in bathorhodopsin is more problematic, because its structure is less well defined experimentally. Car–Parinello *ab initio* calculations, which were very effective in modeling the rhodopsin structure, have also been used to model the batho structure [113,114]. A highly distorted all-*trans* configuration is predicted with minor changes compared to the rhodopsin structure (Fig. 14). The calculated strain energy within just the chromophore is 22 kcal/mol, which compares well with the 32 kcal/mol strain for bathorhodopsin. The indicated structure for bathorhodopsin has been used by Buss et al. [115] to calculate the CD spectrum of bathorhodopsin. It is predicted to be negative in the long wavelength region with an amplitude that is three-fold more intense than rhodopsin. These facts are in agreement with the experimental CD spectra of rhodopsin and bathorhodopsin [116]. Comparison of the rhodopsin and batho structures shows that the main change must be an out-of-plane rotation of the $C_{12}H$ bond converting an 11-*cis* into an 11-*trans* structure. In principle this rotation could occur in either of two directions; however due to the chirality of both chromophores, the rotation with a right sense rotation should be the one with the least motion converting rhodopsin into the bathorhodopsin. The fact that the $C_{12}H$ rotation is an important factor in the isomerization is nicely supported by studies of 11–19 ethanorhodopsin that show unperturbed photochemical conversion to the batho form [117–119].

5. Molecular mechanism of the *cis–trans* isomerization in vision

The information presented here on the ground state structure of the chromophore in rhodopsin, on its photochemical dynamics, and on the chromophore structure in

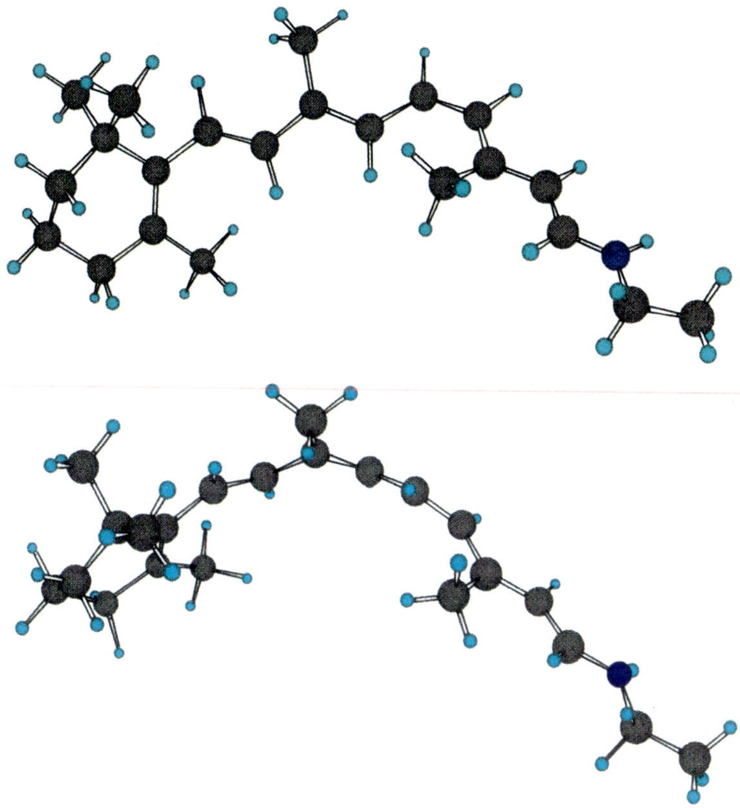

Fig. 14. Comparison of the structures predicted for rhodopsin (top) and bathorhodopsin (bottom) based on Car–Parinello calculations [113,114]. It is striking that a formal *cis–trans* isomerization can be accomplished with such a small change in overall molecular shape.

bathorhodopsin allows us to develop a complete mechanistic understanding of the primary event in vision. This process is most easily understood by discussing the temporal sequence of events that lead to the production of a thermalized bathorhodopsin photoproduct.

The very first step in vision is the vertical electronic excitation to the F–C region of rhodopsin's excited state potential energy surface (PES); the 10 fs time scale of this process is determined by the decay of the optical correlation function [83]. This transition is accompanied by a 12–15 D change in dipole moment which shifts net positive charge toward the ionone ring end of the chromophore [69,120]. The interaction of environmental dipoles with the ground and excited state charge distributions of the chromophore determine the energetics of the optically coupled states and the opsin shift. It should be noted that this charge shift formally consists of an intrinsic change in dipole moment $\Delta\mu$ plus a component due to the interaction of the polarizable retinal chromophore with its protein dielectric environment. The excited state dipole moment polarizes the protein environment in the binding site

molecular energy transfer that thermalizes the excitation within the other retinal normal modes. This is analogous to recent results on the thermalization of stilbene after excitation [124]. With full equipartition there is insufficient thermal population in any one mode to see significant anti-Stokes Raman activity [125]. This 3–5 ps time scale is close to that observed for the photo-to-batho transition, so we suggest that this transition is due to vibrational cooling as well as further low-frequency conformational relaxation of the chromophore–protein complex. The ground state photoproduct is highly twisted in the $C_9 \cdots C_{14}$ region, but the isomerization has rotated the 10, 12 and 14 hydrogens to the same side of the chromophore where they interact strongly with the Glu113 counterion. The anomalous intensities of the HOOPs are produced by the conformational distortion of the photoproduct and the frequencies are a result of the Glu113 perturbation. The batho structure that can be trapped at low temperature has been studied in great detail and is highly relevant for the room temperature process, because time-resolved Raman studies have shown that the molecular structure produced within 1 ps is identical to that trapped at low temperatures [103]. This conformational and electronic distortion is the most likely explanation for the significant energy storage in the primary photoproduct that drives the protein conformation to the MetaII form, activating transducin. It is remarkable that nature has choreographed such complex and interesting molecular and electronic dynamics that enable rhodopsin to do so much in so little time.

Abbreviations

cGMP, 3′,5′-cyclic GMP
RR, resonance Raman
PSB, protonated Schiff base
HOOP, hydrogen out-of-plane
MASS, magic angle sample spinning
CD, circular dichroism
PES, potential energy surface
F–C, Franck–Condon
CCD, charge coupled device
$\Delta\mu$, change in dipole moment
λ, reorganization energy
Δ, dimensionless normal coordinate displacement parameter
S, Franck–Condon displacement parameter where $S_i = \Delta_i^2/2$

Note added in proof

Very recently Palczewski et al. obtained a 2.8 Å X-ray diffraction structure of rhodopsin that shows the 6-s-*cis*, 11-*cis*, 12-s-*trans*, C=N anti-chromophore structure with right-handed helicity as depicted in Fig. 7 [126].

References

1. Goldsmith, T.H. (1972) in: Photochemistry of Vision, ed H.J.A. Dartnall, VII/1, pp. 685–719, Springer, Berlin.

2. Wald, G. (1968) Science **162**, 230–239.
3. Rando, R.R. (1991) Biochem. **30**, 595–602.
4. Stryer, L. (1987) Sci. Amer. **257**, 42–50.
5. Hecht, S., Shlaer, S. and Pirenne, M.H. (1942) J. Gen. Physiol. **25**, 819–840.
6. Hoffman, K.P., Jager, S. and Ernst, O.P. (1995) Isr. J. Chem. **35**, 339–355.
7. Rang, H.P., Dale, M.M. and Ritter, J.M. (1990) in: Pharmacology, pp. 22–26. Churchill Livingstone, London.
8. Zhukovsky, E.A., Robinson, P.R. and Oprian, D.D. (1991) Science **251**, 558–560.
9. Morton, R. (1955) Biochem. **59**, 128–134.
10. Oseroff, A.R. and Callender, R.H. (1974) Biochem. **13**, 4243–4248.
11. Mathies, R., Oseroff, A.R. and Stryer, L. (1976) Proc. Natl. Acad. Sci. USA **73**, 1–5.
12. Loppnow, G.R. and Mathies, R.A. (1989) Rev. Sci. Inst. **60**, 2628–2630.
13. Lin, S.W., Kochendoerfer, G.G., Carroll, K.S., Wang, D., Mathies, R.A. and Sakmar, T.P. (1998) J. Biol. Chem. **273**, 24583–24591.
14. Callender, R.H., Doukas, A., Crouch, R. and Nakanishi, K. (1976) Biochem. **15**, 1621–1629
15. Mathies, R., Freedman, T.B. and Stryer, L. (1977) J. Mol. Biol. **109**, 367–372.
16. Curry, B., Broek, A., Lugtenburg, J. and Mathies, R. (1982) J. Am. Chem. Soc. **104**, 5274–5286.
17. Curry, B., Palings, I., Broek, A.D., Pardoen, J.A., Lugtenburg, J. and Mathies, R. (1985) Adv. Infrared Raman Spectrosc. **12**, 115–178.
18. Smith, S.O., Myers, A.B., Mathies, R.A., Pardoen, J.A., Winkel, C., van den Berg, E.M.M. and Lugtenburg, J. (1985) Biophys. J. **47**, 653–664.
19. Smith, S.O., Braiman, M.S., Myers, A.B., Pardoen, J.A., Courtin, J.M.L., Winkel, C., Lugtenburg, J. and Mathies, R.A. (1987) J. Am. Chem. Soc. **109**, 3108–3125.
20. Eyring, G., Curry, B., Mathies, R., Fransen, R., Palings, I. and Lugtenburg, J. (1980) Biochem. **19**, 2410–2418.
21. Palings, I., Pardoen, J.A., van den Berg, E., Winkel, C., Lugtenburg, J. and Mathies, R.A. (1987) Biochem. **26**, 2544–2556.
22. Mathies, R.A., Smith, S.O. and Palings, I. (1987) in: Biological Applications of Raman Spectroscopy: Vol. 2 – Resonance Raman Spectra of Polyenes and Aromatics, ed T.G. Spiro. pp. 59–108, Wiley, New York.
23. Palings, I., van den Berg, E.M.M., Lugtenburg, J. and Mathies, R.A. (1989) Biochem. **28**, 1498–1507.
24. Groesbeek, M. and Lugtenburg, J. (1992) Photochem. Photobiol. **56**, 903–908.
25. Lin, S.W., Groesbeek, M., van der Hoef, I., Verdegem, P., Lugtenburg, J. and Mathies, R.A. (1998) J. Phys. Chem. **102**, 2787–2806.
26. Myers, A.B., Mathies, R.A., Tannor, D.J. and Heller, E.J. (1982) J. Chem. Phys. **77**, 3857–3866.
27. Verdegem, P.J.E., Lugtenburg, J. and de Groot, H.J.M. (1997) in: Stable Isotopes in Pharmaceutical Research, ed T.R. Browne, pp. 203–217. Elsevier, Amsterdam.
28. de Lange, F., Klaassen, C.H.W., Wallace-Williams, G.E., Bovee-Geurts, P.H.M., Luix, X.M. and DeGrip, W.J. (1998) J. Biol. Chem. **273**, 23735–23739.
29. Creemers, A.F.L., Klaassen, C.H.W., Bovee-Geurts, P.H.M., Keller, K., Raap, J., de Grip, W.J., Lugtenburg, J. and de Groot, H.J.M. (1999) Biochem. **38**, 7195–7199.
30. Eilers, M., Reeves, P.J., Ying, W.W., Khorana, H.G. and Smith, S.O. (1999) Proc. Natl. Acad. Sci. USA **96**, 487–492.
31. Lugtenburg, J. and de Groot, H.J.M. (1998) Photosynthesis Res. **55**, 241–245.
32. Mehring, M. (1985) Principles of High Resolution NMR in Solids, Springer, Berlin.
33. Smith, S.O., Palings, I., Miley, M.E., Courtin, J., de Groot, H.J.M., Lugtenburg, J. and Mathies, R.A. (1990) Biochem. **29**, 8158–8164.
34. Shriver, J.W., Mateescu, G., Fager, R., Torchia, D. and Abrahamson, E.W. (1979) Biochem. **18**, 4783–4792.
35. Santarsiero, B.D., James, M.N.G., Mamendaan, M. and Childs, R.F. (1990) J. Am. Chem. Soc. **112**, 9416–9418.
36. Herzfeld, J. and Berger, A.E. (1980) J. Chem. Phys. **73**, 6021–6030.

37. Mollevanger, L., Kentgens, A., Pardoen, J., Courtin, J., Veeman, W., Lugtenburg, J. and de Grip, W. (1987) Eur. J. Biochem. **163**, 9–14.
38. Makino, C.L., Kraft, T.W., Mathies, R.A., Lugtenburg, J., Miley, M.E., van der Steen, R. and Baylor, D.A. (1990) J. Physiology **424**, 545–560.
39. Levitt, M.H., Raleigh, D.P., Creuzet, F. and Griffin, R.G. (1990) J. Chem. Phys. **92**, 6347–6364.
40. Creuzet, F., McDermott, A., Gebhard, R., van der Hoef, K., Spijker-Assink, M.B., Herzfeld, J., Lugtenberg, J., Levitt, M.H. and Griffin, R.G. (1991) Science **251**, 783–786.
41. Raleigh, D.P., Levitt, M.H. and Griffin, R.G. (1988) Chem. Phys. Lett. **146**, 71–76.
42. Verdegem, P.J.E., Bovee-Geurts, P.H.M., de Grip, W.J., Lugtenburg, J. and de Groot, H.J.M. (1999) Biochem. **38**, 11316–11324.
43. Verdegem, P.J.E., Helmle, M., Lugtenburg, J. and deGroot, H.J.M. (1997) J. Am. Chem. Soc. **119**, 169–174.
44. Feng, X., Verdegem, P.J.E., Lee, Y.K., Sandström, D., Edén, M., Bovee-Geurts, P., de Grip, W., Lugtenburg, J., de Groot, H.J.M. and Levitt, M.H. (1997) J. Am. Chem. Soc. **119**, 6853–6857.
45. Watts, A., Ulrich, A.S. and Middleton, D.A. (1999) Mol. Memb. Biol. **12**, 233–246.
46. Ketchem, R.R., Hu, W. and Cross, T.A. (1993) Science **261**, 1457–1460.
47. Ulrich, A.S., Heyn, M.P. and Watts, A. (1992) Biochem. **31**, 10390–10399.
48. Gröbner, G., Choi, G., Burnett, I.J., Glaubitz, C., Verdegem, P.J.E., Lugtenburg, J. and Watts, A. (1998) FEBS Lett. **422**, 201–204.
49. Matsumoto, H. and Yoshizawa, T. (1975) Nature **258**, 523–526.
50. Daemen, F.J.M. (1978) Nature **276**, 847–848.
51. Crouch, R.K., Veronee, C.D. and Lacy, M.E. (1982) Vision Res. **22**, 1451–1456.
52. Towner, P., Gaertner, W., Walckhoff, B., Oesterhelt, D. and Hopf, H. (1981) Eur. J. Biochem. **117**, 353–359.
53. Gaertner, W., Oesterhelt, D., Seifert-Schiller, E., Towner, P., Hopf, H. and Bohm, I. (1984) J. Am. Chem. Soc. **106**, 5654–5659.
54. Koshland, D.E. (1995) Angew. Chem. **33**, 2475.
55. Buss, V., Kolster, K., Terstegen, F. and Varenhorst, R. (1998) Angew. Chem. Int. **37**, 1893–1896.
56. Kleinschmidt, J. and Harosi, F.I. (1992) Proc. Natl. Acad. Sci. USA **89**, 9181–9185.
57. Baasov, T., Friedman, N. and Sheves, M. (1987) Biochem. **26**, 3210–3217.
58. Kropf, A. and Hubbard, R. (1958) Ann. N. Y. Acad. Sci. **74**, 266–280.
59. Neitz, M., Neitz, J. and Jacobs, G.H. (1991) Science **252**, 971–974.
60. Asenjo, A.B., Rim, J. and Oprian, D. (1994) Neuron **12**, 1131–1138.
61. Beppu, Y. and Kakitani, T. (1994) Photochem. Photobiol. **59**, 660–669.
62. Honig, B., Greenberg, A.D., Dinur, U. and Ebrey, T.G. (1976) Biochem. **15**, 4593–4599.
63. Irving, C.S., Byers, G.W. and Leermakers, P.A. (1970) Biochem. **9**, 858–864.
64. Blatz, P.E. and Mohler, J.H. (1975) Biochem. **14**, 2304–2309.
65. Blatz, P.E. and Liebman, P. (1973) Exp. Eye Res. **17**, 573–580.
66. Kakitani, H., Kakitani, T., Rodman, H. and Honig, B. (1985) Photochem. Photobiol. **41**, 471–479.
67. Honig, B., Dinur, U., Nakanishi, K., Balogh-Nair, V., Gawinowicz, M.A., Arnaboldi, M. and Motto, M.G. (1979) J. Am. Chem. Soc. **101**, 7084–7086.
68. Kochendoerfer, G.G., Lin, S.W., Sakmar, T.P. and Mathies, R.A. (1999) Trends Biochem. Sci. **24**, 300–305.
69. Ponder, M. and Mathies, R. (1983) J. Phys. Chem. **87**, 5090–5098.
70. Busch, G.E., Applebury, M.L., Lamola, A.A. and Rentzepis, P.M. (1972) Proc. Natl. Acad. Sci. USA **69**, 2802–2806.
71. Shichida, Y., Matuoka, S. and Yoshizawa, T. (1984) Photobiochem. Photobiophys. **7**, 221–228.
72. Schoenlein, R.W., Peteanu, L.A., Mathies, R.A. and Shank, C.V. (1991) Science **254**, 412–415.
73. Peteanu, L.A., Schoenlein, R.W., Wang, Q., Mathies, R.A. and Shank, C.V. (1993) Proc. Natl. Acad. Sci. USA **90**, 11762–11766.
74. Taiji, M., Bryl, K., Nakagawa, M., Tsuda, M. and Kobayashi, T. (1992) Photochem. Photobiol. **56**, 1003–1011.

75. Yan, M., Manor, D., Weng, G., Chao, H., Rothberg, L., Jedju, T.M., Alfano, R.R. and Callender, R.H. (1991) Proc. Natl. Acad. Sci. USA **88**, 9809–9812.
76. Wang, Q., Schoenlein, R.W., Peteanu, L.A., Mathies, R.A. and Shank, C.V. (1994) Science **266**, 422–424.
77. Doukas, A.G., Junnarkar, M.R., Alfano, R.R., Callender, R.H., Kakitani, T. and Honig, B. (1984) Proc. Natl. Acad. Sci. USA **81**, 4790–4794.
78. Kochendoerfer, G.G. and Mathies, R.A. (1996) J. Phys. Chem. **100**, 14526–14532.
79. Chosrowjan, H., Mataga, N., Shibata, Y., Tachibanaki, S., Kandori, H., Shichida, Y., Okada, T. and Kouyama, T. (1998) J. Am. Chem. Soc. **120**, 9706–9707.
80. Kobayashi, T., Kim, M. and Taiji, M. (1998) J. Phys. Chem. B **102**, 272–280.
81. Haran, G., Morlino, E.A., Matthes, J., Callender, R.H. and Hochstrasser, R.M. (1999) J. Phys. Chem. A **103**, 2202–2207.
82. Myers, A.B. and Mathies, R.A. (1987) in: Biological Applications of Raman Spectroscopy: Vol. 2 – Resonance Raman Spectra of Polyenes and Aromatics, ed T.G. Spiro. pp. 1–58, Wiley, New York.
83. Loppnow, G.R. and Mathies, R.A. (1988) Biophys. J. **54**, 35–43.
84. Song, L. and El-Sayed, M.A. (1998) J. Amer. Chem. Soc. **120**, 8889–8890.
85. Myers, A.B., Harris, R.A. and Mathies, R.A. (1983) J. Chem. Phys. **79**, 603–613.
86. Myers, A.B., Trulson, M.O., Pardoen, J.A., Heeremans, C., Lugtenburg, J. and Mathies, R.A. (1986) J. Chem. Phys. **84**, 633–640.
87. Garavelli, M., Negri, F. and Olivucci, M. (1999) J. Am. Chem. Soc. **121**, 1023–1029.
88. Schoenlein, R.W., Peteanu, L.A., Wang, Q., Mathies, R.A. and Shank, C.V. (1993) J. Phys. Chem. **97**, 12087–12092.
89. Wang, Q., Kochendoerfer, G.G., Schoenlein, R.W., Verdegem, P.J.E., Lugtenburg, J., Mathies, R.A. and Shank, C.V. (1996) J. Phys. Chem. **100**, 17388–17394.
90. Lawless, M.K., Wickham, S.D. and Mathies, R.A. (1995) Accts. Chem. Res. **28**, 493–502.
91. Rodier, J. and Myers, A.B. (1993) J. Am. Chem. Soc. **115**, 10791–10795.
92. Kakitani, T., Akiyama, R., Hatano, Y., Imamoto, Y., Shichida, Y., Verdegem, P.J.E. and Lugtenburg, J. (1998) J. Phys. Chem. B **102**, 1334–1339.
93. DeLange, F., Bovee-Geurts, P.H.M., Van Oostrum, J., Daniel Portier, M., Verdegem, P.J.E., Lugtenburg, J. and de Grip, W.J. (1999) Biochem. **37**, 1411–1420.
94. Vroegop, P.J., Lugtenburg, J. and Havinga, E. (1973) Tetrahedron **29**, 1393–1398.
95. Yoshizawa, T. and Wald, G. (1963) Nature **197**, 1279–1286.
96. Cooper, A. (1979) Nature **282**, 531–533.
97. Peters, K., Applebury, M.L. and Rentzepis, P.M. (1977) Proc. Natl. Acad. Sci. USA **74**, 3119–3123.
98. Fransen, M.R., Luyten, W.C.M.M., van Thuijl, J., Lugtenburg, J., Jansen, P.A.A., van Breugel, P.J.G.M. and Daemen, F.J.M. (1976) Nature **260**, 726–727.
99. Eyring, G. and Mathies, R. (1979) Proc. Natl. Acad. Sci. USA **76**, 33–37.
100. Honig, B., Ebrey, T., Callender, R.H., Dinur, U. and Ottolenghi, M. (1979) Proc. Natl. Acad. Sci. USA **76**, 2503–2507.
101. Eyring, G., Curry, B., Mathies, R., Broek, A. and Lugtenburg, J. (1980) J. Am. Chem. Soc. **102**, 5390–5392.
102. Eyring, G., Curry, B., Broek, A., Lugtenburg, J. and Mathies, R. (1982) Biochem. **21**, 384–393.
103. Zhu, L., Kim, J. and Mathies, R.A. (1999) J. Raman Spectroscopy **30**, 777–783.
104. Hayward, G., Carlsen, W., Siegman, A. and Stryer, L. (1981) Science **211**, 942–944.
105. Siebert, F., Mantele, W. and Gerwert, K. (1983) Eur. J. Biochem. **136**, 119–127.
106. Bagley, K.A., Balogh-Nair, V., Croteau, A.A., Dollinger, G., Ebrey, T.G., Eisenstein, L., Hong, M.K., Nakanishi, K. and Vittitow, J. (1985) Biochem. **24**, 6055–6071.
107. Kandori, H. and Maeda, A. (1995) Biochem. **34**, 14220–14229.
108. Smith, S.O., Courtin, J., de Groot, H.J.M., Gebhard, R. and Lugtenburg, J. (1991) Biochem. **30**, 7409–7415.
109. Warshel, A., Chu, Z.T. and Hwang, J.-K. (1991) Chem. Phys. **158**, 303–314.
110. Tallent, J.R., Hyde, E.W., Findsen, L.A., Fox, G.C. and Birge, R.R. (1992) J. Am. Chem. Soc. **114**, 1581–1592.

111. Garavelli, M., Vreven, T., Celani, P., Bernardi, F., Robb, M.A. and Olivucci, M. (1998) J. Am. Chem. Soc. **120**, 1285–1288.
112. Terstegen, F. and Buss, V. (1998) J. Mol. Struct. (Theochem) **430**, 209–218.
113. Bifone, A., de Groot, H.J.M. and Buda, F. (1997) Pure Appl. Chem. **69**, 2105–2110.
114. Bifone, A., de Groot, H.J.M. and Buda, F. (1997) J. Phys. Chem. B **101**, 2954–2958.
115. Buss, X. (1999) Unpublished.
116. Yoshizawa, T. and Shichida, Y. (1982) Methods Enzymol. **81**, 634–641.
117. Sheves, M., Albeck, A., Ottolenghi, M., Bovee-Geurts, P., De Grip, W., Einterz, C., Lewis, J., Schaechter, L. and Kliger, D. (1986) J. Am. Chem. Soc. **108**, 6440–6441.
118. Asato, A.E., Denny, M. and Liu, R.S.H. (1986) J. Am. Chem. Soc. **108**, 5032–5033.
119. Lewis, J.W., Pinkas, I., Sheves, M., Ottolenghi, M. and Kliger, D.S. (1995) J. Am. Chem. Soc. **117**, 918–923.
120. Mathies, R. and Stryer, L. (1976) Proc. Natl. Acad. Sci. USA **73**, 2169–2173.
121. Wexler, D. and Mathies, R.A., Unpublished.
122. Humphrey, W., Lu, H., Logunov, I., Werner, H.J. and Schulten, K. (1998) Biophys. J. **75**, 1689–1699.
123. Doig, S.J., Reid, P.J. and Mathies, R.A. (1991) J. Phys. Chem. **95**, 6372–6379.
124. Nakabayashi, T., Okamoto, H. and Tasumi, M. (1998) J. Phys. Chem. A **102**, 9686–9695.
125. Shreve, A.P. and Mathies, R.A. (1995) J. Phys. Chem. **99**, 7285–7299.
126. Palczewski, K., Kumasaka, T., Hori, T., Behnke, C.A., Motoshima, H., Fox, B.A., Le Trong, I., Teller, D.C., Okada, T., Stenkamp, R.E., Yamamoto, M. and Miyano, M. (2000) Science **289**, 739–745.

CHAPTER 3

Late Photoproducts and Signaling States of Bovine Rhodopsin

K.P. HOFMANN

Institute for Medical Physics and Biophysics, Humboldt University, Berlin

Handbook of Biological Physics
Volume 3, edited by D.G. Stavenga, W.J. DeGrip and E.N. Pugh Jr

Contents

1. Introduction

Rhodopsin is the visual pigment of the vertebrate rod. Its bleaching by light, which is the theme of this chapter, has been known since the early investigations on the "Sehpurpur" by Karl Boll and Willy Kühne (see Ref. [1]). Today, we understand that the bleaching reaction enables the interaction of rhodopsin with the G-protein transducin (G_t), which in turn initiates the intermolecular transduction of the light signal by catalyzing GDP/GTP exchange in the nucleotide binding site of the G-protein. Concurrent interaction with rhodopsin kinase leads to the activation of the enzyme and eventually to phosphorylation of the receptor, thus enabling tight interaction with arrestin and shut-off of the signal for the G-protein. One molecule of photoactivated rhodopsin can activate many copies of G_t and of the G_t-coupled effector, a cyclic GMP phosphodiesterase (PDE), leading to the closure of cGMP-dependent channels and the receptor potential (for recent reviews see [2] and Chapters 4 and 5 in this volume).

Rhodopsin is an integral membrane protein of the disc membrane stacks filling the rod outer segment. By its lipid composition (which includes large amounts of polyunsaturated fatty acids, see below), the disc membrane is extremely fluid. This property is crucially important for signal transduction, because it favors both the formation of rhodopsin's active state and the rapid lateral and rotational diffusion of the receptor, thus providing opportunities for collisional coupling in the "random walk amplifier" mechanism [3].

By its dense packing and exact orientation in the membrane, rhodopsin provides an effective target for the light passing through the disc stack. Vertebrate rhodopsins are intrinsic membrane proteins of approximately 40 kDa molecular weight [4]. They share their seven transmembrane helix structure with other G-protein coupled receptors and retinal proteins, such as the proton pump bacteriorhodopsin [5–7]. Bovine rhodopsin, to which this chapter predominantly refers, is 348 amino acid residues in length, identical to human rhodopsin at all but 23 positions. The disposition of the helical transmembrane stretches became best visible by cryo-electron microscopy of two-dimensional crystals of frog rhodopsin [8].

About one-half of rhodopsin's protein mass is buried in the lipid bilayer, the other half is exposed in equal portions to the cytoplasmic and intradiscal surfaces. Part of the carboxyl terminus is anchored to the bilayer by palmitoyl-cysteines, so that four "loop" domains and the carboxyl terminus face the rod cytoplasm [2,6]. The characteristics rhodopsin has in common with other G-protein coupled receptor include an amino terminal glycosylation site and multiple carboxy terminal phosphorylation sites for a receptor kinase [9,10]. The cytoplasmic loop regions are likely to provide loci of interaction with transducin, G_t, rhodopsin kinase, RK, and arrestin.

investigate, in digitonin micelles, a pH- and temperature-dependent equilibrium between the intermediate metarhodopsin known from low temperature experiments and the new photoproduct, now termed Meta II. It was noticeable by its maximal absorption at 380 nm, similar to the final product of rhodopsin bleaching, opsin + all-*trans*-retinal.

2.1.1. The Lumi and Meta I intermediates

Relative to the ground state, the Lumi intermediate is formed with a large positive reaction enthalpy and reaction volume [62]. The enthalpy of reaction depends strongly on the hydrophobic environment (90 vs 11 kJ mol^{-1} for washed membranes and dodecyl maltoside solution, respectively [63]). In Lumi, most of the photon energy absorbed by rhodopsin has already transferred to the apoprotein [64]. MI is formed from rhodopsin with a reaction enthalpy of 70 kJ mol^{-1} [65]. The L $\rightarrow$ MI transition [66,67] and infrared bands typical of the MI state [68] are influenced by the lipid environment.

Although the majority of the available biochemical and biophysical data argues for a dominant role of MII in the function of rhodopsin as a receptor, there is evidence from physiological investigations for an involvement of early transformations, on the time scale of metarhodopsin I, in regulating visual function (see, for example, Refs. [69,70]). The direct evidence for a role of MI in protein interactions of rhodopsin [71,72] will be discussed in Section 3.3.

2.1.2. Metarhodopsin II

In MII, the Schiff base bond of the chromophore to the apoprotein, which is protonated in all other intermediates, becomes deprotonated [73]. When Schiff base deprotonation is blocked by methylation of the active site, Lys^{296}, photolysis leads to formation of a stable MI-like photoproduct with no formation of MII- or MIII-like intermediates [74]. According to FTIR analysis, Schiff base deprotonation is mechanistically coupled to Glu^{113} protonation [75]. Based on the "naive" definition of MII as any form that carries the chromophore bound via a deprotonated Schiff base, the intermediate is characterized by a number of special properties. They are summarized in Table 1 and will be discussed as they appear in the table. A discussion of subforms and activities will be provided in later sections.

MII is formed with a large activation energy, both from the ground state [44,45,53,54,67] and from MI by pressure jump [76]. The reaction enthalpy was either directly determined [65] or derived from the van't Hoff equation [45]. The MI $\rightarrow$ MII reaction volume is 108 ml mol^{-1} [76], in G-protein-free, osmotically active disc vesicles, prepared according to Ref. [77]. In native (sonicated) rod outer segments, the reaction volume was reported to depend on pressure, with ca. 60 ml mol^{-1} at atmospheric pressure and ca. 10 ml mol^{-1} at high pressure. For digitonin-solubilized rhodopsin, the latter small value was found to be about constant over the whole range of pressure [78].

Changes of interfacial potential on membranes incorporating rhodopsin were identified by fluorescence [79] and spin label [80] techniques. In the latter case, spin-labeled hydrophobic ions reported a potential change, positive relative to the

Table 1
Properties of metarhodopsin II

Category	Leads to	Meta II (MII)	Depends on
Thermodynamic		R → MII or MI → MII $\Delta H^{\#} = 160$ kJ mol^{-1} MI → MII $\Delta H^{0} = 40$ kJ mol^{-1} $\Delta V = 108$ ml mol^{-1} (in membrane) 63 ml mol^{-1} (solubilized)	
Electric		Change in the interfacial potential Dielectric charge displacements	
Chemical		Deprotonation of the retinal Schiff base	
	Uptake of H^+		pH Lonic strength
	Access of water-soluble substances to chromophore		Aqueous milieu
	Enhanced digestion of cytoplasmic loops		
			Osmotic pressure Hydrophobic environment
Spectroscopic/biophysical	Changes in: UV-spectrum Lin./circular dichroism Birefringence NIR light scattering FTIR and Raman spectrum EPR spectrum (labeled sites) Surface plasmon resonance		
Biochemical/functional	Exposure of binding sites for protein–protein interaction		Mutations of charged residues
Physiological	Role in signal transduction		

aqueous exterior of the disk, within a boundary region of the native isolated disc membrane near the cytoplasmic surface. Electrical effects can be measured *in situ*, namely as a component of the 'early receptor potential' [81,82] on lipid impregnated filter materials [83], and on lipid bilayers having photoreceptor membranes attached at one side [84]. Not all components of the measured electrical effect are kinetically related to the formation of spectroscopically distinguishable photoproducts [85]. The recent introduction of the 'early receptor current' technique, realized in a unicellular expression system [86], may offer new approaches to the relationships between electrostatics and structure of the receptor protein [87]. The dependence on ionic strength is found in native membranes [88], as well as in detergent, where it can be specified for the active conformation of MII (MII_b, see below).

MII formation depends on the presence of an aqueous milieu [60]. The role of water molecules in the formation of MII was recently specified by FTIR spectroscopy [89,90] (see Section 3.1). Small soluble molecules, which can reach the chromophore after MII has formed, include hydroxylamine and sodium borohydride [92–94]. An intriguing recent observation is that the formation of MII from MI depends on osmotic pressure [95]; analysis of different neutral solutes and separation of the effect on acyl chain packing in the membrane environment led to the estimation that the MI/MII conversion is accompanied by the release of 34 water molecules per MII formed (at 20°C, but almost independent of temperature). The data obtained with solutes of different size were interpreted to mean that the osmotically sensitive regions of rhodopsin, which change their hydration during MI/MII conversion, are narrow crevices or pockets [95], in remarkable similarity to the ideas derived from peptide stabilization (see Section 3.1.5).

The enhanced susceptibility, as compared to the ground state, of MII to partial digestion was first studied in the context of the binding and activation of G_t [96]; it was later related to changes in FTIR difference spectra [97]. Perturbations of the UV-spectrum were interpreted as an indication of a more hydrophilic environment – as compared to the ground state – of aromatic amino acids [98], and changes in linear and circular dichroism were interpreted as a rotation of the chromophore relative to the plane of the disc membrane [27,99]. Birefringence changes were kinetically related to the formation of MII in intact rod outer segments [100,101].

The so-called N-signal [102] or rhodopsin signal [103], a fast change in the near-infrared light scattering of the disc membranes, is kinetically and stoichiometrically related to MII-formation. It reflects the strong shift of the absorption maximum and an as yet unidentified structural change [104].

Vibrational spectroscopy indicates considerable alterations of the apoprotein structure in MII [105–107]. Some bands in the infrared spectrum can be specifically assigned to carboxyl groups [109,110]. Changes at labeled sites in the electron paramagnetic spectrum were interpreted as indicating small movements near the second cytoplasmic loop, and were kinetically correlated with MII-formation [111,112]. The functional significance of these findings will be discussed in Section 3.1. Alterations in the surface plasmon resonance spectrum have been interpreted as an increase in the average thickness of the lipid–protein bilayer [113]. Mutations that affect the formation of Meta II as a spectrally defined species include such of the Schiff base counterion [13] and of certain histidines [47,114]. Much larger is the class of mutations that have been shown to affect the functional activity of rhodopsin, and thus should reflect the modified MII.

Taken together, the available data argue for substantial conformational changes accompanying MII-formation. Most important is the property of MII to provide the conditions for the formation of the active state(s) of rhodopsin [74,115,116]. The virtually complete shift to MII of the MI/MII-equilibrium by G_t or arrestin indicates that these proteins can only bind after MII has formed (discussed in detail in Section 3.1).

2.1.3. Role of the hydrophobic environment

In the early experiments on digitonin-solubilized rhodopsin [24], the metarhodopsin equilibrium was shifted to MII by protons, neutral salts and glycerol. Later, it was found that the precise lipid composition of the disc membrane is crucial for the MII/

MI-eqilibrium. According to recent investigations [117], the phospholipid content of native membrane is roughly equimolar phosphatidylcholine and phosphatidylethanolamine, with a lower proportion of (ca. 14%) of phosphatidylserine and of phosphatidylinositol (2.5%), and a low percentage of cholesterol, which varies from bottom to top of the rod outer segment. The fatty acids of the phospholipids are mainly 22 in length and polyunsaturated to a high percentage (22:6, but no 22:3 in the dominant species) [118–120]. The lipids of the disc membrane undergo rapid (half mean time < 5 min) flip–flop between outer and inner leaflet; in the resulting equilibrium, phosphatidylserine has a distinct preference for the outer leaflet (see [58,121] and references therein). In this native hydrophobic environment, MII is formed on a 1 ms time scale, and the equilibrium is well on the side of MII (Fig. 1a).

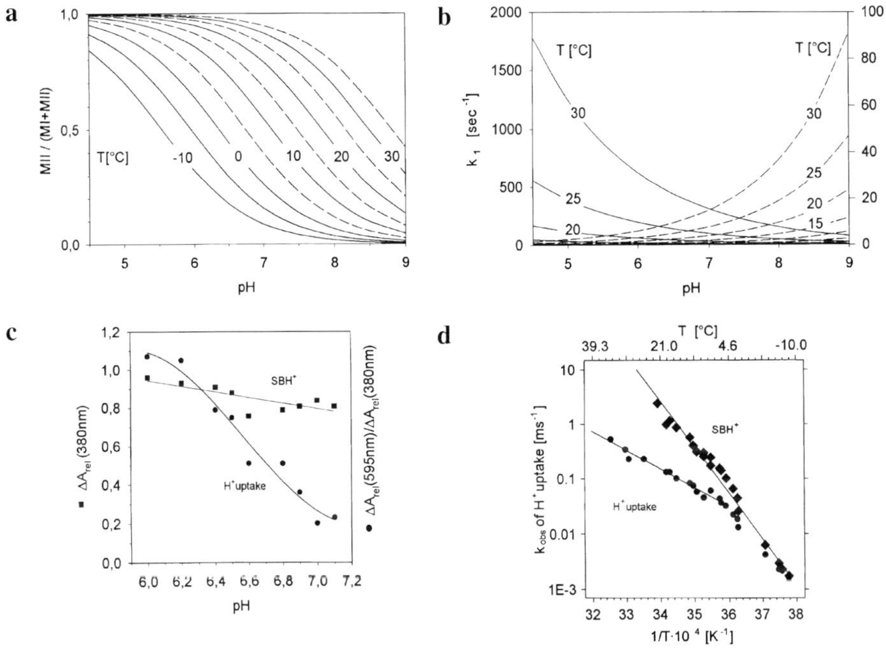

Fig. 1. pH-dependence of Metarhodopsin II. (**a**) Plot of the relative amount of metarhodopsin II (MII) formed after a green flash, as a function of pH; the plot is a representation of Eq. (13b) in Ref. [46]: $\ln([MII]/[MI]) = -9464(1/T) - 1.608$ pH. (**b**) Plot of the forward (k_1) and backward (k_{-1}) rate constant of MII formation; see Ref. [46] for details. (**c**) Plots of the relative amount of total MII (SBH^+, squares) and of the protonated subform MII_b (H^+ uptake, circles), formed in dodecylmaltoside solubilized rhodopsin after a green flash. The data points are the relative absorption change at 380 nm (squares, formation of Meta II, normalized to the maximal amount of Meta II) and at 595 nm (circles, alkalization of the medium from indicator dye bromocresol purple, divided by the relative absorption change at 380 nm), both as a function of pH. The solid lines represent linear regression (squares) and a polynomial fit (circles). Data from Ref. [140]. (**d**) Arrhenius representation of the observed rates of flash-induced formation (see c) of total MII (SBH^+, squares) and of the protonated subform MII_b (H^+ uptake, circles).

Rhodopsin recombined with egg lecithin [122] shows distinct differences to the native environment [123]. The formation rate is highly sensitive even to alterations of the rhodopsin environment which give almost the same CD spectra, thermal stability and regenerability of opsin with 11-*cis*-retinal as those of native rhodopsin in the membrane [124]. It also depends on the presence of unsaturated lipids and on the fluidity of the phospholipid hydrocarbon chains but it does not require a specific phospholipid head group [125]. The enhanced formation of MII in PE or PS bilayers [126] may reflect a surface pH effect [88].

In reconstituted membranes with a reduced lipid content or prepared with short-chain, saturated lecithin the decay of MI is retarded and yields almost no MII but predominantly free retinal plus opsin [38,39,127]. The conclusion was that the ratio between the free energies of MI and MII is modulated by an interfacial tension-like interaction between rhodopsin and the bilayer. In octyl glucoside solubilized discs, the activation free energy of MII-formation is linearly dependent on the level of associated disc phospholipid [34], until the lipid/rhodopsin ratio of the native membrane is reached. This argues against a specific mode of interaction between rhodopsin (and/or MII) and the lipid, as was also verified by the study of alcohol effects [128,129]. This does not exclude the possibility that a very small amount of lipid binds tightly to the receptor. According to recent FTIR data [130], this is indeed the case. The specific effect of the G-protein on the MI/MII-equilibrium (discussed in Section 3.1) is also observed in detergent [33,35,132]. The MI/MII-equilibrium shifts towards MII in proportion to phosphorylation stoichiometry in native disc membranes [133]. This effect, which was not observed with rhodopsin reconstituted into POPC liposomes [134], could now be explained by a change of the membrane surface potential that occurs only when the protein is unidirectionally oriented in the disc membranes carrying a negative surface charge [135].

Both an enhanced rate of MII-formation and a shift of the equilibrium to MII are seen in highly fluid detergent micelles (formed from octyl glucoside or dodecylmaltoside). This means in general terms that the energy barrier for the MI $\rightarrow$ MII conversion is lowered (Fig. 3a), and that photoproducts preceding the conversion must be affected; the effect of detergent may also, at least in part, be assigned to the accumulation of the early MI_{380} product (see below).

2.2. *Reaction schemes*

2.2.1. *Role of proton transfer*

Early investigations on the acid–base properties of rhodopsin in solution [136] showed a light-induced proton uptake with a pK of 6.6. Later kinetic studies compared the pH-dependence in native membranes (e.g. [137]) to proton uptake [43,138,139]. A reaction model for MII must account for the pH-dependence and kinetics of its formation and the properties of proton uptake [140,141].

The "essential proton" concept [46] states that Schiff base deprotonation and proton uptake occur in one and the same concerted reaction, and are both mandatory for the conformational change. This yields a reversible first order process, involving the uptake of *n* protons at constant pH:

$$\mathrm{MI} \underset{k_2}{\overset{k_1'}{\rightleftharpoons}} \mathrm{MII} \qquad \text{where } k_1' = (\mathrm{H}^+)^n k_1 \tag{2}$$

In the approach of scheme (2), changes in protonation and conformation appear as one and the same change of protein free energy [46,65,124]. The conformational change may occur before or after the actual proton transfer reaction, as is expressed in more complex square schemes replacing Eq. (2); (see Refs. [46,141]). However, it is known that proton transfer occurs on a picosecond time scale, as soon as the crucial factors, relative orientation and distance of proton donor and acceptor, and the local electric field, are adjusted [142]. One may thus assume that the conformation of the protein is adjusted in a thermally activated pre-equilibrium, and proton transfer contributes to its stabilization.

Because proton transfer is so rapid, the conformational change must be rate limiting for MII-formation, so that the essential proton reaction mode implies first order kinetics. However, it was noted early on that, in preparations of native disc membranes, the kinetics of MII-formation in general does not fit to a reversible first order process. Characteristic delays of proton uptake with respect to MII-formation [43,143,144] led to the proposal that the formation of the spectroscopic species and the proton uptake reaction occurred in two thermodynamically separate steps, leading to two different isochromic MII species [143]. These early approaches were qualified by the later finding that MII-formation depends on the presence of proteins other than rhodopsin, such as G_t (proteins that are present in the native preparations [145]).

Some progress was made when proton uptake measurements were carried out in detergent purified rhodopsin (for a discussion of technical aspects, see Ref. [146]). Although indicator dyes in solution fail to detect very fast proton movements on membranes [147], the transformations linked to Schiff base deprotonation (proton translocation) and proton uptake could be sufficiently well separated. The resulting subforms $\mathrm{MII_a}$ and $\mathrm{MII_b}$ [140] are distinguished from one another by their dependence on several parameters and are separated by a thermal barrier that takes milliseconds to surpass (see Section 2.2.5 for more details).

2.2.2. *Classical scheme*

To discuss the reaction schemes that have been designed to describe the kinetics of formation of rhodopsin's late photointermediates, we may start with the classical equilibrium scheme introduced by Matthews et al. [24]:

$$\mathrm{R} \rightarrow\rightarrow \mathrm{MI} \overset{\mathrm{pH},T}{\rightleftharpoons} \mathrm{MII} \rightarrow \text{all-}\textit{trans}\text{-retinal} + \text{opsin} \tag{3}$$

Subsequent flash photolysis work has shown that MII is formed in milliseconds and that its formation is related to proton uptake [42,43]. As discussed in the last section, isochromic species were proposed, but a comparison of flash-induced and pressure-induced MII-formation (from the ground state or MI, respectively) in hypotonically washed (i.e. G-protein free) discs yielded a satisfying approximation by a single (pseudo-) first-order reaction, with identical kinetic parameters in both trigger modes [76]. Parkes and Liebman [46] were the first to analyze the MI/MII-transition on the disc membrane in its native condition, but in the absence of G_t [145]. They also provided an explicit kinetic model of MII formation. Based on the "essential proton" concept outlined above, they extracted the true forward and backward rate constants from pH- and temperature-dependent measurements of the apparent kinetics. The pH dependence of the relative amount of MII formed on flash photolysis led them to mathematical expressions for the relative amount of MII and for the rate constants (plotted in Fig. 1a,b).

2.2.3. Parallel pathways

Several investigators applied a biphasic description, with two exponential functions, to describe the kinetics of MII-formation [44,148]. Related schemes were applied to the decay of bathorhodopsin [41] and supported the notion of two parallel pathways from the very beginning of rhodopsin photolysis. It was speculated that rhodopsin itself may exist in its ground state in two different conformations. The dependence of MII-formation on proteins other than rhodopsin [145] may again explain some of these results.

In later investigations, Litman and co-workers performed a detailed investigation on hypotonically washed discs [77]. Temperature-dependent measurements with the analysis of the contribution of each photoproduct, obtained from time-dependent diode-array spectrophotometer measurements led to the following reaction scheme [149]:

$$\mathrm{R} \rightarrow\rightarrow \mathrm{MI} \begin{matrix} \rightleftharpoons \mathrm{MII}_{\mathrm{fast}} \\ \rightleftharpoons \mathrm{MII}_{\mathrm{slow}} \end{matrix} \tag{4}$$

These analyses appreciated the kinetic complexity of the decay of rhodopsin's photointermediates and emphasized that the absorption properties in UV/Vis may not fully determine all the transformations of the receptor. The authors emphasized, however, that in the parallel kinetic model one has to allow the extinction coefficient of Meta I to vary in the 380 nm spectral region, based on a general broadening of the blue side of the absorption band.

2.2.4. Square scheme

The special assumption about Meta I was eliminated in a more recent square scheme approximation [150], derived from the observation of a 380 nm absorption, rising prior to the 'normal' MI-formation [29], and on a careful analysis of Lumi- and Meta I-formation [45]. The kinetic approach employed multi-exponential fitting by global fit analysis [151], and resulted in the following reaction scheme:

$$\begin{matrix} \mathrm{R} \rightarrow\rightarrow & \mathrm{L} & \rightleftharpoons & \mathrm{MI}_{380} & \\ & \downarrow & & \downarrow & \\ & \mathrm{MI}_{480} & \rightleftharpoons & \mathrm{MII} \rightarrow & \end{matrix} \tag{5}$$

The scheme includes MI_{380}, an early intermediate with a deprotonated retinylidene Schiff base, absorbing at 380 nm, which is distinct from the classical MII intermediate. Since it is an early metarhodopsin intermediate, it was termed MI_{380} (with the conventional Meta I designated by MI_{480}). The later deprotonated Schiff base product, MII, is thought to differ from MI_{380} [152] in that it assumes a conformation in which it can interact with transducin [115]. Part of the proton release observed at basic pH in detergent-solubilized rhodopsin [153] may be assigned to MI_{380} [29,154].

2.2.5. *Two-step scheme*

Besides MI_{380} and MII, other isochromic intermediates with absorption maxima at 380 nm have been proposed. Based on the proton uptake measurements mentioned above, a scheme with two sequential isochromic MII species, separated by proton uptake, was derived [140]:

$$R \rightarrow\rightarrow MI \rightleftharpoons MII_a \overset{+H^+}{\rightleftharpoons} MII_b \rightarrow \qquad (6)$$

The scheme is similar to the one proposed in earlier work [143], in that it involves the formation of the spectrophotometrically observable species, taking place prior to proton uptake. MII_a corresponds to a conformational state in which the Schiff base proton has translocated to another internal group within the protein. MII_b corresponds to a conformational state in which an additional spectrally silent net proton uptake from the aqueous phase has occurred. The existence of the two different species was supported by different kinetics (ca. 10 ms delay between MII_a and MII_b at 5°C), different Arrhenius activation energy, different dependence on ionic strength and pH [140] (see Fig. 3a). Properties of the hydrophobic environment (i.e. membranes or detergent) appear to determine the transition to MII_a, while the hydrophilic milieu, including pH and ionic strength, act on the protein during the transition to MII_b. Both proton movements are linked to different conformational alterations of the protein structure, as indicated by their different activation energies (ca. 160 and 60 kJ/mol, respectively [140]). The Arrhenius plots (Fig. 1d) are specifically consistent with the consecutive reaction scheme: at low temperature, the formation of MII_b is slower than that of MII_a, but above 20°C MII_b is rate limited by MII_a. Mutations of the conserved residue Glu^{134} influence proton uptake and thus MII_b, but leave kinetics and extent of MII_a-formation unchanged [153].

Kinetic analyses of UV/Vis absorption changes [154] and of the electrical effect accompanying MII-formation [155] are consistent with the existence of isospectral forms of MII. However, it is important to note that a characterization by molecular spectroscopy is still lacking. The MII_a/MII_b scheme does not exclude more complex reaction schemes, including an inhomogeneous MII_a state [140]. Thus the MII_a/MII_b scheme and the MI_{380} square scheme are not mutually exclusive. The two schemes, each containing two deprotonated Schiff base intermediates, were derived from different types of experimental data. In recent work, the relationships between these intermediates were explored, and the two schemes were combined [154].

At temperatures well below 0°C, two different MI-like intermediates (termed MI_a and MI_b) were recently detected for chicken [156] and bovine [157] rhodopsin, with different suggested roles in G_t binding and activation (see Sections 3.1.2 and 3.1.6). It remains to be explored, how these species fit into the current reaction scheme(s).

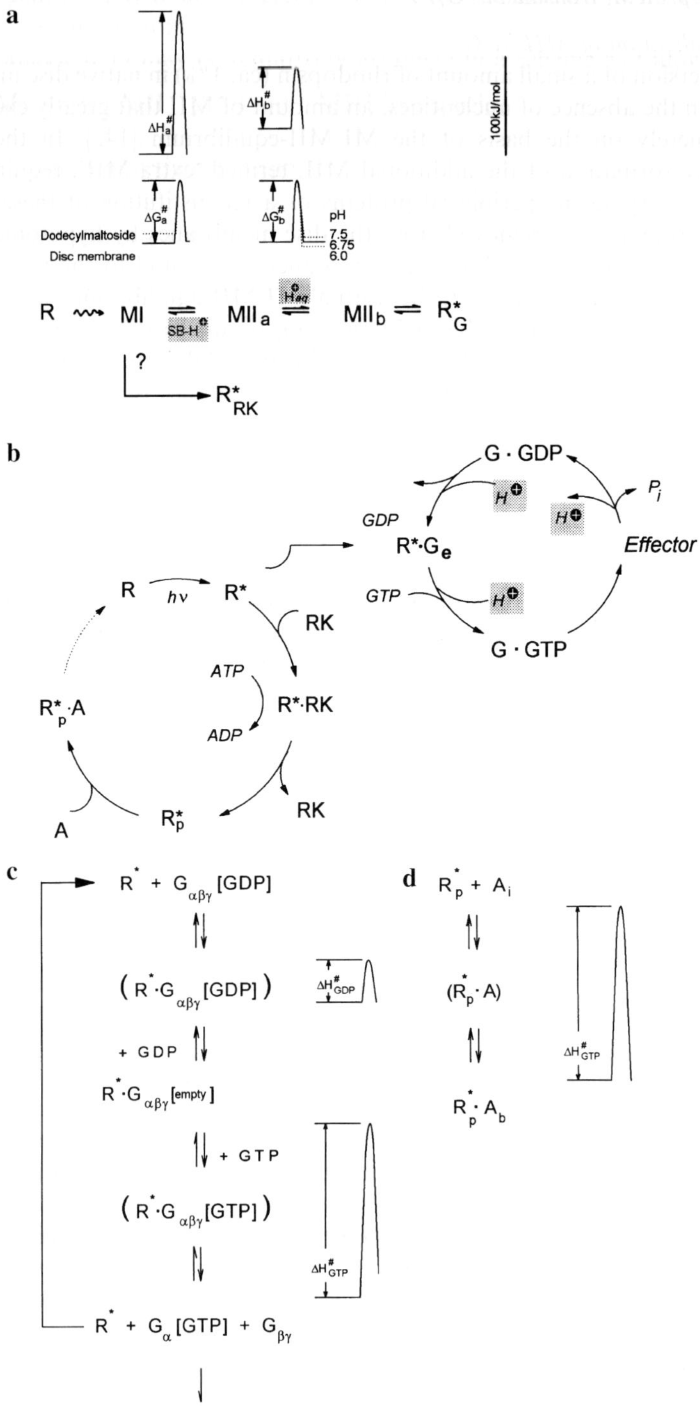
a
100kJ/mol
Dodecylmaltoside
Disc membrane
pH
7.5
6.75
6.0
R
MI
MIIa
MIIb
SB-H
?
b
GDP
Effector
GTP
G · GTP
G · GDP
R
hν
RK
ATP
ADP
A
c
+ GDP
+ GTP
d

reaction enthalpy for the catalytic activation of G_t is ca. 40% of the overall enthalpy of GTP hydrolysis [183]. The stabilization of MII is also observed with arrestin, but not with rhodopsin kinase (Fig. 2d,e; see Sections 3.2 and 3.3). C-terminal synthetic peptides ($G_t\alpha$ (340–350) and $G_t\gamma$ (60–71)farnesyl) can replace the G_t holoprotein in stabilizing MII [181]; this will be dealt with in Section 3.1.5.

3.1.2. Protonated MII interacts with G_t

The Meta I/Meta II conversion is linked to two proton transfer steps, namely deprotonation of the retinal Schiff base bond and proton uptake from solution, with conserved residues Glu^{113} and Glu^{134} as likely proton acceptors (see Section 3.1.6). A recent analysis has suggested that the pK of Glu^{134} may shift from 5.0 in the ground state to 6.3 in the light-activated state [135].

Different experimental approaches have led to a variety of reaction schemes (see Section 2.2 and Ref. [135]), all leading to protonated Meta II as the final product. As discussed in Section 2.2.5, a protonated MII-subform (termed MII_b) was postulated which arises from the isospectral MII_a by silent proton uptake. Photoregeneration experiments have shown that the active conformation arises from protonated MII. The ground state of rhodopsin can be regenerated directly from the Meta II state by flashes of UV-light. The reaction is likely to proceed through chromophore reisomerization, Schiff base reprotonation and proton release; when G_t is present, the regeneration is blocked at pH 6 but not at pH 8, i.e. only when protonated MII has been formed before and has bound the G-protein [168]. This has suggested that protonated MII is required for binding of G_t [168]. The most conclusive data were now obtained by FTIR spectroscopy, which allows to identify individual molecular changes that occur during photoproduct formation [110]. Protonated MII was enriched at low pH and its FTIR difference spectrum analyzed by a global fit procedure. It was then investigated whether the C-terminal synthetic peptides ($G_t\alpha$ (340–350) and $G_t\gamma$ (60–71)farnesyl), which stabilize MII, recognize metarhodopsin conformations other than protonated Meta II. It was found that the spectra reflect

---

Opposite: Fig. 3. Formation and interactions of photoactivated rhodopsin. (**a**) An equilibrium between the Meta states MI, MII_a and MII_b, is attained within a few milliseconds after the absorption of a photon. The MI/MII_a conversion is accompanied with net intramolecular translocation of the Schiff base proton to Glu^{113}. The subsequent MII_a/MII_b conversion is dependent on proton uptake from the aqueous phase. Rhodopsin kinase can bind to all Meta forms (signal for RK binding branches from MI), but catalytically effective binding of G_t requires MII_b. $\Delta G^{\#}_a$, $\Delta G^{\#}_b$, and $\Delta H^{\#}_a$ and $\Delta H^{\#}_b$ illustrate the respective thermodynamic barriers and free energy levels; data from Ref. [140]. (**b**) Reaction cycles of rhodopsin (left) and transducin (right). The receptor cycle comprises the inactive receptor (R), the activated receptor (R*), activated and phosphorylated R (R*p), and the proteins that regulate receptor activity, namely, receptor kinase (RK), arrestin (A). The G-protein cycle includes the inactive GDP-binding G-protein, the empty site complex with R* (R*G_e), and active G-GTP. Both cycles are simplified (see Ref. [2] for details). (**c**) Minimal scheme of nucleotide exchange in G_t, including the central empty site interaction and transient states of collisional interaction; see text for details. (**d**) Minimal scheme for the conformational switch in arrestin.

3.1.5.2. Putative binding regions in G_t
Two synthetic peptides, one containing the 11 C-terminal amino acid residues and one spanning α-residues 311–323 in the α4/β6/α5 regions of the G_t α structure, compete with holo-G_t for binding to photoactivated rhodopsin [212,215]. Residues 320–323 of the G_t-α-sequence contribute directly to nucleotide binding through van der Waals contact with the guanine ring of the nucleotide [231]. This segment may thus be involved in receptor-regulated modulation of nucleotide binding. A peptide from the sequence 8–23 of G_t-α also showed competition and is likely to be involved in R*G_t interaction, possibly by regulating subunit interaction. But not only the α-subunit contains interaction sites for receptor interaction. The C-terminal domain of G_tγ, G_t-γ-(60–71)-farnesylated peptide, was tested extensively in direct stabilization studies. Both the farnesyl moiety and the structure of the G_t-γ subunit C-terminal sequence are specific determinants of receptor–G_t interaction [232]. Thereby the isoprenoid serves not only as a membrane anchor (see [233] and references therein), but co-specifies the protein–protein interaction [221]. Evidence has been provided to show that the C-terminal tail of the γ-subunit is masked in the βγ-complex and becomes exposed on collisional coupling to the receptor, to build up the catalytic interaction [216]. This interaction occurs in the R*G_t empty site interaction characterized above (Fig. 3c), as was confirmed by recent stabilization, proton uptake and competition studies with kinetic resolution [181]. Both C-terminal regions of G_t-α and G_t-γ have been localized on the common surface of the G_t holoprotein by earlier X-ray studies [185]. G_t-α C-terminal mutations at two positions (Leu^{344} and Leu^{349}) prevent activation by rhodopsin [234], indicating that the respective hydrophobic side chains are critical for the activation of G_t. The binding site for this peptide was proposed to consist of CD and EF loops of rhodopsin [235].

Using the transferred nuclear Overhauser technique, the structure of the G_t-α C-terminal peptide, G_t-α (340–350), was studied [236,237], with the result of a different conformation of the peptide when bound to light-activated vs. dark rhodopsin. Evidence was provided [237] that light-activation of rhodopsin causes a shift from a disordered conformation of the peptide to a binding motif with a helical turn followed by an open reverse turn. Docking of the NMR structure to the X-ray structure of G_t-GDP argues for a contact with active rhodopsin which causes helix formation over residues 325–346, terminated by a C-terminal helical cap containing hydrophobic side chains.

The most recent applications of FTIR difference spectroscopy have now been extended to the R*G_t complex [238,239]. Besides alterations in the rhodopsin peptide backbone, a specific protonation, most likely of a Glu residue, was induced by the contact with G_t. Importantly, this effect was also seen with the synthetic G_t-α (340–350) peptide and with the peptide devoid of carboxylic acid groups, arguing for a location of the relevant group at rhodopsin [239].

3.1.5.3. Cytoplasmic loops of rhodopsin
To map the rhodopsin–G_t interface, synthetic peptides with all surface sequences from rhodopsin were used in combination with an assay based on MII stabilization. Only peptides from the second, third and fourth cytoplasmic loop (loops connecting

helices CD, EF, and the sequence between helix G and the palmitoylated Cys sites) of rhodopsin competed for G_t-dependent stabilization of the photoproduct [213]. The participation of loops CD and EF in the interaction with G_t was confirmed by analysis of a large number of site-directed mutants, using MII stabilization [35], G_t activation [241,242], and light scattering binding [40] assays. Replacement of several acidic and basic residues in loop EF by uncharged residues show significant but not critical effects on G_t activation [241], and partial deletions in loop EF or CD were necessary to abolish G_t activation [35]. Such deletion or replacement mutants allowed stabilization of MII by G_t, but failed to transmit the activation signal [35]. No interaction, neither at the MII stabilization nor at the G_t activation level, was seen when the conserved charge pair Glu^{134}/Arg^{135} was reversed [35]. When the EF sequence 237–249 in loop CD is deleted, even the first step of catalytic activation, the release of GDP, is significantly slowed [40]. On the other hand, G_t activation was hardly affected by multiple alanine substitutions in loop AB or CD, but substitutions at the beginning or end of loop EF showed distinct effects [243]. The latter result is consistently found in mutagenesis studies on G-protein coupled receptors (see Refs. [2,57,244], for more details). A specific significant decrease in G_t activation was described for substitution of two threonines (residues 242/243) in the middle of loop EF [241,243], which was explained by the disruption of a putative α-helical secondary structure in this loop [245]. Such defined structures have now been confirmed applying nuclear Overhauser effect NMR spectroscopy to peptides from the cytoplasmic loops and the carboxy terminal domain [246–248]. Introduction of a photoactivatable group in either loop AB, CD, EF or the fourth loop of rhodopsin identified so far only covalent crosslinking for loop EF to G_t; the linkage occurred specifically to the α-subunit [249].

Loop AB alteration by alanine-scanning mutagenesis does not influence light-induced G_t activation, suggesting that this loop plays mainly a structural role. There was however a strongly reduced G_t activation rate by a mutant rhodopsin with two deleted amino acids in this loop [40,250]. It is important to note that even small deletions in loop AB can lead to impaired expression and/or regeneration of the pigment, which indicates perturbations in the helix-arrangement of these mutants [40,250,251]. In a recent spin labeling study (see Section 3.1.6.4 for a detailed discussion of the technique), all first loop amino acids were replaced by single cysteine substitution mutants [252]. EPR labels on these positions reported measurable changes upon light activation, which were, however, much smaller than the ones measured near the sixth helix (see Section 3.1.6.4).

3.1.5.4. Carboxyl terminus and putative fourth loop

Concerning the carboxyl terminus, C-terminal peptides were reported to competitively inhibit R* stimulated GTPase activity in G_t [253–255]. The competing peptide, comprising amino acids 325–338, facilitated the ability of AlF_4^- to activate G_t in the absence of rhodopsin [255]. The interaction of the fourth loop with G_t was specified for the βγ-subunit, based on fluorescence studies with a corresponding loop-peptide [256,257]. On the other hand, even insertion of 22 amino acids at position 333 did not result in a measurable change of G_t activating capability [258], and the study of mutants with C-terminal truncations led to the conclusion that the C-terminal extension of rhodopsin, including the C-terminal half of the putative fourth loop, is not required for G_t activation [259]. Complete removal of the fourth loop led to a loss of the capacity for regeneration. Alanine-scanning mutagenesis of the fourth loop [260] led to mutants that were even more active towards G_t than the wild type. There is an apparent conflict between the effect of peptide competition and of mutational deletions. Possibly, these probes have different effects on the light-dependent proximity [261] between the C-terminal residue 338 (Ser, replaced by a Cys carrying an EPR label) and residues on the extension of helix G, i.e. in the third cytoplasmic loop.

Recent work on the fourth loop is now fully consistent with an important role of this structure in rhodopsin–G_t interaction. It has been convincingly shown that this region is part of a tertiary structure and may undergo microstructural changes, detected by ESR labels, that bring specific amino acids into interaction with G_t [262]. For the same region, reversible conformational changes were described, which are linked to the exposure of an epitope for an antibody specific for the light-activated state [263]. In another approach, chimeric receptors were generated, in which parts of the fourth loop of rhodopsin were replaced by sequences from the β_2 adrenergic receptor. The data yielded strong evidence that the tripeptide sequence $Asn^{310}Lys^{311}Gln^{312}$ at the end of the seventh helix plays an important role in G_t binding and activation [218]. The C-terminal G_t-α and G_t-γ peptides [181] were used to study how they stabilized MII against its photoconversion to the ground state. A stabilizing effect was found as long as an intact fourth loop structure was present, independent of whether the primary sequence was from rhodopsin or the β_2 receptor [226]. These investigations [218,226] also indicate that not only G_t-$\beta\gamma$ [256,257] but also G_t-α interaction is affected by alterations in the fourth loop.

3.1.5.5. Palmitoyl modifications and luminal region

Only minor effects on G_t activation by light-activated rhodopsin were seen when the palmitoyl modifications at cysteines 322 and 323 were removed, either using a high concentration of NH_2OH in the dark [264] or a mutational substitution by serines of the palmitoyl anchoring cysteines [265]. The absence of an observable effect in intrinsically slow assays, such as the commonly used fluorescence assay [266] should not be overestimated, because of saturation effects with respect to the rate of G_t activation (see Refs. [192,267] for a detailed discussion of this technical aspect). However, even at low levels of rhodopsin regeneration, where the G_t activation rate is not saturated [268], depalmitoylation had no effect [269]. The palmitoyl groups are, however, crucially important for another activity of the rhodopsin receptor, which originates from the opsin apoprotein and its reversible non-covalent interaction with all-*trans*-retinal in the dark (see Section 3.4).

Mutations in the luminal region of rhodopsin, opposite to the cytoplasmic interaction domain, can also influence rhodopsin function. Mutants lacking glycosylation sites, show site-specific folding and 11-*cis*-retinal incorporation, but generally severely impaired activity towards G_t [270]. On the other hand, like in other receptors, even assembly of rhodopsin from expressed fragments can lead to a "rhodopsin" with substantial G_t-activating capability [271].

3.1.5.6. Mechanistic conclusions

Either the second or the third cytoplasmic loop in conjunction with the fourth loop appears to be sufficient to maintain the empty site R^*G_t interaction. However, a more complex interaction pattern may be required to allow for the fast release of GDP from the G_t–rhodopsin collisional complex [35,40]. GTP binding to the empty G_t-α site and the subsequent release of GTP bound G_t-α subunit from rhodopsin requires interactions with both the second and the third loop [35,40].

Although many details have been explored, the mechanism of R^*G_t interaction is not yet understood. One interesting aspect is the specific recognition of MII. The stabilization of protonated MII by the G_t-α and G_t-γ peptides implies that these short stretches from the G_t-α and G_t-γ sequences have the capacity to recognize protonated MII and to distinguish it from differently protonated forms of metarhodopsin. The most recent FTIR evidence even indicates that the recognition by both these peptides is specific for the protonated form of MII, MII_b (Bartl, F., Ritter, E., Evers, T. and Hofmann, K.P., unpublished). It is an open question whether a small percentage of the free peptide has a conformation that fits better to protonated MII than to MI and the other intermediates. The free G_t-α-peptide in free solution does not seem to have a specific structure [237]. On the other hand, NMR and CD spectroscopic studies on short peptides have indeed indicated that small populations of β-turn and helical conformations can be present in aqueous solution [272]. Alternatively, the peptides would lack a defined structure with preference for protonated MII; the selectivity could then result from a mechanism, in which the related receptor binding sites are inaccessible in all other conformations [95,263].

Although the microstructural changes detected by EPR spectroscopy may offer possible mechanisms for the exposure of binding sites, we can at present not even answer the innocent question whether only loop structures contribute to the interaction surface. We may have to wait for the co-crystallization of the R^*G_t complex to specify the nature of the interaction sites. However, even the knowledge of all details of the active state would not answer the question how this state is reached, and how it arises from retinal isomerization. In the next section, various mechanisms discussed in the literature will be presented.

3.1.6. Intramolecular signal transmission in rhodopsin

From the energy-rich state reached after retinal isomerization, rhodopsin passes through the thermal intermediate states of chromophore–protein interaction, reflected in the photoproducts, and reaches the active state characterized in the foregoing section. When discussing how this state(s) can form, we will follow the concepts that have been outlined in the literature.

3.1.6.1. Salt bridge breaking

In the ground state, the chromophore 11-*cis*-retinal is bound via a protonated Schiff base to Lys^{296} near the center of the seventh transmembrane stretch. Direct spectroscopic evidence for an interaction of helices C and E in the ground state, mediated by residues Glu^{122} and His^{211}, was provided in a recent FTIR study [68]. Based on the information from constitutively active mutants, Oprian and colleagues [273,274] have suggested that the inactive conformation of rhodopsin is specifically stabilized by a salt bridge between the protonated Schiff base and its counterion, the carboxyl sidechain of Glu^{113} [13]. The required apposition of Glu^{113} and the positive charge of the Schiff base nitrogen is in agreement with the charge distribution according to the spectroscopic data. ^{13}C-magic angle spinning NMR [275] has actually shown that Glu^{113} interacts with the C-12 region of the retinal hydrocarbon chain.

The binding of 11-*cis*-retinal is thought to additionally stabilize the inactive conformation in rhodopsin's ground state [114,276]. Isomerization removes the first barrier to the transition into the active state. A covalent bond of opsin to the 11-*cis*-retinal chromophore is not essential [277] when the chromophore is provided in the form of a Schiff base with an *n*-alkylamine.

Removal of the charge in the side chain of either Lys^{296} or Glu^{113} results in constitutive activation of G_t by mutant opsins [114]. The rate of G_t activation catalysed by different active species, namely light-activated rhodopsin, the constitutively active mutants, and opsin plus all-*trans*-retinal followed a common principle, when plotted as a function of pH. The resulting so-called pH/rate profiles are bell-shaped curves (cf. the Synopsis and Fig. 4), with the left and right wing reflecting the necessary deprotonation and protonation of groups with different pK_a. Evidence was provided that both these groups are localized at the active catalyst of nucleotide

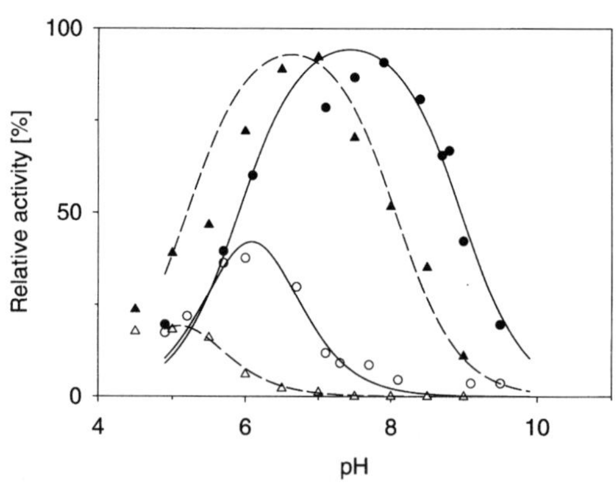

Fig. 4. pH/rate profiles of rhodopsin-induced activities in transducin and rhodopsin kinase. The data points are measured relative rates (in %) of nucleotide exchange by G_t (circles) and of phosphorylation by RK (triangles), both stimulated by interaction with light-activated rhodopsin or the all-*trans*-retinal–opsin complex. Closed symbols, light-induced formation of active rhodopsin; open symbols, light-independent all-*trans*-retinal–opsin complex. Data from Refs. [114,162], respectively. The solid and broken lines are fits with the function % rate $= 100\,(H/(H + K_2))\,(K_1/(H + K_1))$, where H is the free H^+ concentration ($=10^{-pH}$), and K_1 and K_2 are proton dissociation constants ($=10^{pK_a}$) [114]. Solid lines belong to G_t, broken lines to RK.

exchange, i.e. rhodopsin or opsin. Notably, all curves were identical when the right wing of the bell-shaped curves (Fig. 4) was adjusted by a mere change of the relevant pK_a. From this behaviour, the authors derived a mechanism as follows: (i) the salt bridge between Lys^{296} or Glu^{113} is broken, (ii) Lys^{296} becomes deprotonated, and (iii) a proton is taken up from solution. Deprotonation of Lys^{296} is thought to arise from a movement of the residue into a more hydrophobic environment, and proton uptake arises from the exposure to solvent of another group, both in concert with a conformational change. Generally, the active state is thought to arise from deprotonation of a group with pK_{a1} and protonation of another group with pK_{a2} [274], linked by the mechanism described, thus explaining the pH/rate profile. The approach has recently been extended to rhodopsin kinase interaction (Ref. [162], see below).

3.1.6.2. The microswitch approach

Key individual molecular changes were considered in the "weighted sum" approach of on/off switches [211], already mentioned above. These include changes at the chromophore, at the Schiff base imine, at the Schiff base counterion, Glu^{113}, at the other highly conserved Glu in the third helix, Glu^{134}, and in the hydrogen-bonding environments of Asp^{83} and Glu^{122} [210]. Several studies with recombinant-mutant proteins and FTIR difference spectroscopy have investigated the protonation of the Schiff base and its counterion during photoactivation [219,220,278,279]. Strong evidence could be presented that the side chain of Glu^{113} is eventually neutralized through the shift of the Schiff base proton [75], either directly or indirectly via a proton relay system. Using mutants in which the position of the Schiff base was shifted by one helix turn [280–283], it was shown that activity towards G_t is not determined by the protonation state of the Schiff base *per se*, but rather by the mechanistically coupled neutralization of the counterion. The requirement of Schiff base deprotonation to form active rhodopsin [74] is in this view interpreted as the Schiff base proton inducing the protonation of the Glu^{113} side chain.

We have already discussed above that reversal of the Glu^{134}/Arg^{135} in the highly conserved tripeptide GluAsp-Arg-Tyr (rhodopsin residues 134–136) [6] abolishes binding and activation of G_t [35]. The replacement of Glu^{134}, located near the cytoplasmic border of the third transmembrane helix, by neutral Gln makes photoactivation more efficient and changes the pH/rate profile [284]. It also causes constitutive activity in the absence of chromophore [274]. It was further shown that the same replacement abolishes proton uptake [153], and we have seen above that the protonated Meta form of rhodopsin (MII_b) is likely to be the one that eventually develops the conformation for the empty site interaction with G_t.

It is not yet clear whether Glu^{134} may be identified with the group with pK_{a2} in the salt bridge model, which is forced to become protonated. The group cannot be identified with His^{211} [114], which was reported to influence the MI/MII-equilibrium [47]. FTIR spectrocopy could not yet assign a C=O stretching vibration to a protonation of Glu^{134} [220,285], although the most recent FTIR evidence from measurements on the R^*G_t complex support a role of a Glu residue [239]. Also, when the neutralization by protonation of Glu^{134} is mimicked in the double mutant E134Q/K296H, a bell-shaped pH rate profile is still expressed [274], arguing for a more complex charge pattern required for activation of G_t. We should remember, however, that complexity is introduced by the additional proton uptake in the presence of G_t (see above, Ref. [171]).

Open questions about this mechanism include:

(i) How is the proton translocation induced?

(ii) How are the proton transfer reactions at the two Glu residues coupled?

(iii) What is the nature of the titratable group with pK_{a1} that expresses itself in the pH/rate profiles? If it is true that proton transfer, rather than proton release operates the essential "on" switch, no reprotonation from the aqueous phase occurs at this step, and the left wing of the pH/rate profile remains unexplained.

For both the first and second question, interesting solutions have been proposed, associated with the terms "steric trigger" and "solid state coupling" via helix movements. The last aspect (iii) was, to the best of our knowledge, so far not discussed in the literature (see Section 4.4 for a discussion).

What will be discussed first is how the proton translocation to Glu^{113} is brought about.

3.1.6.3. The "steric trigger" problem

The retinal binding site is sensitive to the packing of helices 3 and 6 [210,224,286]. Time resolved UV/Vis spectroscopy of the early intermediates on rhodopsin mutants has shown that Gly^{121} (and to lesser extent, Phe^{261}) have an influence on early photoreactions, such as the decay of the BSI photoproduct [287]. The decisive step, however, in which the proton translocations leading to MII are prepared, is likely to be associated with the Lumi to MI transition.

Maeda and associates have emphasized the role of water in this transition. When the water content in air-dried films is lower than 15%, the conversion from Lumi to MI becomes extremely slow [288]. In the subsequent conversion to MII, a change in the hydrogen bonding at residues Asp^{83} (in helix B) and Glu^{122} (in helix C) is observed, which depends on the presence of internal water molecules. These water-dependent structural changes between helices B and C were proposed to be involved in the generation of the intramolecular transmission signal to the cytoplasmic G_t binding domain [90].

A reaction model which is based on steric interactions, rather than on hydrogen bonding changes, involves Gly^{121} and the 9-methyl group of retinal. It has long been known that, when rhodopsin is regenerated from opsin and the 9-desmethyl analog of 11-*cis*-retinal (9-dm-rhodopsin), its maximal absorption is blue-shifted [289]. Methyl groups extend from both the conjugated chain and from the β-ionone ring of retinal. Single demethylations or insertions of methyl groups show distinct effects on rhodopsin functionality [290,291]. Although an intact ring structure is essential for rhodopsin to reach the active state [292,293] the methyl group at the 9-position of retinal is likely to be the most crucial element. When this group is lacking, the bands in the infrared spectrum typical for MII are not observed, and MI appears to decay directly to an MIII-like product without formation of a deprotonated Schiff base [294]. 9-dm-rhodopsin activates G_t with $<10\%$ efficiency, as compared to normal

MII. Gly^{121}, a highly conserved residue in the middle of the third transmembrane helical stretch appears to play a crucial role as the counterpart of the 9-methyl group. From molecular modeling based on NMR and other evidence, it was concluded that this residue is a determinant of the retinal binding pocket, with a specific interaction between the C_9 methyl group of retinal and Gly^{121} [224,295]. 11-*cis*-retinal acts as an agonist and activates opsin in the dark, when Gly^{121} is replaced with residues with bulkier side chains [296], or when progressively larger substituents (methyl, ethyl, propyl) are present at the retinal 9-position [297]. By the same replacements, the defect seen in G_t activation in 9-dm rhodopsin can be partially restored [297]. It was concluded that the outward movement of transmembrane helix C, relative to F and the helix bundle, a putative requirement for exposing the receptor's interface for G_t interaction [286,298,299], was forced in the dark, or was restored in 9-dm-rhodopsin, by putting protein mass near the position of the 9-methyl group.

It is important to note that the retinal-9-methyl/Gly^{121} switch, although persistently set in its "on"-position by mutation or retinal modification, is likely to be operated in the native system during the Lumi–Meta I transition. From the collected evidence, one may come to the conclusion that the events up to MI-formation are driven by steric interaction, while only later, with the MI/MII-conversion and the exposure of an interface with G_t, electrostatic interactions come into play [211,295,297]. However, there are other determinants, such as the hydrogen bonds of protonated Glu^{122} and Asp^{83}, which are concomitants of the transition into the MII state [107,300], and the water-dependent changes in hydrogen bonding discussed above. Also, the proton release linked to the early MI_{380} product formed on the time scale of the Lumi–Meta I transition [154] indicates that reprotonations may occur on an early time scale.

Steric determinants that come into play in the MI $\rightarrow$ MII time domain are the structure of the β-ionone ring of retinal [293] and the region around the C_{10} position. The introduction of an additional steric constraint at position C_{10} of retinal has clear effects on the formation of MII [291]. With a rhodopsin regenerated from opsin and the analog 10-methyl-retinal (10-methyl-rhodopsin) essentially two effects were seen: it forms normal Batho (which is, however, stable up to 210 K), and it forms normal MII, with the same activation energy, but with ca. six orders of magnitude slower rate, as compared to native rhodopsin. Because the Arrhenius activation energy is the same as in native rhodopsin, the large entropic component of MII-formation, which makes this reaction fast in spite of its large activation enthalpy, is reduced by ca. 40 kJ mol^{-1}. The fact that a single steric hindrance can make such a profound change in the formation of the late product, shows that a steric factor can significantly raise the free energy barrier to this reaction. Thus, although charge interactions may indeed govern all events later than MI [295], the chromophore environment is at least still sensitive to steric factors.

Although important questions remain, we may state that sterically determined transitions precede the reactions involved in Schiff base deprotonation. A similar sequence of events was observed in sensory bacterial rhodopsins and denoted as a "steric trigger" mechanism [294,301]. This terminology may be somewhat misleading in that such steric adjustment may reflect a point of fixation or an "abutment" rather than a trigger, required to adjust the conformation in a way that allows proper proton transfer reactions.

3.1.6.4. Solid state coupling
The approach taken by Hubbell, Khorana and coworkers is to introduce "sensor mutations" [302], i.e. to substitute cysteine residues at appropriate positions and to generate EPR sensitive nitroxide side chains. EPR-measurements allowed to monitor specific changes at labeled Cys-sites and to interpret them as changes in mobility of the respective residue.

Different classes of such changes have been reported, namely:

(i) spin-labeled side chains reflect changes at sites 140 or 316, near the cytoplasmic ends of helices C and G, respectively [112,304,305]. The signal from the site at position 140 was interpreted as a light-induced motion of helix C [304];
(ii) side chains located at sites 227 or 250 in the cytoplasmic domains of helices E and F, respectively, also undergo changes in mobility upon light-activation. These changes were interpreted as a "rigid body" outward motion of helix F in the cytoplasmic domain of the receptor [306];
(iii) spin labels fixed in the two helices C and F allowed to measure their mutual EPR interaction, and to interpret it as a relative motion of the helices, thus supporting the outward movement of helix F [299];
(iv) labeling of residues of the first and putative fourth loops were already discussed in Section 3.1.5; they gave information on tertiary structures in these loops and their light-induced changes.

The interaction of helices C and F was also probed by introducing metal ion binding sites (His residues) between the cytoplasmic surfaces of these helices [298]. It was found that His residues presented at the cytoplasmic terminals of helices C and F led to a rhodopsin that activated G_t in the absence but not in the presence of the metal ion. Ideally, namely if it can be shown that the blockade of movement and not the presence of the ion *per se* is responsible for the absence of activation, this method has the advantage to show a causal relationship between movement and activation. A positive control could indeed be provided by showing that His residues at helices C and G did not produce the effect seen with C and F.

The specific benefit from time-resolved electron-paramagnetic-resonance is that it provides a direct structural and kinetic monitor of microstructural changes accompanying MII-formation. In native disc membranes, the conformational change at Cys^{140} does not appear prior to MII-formation. The EPR (and thus conformational) change appeared with a time constant and activation energy consistent with the optical signal indicating the appearance of MII [111]. Additional evidence was provided by experiments on rhodopsin solubilized in different detergents. When the receptor was solubilized in digitonin, under conditions where MI is the dominant species (see above), almost no change in the specific EPR signals interpreted as changes in label mobility were seen at cysteine residues in selected positions (Cys^{140} and Cys^{316}, respectively) on each of the cytoplasmic loops CD and EF. With detergents allowing formation of MII, the full signals were observed.

The measured movements are certainly suited to transmit chromophore motions to the cytoplasmic surface of the receptor and to cause an increased exposure of residues for the contact with the G-protein and with regulatory proteins. For example, it has been argued that the ionone ring of retinal, a determinant of chromophore–protein interaction (see above), makes contact with Trp^{265} in helix F, and this residue may serve as another abutment for chromophore motion [299]. On the other hand, if we assume that one of the movements measured at the end of helices C and F belongs to the MII_a–MII_b transition, a direct (rigid) coupling between retinal Schiff base deprotonation and helix motions in the vicinity of Glu^{134} is unlikely, considering the two reaction steps in the MI → MII_b transition.

3.1.6.5. Two-step activation scheme

Although the time resolution of the EPR methods would suffice to detect alterations on the surface that occur earlier than the Meta II states, all measured EPR changes have so far been attributed to the MI/MII transformation [111], and no such changes have been observed on a faster time scale. This confirms the notion [168] that the early events, including those related to the steric trigger and the Lumi–Meta I transition, remain decoupled from the cytoplasmic interaction domain.

All the concepts of receptor activation outlined above contain the element of a variable coupling between (at least) two domains of the receptor in the course of its activation. In the MII_a/MII_b model, the proton movements during the lifetime of MII do not occur in one concerted reaction but in two separate transitions, linked to proton transfer from the Schiff base and to proton uptake, respectively. Consistently, MII_a and MII_b are sensitive to quite different influences from the molecular environment (see Section 2.2.5). Because both MII_a and MII_b are sequentially involved in the activation of rhodopsin, at least two steps are necessary to reach the fully active state. The most recent EPR signals recorded in the dark from labels in the vicinity of Glu^{134} have been interpreted to indicate that the movement of helix F can occur independently of those of helices A and E [302]. This could argue for even more complex schemes, in which the cytoplasmic domain is not homogeneous but divided in subdomains, which can react independently (see the Synopsis).

3.1.6.6. Sequential fit scheme

As already discussed in Section 3.1.5.2, peptide stabilization experiments [181] have shown that MII is not only stabilized by the G_t holoprotein, but also by synthetic peptides from the $G_t\alpha$ and $G_t\gamma$ C-terminal sequences. This has suggested that the respective structures on the holoprotein are interaction sites for rhodopsin. Surprisingly however, the X-ray structure of G_t localizes the C-terminal regions of $G_t\alpha$ and $G_t\gamma$ to a common surface but at a large distance, as compared with the dimensions of the rhodopsin projection structure [185]. A two-site sequential fit mechanism provided a minimal model for signal transfer from the activated receptor to the G-protein. It assumes that the contact between a receptor site and one of the sites on G_t leads to a conformational change in the G-protein and/or receptor, with subsequent binding of the second pair of interaction sites. In the light of the two-step activation scheme described in the foregoing section, one may ask whether the two conformations before and after the conversion are – if it occurs in the receptor – related to two different metarhodopsin species, such as MII_a and MII_b. However, recent FTIR data (see Section 3.1.2) show that the two peptides fail to recognize any species other than the fully protonated MII. Properties of Schiff base protonated (i.e. MI-like) species are not detectable. This is interesting in view of the reported interaction of an MI like species (MI_b) that can interact with G_t [156,157]. Although the data do not support a role of intermediates other than protonated MII in the interaction with the two identified G_t sites, the G_t holoprotein may still reserve binding sites other than the α- and γ-C-termini [303], which may interact with earlier intermediates. An earlier exposure of binding sites is at least conceivable, given the intriguing observations on the interactions of rhodopsin kinase (see Section 3.3.2).

3.2. Arrestin

3.2.1. Specificity for Meta II photoproduct

Recent knock-out studies have confirmed the deactivation scheme outlined in the Introduction ([307]; Fig. 3b). Arrestin binds to the phosphorylated form of light-activated rhodopsin. Although preferred sites of rhodopsin kinase mediated phosphorylation include Ser^{334}, Ser^{338}, and Ser^{343}, replacement mutations have suggested that residues Thr^{340} and Ser^{343} play important roles in promoting binding of arrestin [308]. Weak binding of arrestin to both phosphorylated dark rhodopsin and to unphosphorylated, photoactivated rhodopsin appears in immunochemical assays, but has never been observed by biophysical assays (MII stabilization and light scattering binding signals; see Ref. [309] for a discussion of the latter technique). This reflects the fast, flash-induced interaction relevant for signal transduction. Arrestin binds to phosphorylated MII with a K_D of 50 nM [310] and a bimolecular on-rate of about 0.2 $\mu M^{-1}\ s^{-1}$. This mode of binding is specific for MII, as was shown by arrestin-induced stabilization of the photoproduct [310] (Fig. 2d) and by the release of arrestin following hydroxylamine-induced MII decay [311]. The on-rate of binding shows little dependence on concentration [312] (by contrast to the splice variant, see below). At cellular concentration of the protein, a fast binding reaction (in the order of 50 ms) with rhodopsin can be assumed. This means that the interaction with arrestin would not be rate limiting for the overall shut-off reaction sequence (Fig. 3b).

3.2.2. Mapping of the interaction sites

The recognition domain for rhodopsin in the arrestin sequence may overlap with that in G_t [313,314]. It does, however, not appear to include the strongly homologous C-terminal region of arrestin, since the splice variant, p44, binds with the same affinity, although it lacks this part of the sequence (see below). The interaction between arrestin and light-activated phosphorylated rhodopsin (R*p) presumably involves cytoplasmic loop domains of rhodopsin and a phosphorylation site at the C-terminus [315]. Synthetic peptides corresponding to rhodopsin's cytoplasmic loops competed against R*-arrestin interaction in a co-elution assay; the effect was strongest for a peptide from loop CD and much weaker for a peptide from loop AB [314].

Smith et al. [316] determined regions of arrestin that bind to rhodopsin, using a phage-display of arrestin fragments against phosphorylated rhodopsin and PDE activity inhibition by synthetic arrestin peptides. The results indicate an involvement of residues 90–140 in arrestin/rhodopsin interaction. It was found that synthetic peptide 109–130 inhibited the binding of arrestin to rhodopsin completely but with a high IC50 of 1.1 mM. The authors suggested that this portion of arrestin may be only one of several binding sites for rhodopsin [316]. Using the extra MII assay (Pulvermüller, A., Schröder, K. and Hofmann, K.P., unpublished) one finds indeed a similar IC50 in this region but higher competition (IC50 of 50 μM) in arrestin's N-terminal- and C-terminal domains. This would be in agreement with a specific form of a conformational switch (see below), in which a reorientation of the N and C domains presents their surfaces for association with rhodopsin.

3.2.3. Conformational switch

Arrhenius plots of the arrestin binding reaction monitored via Meta II stabilization yielded a large activation energy, which can be interpreted as a conformational transition in domains of arrestin and/or rhodopsin, linked to the interaction [310]. Limited proteolysis of free and bound arrestin have indeed shown that binding to R*p protects arrestin from the attack of the proteolytic enzyme. This was interpreted as an indication that arrestin bound to R*p assumes a conformation (A_b) which is different from that of free, inactive (A_i) arrestin [311,314]. Recent sulfhydryl reactivity studies argue for several different inactive conformations, all of which can attain the active A_b conformation [317]. Heparin [311] or phytic acid [318] mimic the effect of R*p to some degree. A highly cationic region beginning with residue 163 was proposed to mediate the interaction with the negatively charged regions of phosphorylated rhodopsin or heparin [311]. Studies on mutated and truncated arrestins [319,320] have confirmed this hypothesis and localized another major binding site for the phosphorylated region of rhodopsin, heparin and phytic acid to the N-terminus (residues 1–47) of the arrestin sequence. A sequential process was invoked for the conformational transition from free to bound arrestin, involving the contact of the phosphorylated region as a trigger, which switches arrestin into its active conformation and allows it to interact with the binding sites exposed on rhodopsin by photoactivation [320]. Such a mechanism was strongly corroborated by the finding that a synthetic phosphorylated peptide from the C-terminal sequence of rhodopsin induces binding of arrestin to non-phosphorylated R* [321]. A mutant, in which Arg^{175} (within the putative recognition site for the phosphorylated site(s) on rhodopsin) was replaced by Gln, was reported to show enhanced binding to phosphorylated rhodopsin, and the replacement by Glu resulted in an arrestin that does not distinguish between phosphorylated and unphosphorylated rhodopsin [322]. A splice variant of arrestin, p44, in which the C-terminal residues 370–404 are replaced by a single alanine, is apparently only present in rod outer segments [323] and has been reported to inhibit phototransduction similar to the parent protein. However, it interacts not only with phosphorylated but also with unphosphorylated R* [324], although with lower affinity [312]. The lack of the C-terminal sequence also lowers the activation energy of the binding reaction (70 kJ mol^{-1} instead of 140 kJ mol^{-1}), and it removes any specificity for the C-terminal structure of the receptor, so that even C-terminally truncated rhodopsin binds to p44. Thus it appears that, by the lack of the C-terminal structure, the conformational switch between active and inactive conformations of arrestin is absent in the splice variant [312]. This fits nicely to a recent proposal, based on intrinsic fluorescence and circular dichroism data, that the conformational switch involves localized movements of the N- and C-termini of arrestin; these regions may interact in the inactive conformation, and may be separated by interaction with phosphorylated rhodopsin [325].

An interesting role for the arrestin switch was recently discussed by Palczewski et al. [166]. It was found that the visual cycle in arrestin knock-out mice was almost as fast as in normal animals, and did not show the abnormalities in the kinetics of rhodopsin regeneration expected for stable binding of arrestin. Thus the R*-operated switch may in turn lead to a change in R*, enabling decomposition of R*, release of arrestin and all-*trans*-retinal reduction.

3.2.4. Structure

Two X-ray analyses of arrestin are now available [326,327]. Both show an interesting "double-cap" structure, with the larger cap containing the putative binding sites derived in the earlier work from mutagenesis and peptide competition. Differences exist in the location of the functionally important C-terminus. Granzin et al. [326] reported that the C-terminal tail (residues 369–404) was not visible, presumably due to its high flexibility. In this structure, the first nine residues of the N-terminus were assigned positions where Hirsch and coworkers [327] placed the C-terminal residues 375–384. Based on the structural data [327], a mechanism for the $A_i \rightarrow A_b$ conversion [328] was derived. It assumes that interaction of the phosphorylated C-terminus of R*p with the "polar core" (containing Arg^{175}) leads to disruption of a hydrogen-bounded network of buried, charged side chains and structural rearrangement of arrestin.

Although many questions remain, the co-crystallization of arrestin with the phosphorylated peptide and even with R*p is now within the reach of future research, so that the mechanism of the conformational switch can be explored.

3.3. Rhodopsin kinase

3.3.1. Role of photoproducts

Akhtar and coworkers have dissected the action of rhodopsin kinase (RK) into two distinct steps, binding and activation [329,330]. These authors claim that the RK remains active after dissociation from the receptor [331], whereas others assume that it remains only active as long as it is held in an active conformation by interaction with rhodopsin [10,332,333].

In sharp contrast to arrestin and G_t, a preference of RK for any of the Meta states is not measured [72]. Fig. 2e shows experimental data that demonstrates that binding of RK after a flash can destabilize MII by competition of RK with G_t for photoactivated rhodopsin. On the other hand, it was reported that a rhodopsin with chemically blocked Schiff base deprotonation is not a substrate for the kinase [334]. The conflict would be solved by a mechanism, in which the kinase can bind to all Meta forms of rhodopsin but needs the protonated form, MII_b (as discussed in Ref. [162]), to perform the actual phosphorylation step. It is an open question whether constitutively active mutants of rhodopsin, which activate G_t in the absence of the chromophore, can activate the kinase [335,336].

Studies on intact retinae, in which MI (but not MII) was allowed to form by warming up from low temperatures, and subsequently photoregenerated to rhodopsin, confirm the notion that MI provides the trigger for the generation of the kinase substrate [71]. Since the actual phosphorylation reaction was performed at room temperature, the exposure of phosphorylation sites could well have occurred at a stage later than MI. The experiment is also significant in that it shows that photoregeneration restores the spectral identity of rhodopsin but not the conformational changes that trigger phosphorylation (in remarkable analogy to the reverted Meta product discussed above [168]).

3.3.2. Mapping of the interaction sites

Like with arrestin and the G-protein, the mapping of the binding sites of RK to active rhodopsin is not yet complete. Involvement of the cytoplasmic loop EF was concluded from partial digestion data [338], but the effects of amino acid replacements argue for the involvement of more than one loop in the phosphorylation reaction [243]. In a recent report, Khorana and coworkers studied the role of loops CD and EF in rhodopsin–RK interaction, using a solubilized system [337]. The ratio k_{cat}/K_m was used to distinguish whether the mutations affected binding of the RK or whether the enzyme can bind where its activation and/or the catalysis of phosphoryation is impaired. It turned out that both loops are involved in binding, with a specific role of acidic residues in EF but not in CD. The reversal of the Glu^{134}/Arg^{135} charge pair removed any interaction. Generally, a role for the CD loop in RK binding and a role for EF, both in binding and catalysis, was concluded [337]. Combined with the evidence mentioned above that RK can bind to MI, one would have to assume that the relevant interaction sites on loop CD are already exposed in MI.

3.3.3. Kinetics of complex formation

The kinetics of R*–RK complex formation (which precedes phosphorylation) was measured in a reconstituted system of detergent-solubilized rhodopsin kinase and washed disc membranes, yielding a K_D of 0.5 mM and a bimolecular rate constant of 1 μM^{-1} s^{-1}[72]. However, detergent solubilization is likely to affect the kinetic properties of rhodopsin kinase (as to the properties of RK, the reader is referred to Ref. [10]). Moreover, even in a more native membrane preparation system, the bimolecular rate constant measured in dilution cannot be directly extrapolated to the kinetics *in situ*; see Ref. [312] for a detailed investigation of the problem in the case of the arrestin splice variant. A recent study [333] has now attempted to derive kinetic constants of RK–rhodopsin interaction from the behavior of the photoresponse in dependence on the number of R* formed per disc membrane [339]. The

R*–RK interaction was described as a two-step reaction comprising complex formation and phosphorylation/dissociation. In this model, R* binding obeys different kinetics dependent on whether the kinase is substrate saturated or not. Thus properties of the photoresponse saturation function (for warm-blooded animals) could be explained, and reaction times for the binding of the kinase and the phosphorylation step could be given (0.25 and 1 s under conditions *in vivo*, respectively).

Not only arrestin, but also rhodopsin kinase competes with G_t for binding to active rhodopsin [72]. This "pre-arrestin function", which is independent of the cofactor ATP and RK's enzymatic function, is presumably due to direct steric hindrance by the bound protein. It should have direct implications for the understanding of the shut-off mechanism of rhodopsin, because together with the above-mentioned different R* binding kinetics for saturated/subsaturated RK [333], it transcends the classical notion [340,343] of a constant characteristic lifetime of activated rhodopsin. However, rhodopsin shut-off by binding of RK (in the absence of phosphorylation) is, in direct electrophysiological experiments [341], so far not evident. It is not clear how a mechanism of direct inhibition can be absent *in situ*. This is the more surprising since in the parathyroid hormone receptor/inositol phosphate pathway, G-protein coupled receptor kinases can inhibit receptor signaling already under nonphosphorylating conditions [342].

3.3.4. Light-dependent signaling state

Regeneration of the opsin in salamander rods with the retinal analog 11-*cis*-9-desmethyl retinal causes prolonged activation and a smaller quantal response [340], arguing for a delayed action of RK in rhodopsin shut-off. Reduced light-dependent phosphorylation has indeed been shown biochemically [344]. Whether this is due to slower binding of the enzyme or slower phosphorylation of the substrate, once the enzyme is bound, remains to be shown.

In another approach, retinal analogs were regenerated with amidated forms of opsin, in which all exposed Glu residues were converted to Gln [162]. The study comprised light-induced RK–rhodopsin interaction, its comparison with G_t–rhodopsin interaction, and the analysis of light-independent opsin activity (see below), to reach conclusions about the mechanisms of opsin activation. To explain how Glu^{134} can be protonated at neutral pH, Glu–Glu pairing was proposed. The essential result of this study is that, in close analogy to G_t activation models (Section 3.1; Ref. [114]), the energy channeled into the receptor by binding or by light-induced transformation of the retinal ligand forces cytoplasmic Glu sites into the protonated state, thus providing the interface for RK binding.

3.4. Light-independent signaling

3.4.1. Opsin

Besides the activation of G_t in response to the light-triggered activation of rhodopsin, there is a light-independent activation of G_t by opsin with the chromophore site empty. At neutral pH, one measures an activity lower than 1% of that of Meta II [268]; higher activities have been reported [345], which may, however, result from an underestimation of the Meta II activity (for a discussion of this problem, see Ref. [192]). Opsin activity may have an interesting physiological role in maintaining a certain stimulation of the visual cascade, which is one of the potential explanations for "bleaching desensitization" [346]. The level of activity required for this function is in the order of 10^{-6} of the Meta II activity. In recent measurements of PDE effector activity, after removal of any residual retinoids, such a low activity could indeed be measured *in vitro* [276]. It is important to distinguish the desensitizing activity of opsin from the activity expressed in "photon-like noise" [175]; by its sensitivity to hydroxylamine, this latter activity was assigned to Meta II, forming via reversal of phosphorylation and arrestin binding [165].

The low activity of opsin was explained by a salt bridge between Glu^{113} and Lys^{296}, which forces the protein into a low activity conformation [273] (see also the Section 3.1.6). The mechanism was derived from results obtained with mutant opsins, where one or the other charge, Glu^{113} or Lys^{296}, was removed and consequently opsins with much higher activity were generated. 11-*cis*-retinal strengthens this salt bridge, forming the inactive, 11-*cis*-retinal ground state rhodopsin, which can be activated by light. Any low activity can either be explained by a form of the receptor with intrinsically low activity or by an active, Meta II-like conformation, present at very low concentration (as discussed for the β-adrenergic receptor [347].

3.4.2. All-trans-retinal–opsin complexes

Addition of all-*trans*-retinal to opsin enhances the activity [268,348,349]. At saturation, the G_t activation state of the all-*trans*-retinal–opsin complex is at neutral pH by a factor of 1/30 lower, relative to Meta II. A plausible explanation for this activity would be to assume that a small amount of a Meta-II-like active state is formed at equilibrium (see the next section for details). Actually, two different all-*trans*-retinal-opsin complexes are reversibly generated, and none of these complexes involves Schiff base formation with the original Lys^{296} site. There is a non-covalent complex of opsin with all-*trans*-retinal and retinal analogs that can generate a conformation capable of interacting with G_t [114,268] and also with rhodopsin kinase and arrestin [162,350,268].

Schiff base formation between all-*trans*-retinal and opsin does occur, leading to the reversible "pseudo-photoproducts" [351]. Neither in the non-covalent complex nor in the pseudo-photoproduct, all-*trans*-retinal competes with 11-*cis*-retinal when it regenerates rhodopsin by binding to Lys^{296}, suggesting that residues other than Lys^{296} are involved [268]. The pseudo-photoproduct interacts with arrestin and kinase, but an interaction with G_t is not measured [351].

Spontaneous activity of rhodopsin in the ground state, i.e. of the light-sensitive 11-*cis*-retinal–opsin complex, in fully dark adapted rods can arise at a rate of $\leqslant 10^{-9}\ s^{-1}$ per molecule of R* [352]. The Schiff base may spontaneously deprotonate, followed by thermal *cis*/*trans* isomerization, or a G_t interaction domain may even form without thermal *cis*/*trans* isomerization of the chromophore [57]. Recent results, which show a certain degree of decoupling between the chromophore domain and the actual recognition domain [168,302], are consistent with such a mechanism.

We may summarize the results as follows (atr, all-*trans*-retinal):

$$\text{atr} + \text{opsin} \rightleftharpoons \text{opsin-atr-SB} \rightarrow \text{Interaction with RK, arrestin} \qquad (9)$$

$$\text{atr} + \text{opsin} \rightleftharpoons \text{opsin-atr-complex} \rightarrow \text{Interaction with } G_t, \text{RK, arrestin} \qquad (10)$$

where opsin-atr-SB stands for the reversibly formed "pseudo-photoproduct", with the Schiff base formed between all-trans-retinal and site(s) other than Lys^{296}; ops–atr-complex is the complex arising from non-covalent interaction.

3.4.3. *Palmitoylation-dependent binding site for all-*trans*-retinal*

As mentioned above, the low activity of opsin as compared to Meta II could be due to Meta II at low equilibrium concentration. Such a small amount of Meta II-like species would be hard to be determined by direct spectroscopy. However properties of the G_t activation argue for opsin–atr representing a separate form of the active receptor with intrinsically low activity. These include that all-*trans*-retinal recombines not only with native opsin but also with reductively permethylated (PM)-opsin, lacking free lysine side chains with the exception of the original blocked binding site [268], and produces a similar level of G_t activation as with native opsin. More importantly, removal of the palmitoyl anchors from the C-terminal cysteines, which does not measurably affect the activity of Meta II, has a strong inhibitory effect on the activity of opsin–atr [269]. This observation yields strong evidence that the rod photoreceptor is capable of two different types of activity, one Meta II-like, one opsin-like. The second activity arises from a second binding site for atr, which may be prevented from reaching the 11-*cis*-retinal binding pocket in native opsin (but not in distinct mutants of opsin; see [353] and references therein). Interestingly, however, the opsin-like activity is sensitive to competition by the same C-terminal G_t-α and G_t-γ peptides which were found to inhibit the interaction of MII with G_t [181,269].

On the other hand, this finding raises interesting questions as to the role of the palmitoyl modification. A physiological role for the opsin–atr complex as a storage form of the photolysed chromophore, all-*trans*-retinal, was recently found in mice lacking the rim protein of rod outer segment discs, a member of the ABC transporter superfamily [354]. The delay of dark adaptation in these retinae was interpreted as an accumulation of the opsin–atr complex and a reduced transport of all-*trans*-retinal. The rim protein is a likely component of a larger signaling complex organized by glutamic acid rich protein [355]. It will be of interest to analyze if and how the palmitoyl-dependent binding of the retinal to opsin is involved in its translocation to the export machinery of the ABC transporter.

4. Synopsis

4.1. *Signaling states related to Meta intermediates*

After having discussed the photointermediates and signaling states of rhodopsin, we may summarize the results. As Scheme (11) shows, functional roles for all the

		Stabilized by salt bridge & 11-*cis*-retinal	
↓	ps	Isomerization of retinal into strained *trans* conformation	
B ⇔ BSI		Energy stored in strained retinal–opsin interaction Partial relaxation of retinal	
↓	ns	Transfer of absorbed energy to opsin, positive ΔH^0	
L ⇔ MI_{380}		Retinal partially relaxed Partial retinal SB deprotonation/ proton release	
↓ ↓	μs	Specific steric retinal/opsin interactions	
MI	ms	Retinal fully relaxed	RK can bind
⇔		Sensitive to lipid; pos. ΔH^0	
MII_a		SB deprotonated	
⇔		Sensitive to pH, T	
MII_b		Protonated	G_t and RK activated
⇔		Sensitive to pH, T.	
MIII		SB reprotonated	
		↓	
		Opsin + all-*trans*-retinal	
		↓	
	>100 s		
Opsin–all-*trans*-retinal complex		Non-covalent retinal–opsin complex	IA with G_t
Pseudo-MII		Reversible Schiff base	IA RK & arrestin

photointermediates are currently emerging. They all reflect a specific state of chromophore–protein interaction in the time ordered sequence of events that lead to receptor activity, i.e. interaction (IA in Scheme (11) above) with transducer or regulatory signaling proteins, which include the G-protein transducin (G_t), rhodopsin kinase (RK), and arrestin.

All light-induced protein–protein interactions of rhodopsin studied so far are related to the Meta intermediates, which represent distinct conformations of the receptor, formed in thermal equilibrium and on a time scale comparable to that of the physiological rod response. They are distinguished by the state of protonation of the retinal Schiff base and of highly conserved amino acids. It is now emerging that the protein–protein interactions in the Meta intermediates may be more selective than has been assumed before; Meta I is competent to bind RK but formation of Meta II is mandatory to bind G_t or arrestin. Each signaling protein is specifically activated by the contact with the receptor. To operate the conformational switch that leads to (i) activity towards the effector in the case of G_t, (ii) enzymatic kinase activity in RK, and (iii) tight "capping" (and/or as yet hypothetical transformation) of the phosphorylated receptor in arrestin, the protonated form of Meta II, MII_b, appears to be required in all cases.

While proton transfer reactions govern the transitions between the Meta states, steric constraints prevail in the early steps of intramolecular signal transmission from the light-absorbing chromophore into the structure of the protein.

4.2. Steric constraints and Coulomb interactions

In the early Batho stage of intramolecular signal transmission, the chromophore is stabilized in a strained transoid conformation. When the stored energy has dissipated into the apoprotein, specific interactions are selected from the many degrees of freedom in the apoprotein structure. It takes microseconds, i.e. 10^6 times longer than Batho formation, to build up these specific interactions in the Meta I state. Identified crucial contacts include the C_9 methyl group of the retinal and Gly^{121}, an interaction often characterized as a steric trigger. We have argued that the term "steric abutment" might be more adequate, because of the evidence that this state serves to adjust the proton acceptor and donor groups for proton transfer. In this view, the only irreversible trigger element is the photochemical isomerization of the chromophore.

Early proton release on the time scale of Meta I-formation (MI_{380}) does occur and may indicate more than just some irrelevant "leakage"; the early events may well be co-determined by reprotonations. Conceivably, however, the steric contacts are necessary to build up a defined adjustment of donor and acceptor groups for the subsequent proton translocations.

It would be an obvious assumption that the steric adjustment in MI serves to adjust retinal Schiff base and Glu^{113} side chain, and thus proton donor and acceptor, with respect to distance and/or orientation and electric field. The neutralization of Glu^{113} is certainly a key requirement of activity. However, it is not necessarily brought about by direct proton transfer from the retinal Schiff base. Specific reasons are: (i) mutants with the Schiff base counterion, i.e. the putative final proton acceptor, shifted by one helix turn, form a photoproduct in which the Schiff base remains protonated; this mutant is active towards G_t [281]; (ii) formation of the early MI_{380} product (Scheme (8), and accompanying proton release into the aqueous phase (both dominant in detergent solution [154]) indicate leakage of proton transfer from the Schiff base before the steric situation is adjusted by forming MI; yet G_t is activated in detergent, i.e. under conditions of dominant (though transient) MI_{380} formation.

These experimental facts indicate that both the requirements for Schiff base deprotonation and Glu^{113} protonation are not very stringent, and that the events at Glu^{113} and at the Schiff base can be uncoupled from each other without much effect on the activity of the receptor. Available methods do not answer the question where, in the course of light-induced activation, the Schiff base proton goes to and from where the proton that neutralizes Glu^{113} is received. On the other hand, prevention of Schiff base deprotonation in the native structure blocks activity, and the proton movement at the Schiff base is always faster than proton uptake, indicating mechanistic coupling in the native structure. Aspects of coupling and decoupling are best accounted for by the assumption of thermodynamic coupling, for example via conformations of protein subdomains, so that the Schiff base deprotonated conformation enhances the probability of Glu^{113} protonation and vice versa.

These proton translocation events near the retinal Schiff base bond, and at residues Glu^{113} and Glu^{134} are conceivably related to a profound change in the charge pattern within the protein, thus resulting in a broken salt bridge [274] and unlocking the helices, to allow their relative motions [298,299]. Although it seems clear that all the observable events depend on one another, it is less obvious what is the cause and what is the consequence.

4.3. Decoupling from chromophore–protein interaction

The fact that the different Meta II intermediates possess different activities towards G_t implies decoupling between the state of the chromophore and the activities, which are now co-determined by external factors such as pH and lipid environment. In detergent solution, the influences of pH and lipid could be assigned to subforms MII_a and MII_b, respectively [140]. Another example is the form of active rhodopsin apoprotein (opsin) which arises from adding the photolysed chromophore to the apoprotein. It is active without light and depends on the palmitoyl modification [269]. Most extreme examples can be constructed by "mismatch" experiments, as in the photoreverted Meta product, RM, which releases G_t after adding GTP although it shows the absorption spectrum of the ground state. While this argues for a high degree of flexibility of rhodopsin in forming the active state, certain tight constraints are present even after the Meta states are reached, as shown by the strong inhibition of Meta II formation in 10-methyl-rhodopsin [291].

There are indications of not only temporal but also spatial complexity in the sequential interactions of rhodopsin. Subdomains of the cytoplasmic surface are selectively activated in the dark by mutation [302], and loop alterations distinguish between binding and activation [35].

4.4. Convergence in forced protonation

We have seen above that all interactions of the receptor eventually lead to the operation of an activating switch in the partner protein. The concept of "forced protonation" [162] provides a basis for understanding the underlying interactions. This applies for the G-protein [114,153,284] and rhodopsin kinase [162], but it may also be extended to arrestin [351]. The principle elegantly unifies the different activities by stating that they all converge in protonations and deprotonations of relevant groups, with the highly conserved Glu^{134} as a key residue. What is different is solely the degree of coupling between the location where the retinal ligand is bound and the residue(s) which are forced to be protonated. It requires energy to protonate a residue at a pH higher than its endogenous pK_a, and this is expressed in the relative position of the right wing of the bell-shaped pH/activity profiles (Fig. 4).

The light-induced formation of MII_b has an apparent pK_a of 6.7; the higher pK_a as compared to isolated Glu ($pK_a = 4.8$) may be due to a high local negative charge density, as it arises from Glu–Glu pairing [162]). Shift of the pH/rate curve to the right means that excess energy is fed in, which must originate from the interacting protein (the GTP/GDP free energy gap in the case of G_t). Inversely, the relative shift to the left in the retinoid-induced vs. the light-induced activation reflects a different coupling energy. Interestingly, this difference in the apparent pK_a is the same for G_t and RK, arguing for a common coupling mechanism.

An interesting problem is the left wing of the pH/rate profiles, which is seen for G_t and RK, both in membranes and detergent. It also appears in measurements of stable interaction of Meta II with G_t. The corresponding pK_a is in the order of 5.5–6.0, and should indicate proton release, necessary to reach the active state; intramolecular proton translocations do not express themselves in titration curves. It may reflect the measured preference of microscopic (peptide) recognition for basic pH [181]. Together with the preference of the active MII_b-like conformation for acidic pH, the picture arises that the bell shaped pH/rate profile reflects features of catalytic R^*G_t interaction, namely, microscopic recognition in the left and conformational interlocking in the right wing, respectively.

4.5. Open questions

From the considerations above, a variety of interesting topics for future research can be anticipated. They include:

(i) the further elucidation of steric vs Coulomb interactions in light-induced trigger processes, or more generally the question of how an active state can form;

(ii) the structure of the interaction domains that make up the interface between active rhodopsin and its interaction partners, at least for one representative complex, most importantly for the empty site R^*G_t complex;

(iii) the sequential formation of different interactive states in the receptor and in the partner proteins, including the intriguing role of the conformational switch in arrestin;

(iv) the common and varying mode of activation between light-dependent Meta activity and all-*trans*-retinal–opsin complexes;

and the many questions related to physiology, including

(v) what is and how active is the physiologically relevant form of phosphorylated rhodopsin;

(vi) what is the mechanism of the ground state activity;

(vii) what is the activity behind bleaching desensitization;

(viii) what is the potential role of opsin palmitoylation in all-*trans*-retinal metabolism and transport?

The basic problems are similar or analogous for any of the G-protein coupled receptors, and rhodopsin is likely to be among the first to provide answers to these key questions.

Abbreviations

R, rhodopsin
R*, active rhodopsin
R*p, active phosphorylated rhodopsin
B, bathorhodopsin
L, lumirhodopsin
MI, metarhodopsin I
MII, metarhodopsin II
MIII, metarhodopsin III
BSI, blue-shifted intermediate
MII_a, unprotonated subform of MII
MII_b, protonated subform of MII
MI_{380}, early MII-like photoproduct
$MII^{\S}$, late partially deprotonated subform of MII
RM, photoreverted metarhodopsin
atr, all-*trans*-retinal
PDE, cyclic GMP phosphodiesterase
RK, rhodopsin kinase
DDAO, dodecyldimethylamine oxide
G_t, G-protein of the rod cell, transducin
G_t-GDP, GDP-bound form of G_t
G_t-GTP, GTP-bound form of G_t
G_t-α, α-subunit of G_t
G_t-γ, γ-subunit of G_t
FTIR, Fourier Transform Infrared Spectroscopy
EPR, Electron Paramagnetic Resonance
NMR, Nuclear Magnetic Resonance

Acknowledgements

I wish to thank all members of the laboratory for discussions and Franz Bartl, Elke Hessel, Dieter Maretzki, Christoph K. Meyer and Alexander Pulvermüller for providing me with results prior to publication. The work carried out in my laboratory has been supported by grants from the Deutsche Forschungsgemeinschaft, the Human Frontier Science Program, and from the Fonds der Chemischen Industrie.

References

1. Kühne, W. (1978) in: Untersuchungen aus dem Physiologischen Institute der Universität Heidelberg. Carl Winters Universitätsbuchhandlung, Heidelberg, Germany,Vol. I.
2. Helmreich, E.J. and Hofmann, K.P. (1996) Biochim. Biophys. Acta **1286**, 285–322.
3. Liebman, P.A., Parker, K.R. and Dratz, E.A. (1987) Annu. Rev. Physiol. **49**, 765–791.
4. Applebury, M.L. and Hargrave, P.A. (1986) Vision Res. **26**, 1881–1895.

94. Johnson, R.H. (1970) Vision Res. **10**, 897–900.
95. Mitchell, D.C. and Litman, B.J. (1999) Biochemistry **38**, 7617–7623.
96. Kühn, H. and Hargrave, P.A. (1981) Biochemistry **20**, 2410–2417.
97. Ganter, U.M., Charitopoulos, T., Virmaux, N. and Siebert, F. (1992) Photochem. Photobiol. **56**, 57–62.
98. Rafferty, C.N., Muellenberg, C.G. and Shichi, H. (1980) Biochemistry **19**, 2145–2151.
99. Rafferty, C.N., Cassim, J.Y. and McConnell, D.G. (1977) Biophys. Struct. Mech. **2**, 227–320.
100. Liebman, P.A., Jagger, W.S., Kaplan, M.W. and Bargoot, F.G. (1974) Nature **251**, 31–36.
101. Kaplan, M.W. (1982) Biophys. J. **38**, 237–241.
102. Hofmann, K.P., Uhl, R., Hoffmann, W. and Kreutz, W. (1976) Biophys. Struct. Mech. **2**, 61–77.
103. Kühn, H., Bennett, N., Michel Villaz, M. and Chabre, M. (1981) Proc. Natl. Acad. Sci. USA **78**, 6873–6877.
104. Schleicher, A. and Hofmann, K.P. (1983) J. Biochem. Biophys. Methods **8**, 227–237.
105. Siebert, F., Mäntele, W. and Kreutz, W. (1980) Biophys. Struct. Mech. **6**, 139–146.
106. de Grip, W.J. (1988) Photochem. Photobiol. **48**, 799–810.
107. Klinger, A.L. and Braiman, M.S. (1992) Biophys. J. **63**, 1244–1255.
108. Siebert, F., Mäntele, W. and Gerwert, K. (1983) Eur. J. Biochem. **136**, 119–127.
109. DeGrip, W.J., Gillespie, J. and Rothschild, K.J. (1985) Biochim. Biophys. Acta **809**, 97–106.
110. Siebert, F. (1995) Methods Enzymol. **246**, 501–526.
111. Farahbakhsh, Z.T., Hideg, K. and Hubbell, W.L. (1993) Science **262**, 1416–1419.
112. Resek, J.F., Farahbakhsh, Z.T., Hubbell, W.L. and Khorana, H.G. (1993) Biochemistry **32**, 12025–12032.
113. Salamon, Z., Wang, Y., Brown, M.F., Macleod, H.A. and Tollin, G. (1994) Biochemistry **33**, 13706–13711.
114. Cohen, G.B., Oprian, D.D. and Robinson, P.R. (1992) Biochemistry **31**, 12592–12601.
115. Emeis, D., Kühn, H., Reichert, J. and Hofmann, K.P. (1982) FEBS Lett. **143**, 29–34.
116. Kibelbek, J., Mitchell, D.C., Beach, J.M and Litman, B.J. (1991) Biochemistry **30**, 6761–6768.
117. Boesze Battaglia, K. and Albert, A.D. (1992) Exp. Eye Res. **54**, 821–823.
118. Fliesler, S.J. and Anderson, R.E. (1983) Prog. Lipid Res. **22**, 79–131.
119. Miljanich, G.P., Nemes, P.P., White, D.L. and Dratz, E.A. (1981) J. Membr. Biol. **60**, 249–255.
120. Wang, N. and Anderson, R.E. (1992) Curr. Eye Res. **11**, 783–791.
121. Hessel, E., Herrmann, A., Müller, P., Schnetkamp, P.P.M. and Hofmann, K.P. (2000) Eur. J. Biochem. **267**, 1473–1483.
122. Applebury, M.L., Zuckerman, D.M., Lamola, A.A. and Jovin, T.M. (1974) Biochemistry **13**, 3448–3458.
123. Brown, M.F. (1997) Curr. Top. Membr. Transp. **44**, 285–356.
124. Makino, M., Suzuki, T., Ebina, Y. and Nagai, K. (1980) Biochim. Biophys. Acta **600**, 332–342.
125. O'Brien, D.F., Costa, L.F. and Ott, R.A. (1977) Biochemistry **16**, 1295–1303.
126. de Grip, W.J., Olive, J. and Bovee-Geurts, P.H.M. (1983) Biochim. Biophys. Acta **734**, 168–179.
127. VanBreugel, P.J.G.M., Bovee-Geurts, P.H.M., Daemen, F.J.M. and Bonting, S.L. (1978) Biochim. Biophys. Acta **509**, 136–147.
128. Mitchell, D.C. and Litman, B.J. (1994) Biochemistry **33**, 12752–12756.
129. Mitchell, D.C., Lawrence, J.T.R. and Litman, B.J. (1996) J. Biol. Chem. **271**, 19033–19036.
130. Beck, M., Siebert, F. and Sakmar, T.P. (1998) FEBS Lett. **435**, 304–308.
131. Aveldano, M.I. (1995) Arch. Biochem. Biophys. **324**, 331–343.
132. König, B., Welte, W. and Hofmann, K.P. (1989) FEBS Lett. **257**, 163–166.
133. Gibson, S.K., Parkes, J.H. and Liebman, P.A. (1998) Biochemistry **37**, 11393–11398.
134. Mitchell, D.C., Kibelbek, J. and Litman, B.J. (1992) Biochemistry **31**, 8107–8111.
135. Gibson, S.K., Parkes, J.H. and Liebman, P.A. (1999) Biochemistry **38**, 11103–11114.
136. Radding, C.M. and Wald, G. (1956) J. Gen. Physiol. **39**, 909–922.
137. Stieve, H., Wilms, M. and Nöll, G. (1973) Z. Naturforsch., C: Biosci. **28**, 600–602.
138. Wong, J.K. and Ostroy, S.E. (1973) Arch. Biochem. Biophys. **154**, 1–7.
139. Antonov, V.F., Afanasev, A.L., Ostrovsky, M.A. and Feodorovich, I.B. (1971) Biofizika **16**, 78–86.

140. Arnis, S. and Hofmann, K.P. (1993) Proc. Natl. Acad. Sci. USA **90**, 7849–7853.
141. Gutfreund, H. (1995) Kinetics for the Life Sciences – Receptors, Transmitters and Catalysts. Cambridge University Press, Cambridge, UK.
142. Gutman, M. and Nachliel, E. (1990) Biochim. Biophys. Acta **1015**, 391–414.
143. Emrich, H.M. and Reich, R. (1974) Z. Naturforsch., C: Biosci. **29C**, 577–591.
144. Baumann, C. and Zeppenfeld, W. (1981) J. Physiol. **317**, 347–364.
145. Emeis, D. and Hofmann, K.P. (1981) FEBS Lett. **136**, 201–207.
146. Meyer, C.K. and Hofmann, K.P. (2000) Meth. Enzymol. **315**, 377–387.
147. Heberle, J. and Dencher, N.A. (1992) Proc. Natl. Acad. Sci. USA **89**, 5996–6000.
148. Stewart, J.G., Baker, B.N. and Williams, T.P. (1975) Nature **258**, 89–90.
149. Straume, M., Mitchell, D.C., Miller, J.L. and Litman, B.J. (1990) Biochemistry **29**, 9135–9142.
150. Thorgeirsson, T.E., Lewis, J.W., Wallace Williams, S.E. and Kliger, D.S. (1993) Biochemistry **32**, 13861–13872.
151. Szundi, I., Lewis, J.W. and Kliger, D.S. (1997) Biophys. J. **73**, 688–702.
152. Jäger, S., Szundi, I., Lewis, J.W., Mah, T.L. and Kliger, D.S. (1998) Biochemistry **37**, 6998–7005.
153. Arnis, S., Fahmy, K., Hofmann, K.P. and Sakmar, T.P. (1994) J. Biol. Chem. **269**, 23879–23881.
154. Szundi, I., Mah, T.L., Lewis, J.W., Jäger, S., Ernst, O.P., Hofmann, K.P. and Kliger, D.S. (1998) Biochemistry **37**, 14237–14244.
155. Dickopf, S., Mielke, T. and Heyn, M.P. (1998) Biochemistry **37**, 16888–16897.
156. Tachibanaki, S., Imai, H., Mizukami, T., Okada, T., Imamoto, Y., Matsuda, T., Fukada, Y., Terakita, A. and Shichida, Y. (1997) Biochemistry **36**, 14173–14180.
157. Tachibanaki, S., Imai, H., Terakita, A. and Shichida, Y. (1998) FEBS Lett. **425**, 126–130.
158. Bennett, N. (1978) Biochem. Biophys. Res. Commun. **83**, 457–465.
159. Bennett, N. (1980) Eur. J. Biochem. **111**, 99–103.
160. Bennett, N. (1980) Biochem. Biophys. Res. Commun. **96**, 1695–1701.
161. Hartmann, H., Parak, F., Steigemann, W., Petsko, G.A., Ponzi, D.R. and Frauenfelder, H. (1982) Proc. Natl. Acad. Sci. USA **79**, 4967–4971.
162. Buczyłko, J., Saari, J.C., Crouch, R.K. and Palczewski, K. (1996) J. Biol. Chem. **271**, 20621–20630.
163. Hofmann, K.P., Emeis, D. and Schnetkamp, P.P. (1983) Biochim. Biophys. Acta **725**, 60–70.
164. Donner, K.O. and Hemila, S. (1975) Vision Res. **15**, 985–995.
165. Leibrock, C.S. and Lamb, T.D. (1997) J. Physiol. **501**, 97–109.
166. Palczewski, K., Van Hooser, J.P., Garwin, G.G., Chen, J., Liou, G.I. and Saari, J.C. (1999) Biochemistry **37**, 12012–12019.
167. Pfister, C., Kühn, H. and Chabre, M. (1983) Eur. J. Biochem. **136**, 489–499.
168. Arnis, S. and Hofmann, K.P. (1995) Biochemistry **34**, 9333–9340.
169. Ostroy, S.E. (1974) Arch. Biochem. Biophys. **164**, 275–284.
170. Watanabe, M. and Asai, H. (1980) Biochem. Biophys. Res. Commun. **94**, 529–534.
171. Schleicher, A. and Hofmann, K.P. (1985) Z. Naturforsch., C: Biosci. **40**, 400–405.
172. Vuong, T.M., Chabre, M. and Stryer, L. (1984) Nature **311**, 659–661.
173. Kahlert, M. and Hofmann, K.P. (1991) Biophys. J. **59**, 375–386.
174. Pugh, E.N. Jr and Lamb, T.D. (1993) Biochim. Biophys. Acta **1141**, 111–419.
175. Leibrock, C.S., Reuter, T. and Lamb, T.D. (1994) Vision Res. **34**, 2787–2800.
176. Kühn, H. (1980) Nature **283**, 587–589.
177. Bennett, N., Michel Villaz, M. and Kühn, H. (1982) Eur. J. Biochem. **127**, 97–103.
178. Parkes, J.H., Gibson, S.K. and Liebman, P.A. (1999) Biochemistry **38**, 6862–6878.
179. Hofmann, K.P. (1985) Biochim. Biophys. Acta **810**, 278–281.
180. Kohl, B. and Hofmann, K.P. (1987) Biophys. J. **52**, 271–277.
181. Kisselev, O.G., Meyer, C.K., Heck, M., Ernst, O.P. and Hofmann, K.P. (1999) Proc. Natl. Acad. Sci. USA **96**, 4898–4903.
182. DeLange, F., Bovee-Geurts, P.H., Pistorius, A.M., Rothschild, K.J. and de Grip, W.J. (1999) Biochemistry **38**, 13200–13209.
183. Vuong, T.M. and Chabre, M. (1991) Proc. Natl. Acad. Sci. USA **88**, 9813–9817.

261. Cai, K., Langen, R., Hubbell, W.L. and Khorana, H.G. (1997) Proc. Natl. Acad. Sci. USA **94**, 14267–14272.
262. Cai, K., Klein-Seetharaman, J., Farrens, D., Zhang, C., Altenbach, C., Hubbell, W.L. and Khorana, H.G. (1999) Biochemistry **38**, 7925–7930.
263. Abdulaev, N.G. and Ridge, K.D. (1989) Proc. Natl. Acad. Sci. USA **95**, 12854–12859.
264. Morrison, D.F., O'Brien, P.J. and Pepperberg, D.R. (1991) J. Biol. Chem. **266**, 20118–20123.
265. Karnik, S.S., Ridge, K.D., Bhattacharya, S. and Khorana, H.G. (1993) Proc. Natl. Acad. Sci. USA **90**, 40–44.
266. Higashijima, T., Ferguson, K.M., Sternweis, P.C., Ross, E.M., Smigel, M.D. and Gilman, A.G. (1987) J. Biol. Chem. **262**, 752–756.
267. Sachs, K., Maretzki, D. and Hofmann, K.P. (2000) Meth. Enzymol. **315**, 238–251.
268. Jäger, S., Palczewski, K. and Hofmann, K.P. (1996) Biochemistry **35**, 2901–2908.
269. Sachs, K., Maretzki, D., Meyer, C.K and Hofmann, K.P. (2000) J. Biol. Chem. **275**, 6189–6194.
270. Kaushal, S., Ridge, K.D. and Khorana, H.G. (1994) Proc. Natl. Acad. Sci. USA **91**, 4024–4028.
271. Ridge, K.D., Lee, S.S. and Yao, L.L. (1995) Proc. Natl. Acad. Sci. USA **92**, 3204–3208.
272. Dyson, H.J. and Wright P.E. (1995) FASEB Journal **9**, 37–42.
273. Robinson, P.R., Cohen, G.B., Zhukovsky, E.A. and Oprian, D.D. (1992) Neuron **9**, 719–725.
274. Cohen, G.B., Yang, T., Robinson, P.R. and Oprian, D.D. (1993) Biochemistry **32**, 6111–6115.
275. Han, M., DeDecker, B.S. and Smith, S.O. (1993) Biophys. J. **65**, 899–906.
276. Melia, T.J.J., Cowan, C.W., Angleson, J.K. and Wensel, T.G. (1997) Biophys. J. **73**, 3182–3191.
277. Zhukovsky, E.A., Robinson, P.R. and Oprian, D.D. (1991) Science **251**, 558–560.
278. Siebert, F. (1992) in: Biomolecular Spectroscopy. Advances in Spectroscopy, eds R.J.H. Clark and R.E. Hesters. pp. 1–54, Wiley, New York.
279. Jansen, J.J., Mulder, W.R., De Caluwe, G.L., Vlak, J.M. and de Grip, W.J. (1991) Biochim. Biophys. Acta **1089**, 68–76.
280. Zhukovsky, E.A., Robinson, P.R. and Oprian, D.D. (1992) Biochemistry **31**, 10400–10405.
281. Zvyaga, T.A., Min, K.C., Beck, M. and Sakmar, T.P. (1993) J. Biol. Chem. **268**, 4661–4667.
282. Zvyaga, T.A., Fahmy, K. and Sakmar, T.P. (1994) Biochemistry **33**, 9753–9761.
283. Fahmy, K. and Sakmar, T.P. (1993) Biochemistry **32**, 9165–9171.
284. Fahmy, K. and Sakmar, T.P. (1993) Biochemistry **32**, 7229–7236.
285. Sakmar, T.P. and Fahmy, K. (1995) Isr. J. Chem. **35**, 325–337.
286. Han, M., Lin, S.W., Minkova, M., Smith, S.O. and Sakmar, T.P. (1996) J. Biol. Chem. **271**, 32337–32342.
287. Jäger, S., Han, M., Lewis, J.W., Szundi, I., Sakmar, T.P. and Kliger, D.S. (1997) Biochemistry **36**, 11804–11810.
288. Nishimura, S., Sasaki, J., Kandori, H., Lugtenburg, J. and Maeda, A. (1995) Biochemistry **34**, 16758–16763.
289. Kropf, A., Whittenberger, B.P., Goff, S.P. and Waggoner, A.S. (1973) Exp. Eye Res. **17**, 591–606.
290. Corson, D.W. and Crouch, R.K. (1996) Photochem. Photobiol. **63**, 595–600.
291. Delange, F., Boveegeurts, P.H.M., Vanoostrum, J., Portier, M.D., Verdegem, P.J.E., Lugtenburg, J. and de Grip, W.J. (1998) Biochemistry **37**, 1411–1420.
292. Renk, G. and Crouch, R.K. (1989) Biochemistry **28**, 907–912.
293. Jäger, F., Jäger, S., Kraütle, O., Friedman, N., Sheves, M., Hofmann, K.P. and Siebert, F. (1994) Biochemistry **33**, 7389–7397.
294. Ganter, U.M., Schmid, E.D., Perez Sala, D., Rando, R.R. and Siebert, F. (1989) Biochemistry **28**, 5954–5962.
295. Shieh, T., Han, M., Sakmar, T.P. and Smith, S.O. (1997) J. Mol. Biol. **269**, 373–384.
296. Han, M., Lou, J., Nakanishi, K., Sakmar, T.P. and Smith, S.O. (1997) J. Biol. Chem. **272**, 23081–23085.
297. Han, M., Groesbeek, M., Sakmar, T.P. and Smith, S.O. (1997) Proc. Natl. Acad. Sci. USA **94**, 13442–13447
298. Sheikh, S.P., Zvyaga, T.A., Lichtarge, O., Sakmar, T.P. and Bourne, H.R. (1996) Nature **383**, 347–350.

299. Farrens, D.L., Altenbach, C., Yang, K., Hubbell, W.L. and Khorana, H.G. (1996) Science **274**, 768–770.
300. Zvyaga, T.A., Fahmy, K., Siebert, F. and Sakmar, T.P. (1996) Biochemistry **35**, 7536–7545.
301. Yan, B., Nakanishi, K. and Spudich, J.L. (1991) Proc. Natl. Acad. Sci. USA **88**, 9412–9416.
302. Kim, J.M., Altenbach, C., Thurmond, R.L., Khorana, H.G. and Hubbell, W.L. (1997) Proc. Natl. Acad. Sci. USA **94**, 14273–14278.
303. Bae, H., Carbrera Vera, T.M., Depree, K.M., Graber, S.G. and Hamm, H.E. (1999) J. Biol. Chem. **274**, 14963–14971.
304. Farahbakhsh, Z.T., Ridge, K.D., Khorana, H.G. and Hubbell, W.L. (1995) Biochemistry **34**, 8812–8819.
305. Yang, K., Farrens, D.L., Altenbach, C., Farahbakhsh, Z.T., Hubbell, W.L. and Khorana, H.G. (1996) Biochemistry **35**, 14040–14046.
306. Altenbach, C., Yang, K., Farrens, D.L., Farahbakhsh, Z.T., Khorana, H.G. and Hubbell, W.L. (1996) Biochemistry **35**, 12470–12478.
307. Xu, J., Dodd, R.L., Makino, C.L., Simon, M.I., Baylor, D.A. and Chen, J. (1997) Nature **389**, 505–509.
308. Zhang, L., Sports, C.D., Osawa, S. and Weiss, E.R. (1997) J. Biol. Chem. **272**, 14762–14768.
309. Heck, M., Pulvermüller, A. and Hofmann, K.P. (2000) Methods Enzymol. **315**, 329–347.
310. Schleicher, A., Kühn, H. and Hofmann, K.P. (1989) Biochemistry **28**, 1770–1775.
311. Palczewski, K., Pulvermüller, A., Buczylko, J. and Hofmann, K.P. (1991) J. Biol. Chem. **266**, 18649–18654.
312. Pulvermüller, A., Maretzki, D., Rudnicka Nawrot, M., Smith, W.C., Palczewski, K. and Hofmann, K.P. (1997) Biochemistry **36**, 9253–9260.
313. Wistow, G.J., Katial, A., Craft, C. and Shinohara, T. (1986) FEBS Lett. **196**, 23–28.
314. Krupnick, J.G., Gurevich, V.V., Schepers, T., Hamm, H.E. and Benovic, J.L. (1994) J. Biol. Chem. **269**, 3226–3232.
315. Palczewski, K. (1994) Protein Sci. **3**, 1355–1361.
316. Smith, W.C., McDowell, J.H., Dugger, D.R., Miller, R.L., Arendt, A., Popp, M.P. and Hargrave, P.A. (1999) Biochemistry **38**, 2752–2761.
317. McDowell, J.H., Smith, W.C., Miller, R.L., Popp, M.P., Arendt, A., Abdulaeva, G. and Hargrave, P.A. (1999) Biochemistry **38**, 6119–6125.
318. Palczewski, K., Pulvermüller, A., Buczylko, J., Gutmann, C. and Hofmann, K.P. (1991) FEBS Lett. **295**, 195–199.
319. Gurevich, V.V. and Benovic, J.L. (1993) J. Biol. Chem. **268**, 11628–11638.
320. Gurevich, V.V., Chen, C.Y., Kim, C.M. and Benovic, J.L. (1994) J. Biol. Chem. **269**, 8721–8727.
321. Puig, J., Arendt, A., Tomson, F.L., Abdulaeva, G., Miller, R., Hargrave, P.A. and McDowell, J.H. (1995) FEBS Lett. **362**, 185–188.
322. Gurevich, V.V. and Benovic, J.L. (1997) Mol. Pharmacol. **51**, 161–169.
323. Smith, W.C., Milam, A.H., Dugger, D., Arendt, A., Hargrave, P.A. and Palczewski, K. (1994) J. Biol. Chem. **269**, 15407–15410.
324. Palczewski, K., Buczylko, J., Ohguro, H., Annan, R.S., Carr, S.A., Crabb, J.W., Kaplan, M.W., Johnson, R.S. and Walsh, K.A. (1994) Protein Sci. **3**, 314–324.
325. Wilson, C.J. and Copeland, R.A. (1997) J. Protein Chem. **16**, 755–763.
326. Granzin, J., Wilden, U., Choe, H.W., Labahn, J., Krafft, B. and Büldt, G. (1998) Nature **391**, 818–921.
327. Hirsch, J.A., Schubert, C., Gurevich, V.V. and Sigler, P.B. (1999) Cell **97**, 257–269.
328. Vishnivetskiy, S.A., Paz, C.L., Schubert, C., Hirsch, J.A., Sigler, P.B. and Gurevich, V.V. (1999) J. Biol. Chem. **274**, 11451–11454.
329. Brown, N.G., Fowles, C., Sharma, R. and Akhtar, M. (1992) Eur. J. Biochem. **208**, 659–667.
330. Brown, N.G., Fowles, C., Sharma, R. and Akhtar, M. (1993) Eur. J. Biochem. **212**, 840.
331. Dean, K.R. and Akhtar, M. (1996) Biochemistry **35**, 6164–6172.
332. Rim, J., Faurobert, E., Hurley, J.B. and Oprian, D.D. (1997) Biochemistry **36**, 7064–7070.

333. Laitko, U. and Hofmann, K.P. (1998) Biophys. J. **74**, 803–815.
334. Seckler, B. and Rando, R.R. (1989) Biochem. J. **264**, 489–493.
335. Rim, J. and Oprian, D.D. (1995) Biochemistry **34**, 11938–11945.
336. Robinson, P.R., Buczylko, J., Ohguro, H. and Palczewski, K. (1994) Proc. Natl. Acad. Sci. USA **91**, 5411–5415.
337. Thurmond, R.L., Creuzenet, C., Reeves, P.J. and Khorana, H.G. (1997) Proc. Natl. Acad. Sci. USA **94**, 1715–1720.
338. Palczewski, K., Buczylko, J., Kaplan, M.W., Polans, A.S. and Crabb, J.W. (1991) J. Biol. Chem. **266**, 12949–12955.
339. Pepperberg, D.R., Birch, D.G., Hofmann, K.P. and Hood, D.C. (1996) J. Opt. Soc. Am. A **13**, 586–600.
340. Corson, D.W., Cornwall, M.C. and Pepperberg, D.R. (1994) Vis. Neurosci. **11**, 91–98.
341. Chen, J., Makino, C.L., Peachey, N.S., Baylor, D.A. and Simon, M.I. (1995) Science **267**, 374–377.
342. Dicker, F., Quitterer, U., Winstel, R., Honold, K. and Lohse, M.J. (1999) Proc. Natl. Acad. Sci. USA (1999) **96**, 5476–5481.
343. Felber, S., Breuer, H.P., Petruccione, F., Honerkamp, J. and Hofmann, K.P. (1996) Biophys. J. **71**, 3051–3063.
344. Morrison, D.F., Ting, T.D., Vallury, V., Ho, Y.K., Crouch, R.K., Corson, D.W., Mangini, N.J. and Pepperberg, D.R. (1995) J. Biol. Chem. **270**, 6718–6721.
345. Surya, A., Foster, K.W. and Knox, B.E. (1995) J. Biol. Chem. **270**, 5024–5031.
346. Jin, J., Crouch, R.K., Corson, D.W., Katz, B.M., MacNichol, E.F. and Cornwall, M.C. (1993) Neuron **11**, 513–522.
347. Samama, P., Cotecchia, S., Costa, T. and Lefkowitz, R.J. (1993) J. Biol. Chem. **268**, 4625–4636.
348. Yoshizawa, T. and Fukada, Y. (1983) Biophys. Struct. Mech. **9**, 245–258.
349. Fukada, Y. and Yoshizawa, T. (1982) Biochim. Biophys. Acta **708**, 112–117.
350. Palczewski, K., Jäger, S., Buczylko, J., Crouch, R.K., Bredberg, D.L., Hofmann, K.P., Asson Batres, M.A. and Saari, J.C. (1994) Biochemistry **33**, 13741–13750.
351. Hofmann, K.P., Pulvermüller, A., Buczylko, J., Van Hooser, P. and Palczewski, K. (1992) J. Biol. Chem. **267**, 15701–15706.
352. Barlow, R.B., Birge, R.R., Kaplan, E. and Tallent, J.R. (1993) Nature **366**, 64–66.
353. Han, M. and Sakmar, T.P. (2000) Methods Enzymol. **315**, 251–267.
354. Weng, J., Mata, N.L., Azarian, S.M., Tzekov, R.T., Birch, D.G. and Travis, G.H. (1999) Cell **98**, 13–23.
355. Körschen, H.G., Beyermann, M., Müller, F., Heck, M., Koch, K.-W., Kellner, R., Hofmann, K.P. and Kaupp, U.B. (1999) Nature **400**, 761–766.

CHAPTER 4

Ion Channels of Vertebrate Photoreceptors

R.S. MOLDAY

Department of Biochemistry and Molecular Biology, The University of British Columbia, Vancouver

U.B. KAUPP

Forschungszentrum Jülich, Institut für Biologische Informationsverarbeitung

Handbook of Biological Physics
Volume 3, edited by D.G. Stavenga, W.J. DeGrip and E.N. Pugh Jr

Contents

1. Introduction

Rod and cone photoreceptors of the vertebrate retina respond to light by a brief and transient hyperpolarization of the cell membrane, due to the closure of light-dependent channels in the surface membrane of the outer segment. The electrical signal is propagated to the inner segment and the synaptic ending of the photoreceptor cell. Several different ion channels are involved in the generation and shaping of the light response and in the control of transmitter release from the photoreceptor synapse. In this chapter, we will review the electrical and molecular properties of these ion channels and the specific function they subserve during the light response.

2. Cyclic GMP-gated channels of photoreceptor cells

2.1. Role of the channel in phototransduction

The cGMP-gated channels of vertebrate rod and cone cells play a central role in phototransduction by controlling the flow of Na^+ and Ca^{2+} ions into the photoreceptor outer segment (Fig. 1; for review [1]). A relatively high concentration of cGMP in the outer segments of dark-adapted photoreceptor cells maintains a significant number of channels in their open state by the direct binding of cGMP to the channels. The influx of Na^+ and Ca^{2+} ions through these channels maintains the cell in a depolarized state. Low intracellular Na^+ concentration is retained by the active transport of Na^+ from the photoreceptor cell by Na, K ATPase located in the plasma membrane of the inner segment. The balanced efflux of Ca^{2+} through the Na/Ca–K exchanger in the plasma membrane of outer segment maintains an intracellular Ca^{2+} concentration of ~400 nM (for review [2]).

Photoexcitation of the rod cell is initiated when a photon is captured by rhodopsin resulting in the conversion of the 11-*cis* retinal chromophore to its all-*trans* isomer. The resulting formation of Meta II rhodopsin triggers the activation of the visual enzyme cascade system and an amplification of the signal (for review [3–7]. Briefly, Meta II rhodopsin catalyzes the exchange of GDP for GTP on the α-subunit of transducin. Interaction of this subunit with inhibitory subunit of phosphodiesterase (PDE) activates PDE leading to the hydrolysis of cGMP. As the intracellular concentration of cGMP decreases, the cGMP-gated channels close causing a hyperpolarization of the rod cell. This in turn results in an inhibition of glutamate transmitter release at the synapse and the transmission of the signal to other retinal neurons. Since the Na/Ca–K exchanger continues to extrude Ca^{2+} from the outer segment, a marked decrease in intracellular Ca^{2+} concentration (below 100 nM) also occurs. This reduction in Ca^{2+} plays a central role in photorecovery and light adaptation [4,8–11].

that the channel and exchanger distribute together into the same vesicles when ROS are subjected to freezing and thawing or after reconstitution of solubilized ROS proteins into liposomes. More recently, the Na/Ca–K exchanger has also been detected in immunoaffinity purified channel preparations by Western blotting [57]. In addition to the exchanger, it is possible that other ROS proteins may interact with the channel in a transient or stable manner.

2.5. Molecular cloning and expression

2.5.1. Rod cGMP-gated channel

Molecular characterization of CNG channels was initiated when Kaupp et al. [40] first cloned the cDNA for the α-subunit of the bovine rod channel. The full-length cDNA codes for a polypeptide of 690 amino acids with a predicted molecular mass of 79.6 kDa. This size is in good agreement with the molecular mass (~78 kDa) of the heterologously expressed polypeptide measured by SDS gel electrophoresis, but considerably larger than the apparent molecular mass of 63 kDa observed in bovine ROS and purified channel preparations [32]. The discrepancy in size is primarily due to the absence of the N-terminal 92 amino acids. Immunocytochemical labeling studies of retinal tissue has firmly established that this truncated 63 kDa subunit is the predominant form of the α-subunit in bovine rod photoreceptor cells. The truncated α-subunit is also present in ROS from human, mouse, rat, and pig retinas [32]. Shortened forms of the rod and cone α-subunits have also been detected in chicken retinal extracts [58]. These results lead to the view that the N-terminus of the α-subunit undergoes a photoreceptor-specific post-translational cleavage reaction. The role for this proteolytic processing is not known at the present time, but it may be required for targeting of the channel to the outer segment plasma membrane or for interaction with other photoreceptor proteins.

The α-subunit of the rod cGMP-gated channel has been cloned for a number of vertebrate species including human, cow, mouse, rat, dog and chicken [58–62]. The subunits are highly conserved, typically being 85–90% identical in amino acid sequence between mammalian polypeptides and 76% identical between cow and chicken rod subunits. The greatest variation in sequence occurs near the N and C termini.

The rod α-subunit assembles into a functional channel when expressed in *Xenopus* oocytes or HEK 293 cells [40,59]. Channel recordings have shown that the heterologously expressed channel composed of α-subunits has many properties in common with the native channel from ROS. In particular, it is cooperatively activated by cGMP with a $K_{1/2}$ of ~40–80 μM and a Hill coefficient of about 2, has a single channel conductance of 20–30 pS, and exhibits a current–voltage relationship and broad ion selectivity similar to the native channel in excised patches of ROS plasma membranes. These findings led to the early view that the rod cGMP-gated channel is an oligomeric complex consisting of identical subunits.

The first indication that the channel contains a second subunit came from the cloning and expression studies of Chen et al. [34]. In this study, two cDNAs coding for polypeptides of 71 and 102 kDa were isolated from a human retinal cDNA library. These two forms (referred to as subunit 2a and 2b, respectively) are identical in sequence, but differ in length at the N-terminus. Subunit 2 is about 30% identical in sequence to the α-subunit of the rod channel and contains similar structural features. However, unlike the α-subunit, it does not assemble into a functional channel when expressed by itself in HEK 293 cells. Co-expression of subunit 2 with the rod α-subunit, however, results in a heteromeric channel that displays a number of properties characteristic of the native channel that are not reproduced when the α-subunits are expressed individually. These include the rapid channel opening and closing or flickering behavior, inhibition by micromolar concentrations of l-*cis* diltiazem and modulation by Ca-calmodulin [34,50].

Questions were raised, however, regarding the true nature of the second subunit of the rod cGMP-gated channel, since purified channel preparations from ROS do not contain polypeptides in the size range expected for either subunits 2a or 2b [49,50]. This issue was resolved when Körschen et al. [35] showed by peptide sequencing, molecular cloning and heterologous expression that subunit 2 is part of a longer polypeptide corresponding to the 240 kDa protein present in purified channel preparations from ROS. This subunit is now generally referred to as the β-subunit of the rod cGMP-gated channel. The human cDNA clones coding for the subunit 2 variants most likely represent either incomplete cDNA clones or spliced variants of the β-subunit expressed in low abundance in other retinal cells [36,63].

2.5.2. Cone cGMP-gated channel

The α-subunits of the chicken, bovine and human cone CNG channels have been cloned and shown to be approximately 60% identical in amino acid sequence to the rod subunit [58,64–66]. The heterologously expressed cone α-subunit assembles into a homomeric channel that exhibits a cyclic nucleotide specificity and monovalent ion selectivity similar to that of the rod subunit. However, the expressed cone channel is 10–15-fold less sensitive to external Ca^{2+} blockage and is significantly more permeable to calcium than the rod channel [67]. Similar differences in calcium permeability have been reported for the native cone and rod channels [68]. Higher calcium permeability of the cone channel has been suggested to play a role in the photoresponse of cone cells [67,69].

A β-subunit of the cone channel has not yet been identified. Site-directed antibodies against the N-terminal and C-terminal segments of the rod β-subunit do not stain cone cells by immunofluorescence techniques [34–36]. If a cone β-subunit exists, it is either a spliced variant of the rod β-subunit or a polypeptide encoded by a distinct gene.

2.5.3. Olfactory cyclic nucleotide-gated channel

The α-subunit of olfactory CNG channels is similar to the rod and cone subunits in both size and amino acid sequence [70,71]. The bovine olfactory subunit consists of 663 amino acids (Mr ~ 76,000) and is 57% identical in sequence to the bovine rod α-subunit and 76% similar in sequence when conserved amino acid substitutions are included.

The photoreceptor and olfactory channels, however, differ in their sensitivity to cyclic nucleotides. The heterologously expressed rat and bovine olfactory homomeric channels exhibit a $K_{1/2}$ of 1–2 μM for cGMP and 55 μM for cAMP compared to a $K_{1/2}$ of 60 μM for cGMP and ~1500 μM for cAMP for the expressed rat and bovine rod homomeric channel [71–73]. Unlike the expressed channel, however, the native olfactory channel is activated almost equally well by cGMP ($K_{1/2} \sim 1.5$ μM) and cAMP ($K_{1/2} \sim 4$ μM) [74,75]. Differences in the functional properties of the native olfactory channel and the heterologously expressed α-subunit appear to be due to the presence of two additional subunits [76,77]. One subunit is more akin to the α-subunit [76,77]; the other subunit is a shorter splice variant of the rod β-subunit [78,79]. Co-expression of the two α-like subunits and the β-subunit produces channels with properties that are largely similar if not identical with those of the native channel. Using subtype-specific antibodies, Bönigk et al. [79] demonstrate that all three subunit types are localized to olfactory cilia. In conclusion, unlike the CNG channel from rod photoreceptors, the native CNG channel in chemo-sensory cilia is composed of three distinct subunits of unknown stoichiometry.

2.6. *Structural analysis*

2.6.1. *Subunit structure*

The current models for the membrane topology of α- and β-subunits of the rod cGMP-gated channel are illustrated in Figs. 3 and 4 [58,80,81]. The core structural unit consists of six putative membrane-spanning segments, designated S1–S6, followed by a cyclic GMP-binding domain near the C-terminus. The S4 segment contains four positively charged residues, each separated by two hydrophobic residues, and resembles the voltage-sensor motif found in corresponding transmembrane segment of voltage-gated cation channels [82]. A pore region of about 20–30 amino acids is located between the S5 and S6 transmembrane segments. Since the S4 voltage sensor-like motif and the pore region are also characteristic features of voltage-gated cation channels, it has been suggested that CNG channels and voltage-gated channels are members of a superfamily of cation channels, which have evolved from a common primordial channel [83].

Experimental evidence in support of the topological model of the rod cGMP-gated subunit was first obtained from immunogold labeling for electron microscopy. In these studies both the N- and C-termini of the rod α-subunit were localized to the cytoplasmic side of the ROS plasma membrane [32], and a glycosylated segment linking the S5 transmembrane segment to the pore region was localized to the extracellular side [84]. Additional support for the transmembrane topology of CNG channels has been obtained from gene fusion techniques using enzyme reporters [85].

The β-subunit is considerably longer than the α-subunit (Fig. 4). The bovine β-subunit consists of 1394 amino acids and has a calculated molecular mass of 155 kDa [35]; the human orthologue is slightly shorter, containing 1251 amino acids and having a calculated molecular mass of 140 kDa [36,63]. In either case, the calculated mass is significantly less than the apparent size of 220–240 kDa for the native and heterologously expressed β-subunit determined by SDS polyacrylamide gel electrophoresis [35,36]. This difference has been attributed to the anomalous migration

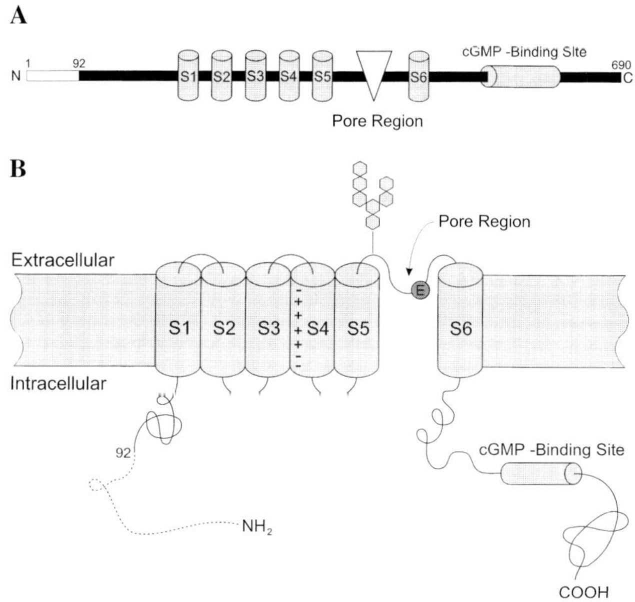

Fig. 3. Linear diagram (A) and the proposed topological model (B) for the α-subunit of the rod cGMP-gated channel. The cloned bovine α-subunit consists of 690 amino acids, the first 92 amino acids of which are absent in the native channel in rod photoreceptors [40]. Structural features include six putative transmembrane segments (S1–S6), a voltage-sensor-like motif comprising the S4 segment, a pore region containing a negatively charged glutamate residue (E), an N-linked glycosylation site (hexagons), and a cyclic GMP binding site.

of the β-subunit on SDS polyacrylamide gels due to its high content of proline and glutamic acid residues.

The β-subunit has an unusual bipartite structure. The C-terminal region or β′ part of approximately 800 amino acids contains the core structural unit of six putative transmembrane segments (S1–S6), an S4 voltage sensor-like motif, a pore region and a cyclic nucleotide binding domain (Fig. 4). This part is highly conserved between the bovine and human orthologues (~86% identical in amino acid sequence), but is only 30% identical in amino acid sequence to the α-subunit. This sequence identity between the α- and β-subunits rises to ~50% in the cyclic nucleotide binding domain (see Fig. 7). When the β′ part of the β-subunit is co-expressed with the α-subunit, a hetero-oligomeric channel is formed that has electrophysiological properties identical to the heterologously expressed channel containing the full-length β-subunit and properties that closely resemble the native channel of ROS [34,35,50]. These include the rapid opening and closing or flickering behavior of the channel and micromolar sensitivity to l-*cis* diltiazem (Fig. 5), relative monovalent ion selectivity, divalent cation blockage and calmodulin modulation of the channel. These results indicate that all the information required for hetero-oligomeric channel assembly and function is inherent in the β′ part of this subunit.

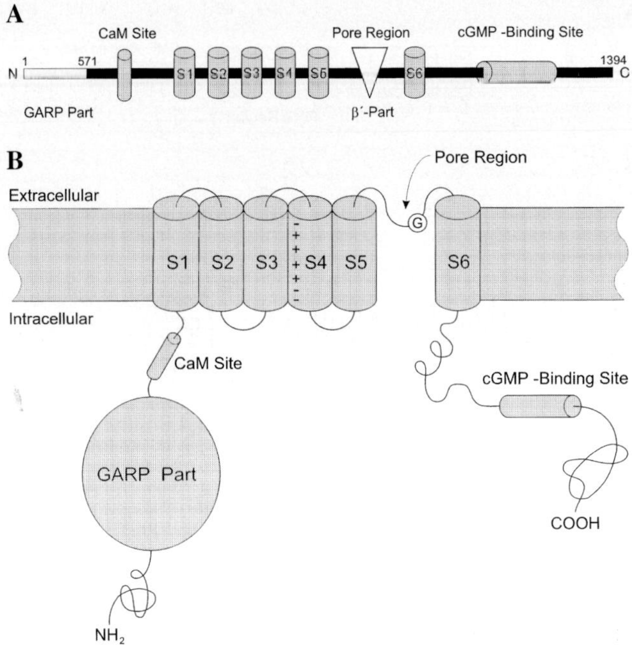

Fig. 4. Linear diagram (A) and the proposed topological model (B) for the β-subunit of the rod cGMP-gated channel. The bovine polypeptide consists of 1394 amino acids and has a bipartite structure. The β′-part contains the characteristic structural features of other CNG subunits including six transmembrane segments (S1–S6), a voltage sensor-like motif comprising the S4 segment, a pore region, and a cyclic GMP binding domain. In addition, it contains the calmodulin binding site. The GARP part consisting of the first 571 amino acids is rich in proline and glutamate residues.

The N-terminal region of the bovine β-subunit, called the glutamic acid rich protein (GARP) part, consists of 571 amino acids [35] and is essentially identical in sequence to a GARP cloned previously by Sugimoto et al. [86]. This protein contains a high content of glutamic acid and proline residues and two multiple repeat segments near its C-terminal region. The GARP part of the human β-subunit is also rich in glutamic acid and proline residues, but unlike its bovine counterpart, it does not contain repeat segments [36].

Two additional spliced forms of GARP have been cloned from a bovine retinal cDNA library (Fig. 4). One spliced variant codes for a full-length form of the protein, called f-GARP, is identical to the GARP part of the β-subunit but contains an additional 19 amino acids at its C-terminus [35,86]. A second, truncated form referred to as t-GARP contains the first 291 amino acids of f-GARP and an additional 8 amino acid C-terminal extension [36,87]. Western blotting and immunofluorescence microscopy have confirmed that these GARP forms are exclusively localized in the outer segments of rod photoreceptors. The f-GARP and t-GARP variants do not co-purify with the channel, indicating that they are not subunits of

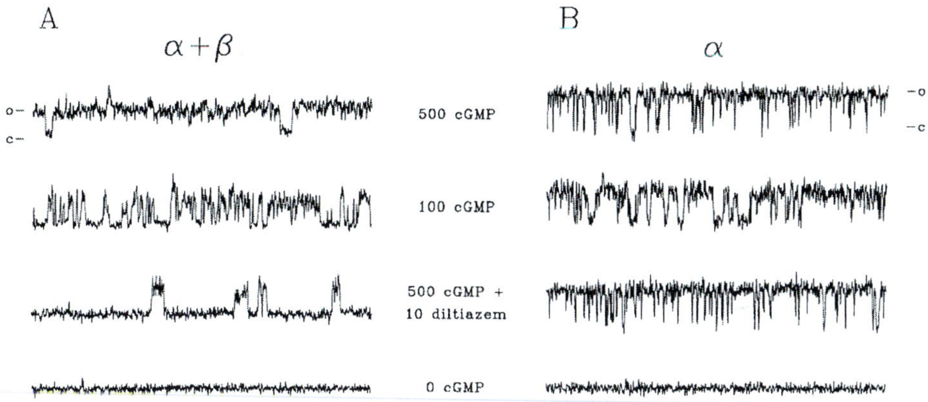

Fig. 5. Single-channel activity of the hetero-oligomeric (α + β) and homo-oligomeric (α) channels. Single channel currents from an inside-out patch of HEK 293 cells expressing the hetero-oligomeric channel (A) and the homo-oligomeric channel (B) at the indicated cGMP concentrations in the presence and absence of l-*cis*-diltiazem. The holding potential was +50 mV. c – closed channel; o – open channel (from Ref. [35]).

the channel. Related GARP variants are also present in ROS from other mammalian retinas [36,87,88].

The high content of proline residues suggests that GARP may be involved in protein–protein interactions. In the case of the channel, the GARP part of the β-subunit may be involved in interactions that maintain the spatial distribution of the channel in the plasma membrane or bind enzymes involved in the phototransduction process in order to bring them in close proximity to the channel. The smaller GARP variants may also interact with other proteins of ROS including enzymes of the visual cascade system.

In fact, Körschen and coworkers [89] identified GARPs as multivalent proteins that interact with the key players of cGMP-signalling, phosphodiesterase (PDE) and guanylate cyclase, and with a retina-specific ATP binding cassette transporter (ABCR) [90,91] through four short, repetitive sequences. In electron micrographs, GARPs are restricted to the rim region and incisures of discs in close proximity to the ABCR and probably, guanylate cyclase, whereas the PDE is randomly distributed. The most abundant splice form, t-GARP, associates more strongly with light-activated than with inactive PDE, and t-GARP potently inhibits PDE activity. These observations are consistent with the idea that GARPs organize a dynamic protein complex near the disc rim. Future work needs to elucidate the precise composition and dynamics of this macromolecular complex and the molecular nature of the interactions.

What is the function of GARPs? Active PDE is turned off by the endogenous GTPase activity of T_a-GTP involving RGS9 [92] – a member of the family of GAP proteins – and the β-subunit of type-5 G protein [93]. The powerful inhibition of active PDE by recombinant t-GARP suggests an independent mechanism of PDE inactivation that might be important for termination and adaptation of the light response: active PDE molecules that reach the disc rim by diffusion will be deactivated by t-GARP or the GARP part of the channel. An intriguing scenario is that the two inactivation mechanisms operate in different light regimes. Sequestration by GARPs may become the dominant inactivation mechanism at high light intensities, when it is highly probable that a significant fraction of active PDE is reaching the disc margins [6]. In this respect, GARPs may organize an 'adaptation' signalling complex at the disc rim that downregulates the high cGMP turnover during daylight, when rod function is saturated.

2.6.2. Subunit stoichiometry

The CNG channels, like the voltage-gated K^+ channels, are tetrameric complexes [94]. This has been established by co-expression of the wild-type rod α-subunit with a chimeric rod α-subunit in which the pore region of the rod subunit is replaced with that of the olfactory subunit. In addition to the 30 and 85 pS conductances characteristic of the wild-type rod and chimeric homomeric channels, respectively, four intermediate conductance states were also obtained. The four intermediate conductances could arise from the four different combinations of two subunits, which give rise to pentameric species. However, using dimer constructs of wild-type and chimeric subunits, Liu et al. [94] showed that two of the intermediate conductances arise from different arrangements of two wild-type and two chimeric subunits in a tetrameric channel, i.e. the same subunits can be adjacent or opposite to each other in a tetramer. On this basis, they have concluded that the channel is a tetramer with 4-fold symmetry in which the subunits are arranged in a head-to-tail configuration around the central axis. Evidence for the tetrameric subunit composition of the heterologously expressed channel has also been supported from Ni^{2+} potentiation studies of Gordon and Zagotta [95]. More recently, Ni^{2+} studies have been carried out for the heterologously expressed heteromeric rod channel [96]. These studies confirm the tetrameric subunit composition of the rod channel and further suggest a paired α- and β-subunit arrangement (α/α/β/β) as shown in Fig. 6. This conclusion, however, has been questioned recently by He and coworkers [97]. By constraining the subunit order of the rod CNG channel, these authors conclude that the subunits are diagonally arranged in the channel complex.

The subunit composition of the native rod cGMP-gated channel is yet to be determined experimentally. However, since the purified channel contains roughly similar amounts of the α- and β-subunits, it is likely that the native rod channel consists of 2α- and 2β-subunits with either 2-fold or 4-fold symmetry (Fig. 6).

2.7. Structure–function relationships

2.7.1. Cyclic nucleotide binding domain and nucleotide selectivity

The cyclic nucleotide binding domain of CNG channels contain approximately 120 amino acids and show significant homology to the nucleotide binding domain of cGMP-dependent protein kinases and the cAMP-binding domain of the *E. coli* catabolite gene activator protein called CAP (Fig. 7; [40]). The 3-dimensional structure of the latter consists of a short α-helix (helix A) followed by a set of 8

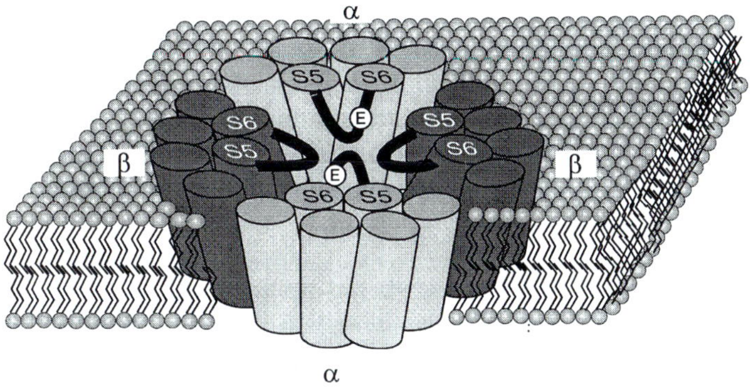

Fig. 6. Schematic model of the rod cGMP-gated channel showing the tetrameric arrangement of the subunits. In this model the S6 and S5 segments are visualized to line the central cavity of the channel. The pore regions joining the S5 and S6 segment near the extracellular surface of the membrane extend toward the center of the cavity where they can function as a gate and selectivity filter. The glutamate residue (E) in the pore region of the α-subunit that is responsible for external divalent cation blockage of the channel is indicated. Although the channel is shown with identical subunits across from each other, it is possible that the two identical subunits may be adjacent to each other to form a channel with 2-fold symmetry.

β-strands and two additional α-helices (helices B and C) [98]. The bound cyclic nucleotide is buried in a hydrophobic pocket between the β-roll and helix C. This structure has been used to develop a model for the cGMP binding domain of the rod GMP-gated channel [99]. In this model, hydrogen bonds are thought to form between the Glu 543 and Glu 544 and the 2′ hydroxyl group of ribose. Binding is further stabilized by formation of a hydrogen bond between Thr560 and an exocyclic phosphate oxygen of the cyclic nucleotide, and ionic interactions between Arg559 and an exocyclic phosphate oxygen of the ligand (for review see Ref. [100]). By analogy with molecular modeling of cGMP-dependent protein kinases, an additional hydrogen bond between Thr560 and the 2-amino group of cGMP has been suggested to account for the selectivity of the rod channel for cGMP over cAMP [72,99]. Cyclic GMP-dependent protein kinases invariantly have a Thr residue in this position, whereas cAMP-dependent protein kinases most often contain an Ala residue which cannot form side-chain hydrogen bonds [101]. Indeed, substitution of an Ala with a Thr in a cAMP-dependent protein kinase was found to increase the affinity of this kinase for cGMP without significantly affecting the cAMP binding [102]. The putative role of Thr560 in nucleotide selectivity of the rod channel was examined by substituting this residue with an alanine [72]. This mutation caused a significant reduction in the apparent affinity of the rod channel for cGMP with only a minimal reduction in the cAMP affinity. However, Thr560 is not the only amino acid that determines nucleotide selectivity of the channel since the principal α-subunit of the catfish olfactory channel contains a Thr at this position, but exhibits a different selectivity for cGMP and cAMP [103].

More recent studies indicate that the C helix plays a key role in conferring nucleotide selectivity to CNG channels. In one study, the high degree of selectivity

nism, the channel is considered to exist in five states (an unliganded and four liganded states), each capable of undergoing a conformational transition from a closed to an open state with different probabilities [100,119]. Evidence for this concerted model has come from the findings that spontaneous opening of the channel can occur in the absence of ligand, but with low probability [120]. Furthermore, Ruiz and Karpen [121] have also reported that channels maintained in their partially liganded states through covalently bound cGMP analogues also open. Their results indicate that channels can open when a single cGMP is bound, but with an opening probability ($i/i_{max} \sim 1 \times 10^{-5}$) not significantly different from spontaneous opening of the unliganded channel. A marked increase in the opening probability is observed when the channel contains two ($i/i_{max} \sim 0.01$) and three ($i/i_{max} \sim 0.33$) bound ligands. Thus, the transition from a closed (C) state to an open (O) state is determined by the number of bound ligands and is favored at high ligand occupancy:

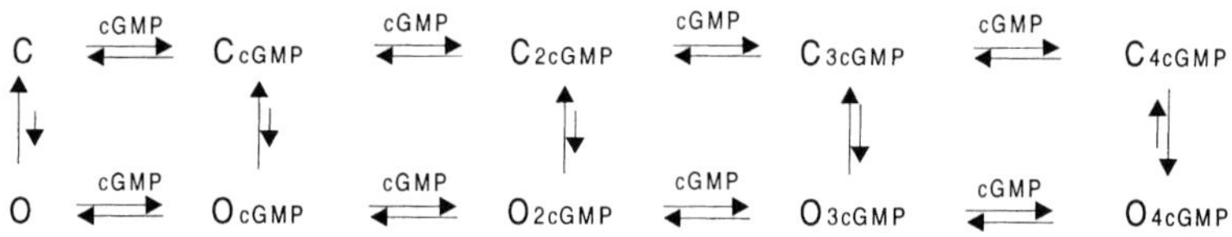

The effect of individual binding events on activation has also been studied with a channel containing less than four functional cGMP-binding sites [122]. Binding sites were disabled by mutating two residues in the cNMP-binding domain known to be crucially important for ligand binding and/or gating. This study confirmed that the binding of a single ligand significantly enhanced open probability, but four ligands are required to fully activate the channel [122]. Although the studies by Ruiz and Karpen [121] and by Liu and coworkers [122] are in qualitative agreement, they differ in important quantitative aspects. Notably, the conclusions by Ruiz and Karpen [121] rest on the analysis of subconductance states that are routinely observed, whereas other studies, also using homomeric α-subunit channels, could not detect a sizeable incidence of sublevels [122,123].

Various regions on the CNG channel that contribute to the allosteric conformational transition between the open and closed states have been identified. These studies have taken advantage of the property that the closed-to-open transition of the rod channel is energetically less favorable than for that of the olfactory channel. Using chimeric channels constructed from various regions of the rod and olfactory, several groups have identified the amino terminal domain of CNG channels as essential in the gating process [104,124,125]. Replacement of the amino terminal segment of the rod channel with that of the olfactory channel produces a chimeric channel having gating properties and spontaneous openings characteristic of the olfactory channel. Analysis of Ca-calmodulin binding and modulation of the olfactory channel implicate a domain near the N-terminus of the olfactory channel in the gating process ([126,127], see below).

Varnum and Zagotta [127] have examined the interaction of the amino terminal region with other domains of the CNG channel using GST fusion constructs. The amino terminus of the olfactory channel has been shown to interact strongly with the carboxyl terminal region of the olfactory and rod channels. Deletion of a 30 amino acid segment (aa 62–91) of the olfactory channel disrupts this interaction and reduces both the apparent affinity for cyclic nucleotides and the maximum current elicited at saturating cAMP concentrations. These results suggest that this segment plays a crucial role in the gating process. The amino terminal region interacts with the carboxyl terminal domain on either the same subunit or an adjacent subunit. The latter would implicate the amino terminal domain in subunit–subunit interactions [127]. The N-terminal region of the Shaker K^+ channel has previously been shown to be involved in subunit assembly [128]. In addition to the N-terminal region, other segments of the channel also appear to be involved in gating. Recent site-directed mutagenesis studies [129] indicate that selected amino acids in a segment of the cone α-subunit that links the S6 transmembrane segment to the cyclic nucleotide binding domain alter channel gating properties.

The picture is emerging whereby cyclic nucleotide binding is coupled to the transition from the closed to the open state of CNG channels. Cyclic nucleotides preferentially bind and stabilize the open state of the channel. Deletion of the amino terminal domain increases the energy of this closed-to-open transition as reflected in a decrease in the apparent affinity of the channel for cyclic nucleotides. Other segments of the channel also affect the ease of transition from the closed to open states.

2.8. Modulation of channel activity

Although CNG channels are activated by the binding of cyclic nucleotides, the concentration range over which this activation occurs can be regulated by other biochemical processes. Some factors that have been reported to modulate the sensitivity of rod channel to cGMP include Ca-calmodulin and unidentified endogeneous Ca-binding proteins, phosphorylation, diacylglyceride, sulfhydryl reagents and divalent transition metal ions (see Ref. [130]).

2.8.1. Ca-calmodulin modulation

Regulation of the photoreceptor and olfactory channel by Ca-calmodulin has been extensively studied due to the important physiological role that Ca^{2+} plays in photoreceptor and olfactory receptor signal transduction and adaptation mechanisms (for review see Refs. [4,11,16,81,131]). Ca-calmodulin was first shown to reduce the apparent affinity of the rod cGMP-gated channel for cGMP in ROS vesicles [46]. Subsequently, this effect has been reproduced in a number of experimental systems including excised patches from ROS, native and reconstituted membrane vesicles, and heterologously channel expressing cells [35,47,50,132,133]. Ca-calmodulin binds tightly to the rod channel with an apparent dissociation constant of 1–2 nM. This interaction results in a shift of the dose response curve of the channel to higher cGMP concentration corresponding to a ~2-fold decrease in the apparent affinity of the channel for cGMP. The olfactory channel is also modulated by Ca-calmodulin, but in a more robust manner [126]. Ca-calmodulin reduces the apparent affinity of the olfactory channel for both cAMP and cGMP by 10–20-fold. Binding and modulation of the channel occurs over physiologically relevant Ca^{2+} concentrations (50–300 nM for the rod channel and 2–20 μM for the olfactory channel). Modulation of cone channels by Ca-calmodulin has also been examined. Although Ca-calmodulin was found to modulate the catfish rod channel, no modulation was observed for the cone channel [134]. Recent studies of bass cone cells, however, suggest that an endogeneous Ca-binding protein distinct from calmodulin does modulate the sensitivity of the cone channel for cGMP [135]. In contrast, heterologously expressed chicken cone α-subunit and two short spliced variants found in chicken pineal gland have been reported to be modulated by Ca-calmodulin [136].

The Ca-calmodulin binding site has been localized to the β-subunit of the rod channel [35,46,50] and α-subunit of the olfactory channel [126]. In the olfactory channel, the binding site is located near the N-terminus of the α-subunit and consists of a stretch of about 14 amino acids that can form a basic amphipathic helical structure characteristic of other calmodulin recognition sites [126]. The Ca-calmodulin binding site of the rod channel has been recently localized to a segment on the N-terminal side of the membrane-spanning domain of the β-subunit [88,137] (also see Fig. 4). Interestingly, this site does not contain a common calmodulin binding motif. Instead, this 15 amino acid segment contains a number of negatively charged residues not generally found in calmodulin binding sites and does not fold into a basic amphipathic structure when represented as an α-helical wheel [137]. High affinity binding to Ca-calmodulin is only obtained when much longer peptides derived from the rod β-subunit that encompass this site are used in binding studies. Apparently, flanking regions of the calmodulin binding site are needed to properly fold this segment into this unconventional high affinity binding site.

The mechanism by which Ca-calmodulin reduces the sensitivity of CNG channels for cyclic nucleotides has been investigated [126,127]. In the case of the olfactory channel, deletion of the Ca-calmodulin binding site results in a channel that exhibits the same apparent affinity for cyclic nucleotides as the non-mutated (wild-type) channel in the absence of Ca-calmodulin. Deletion of this site also abolishes interaction of the amino-terminal and carboxyl-terminal regions of the channel [127]. These and other studies lead to the view that Ca-calmodulin alters the gating of the channel by cyclic nucleotides. This occurs through the binding of Ca-calmodulin to the amino-terminal segment of the α-subunit, thereby altering the interaction of this domain with the carboxyl domain and decreasing the ease of transition from the closed to the open state. It is likely that a similar mechanism may occur for the rod channel, although this remains to be investigated in detail.

Although calmodulin is known to be present in ROS, several studies indicate that another calcium binding protein endogeneous to ROS may bind to the channel and modulate its sensitivity to cGMP [130,132]. The identity of such an endogeneous calcium binding protein, however, remains to be determined.

Modulation of CNG channels by Ca-calmodulin plays an important role in sensory neurons. In olfactory neurons, activation of the signal transduction pathway results in the opening of CNG channels to the entry of Ca^{2+}, as well as Na^{+} ions. As the intracellular concentration of Ca^{2+} rises, the Ca-calmodulin complex interacts with the olfactory CNG channel to reduce its sensitivity for cAMP. This negative feedback mechanism contributes to olfactory adaptation [16,18,19]. In the case of rod cells, activation of the visual cascade system results in closure of the rod channel and a decrease in intracellular Ca^{2+}. The resulting dissociation of Ca-calmodulin from the rod channel gives rise to an increased sensitivity of the channel for cGMP. This mechanism is thought to facilitate the recovery of the photoreceptors following bleaching [46,81]. Recent quantitative analysis of this and other Ca-dependent feedback mechanisms, however, suggests that calmodulin modulation of the rod channel may play a relatively small role in the recovery process [131,138].

2.8.2. Modulation by other factors

Several indirect studies suggest that the rod channel can be modulated by protein phosphorylation. Using excised patches from frog ROS, Gordon et al. [139] have reported that the sensitivity of the channel to cGMP gradually increases with time. This effect is blocked by serine/threonine phosphatase inhibitors. Addition of an exogenous phosphatase was also observed to increase the sensitivity of the channel for cGMP. Molokanova et al. [140] have also reported that the native and expressed rod channel is modulated by tyrosine phosphorylation. Although both these studies suggest that protein phosphorylation can regulate the activity of the channel, it is unclear if this modulation occurs through the direct phosphorylation of the channel or through phosphorylation of other cellular proteins which in turn interact with the channel and alter its activity.

Transition divalent ions such as Zn^{2+}, Ni^{2+}, Cd^{2+}, Mn^{2+} and Co^{2+} can also affect the gating properties of the native rod channels [119,141]. Studies of Gordon and Zagotta [124] using the expressed homo-oligomeric rod channel indicate that Ni^{2+} increases the apparent affinity of the channel for cGMP through its interaction with His residue (His420) located near the intracellular mouth of the channel. In contrast, Ni^{2+} binding to a related His residue (His396) on the olfactory channel decreases the apparent affinity of this channel. Evidently, for the rod channel Ni^{2+} binding favors the open state of the channel, whereas in the olfactory channel Ni^{2+} binding favors the closed state.

Other agents, including diacylglycerol and sulfhydryl modifying reagents, have also been reported to alter the activity of the rod channel [142,143].

2.9. Related cyclic nucleotide-gated channels in other tissues

CNG channels have been found in a wide variety of neuronal and non-neuronal tissues in addition to photoreceptor and olfactory neurons (for review see Refs. [80,144]). In some cases the CNG channels present in other tissues are encoded by the gene for the rod, cone or olfactory channel, but may be expressed as spliced variants [145]. For example, the α-subunit of the rod channel has been detected in chicken pineal gland [136], kidney cells [146] and hippocampal neurons [147]. The olfactory α-subunit has been cloned from aorta tissue [148] and hippocampal neurons [147]. The cone α-subunit has been cloned from spermatozoa [64] as well as kidney and heart [149], and spliced variants of this subunit have been found in chicken pineal gland [136]. In addition, a CNG channel has been recently cloned from rat taste buds that is 82% similar in amino acid sequence to the bovine and human cone α-subunit [150].

Less is known about the tissue distribution of the β-subunit. Shorter spliced variants of the bovine rod β-subunit have been cloned from bovine testes and spliced variants of the rat β-subunit have been found in rat pinealocytes [151,152] and in olfactory sensory neurons [78,79].

In addition to the rod, cone and olfactory CNG channels, a number of other channels that are activated or inhibited by cyclic nucleotides have been detected in a variety of different types of vertebrate and invertebrate cells, as recently reviewed [80,144].

3. Ion channels in the inner compartments

Rod and cone photoreceptors consist of an outer segment and inner compartments. The inner compartment comprises the inner segment proper, the soma, a short axon, and a nerve ending. In what follows, the inner compartment will be referred to as inner segment for simplicity.

The light response in rod and cone photoreceptors is engendered by the collaboration of ionic mechanisms in both the outer and inner segments. As the light-evoked hyperpolarization generated in the outer segment spreads to the inner segment and across gap junctions to neighboring cells, it is modified by voltage-dependent currents in the inner segment.

Vertebrate photoreceptors have a unique spatial disposition of ionic conductances on the cell surface. The outer segment membrane exclusively harbors the cGMP-gated channels that are closed by light [153], whereas the inner segment is furnished with at least six different channel types. The various ionic conductances were dissected by means of channel blockers in solitary photoreceptors obtained by enzymatic dissociation of the retina. Using solitary photoreceptors either with or without outer segments facilitated the interpretation of ionic currents in voltage-clamp studies, because in the retinal network, rod and cone cells are electrically coupled through gap junctions [154]. Ion channels identified in the inner segment include: (1) a weakly selective cation channel, activated by hyperpolarization commonly referred to as I_h channels; (2) a non-inactivating voltage-gated K^+ channel; (3) a voltage-gated Ca^{2+}-channel; (4) a Ca^{2+}-activated K^+ channel; (5) a Ca^{2+}-activated Cl^- channel; and (6) a cGMP-gated channel. The channel inventory of the inner segments of rod and cone photoreceptors is similar (Fig. 10), except for the cGMP-gated channels that only exist in the presynaptic terminals of the cone synapse. While the function of these channels in rod and cone photoreceptors has been known for some time, it was only recently that the molecular identity of the channel polypeptides has been revealed by cloning. In the following, we will review the physiological and molecular properties of these channels and their function in shaping the light response.

3.1. Hyperpolarization-activated channel (I_h)

A conductance activated by hyperpolarization in rod photoreceptors of the intact retina has been first described by Fain and coworkers [155]. For light flashes of high

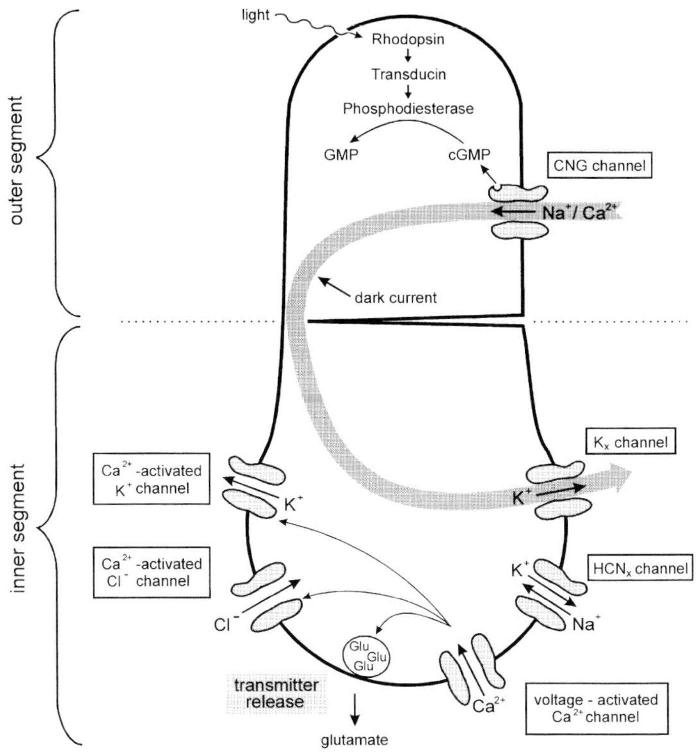

Fig. 10. Channel inventory of the outer and inner segments of rod photoreceptor. CNG, cyclic nucleotide-gated channel; HCN_x, hyperpolarization-activated cyclic nucleotide-sensitive channel of unknown subtype (after [213]). Although the precise distribution of ion channels has not yet been experimentally determined, it is plausible that K_x and HCN_x channels are located in the membrane of the inner segment, whereas the Ca^{2+} channels, and the Ca^{2+}-activated Cl^- and K^+ channels are located in the synaptic terminal.

intensities, the rapid hyperpolarization is followed by relaxation to a less negative plateau (Fig. 11). The initial spike is absent at lower light intensities that cause smaller hyperpolarizations. In the presence of extracellular Cs^+, the voltage sag is abolished and a saturating response hyperpolarizes the cell almost towards the potassium equilibrium potential (E_K), whereas at low light intensities, Cs^+ has no effect on the shape of the voltage response. Because the photocurrent measured with a suction electrode shows no relaxation [156], Fain and coworkers [155] correctly concluded that Na^+ ions pass through a Cs^+-sensitive conductance that is activated by the light-evoked hyperpolarization. Subsequent voltage-clamp studies on solitary rod photoreceptors either with or without outer segment revealed that hyperpolarization activates an inward current in the inner segment and thereby causes the depolarizing sag in the voltage response [157–160]. A similar hyperpolarization-activated current has also been characterized in the inner segment of cone photoreceptors [161–164].

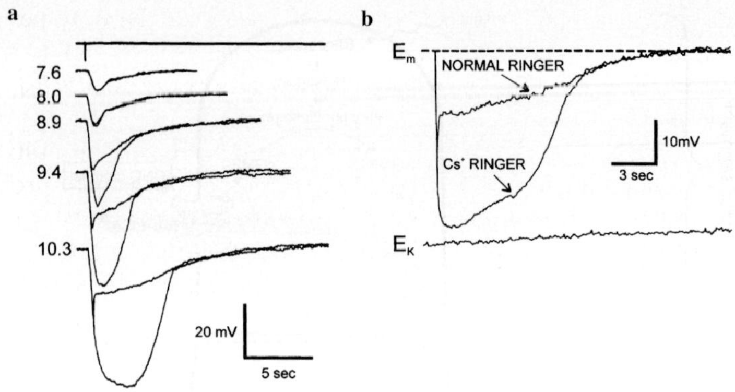

Fig. 11. Effect of Cs^+ on the waveform of rod photoresponses. (a) Responses of toad rod in normal and 2 mM CsCl Ringer. Responses in the two Ringers have been superimposed at five light intensities (numbers refer to log quanta cm^{-2} per flash) for 100 ms, 501 nm flashes of full-field illumination. Responses at intensities 7.6 and 8.0 in normal and CsCl represent averages of four responses. Other traces are single responses. At 7.6 the waveforms are nearly identical, but, in brighter light, Cs^+ responses are always larger and lack fast decay. Dark membrane potential (in both Ringers), −35 mV. (b) Comparison of response to saturating light intensities in normal and 2 mM CsCl Ringers with the membrane potential in zero Na^+ Ringer. The three traces in this figure were all recorded intracellularly from the same toad rod. E_m (dotted line) indicates the dark resting membrane potential in normal (and CsCl) Ringer (−30 mV). The trace labeled *Normal Ringer* shows the response to a 100 ms, full-field illumination of intensity 11.8 log quantum cm^{-2} per flash, at 501 nm. This intensity was bright enough to produce amplitude saturation. The retina was then superfused with 2 mM CsCl Ringer. The response labeled *Cs Ringer* shows the response in CsCl to a flash of the same intensity as for the one shown above in normal Ringer. Finally, the retina was superfused with a Ringer containing 2 mM CsCl but for which all of the Na^+ was replaced with choline. The trace labeled E_K shows the membrane potential of the rod in this Ringer at steady-state. Reprinted by permission from [214].

Currents activated by hyperpolarization exist in a host of various neurons, the best studied examples being I_h channels in the sino-atrial node of the heart and in cortico-thalamic neurons. Because of their unusual features, these currents have also been designated "queer" or "funny" currents (I_q and I_f) (for review see Ref. [165]). Many of the unusual properties of cardiac and neuronal I_h are also shared by I_h from rod and cone photoreceptors. First, the underlying channels are unique among the family of voltage-gated channels in that they appear to become activated by hyperpolarization instead of depolarization like many K^+, Ca^{2+}, or Na^+ channels. The mid-point voltage of activation ($V_{1/2}$) in the rod is roughly −67 mV [160], suggesting that these channels are closed in the dark and do not contribute to the V_m of rod photoreceptors at rest (−30 to −40 mV).

Second, I_h channels are weakly selective for K^+ ions. Relative ion permeabilities P_x/P_K are: Tl^+ (1.55) > K^+ (1) > Rb^+ (0.55) > Na^+ (0.33) > Li^+ (0.02) [166]. As a consequence, at physiological ion concentrations, currents reverse at V_{rev} of −30 mV and for $V_m \geqslant -60$ mV, i.e. during a saturating light response, I_h channels predominantly carry a Na^+ inward current. When K^+ is the only permeant ion species and when its extracellular concentration is changed, V_{rev} of I_h follows the Nernst potential for a K^+ electrode [166]. The relative permeability P_{Na}/P_K is not constant, but changes from 0.2 to 0.3 upon increasing the $[K^+]_e$ from 2 to 10 mM, respectively [160]. Thus, I_h channels display features diagnostic of multi-ion pores that do not accommodate ions independent of each other. A peculiar property of I_h channels is that Na^+ carries little or no inward current in the absence of extracellular K^+ [160,166]. As discussed by Wollmuth [167], the gating of I_h channels is normal in Na^+ solutions lacking K^+, and, under these conditions, Na^+ probably permeates only very slowly. The K^+ sensitivity of the Na^+ inward current may be of some physiological significance, because the extracellular $[K^+]$ near the inner segment is reduced in the light [168].

Third, an important feature of I_h channels is their modulation by cyclic nucleotides. While this property has been extensively studied in I_h channels from other tissues, little is known about I_h regulation in rod [169] and nothing in cone photoreceptors. Cyclic AMP enhances I_h channel activity by shifting the mid-point voltage of activation ca. 8–15 mV to more positive values [170]. In I_h channels of the sino-atrial node, this shift is accomplished directly without involving a phosphorylation mechanism [171]. Therefore, agonists that either increase or decrease the cAMP concentration will augment or suppress I_h currents, respectively. The cardiac and cortico-thalamic I_h channels are also modulated by cGMP [171,172]. For I_h currents in rod and cone photoreceptors, modulation by either nucleotide has not yet been reported. Dopamine reversibly reduced I_h in a dose-dependent manner. The dopamine-mediated inhibition is blocked by antagonists for D2 dopamine receptors [169]. Unexpectedly, the inhibition did not appear to involve changes in intracellular cAMP or Ca^{2+} levels. However, the D2 agonist quinpirole failed to inhibit I_h when the patch pipette contained 10 mM of the Ca^{2+} buffer BAPTA. In conclusion, dopamine inhibits I_h in the inner segment of rod photoreceptors by an unknown mechanism that appears to be different from modulation of I_h channels in other tissues.

Recently, the cDNA of several I_h channels have been cloned from sea urchin sperm [173] and vertebrate heart and brain tissue [174,175] (for brief accounts of this work, see Refs. [176,177,223]) When functionally expressed in *Xenopus* oocytes or cell lines, the cloned cDNA gives rise to currents that display the characteristic features of I_h channels. So far, four different genes encoding vertebrate I_h channels have been identified [174,175,178–180]. The encoded polypeptides have been designated HCN1 to HCN4 [176]. Which of the four subtypes of HCN channels is expressed in rod and cone photoreceptors is unknown. cAMP shifts the $K_{1/2}$ of the heterologously expressed subtypes HCN1 only 2 mV towards more positive potentials – if at all [175]. This subtype, therefore, might be a good candidate for the I_h channels in the inner segment of rod photoreceptors, that seem to operate independent of cAMP [169]. The I_h channels belong to the family of CNG channels and EAG/HERG channels. The channel sequences encompass a full-blown voltage sensor motif (S4), a pore motif that is characteristic of the signature sequence of K^+-selective channels, and a cyclic nucleotide-binding domain (Fig. 12).

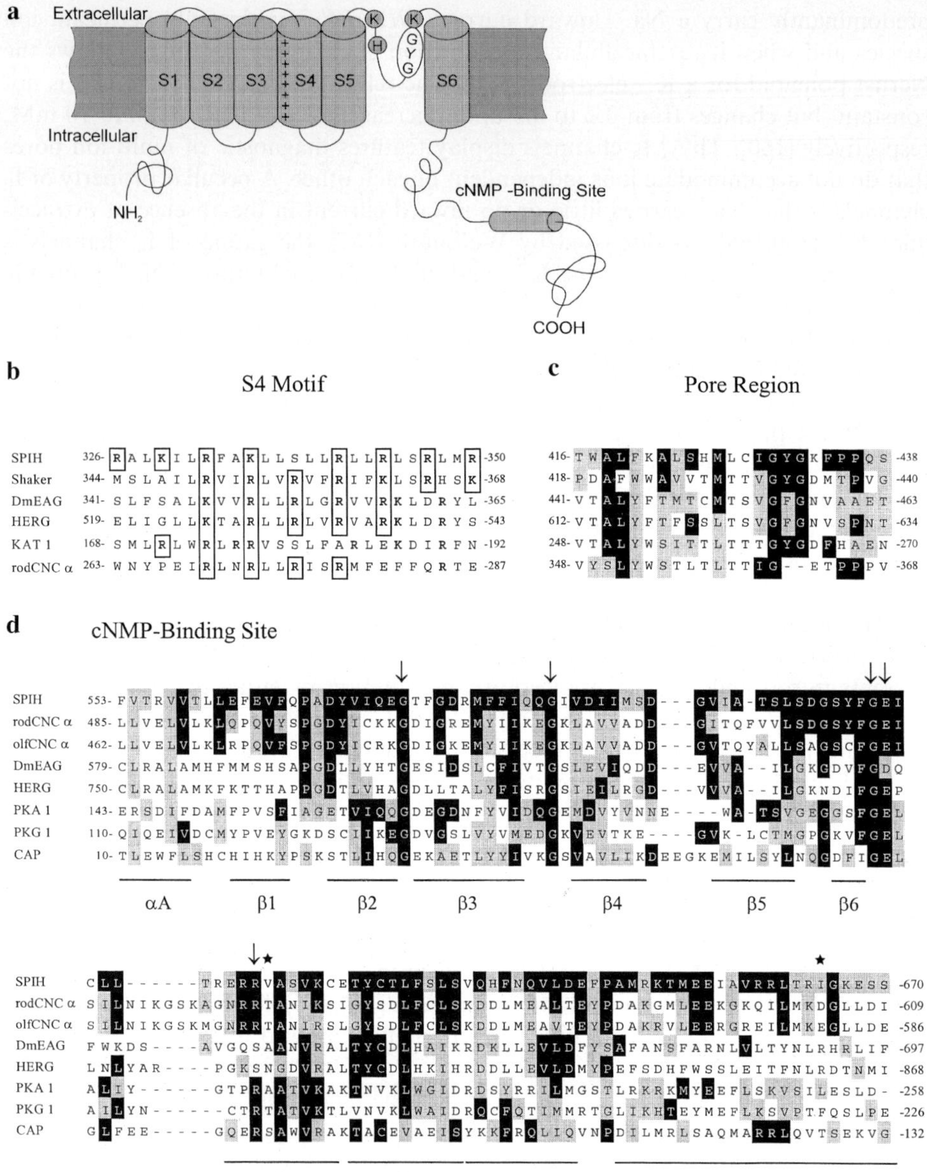

Fig. 12. Structural features of hyperpolarization-activated channels. (a) Transmembrane topology of a HCN channel cloned from sea urchin *Strongylocentrotus purpuratus* (SPIH). K and H refer to positively charged residues in the pore region; GYG refers to the signature sequence of K^+ selective channels [215]; cNMP, binding site for cyclic nucleotides. (b) Comparison of the voltage-sensor (S4) motif of SPIH with that of other channels. Shaker K^+ channel encoded by the *Drosophila Shaker b* gene [216], DmEAG, *Drosophila* ether-á-gogo (eag) channel [217]; HERG, channel encoded by the human eag-related gene [218]; KAT1, K^+ channel from *Arabidopsis thaliana* [219]; rod-CNCα, α-subunit of the CNG channel from bovine rod photoreceptors [40]. Regularly spaced arginine or lysine residues are boxed. Other

3.2. K_x channels

The dark inward current through cGMP-gated channels in the surface membrane of the outer segment is balanced by a K^+ outward current in the inner segment. This current and the underlying channels have been dubbed I_{Kx} and K_x channels, respectively [181]. The I_{Kx} displays some similarity with M-currents observed in other cell types. Like the K^+ channels underlying M-currents, the K_x channels open relatively slowly at $V_m \geqslant -50$ mV, do not inactivate, and are set apart from Shaker type K^+ channels by a modestly different pharmacology [181]. However, while M-currents are suppressed by acetylcholine, I_{Kx} is not. In addition, external Ba^{2+} shifts the voltage-dependence of K_x activation to much more positive values, but has only minor effects on M-currents. A similar current has also been identified in the inner segments of cones [163,182].

Frings and coworkers [183] provided initial evidence that K_x channels belong to the ether-á-gogo (EAG) subfamily of K^+ channels. These authors showed by *in situ* hybridization that transcripts encoding EAG channels are expressed in retinal ganglion cells and rod photoreceptors. Several functional similarities between native K_x channels and heterologously expressed EAG channels suggest that EAG polypeptides might be part of the K_x channels. Both channels are resistant to blockage by extracellular Cs^+/Cd^{2+}, activate at $V_m > -45$ to -50 mV, do not inactivate, and are sensitive to changes in extracellular pH [181,184]. However, the similarity is not perfect. EAG and K_x channels differ significantly in their deactivation kinetics and some aspects of Ba^{2+} sensitivity and modulation by extracellular Mg^{2+} [183,185]. The native K_x channels have been studied in the salamander retina, whereas retinal EAG channels were cloned from bovine retina. Therefore, variations among species could account for the differences between salamander I_{Kx} and bovine EAG. Furthermore, mutations in the *eag* gene of *Drosophila* affect four different K^+ channels [186,187] and coexpression of *Drosophila* EAG with Shaker K^+ channels alters the kinetics of EAG channel gating [188]. Both observations support the possibility that native K_x channels are composed of several distinct subunits, one of which is EAG.

More recently, several novel genes of the KCNQ family of K^+ channels have been identified [189,190]. The subunit species KCNQ2 and KCNQ3 can co-assemble to produce K^+ selective currents with essentially identical biophysical properties and pharmacological sensitivities to the native M current. These findings suggest the

---

positively charged residues are in bold. (c) Comparison of the pore motif of SPIH with that of other channels. Residues identical or similar to the corresponding positions in SPIH are highlighted by a black or grey background, respectively. (d) Comparison of cNMP-binding domains. Olf-CNCα, α-subunit of the CNG channel from bovine olfactory neurons [70]; PKA1, cAMP-binding site 1 of protein kinase A [220]; PKG1, cGMP-binding site 1 of protein kinase G [221]; CAP, catabolite gene activator protein [222]. Residues important for the structure of cNMP-binding domains and for the binding of the ligand are marked by arrows; residues that co-determine the ligand selectivity in bCNCα are marked by an asterisk. Secondary-structure predictions derived from CAP are shown below the sequences.

possibility that KCNQ channels may also participate in forming the K_x channel in the inner segment.

3.3. Physiological function of I_h and I_{Kx}

The physiological function of hyperpolarization-activated currents and K_x currents is illustrated in Fig. 13. The activation curves and the respective reversal voltages V_{rev} of both currents are positioned at opposite ends of the operating range of rod photoreceptors. The I_h is half-maximally activated at −67 mV and reverses at −30 to −35 mV; thus, it is closed in the dark and does not contribute to the resting voltage of roughly −35 mV. Due to the steep voltage dependence of I_h activation, small hyperpolarizations from the resting voltage do not activate I_h channels and in this voltage range there is little driving force for I_h. These considerations provide an explanation why extracellular Cs^+ has no effect on small hyperpolarizing responses [155]. For larger hyperpolarizing responses, I_h becomes activated and carries a depolarizing inward current. Conversely, in the dark, I_{Kx} is activated to a significant extent and the driving force for the outward K^+ current is sizeable. Therefore, currents flowing through cGMP-gated channels in the outer segment and K_x channels in the inner segment set the resting voltage in the dark. Small hyperpolarizing responses cause deactivation of K_x channels. In the voltage range 5–15 mV negative of the resting value, changes in the open probability of K_x channels are particularly effective in opposing the light-evoked hyperpolarization, because the driving force on K^+ currents is large. At larger hyperpolarizations ($\leqslant -60$ mV), I_{Kx} disappears because K_x channels close and V_m is near V_{rev} of I_{Kx}. In conclusion, two sets of channels with opposite voltage dependence and V_{rev} values at juxtapositioned ends of the operating range of the rod cell cooperate to resist the hyperpolarizing response to light.

All three ion channels that produce and shape the light response, i.e. the cGMP-gated channel in the outer segment and the K_x and I_h channels in the inner segment,

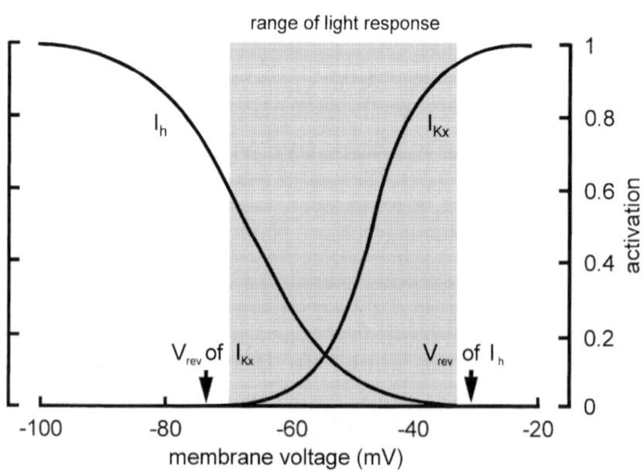

Fig. 13. Comparison of the activation curves of K_x and HCN_x channels with the operating range of light responses (for explanation see text) (after Ref. [213]).

belong to the superfamily of cyclic nucleotide-sensitive channels. In particular, all three channel types feature a cyclic nucleotide-binding motif (Fig. 12d). It is tempting, therefore, to speculate that the activity of I_h and I_{Kx} in the inner segment, like those of cGMP-gated channels, depends on the concentrations of cAMP or cGMP. So far, neither the K_x nor the I_h channels have been shown to be modulated by cyclic nucleotides. An NO-generating compound, S-nitrocysteine, enhanced the activity of Ca^{2+} channels and cGMP-gated channels. Both channel types are probably located in the synaptic ending of the inner segment (see below), whereas currents carried by K_x and HNC_x channels were unaffected by this treatment [191]. Moreover, the I_h amplitude and kinetics were not significantly altered by 200 μM cAMP in the patch pipette [169]. Because I_h channels in other tissues are regulated by receptors or agents that either enhance or lower cAMP or cGMP levels, it will be necessary to re-examine I_h modulation by cyclic nucleotides.

3.4. Voltage-activated Ca^{2+} channels

Light produces a graded hyperpolarization in rod and cone photoreceptors up to 25 mV in amplitude. However, not all of this response range is effectively transmitted to the post-synaptic bipolar cells and horizontal cells. The rod synapse transmits only small deviations up to 5 mV from the resting voltage in the dark and synaptic transmission ceases when a rod is hyperpolarized beyond −45 mV [192,193]. The highly non-linear input–output relation of the rod synapse is largely accounted for by the voltage dependence of presynaptic Ca^{2+} channels. At the dark resting voltage of −35 mV, a small proportion of the Ca^{2+} channels are open. The continuous Ca^{2+} entry sustains a tonic release of the neurotransmitter glutamate from the synaptic terminal. The Ca^{2+} channels are characterized by an activation threshold of ~−45 mV [158]. Therefore, when a rod is hyperpolarized to −45 mV, the Ca^{2+} channels will close and synaptic transmission ceases [192,193]. The fraction of open Ca^{2+} channels changes e-fold with each 6 mV [158]; by comparison, the exponential input–output relation for rod to horizontal cell transmission increases e-fold with each 2.1 mV of depolarization [192]. This difference in the voltage-dependence of Ca^{2+} current and the postsynaptic response in rod photoreceptors is reminiscent of a similar difference in the synapse of the squid giant axon [194,195]. The steep input–output relationship reflects the requirement for the binding of 3 or 4 Ca^{2+} ions to trigger the exocytotic events in the presynaptic terminal. In conclusion, much of the non-linearity of synaptic transmission can be accounted for by the voltage-dependence of Ca^{2+} channel activation. Small deviations from the resting voltage modulate glutamate release, while signals more negative than −45 mV are clipped. In some instances, the voltage range of rod outputs is as wide as 10–15 mV [196,197] and is modulated by light [198], suggesting that the input–output relation of the synaptic gain is not static, but exhibits some degree of plasticity (see Ref. [199] for review).

Ca^{2+} channels in both, rod and cone photoreceptors have been identified as L-type channels on the basis of activation range, gating kinetics, slow inactivation and sensitivity to blockage by dihydropyridines and divalent cations [158,163,

90. Illing, M., Molday, L.L. and Molday, R.S. (1997) J. Biol. Chem. **272**, 10303–10310.
91. Allikmets, R., Singh, N., Sun, H., Shroyer, N.F., Hutchinson, A., Chidambaram, A., Gerrard, B., Baird, L., Stauffer, D., Pfeiffer, A., Rattner, A., Smallwood, P., Li, Y., Anderson, K.L., Lewis, R.A., Nathans, J., Leppert, M., Dean, M. and Lupski, J.R. (1997) Nat. Gen. **15**, 236–246.
92. He, W., Cowan, C.W. and Wensel, T.G. (1998) Neuron **20**, 95–102.
93. Makino, E.R., Handy, J.W., Li, T. and Arshavsky, V.Y. (1999) Proc. Natl. Acad. Sci. USA **96**, 1947–1952.
94. Liu, D.T., Tibbs, G.R. and Siegelbaum, S.A. (1996) Neuron **16**, 983–990.
95. Gordon, S.E. and Zagotta, W.N. (1995) Proc. Natl. Acad. Sci. USA **92**, 10222–10226.
96. Shammat, I.M. and Gordon, S.E. (1999) Neuron **23**, 809–819.
97. He, Y., Ruiz, M.L. and Karpen, J.W. (2000) Proc. Natl. Acad. Sci. USA **97**, 895–900.
98. Weber, I.T. and Steitz, T.A. (1987) J. Mol. Biol. **198**, 311–326.
99. Kumar, V.D. and Weber, I.T. (1992) Biochemistry **31**, 4643–4649.
100. Zagotta, W.N. and Siegelbaum, S.A. (1996) Annu. Rev. Neurosci. **19**, 235–263.
101. Weber, I.T., Shabb, J.B. and Corbin, J.D. (1989) Biochemistry **28**, 6122–6127.
102. Shabb, J.B., Ng, L. and Corbin, J.D. (1990) J. Biol. Chem. **265**, 16031–16034.
103. Goulding, E.H., Ngai, J., Kramer, R.H., Colicos, S., Axel, R., Siegelbaum, S.A. and Chess, A. (1992) Neuron **8**, 45–58.
104. Goulding, E.H., Tibbs, G.R. and Siegelbaum, S.A. (1994) Nature **372**, 369–374.
105. Varnum, M.D., Black, K.D. and Zagotta, W.N. (1995) Neuron **15**, 619–625.
106. Karpen, J.W. and Brown, R.L. (1996) J. Gen. Physiol. **107**, 169–181.
107. Kramer, R.H., Goulding, E. and Siegelbaum, S.A. (1994) Neuron **12**, 655–662.
108. Zagotta, W.N., Hoshi, T. and Aldrich, R.W. (1990) Science **250**, 568–571.
109. Goulding, E.H., Tibbs, G.R., Liu, D. and Siegelbaum, S.A. (1993) Nature **364**, 61–64.
110. Root, M.J. and MacKinnon, R. (1993) Neuron **11**, 459–466.
111. Eismann, E., Müller, F., Heinemann, S.H. and Kaupp, U.B. (1994) Proc. Natl. Acad. Sci. USA **91**, 1109–1113.
112. Bucossi, G., Eismann, E., Sesti, F., Nizzari, M., Seri, M., Kaupp, U.B. and Torre, V. (1996) J. Physiol. **493**, 409–418.
113. Root, M.J. and MacKinnon, R. (1994) Science **265**, 1852–1856.
114. MacKinnon, R. (1995) Neuron **14**, 889–892.
115. Lu, Q. and Miller, C. (1995) Science **268**, 304–307.
116. Sun, Z.-P., Akabas, M.H., Goulding, E.H., Karlin, A. and Siegelbaum, S.A. (1996) Neuron **16**, 141–149.
117. Doyle, D.A., Cabral, J.M., Pfuetzner, R.A., Kuo, A., Gulbis, J.M., Cohen, S.L., Chait, B.T. and MacKinnon, R. (1998) Science **280**, 69–77.
118. Karpen, J.W., Zimmerman, A.L., Stryer, L. and Baylor, D.A. (1988) Proc. Natl. Acad. Sci. USA **85**, 1287–1291.
119. Ildefonse, M. and Bennett, N. (1991) J. Membr. Biol. **123**, 133–147.
120. Picones, A. and Korenbrot, J.I. (1995) J. Physiol. **485**, 699–714.
121. Ruiz, M.L. and Karpen, J.W. (1997) Nature **389**, 389–392.
122. Liu, D.T., Tibbs, G.R., Paoletti, P. and Siegelbaum, S.A. (1998) Neuron **21**, 235–248.
123. Benndorf, K., Koopmann, R., Eismann, E. and Kaupp, U.B. (1999) J. Gen. Physiol. **114**, 477–489.
124. Gordon, S.E. and Zagotta, W.N. (1995) Neuron **14**, 857–864.
125. Tibbs, G.R., Goulding, E.H. and Siegelbaum, S.A. (1997) Nature **386**, 612–615.
126. Liu, M., Chen, T.-Y., Ahamed, B., Li, J. and Yau, K.-W. (1994) Science **266**, 1348–1354.
127. Varnum, M.D. and Zagotta, W.N. (1997) Science **278**, 110–113.
128. Li, M., Jan, Y.N. and Jan, L.Y. (1992) Science **257**, 1225–1230.
129. Zong, X., Zucker, H., Hofmann, F. and Biel, M. (1998) EMBO J. **17**, 353–362.
130. Zimmerman, A. (1995) Curr. Opinion Neurobiol. **5**, 296–303.
131. Koutalos, Y. and Yau, K.-W. (1996) Trends Neurosci. **19**, 73–81.
132. Gordon, S.E., Downing-Park, J. and Zimmerman, A.L. (1995) J. Physiol. **486**, 533–546.
133. Bauer, P.J. (1996) J. Physiol. **494**, 675–685.

134. Haynes, L.W. and Stotz, S.C. (1997) Vis. Neurosci. **14**, 233–239.
135. Hackos, D.H. and Korenbrot, J.I. (1997) J. Gen. Physiol. **110**, 515–528.
136. Bönigk, W., Müller, F., Middendorff, R., Weyand, I. and Kaupp, U.B. (1996) J. Neurosci. **16**, 7458–7468.
137. Weitz, D., Zoche, M., Müller, F., Beyermann, M., Körschen, H.G., Kaupp, U.B. and Koch, K.-W. (1998) EMBO J. **17**, 2273–2284.
138. Nakatani, K., Koutalos, Y. and Yau, K.-W. (1995) J. Physiol. **484**, 69–76.
139. Gordon, S.E., Brautigan, D.L. and Zimmerman, A.L. (1992) Neuron **9**, 739–748.
140. Molokanova, E., Trivedi, B., Savchenko, A. and Kramer, R.H. (1997) J. Neurosci. **17**, 9068–9076.
141. Karpen, J.W., Brown, R.L., Stryer, L. and Baylor, D.A. (1993) J. Gen. Physiol. **101**, 1–25.
142. Gordon, S.E., Downing-Park, J., Tam, B. and Zimmerman, A.L. (1995) Biophys. J. **69**, 409–417.
143. Serre, V., Ildefonse, M. and Bennett, N. (1995) J. Membr. Biol. **146**, 145–162.
144. Finn, J.T., Grunwald, M.E. and Yau, K.-W. (1996) Annu. Rev. Physiol. **58**, 395–426.
145. Ahmad, I., Redmond, L.J. and Barnstable, C.J. (1990) Biochem. Biophys. Res. Commun. **173**, 463–470.
146. Ahmad, I., Korbmacher, C., Segal, A.S., Cheung, P., Boulpaep, E.L. and Barnstable, C.J. (1992) Proc. Natl. Acad. Sci. USA **89**, 10262–10266.
147. Kingston, P.A., Zufall, F. and Barnstable, C.J. (1996) Proc. Natl. Acad. Sci. USA **93**, 10440–10445.
148. Biel, M., Altenhofen, W., Hullin, R., Ludwig, J., Freichel, M., Flockerzi, V., Dascal, N., Kaupp, U.B. and Hofmann, F. (1993) FEBS Lett. **329**, 134–138.
149. Biel, M., Zong, X., Distler, M., Bosse, E., Klugbauer, N., Murakami, M., Flockerzi, V. and Hofmann, F. (1994) Proc. Natl. Acad. Sci. USA **91**, 3505–3509.
150. Misaka, T., Kusakabe, Y., Emori, Y., Gonoi, T., Arai, S. and Abe, K. (1997) J. Biol. Chem. **272**, 22623–22629.
151. Biel, M., Zong, X., Ludwig, A., Sautter, A. and Hofmann, F. (1996) J. Biol. Chem. **271**, 6349–6355.
152. Sautter, A., Biel, M. and Hofmann, F. (1997) Molec. Brain Res. **48**, 171–175.
153. Baylor, D.A. and Lamb, T.D. (1982) J. Physiol. **328**, 49–71.
154. Attwell, D. and Wilson, M. (1980) J. Physiol. **309**, 287–315.
155. Fain, G.L., Quandt, F.N., Bastian, B.L. and Gerschenfeld, H.M. (1978) Nature **272**, 467–469.
156. Yau, K.-W., Lamb, T.D. and Baylor, D.A. (1977) Nature **269**, 78–80.
157. Bader, C.R., MacLeish, P.R. and Schwartz, E.A. (1979) J. Physiol. **296**, 1–26.
158. Bader, C.R., Bertrand, D. and Schwartz, E.A. (1982) J. Physiol. **331**, 253–284.
159. Baylor, D.A., Matthews, G. and Nunn, B.J. (1984) J. Physiol. **354**, 203–223.
160. Hestrin, S. (1987) J. Physiol. **390**, 319–333.
161. Baylor, D.A., Hodgkin, A.L. and Lamb, T.D. (1974) J. Physiol. **242**, 759–791.
162. Attwell, D., Werblin, F.S. and Wilson, M. (1982) J. Physiol. **328**, 259–283.
163. Barnes, S. and Hille, B. (1989) J. Gen. Physiol. **94**, 719–743.
164. Maricq, A.V. and Korenbrot, J.I. (1990) J. Neurophysiol. **64**, 1917–1928.
165. Pape, H.-C. (1996) Annu. Rev. Physiol. **58**, 299–327.
166. Wollmuth, L.P. and Hille, B. (1992) J. Gen. Physiol. **100**, 749–765.
167. Wollmuth, L.P. (1995) Pflügers Arch. **430**, 34–43.
168. Oakley, B.I., Flaming, D.G. and Brown, K.T. (1979) J. Gen. Physiol. **74**, 713–737.
169. Akopian, A. and Witkovsky, P. (1996) J. Neurophysiol. **76**, 1828–1835.
170. DiFrancesco, D. (1993) Annu. Rev. Physiol. **55**, 455–472.
171. DiFrancesco, D. and Tortora, P. (1991) Nature **351**, 145–147.
172. Pape, H.-C. and Mager, R. (1992) Neuron **9**, 441–448.
173. Gauss, R., Seifert, R. and Kaupp, U.B. (1998) Nature **393**, 583–587.
174. Ludwig, A., Zong, X., Jeglitsch, M., Hofmann, F. and Biel, M. (1998) Nature **393**, 587–591.
175. Santoro, B., Liu, D.T., Yao, H., Bartsch, D., Kandel, E.R., Siegelbaum, S.A. and Tibbs, G.R. (1998) Cell **93**, 717–729.
176. Clapham, D.E. (1998) Neuron **21**, 5–7.
177. Lüthi, A. and McCormick, D.A. (1998) Neuron **20**, 553–563.

Contents

1. Introduction

1.1. Phototransduction is a prototypical example of a G-protein signaling cascade

Phototransduction is the process by which light, captured by a visual pigment molecule in a photoreceptor cell, generates an electrical response. In all vertebrate and invertebrate photoreceptor cells that have been investigated (including, of course, retinal rods and cones), phototransduction is effected by a 'G-protein cascade' – a sequence of reactions initiated by a G-protein-coupled receptor (GPCR) protein.

A generic G-protein cascade is illustrated in Fig. 1. In the first step of the cascade the binding of a specific ligand, L, induces the receptor, R, to undergo a conformational change to a state, R*, in which it is enzymatically active. In the second step, R* catalyzes the exchange of GTP for GDP on the α-subunit (Gα) of a specific heterotrimeric 'G-protein' – so-named because of its guanine nucleotide exchange function. In the third step the activated subunit of the G-protein, G* (≡Gα–GTP) conveys the signal onward, by binding to a specific 'effector protein', E, which is thereby converted to its active form, E*. A few examples of the vast variety of G-protein cascades are identified in Fig. 1.

The cascade of phototransduction may appear to be the 'odd man out', in the sense that the stimulus to activation is a photon of light rather than a chemical substance. Nevertheless, it does conform to the general pattern of ligand activation, because in the dark state the ligand (11-*cis* retinal) acts as a powerful antagonist, preventing activation. Absorption of a photon of light isomerizes the ligand to all-*trans* retinal, which acts instead as a powerful agonist to activation.

Receptor protein. Each GPCR is an integral membrane protein, comprising seven *trans*-membrane helices linked by six extra-membrane segments. GPCRs are members of a super-family of proteins which includes at least five main families, and GPCR genes are thought to constitute 3–5% of the human genome (cf. [12,13]). Numerically, the olfactory receptor proteins appear to comprise the largest group of GPCRs, with more than 1000 members [14,15]. To date the genes of several thousand GPCRs from many species have been sequenced, and characterized with respect to their ligand- and G-protein specificity, though many remain 'orphans'.

G-protein. In the second stage of the phototransduction cascade (Fig. 1), the signal (that a photon of light has been captured by the GPCR) is not simply transmitted, but is greatly amplified, by the catalytic activation of a heterotrimeric G-protein. Heterotrimeric G-proteins comprise one division of the super-family of G-proteins, all of which switch between two signaling states. The inactive state is that with a molecule of GDP bound; transition to the active state occurs upon

A

B

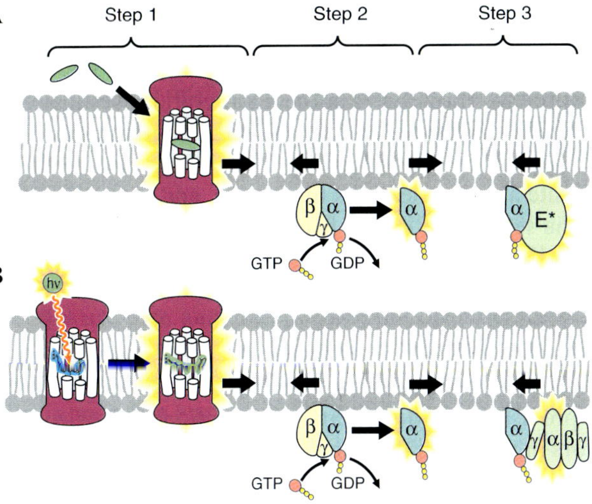

C

	S	R*	G G*	E*
	Stimulus	*Receptor*	*G-protein*	*Effector*
1	photon	rhodopsin	G_T	PDE
2	photon	invert. rhodopsin	G_q	PLC
3	odorant	vert. olfactory	G_{olf}	AC
4	odorant	invert. chemosensory	G_q,$G_{i/o}$	PLC
5	dopamine	D_2-dopamine	$G_{i/o}$	AC
6	NAdr	β-adrenergic	G_S	AC
7	NAdr	β-adrenergic	$G_{\beta\gamma}$	N-type Ca^{2+} chan.
8	ACh	M_2-muscarinic	$G_{\beta\gamma}$	K^+ channel

Fig. 1. Generic G-protein signal transduction cascade. (A) The cascade comprises three proteins, a G-protein-coupled receptor, R, a G-protein, G, and an effector protein, E, which are activated in three steps. In the first step, the receptor R is activated to R*, and in most cases this is brought about by the binding of a ligand. In the second step, R* activates a specific G-protein by catalyzing the release of GDP from the inactive form (G–GDP), and thereby permitting the binding of GTP, to create the active form G* (=Gα–GTP). A single R* can activate many molecules of G-protein, because it is released unaltered after the interaction. In the third step, the usual mechanism is that the G* binds to the effector protein E, causing it to switch states to an active form E*. (B) The G-protein transduction cascade of vertebrate photoreceptors follows the general pattern shown in A, except that activation of R to R* is caused by the photoisomerization of a ligand, 11-*cis* retinal, that is already attached covalently to the receptor protein. (C) Identification of the components in a few examples of the many families of heterotrimeric G-protein signaling cascades, with sample references as follows. (1) Pugh and Lamb (this chapter); (2) [1]; (3) [2–4]; (4) [5]; (5) [6]; (6) [7,8]; (7) and (8) the Gβγ subunit serves to carry the signal modulating the conductance of subclasses of K^+ channels and Ca^{2+} channels [9–11].

exchange of this GDP for a GTP; transition back to the inactive state results from hydrolysis of the terminal phosphate of the GTP, converting it back to GDP. Because of the nature of the state-switch, the super-family is also known as the GTPase super-family [16].

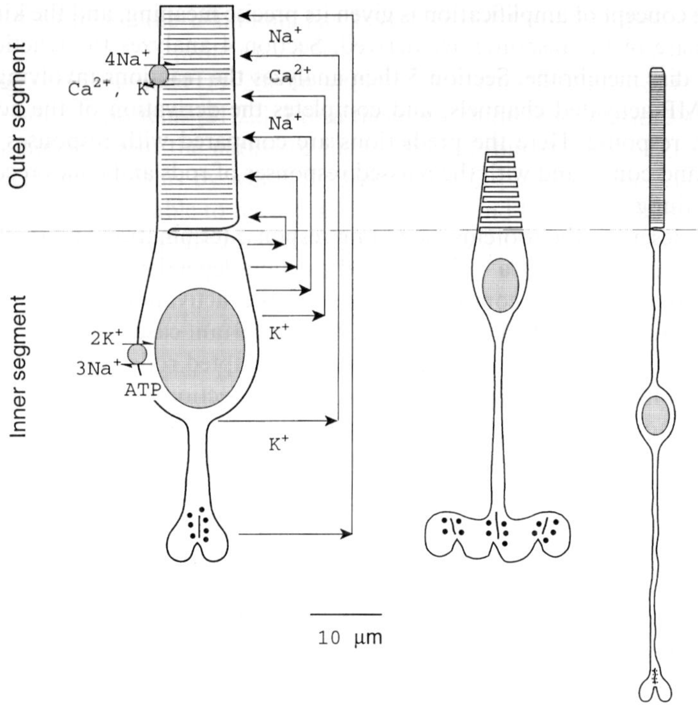

Fig. 2. Structure of rods and cones, and nature of the circulating current. A salamander rod, a salamander red-sensitive cone, and a mammalian rod are shown approximately to scale. The outer segment of each cell, which contains the lamellar membranes in which the visual pigment molecules are embedded, and where phototransduction occurs, is situated above the dotted line, while the inner segment, which contains the mitochondria, nucleus and synaptic region, is below the dotted line. The circulating current is illustrated for the salamander rod only (but applies to each cell).

Beer's Law to be $p = 1 - 10^{-\alpha L} \approx 89\%$. Here α is the axial pigment density, which has been measured as 0.016 μm^{-1} [32], and which can also be obtained as $\alpha = \varepsilon_{max} C$, where $\varepsilon_{max} \approx 60{,}000$ cm^2 $mmole^{-1}$ is the decadic molar absorbance coefficient of rhodopsin in the disc membrane for axially propagating light [33]. Were the outer segment only 1 μm in length (36 discs), the photon capture probability would be $p \approx 4\%$, drastically lower than the actual probability in a real photoreceptor.

Physiological recordings have established the remarkable finding that photoisomerization of any one of the 3×10^9 rhodopsin molecules assembled into the toad rod disc stack will generate a reliable electrical signal [34–36]. Consistent with this physiological finding, behavioral experiments have demonstrated that a toad can detect and capture prey with 100% accuracy, under illumination conditions that generate no more than 1 photoisomerization per rod every 10 s [37]. Single photon responses have also been recorded from primate rods [38], and from mouse rods [39,40]. Thus, a universal (and virtually defining) characteristic of vertebrate rods is their ability to provide reliable single-photon detection. Characterization of the molecular mechanisms underlying this ability has long posed a central challenge to research in phototransduction.

2.3. The circulating electrical current

The protein composition of the surface membrane of photoreceptors differs greatly between the inner and outer segments. With regard to ion permeation, the outer segment membrane contains only two contributing classes of protein: the cGMP-gated channel (or cyclic-nucleotide gated channel, CNGC), and the electrogenic Na^+/Ca^{2+}, K^+ exchanger (NCKX) [41–46]. The membrane of the inner segment, on the other hand, contains K^+ channels of two main varieties (the so-called I_{Kx} and I_h channels) as well as channels permeable to other cations, including Ca^{2+} [46]. In addition, the inner segment membrane contains the Na^+/K^+ pump (or Na^+/K^+-ATPase) that maintains the ionic concentration gradients between the inside and outside of the cell.

Under resting dark conditions the cytoplasmic concentration of free cGMP is several μM (see Section 3), so that a small proportion of the cGMP-gated channels are held open. Since these channels are relatively non-specific in their permeability for different monovalent cations, the net current flowing through them is inward (at the normal resting potential of vertebrate rods and cones, ca. −35 to −45 mV). About 85–90% of this inward current is carried by Na^+, simply because Na^+ is the predominant external cation. And since the channels are highly permeable to Ca^{2+} ions, most of the remaining 10–15% of the current is carried by Ca^{2+}, with an additional contribution from Mg^{2+} that normally is quite small [47].

At rest, a balancing current flows outwards across the inner segment membrane, carried primarily by the I_{Kx} channels. Together, the influx of current into the outer segment and the balancing efflux of current from the inner segment create a loop (Fig. 2) known as the 'circulating current' – or, in darkness, as the 'dark current'. In different vertebrate rods, whether of the large amphibian variety or of the smaller mammalian type, the magnitude of the dark current ranges from 20 to 70 pA per rod. Table 1 summarizes the magnitude of the dark current, and gives other electrical properties, for rods and cones of a number of species that have been investigated extensively.

2.4. The photoresponse and its underlying molecular mechanism

The primary electrical event in vertebrate phototransduction is a transient suppression of the circulating current, that results from closure of cGMP-gated channels in the outer segment. Characteristic families of electrical responses are illustrated in Fig. 3, for both a salamander rod and a mammalian rod, exposed to brief flashes of progressively greater intensity. The ordinate of Fig. 3 is plotted in terms of the circulating current, $J(t)$, and the photocurrent response $r(t)$ is the change in current induced by the flash: for flashes presented in darkness the photocurrent response is simply $r(t) = J_{dark} - J(t)$.

The photocurrent families exhibit a number of well-known properties. For example: each response reflects a graded suppression of the circulating current; the peak amplitude increases monotonically as a function of flash intensity, until saturation

Table 1
Typical outer segment parameters, and dark resting electrical properties, of vertebrate rods and cones[a]

Quantity	Symbol	Unit	Rods				Cones		
			Salamander	Toad	Primate	Mouse	Salamander	Turtle	Primate
Outer segment parameters									
Wavelength of maximum absorption	λ_{max}	nm	500	500	500	500	610	650	565
Length	L	μm	22	60	25	20	8	15	13
Diameter (base, tip)	d	μm	11	6	2	~2	4, 2.5	2.5, ~1	3, ~1
Envelope volume	V_{tot}	μm^3	2000	1800	40	30	70	30	30
Cytoplasmic volume	V_{cyto}	μm^3	1000	900	20	15	35	30	15
Temperature	T	C	22	22	37	37	22	22	37
cGMP-activated current (in dark)	J_{cG}^{D}	pA	−70	−40	−50	−20	−50		−40
$Na^+/Ca^{2+},K^+$ exchange current (in dark)	J_{ex}^{D}	pA	−4	−3	−2	−1	2		
Typical whole-cell electrical properties									
Resting potential (in dark)	V_m	mV	−40		−37		−40	−40	−45
Time-to-peak (dim flashes)	t_{peak}	ms	700	1500	200	240	170	100	50
Flash sensitivity	S_F^D	pA/Φ	0.2	1	0.7	0.4	0.03		0.02
(at t_{peak}, in darkness)		mV/Φ	0.06		1.2		0.03	0.025	0.005
Capacitance	C	pF	20	20		1	70	40	
Time constant (dark)	τ_m	ms	20	20	1	1	50	5	3
References			[48–50]	[34,51]	[38,52,53]	[39,40,54]	[50,55–57]	[58–60]	From [61] as cited in [62]; [63,64]

[a] *Notes*: 1. Values given are representative of measurements in the literature.
2. Where whole-cell recording data are not available, we have used estimates based on suction-electrode recordings, and assumed a suction electrode collection efficiency of 2/3.
3. Primate cone anatomical properties are systematically dependent on distance from the fovea; the parameters given are applicable for the shorter but wider peripheral cones. The cited references illustrate and discuss this variation in cone anatomy.

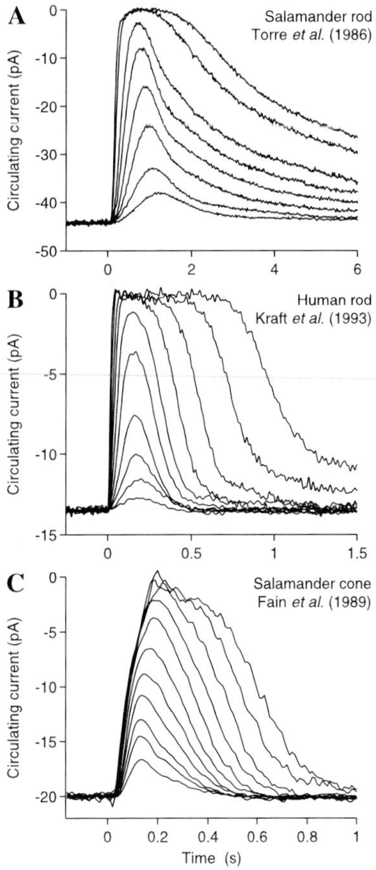

Fig. 3. Families of flash responses for three types of photoreceptor, recorded using the suction pipette method. In each experiment, flashes of progressively greater intensity were delivered at time zero. (A) Salamander rod, exposed to flashes estimated to deliver from 10 to 2000 photoisomerizations; ca. 22°C; data from Fig. 2A of Ref. [65]. (B) Human rod, for flashes delivering from 12 to 5200 photoisomerizations; 37°C; data from Fig. 1 of [53], kindly supplied by Dr. J.L. Schnapf. (C) Salamander red-sensitive cone; flashes delivering 1400 to 6×10^5 photoisomerizations; ca. 22°C; data from Fig. 6A of Ref. [56].

is reached, when the circulating current declines to zero; the dim-flash responses reach peak in roughly 1 s (at room temperature), or roughly 200 ms (at mammalian body temperature), while the bright-flash responses reach peak earlier.

The molecular mechanisms underlying activation of the electrical response (along with some of the termination reactions) are illustrated in Fig. 4A. Activation of the G-protein cascade causes stimulation of the cGMP phosphodiesterase and hence increased hydrolysis of cGMP, so that the cytoplasmic concentration of cGMP is reduced in the vicinity of the absorbed photon. As a result, the cGMP-gated channels in this region are no longer held open, and a localized reduction occurs in the influx of cations (primarily Na^+ and Ca^{2+}) into the outer segment.

Table 4
Definitions of variables and parameters of the modeling, with representative values[a]

Symbol	Units	Value (in dark)	Refs.	Definition	Used in Eqs.
				A. Variables	
Φ	Isomerizations			Number of photoisomerizations per rod per flash	1, 4, 5, 19, 20, 25, 29
R^*	Molecules			Number of activated rhodopsin molecules per rod	1
G^*	Molecules			Number of activated G-protein molecules per rod	4
E^*	Subunits			Number of activated PDE subunits per rod	5, 29
α	$\mu M\ s^{-1}$	1.2–2.4		Rate of synthesis of cGMP by guanylyl cyclase	13–16, 33, 34
β	s^{-1}	0.6–1.2	[118,119]	Rate constant of cGMP hydrolysis by PDE	13–16, 34
cG	μM	2	[46,82]	Free concentration of cGMP in the outer segment	9, 10, 13, 14, 16–18, 20, 22
J_{cG}	pA	Table 1		Current carried by cGMP-gated channels	21, 31
J_{ex}	pA	Table 1		Electrogenic current carried by exchanger	31, 32
J	pA			Total outer segment current; $J = J_{cG} + J_{ex}$	
F	–			Normalized circulating current; $F = J/J_0$	23, 25
Ca	μM	Table 2		Free concentration of Ca^{2+} in the outer segment	31
				B. Activation parameters	
ν_{RG}	s^{-1}	~150	[127]	Rate of G* formation per fully activated R*	6
ν_{RE}	s^{-1}	~150	[127]	Rate of E* formation per fully activated R*	5, 6, 26, 29
c_{GE}	–	< 1		Coupling coefficient from G* to E*; $c_{GE} = \nu_{RE}/\nu_{RG}$	6
k_{cat}	s^{-1}	4400	[78]	Turnover rate of doubly-activated PDE holomer	9
K_m	μM	10	[127]	Michaelis constant of cGMP hydrolysis by PDE	9
V_{cyto}	Pl	1		Cytoplasmic volume of the outer segment	9, 31
B_{cG}	–	~2		Buffering power of the cytoplasm for cGMP	9
β_{sub}	s^{-1}	$2–4 \times 10^{-4}$	[127]	Rate constant of cGMP hydrolysis per E* subunit	11, 26

Table 4 (*continuation*)

Symbol	Units	Value (in dark)	Refs	Definition	Used in Eqs
K_{cG}	μM	20		cGMP conc. for half-maximal channel opening	21
n_{cG}	–	2–3	[82, 115]	Hill coefficient of the cGMP channel activation	21–23, 26, 34
A	s^{-2}	0.1	[26, 27]	Amplification constant; $A = \nu_{RE}\ \beta_{sub}\ n_{cG}$	25, 26
t_R, t_{RG}, etc.	s			Short delays in various activation steps	1, 4, 5, 19 20, 24, 29
t_{eff}	s	0.02	[26, 27]	Effective delay contributed by all short steps	21–23
				C. Inactivation parameters	
τ_R	s	0.4	[120] non-dominant	Time constant for inactivation of R* activity	29, 30
τ_E	s	1.5	[120] dominant	Time constant for inactivation of G*–E* complex	29, 30
$1/\beta_0$	s	1		Time constant for decay of cGMP (Ca clamped)	
f_{Ca}	–	0.1–0.15	[46]	Fraction of cGMP-activated current carried by Ca^{2+}	31
K_{ex}	nM	1600	[70]	$[Ca^{2+}]_i$ giving half-maximal exchange current	32
$J_{ex,\ sat}$	pA	−20	[70]	Saturated exchange current at high $[Ca^{2+}]_i$	32
K_{cyc}	nM	100–230	[81, 121, 122]	$[Ca^{2+}]_i$ giving half-maximal cyclase activity	33
n_{cyc}	–	2	[81, 121, 122]	Hill coefficient for cyclase activation by $[Ca^{2+}]_i$	33
α_{min}	$\mu M\ s^{-1}$	~0	[81, 121, 122]	Min. value of α, at high $[Ca^{2+}]_i$	33
α_{max}	$\mu M\ s^{-1}$	~40	[81, 121, 122]	Max. value of α, at low $[Ca^{2+}]_i$	33

[a] *Notes*: 1. The variables listed in part A of the Table can in general take several forms. For example, the concentration of cGMP (represented by the variable *cG*) can be given in the forms: $cG(t)$, as a function of time; cG_0, at some resting steady level; cG_{dark}, in darkness; and $\Delta cG(t)$, as the incremental change from the resting level cG_0. (One exception is Φ, which has the single meaning of the number of photoisomerizations delivered to the outer segment at time zero.).
2. A dash "–" in the units column signifies a dimensionless quantity.

$$\nu_{enc}(t) = \frac{4\pi(D_R + D_G)C_G}{\ln[4(D_R + D_G)t/\rho^2] - 2\gamma} \qquad (2)$$

where D_R and D_G are the lateral diffusion coefficients for R* and G–GDP respectively, while $\rho \approx 5$ nm is the encounter radius for contact between the two molecules, and $\gamma \approx 0.57722$ is Euler's constant. In amphibian rods at room temperature, we may take $D_R = 0.7\ \mu m^2\ s^{-1}$, and $D_G = 1.2\ \mu m^2\ s^{-1}$ [26,27].

Equation (2) automatically takes account of the local depletion of G–GDP that occurs near the R*, and accordingly it predicts that the encounter rate ν_{enc} should not be constant, but should instead decline with time according to the logarithmic term in the denominator. Evaluation of the denominator during the interval from 5–500 ms after formation of the R* shows that the decline is quite modest, though, and that for amphibian rods a good approximation may be provided by the rate determined at 50 ms, i.e.,

$$\nu_{enc} \approx 1.3(D_R + D_G)C_G \tag{3}$$

The value of ν_{enc} estimated from (3) for an amphibian rod at room temperature is 6000 s^{-1}, and for a mammalian rod at body temperature (at an appropriately early time) 16,000 s^{-1} [26,27].

These calculated rates of encounter represent the diffusion-limit to the possible rate of reaction, since reaction can only occur upon encounter; hence, if every encounter were successful, then the reaction rate ν_{RG} would equal the encounter rate ν_{enc}. However, in the likely scenario that R* fails to trigger activation at every encounter with a G–GDP, then first the reaction rate will be lower than this diffusion limit, and second it will exhibit even less time-dependence (since the pool of inactive G–GDPs will suffer less local depletion in the neighborhood of the R*). Both these theoretical predictions are borne out by numerical simulations [110]. In sum, then, it is expected that a single fully-activated R* will generate G*s at a nearly constant rate, which we shall denote by the symbol ν_{RG}. Therefore, provided we can ignore inactivation of R* and of G*, we predict the number of activated G*s in a rod stimulated at $t = 0$ by a flash producing Φ photoisomerizations should be well approximated by a 'ramp' function of time

$$G^*(t) = \Phi\, \nu_{RG}(t - t_{RG}), \quad t > t_{RG} \tag{4}$$

Here $t_{RG} = t_R + t_G$,where t_R is as defined above, and t_G is a delay no greater than the time between successful encounters between R* and G–GDP.

4.3. E production*

As G* (=Gα–GTP) is acylated, it is likely that after separation from R* and Gβγ the G* continues to move at the surface of the disc membrane. Since the PDE effector protein is firmly anchored to the membrane, interaction between G* and the PDE γ-subunit requires the molecules to encounter each another by lateral diffusion. Quantitative analysis of this step is complicated by the fact that the G*s are produced not at a fixed location, but at whatever location the R* has moved to at the particular instant in question. Nevertheless a simplified analytical treatment was developed [26], and its accuracy was subsequently investigated quantitatively using numerical simulation [110].

The bottom line of the latter investigation was that over a wide range of plausible values of the parameters that affect ν_{RG} (such as the diffusion coefficients, and the probability of successful encounter), the generation of E* (=PDE*) was again predicted to follow a ramping function of time

$$E^*(t) = \Phi\, \nu_{RE}(t - t_{RGE}), \quad t > t_{RGE} \tag{5}$$

Here ν_{RE} is the effective rate with which a single R* triggers activation of E*s, and t_{RGE} is given by $t_{RGE} = t_R + t_G + t_E$, where t_E is a brief delay that arises between buildup of G* and buildup of E* [110]. Thus, t_{RGE} is a cumulative delay that combines the 'micro' delays introduced by the three steps.

The form of the kinetic predictions embodied in Eqs. (1), (4) and (5) are illustrated in Figs. 5A and 5B.

Since, in the absence of inactivation reactions, $G^*(t)$ and $E^*(t)$ are predicted to follow an identical time-course (a ramp with time), it is convenient to define the 'coupling efficiency' of the step from G* to E* as the ratio of their slopes of activation. Thus we denote the coupling efficiency c_{GE} as

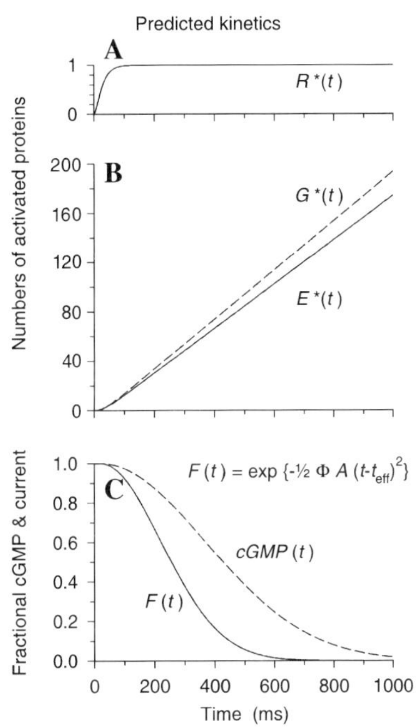

Fig. 5. Predicted kinetics of activation, for the disc membrane proteins, cGMP, and electrical response. Inactivation reactions have been ignored at this stage. (A) Following photoisomerization, a single R* is activated with a time constant t_R; see Eq. (1). (B) According to Eqs. (4) and (5), the quantities $G^*(t)$ and $E^*(t)$ of activated G-protein and PDE increase linearly with time after R* activation; the slope is estimated to be around 150 s^{-1} per R*, see Section 5.9. (C) Activation of E* causes the cGMP concentration to decline, according to Eq. (20). Consequently cGMP-gated channels close, and the fractional circulating current $F(t)$ declines according to Eq. (25). Responses in C are shown for a flash delivering $\Phi = 250$ photoisomerizations, with $A = 0.1$ s^{-2}.

$$\frac{dcG}{dt} = \alpha(t) - \beta(t)cG \tag{13}$$

where $\alpha(t)$ denotes the rate of synthesis by guanylyl cyclase (in $\mu M\ s^{-1}$) and $\beta(t)$ denotes the rate constant of hydrolysis (in s^{-1}). During the light response these parameters will in general both be time-varying functions; we nevertheless employ the conventional term rate 'constant' for $\beta(t)$.

Under steady-state conditions (either in darkness, or during steady illumination) both sides of Eq. (13) are zero, so that

$$\alpha_0 = \beta_0 cG_0 \tag{14}$$

where the subscript '0' indicates the initial steady-state condition; in the special case where this state is darkness we shall employ the subscript 'dark'. It is convenient to consider increments from the steady level, by defining

$$\alpha(t) = \alpha_0 + \Delta\alpha(t), \quad \beta(t) = \beta_0 + \Delta\beta(t) \tag{15}$$

Substitution of these definitions into Eq. (13) yields

$$\begin{aligned}\frac{dcG}{dt} &= \alpha_0 + \Delta\alpha(t) - (\beta_0 + \Delta\beta(t))cG \\ &= \Delta\alpha(t) - \Delta\beta(t)cG + \beta_0(cG_0 - cG)\end{aligned} \tag{16}$$

where the second line follows from the equality in (14).

We now make two approximations that restrict analysis to early times or small responses. Any increment $\Delta\alpha(t)$ in cGMP synthesis during the light response will result from the decline of $[Ca^{2+}]_i$ that occurs as a consequence of the closure of cGMP-gated channels and the continued extrusion of Ca^{2+} by the exchanger. Such changes will take time to develop, so that at sufficiently early times in the light response we can make the approximation that $\Delta\alpha(t) \approx 0$. Since we are limited to early times, where $cG \approx cG_0$, the final term $\beta_0(cG_0 - cG)$ will also approximate to zero. Thus the first and third terms in the second line of Eq. (16) disappear and, within the validity of this approximation, the differential equation simplifies to

$$\frac{dcG}{dt} \approx -\Delta\beta(t)cG \tag{17}$$

Thus, when synthesis occurs at a fixed rate, the differential equation approximates to the exact form in Eq. (10), which applies in the absence of synthesis.

5.5. *Solution for the light-induced change in cGMP concentration*

Equation (17) has the general solution

$$\frac{cG(t)}{cG_0} = \exp\left[-\int_0^t \Delta\beta(t')dt'\right] \tag{18}$$

We now require an expression for $\Delta\beta(t)$, which, in the case of a brief flash delivering Φ photoisomerizations to the outer segment at time zero, we obtain from Eqs. (5) and (12), as

$$\Delta\beta(t) = \Phi \nu_{RE} \beta_{sub}(t - t_{RGE}) \quad \text{for } t > t_{RGE} \tag{19}$$

Substitution of (19) into (18) yields the solution

$$\frac{cG(t)}{cG_0} = \exp\left[-\tfrac{1}{2}\Phi \nu_{RE} \beta_{sub}(t - t_{RGE})^2\right] \quad \text{for } t > t_{RGE} \tag{20}$$

which indicates that the cGMP concentration should decline according to a 'delayed Gaussian' function of time following a brief flash (Fig. 5c). The range of validity of the approximations underlying Eqs. (17) and (19) will be investigated after the electrical response has been derived.

5.6. *Solution for the electrical response to a flash of light*

The change in intracellular concentration of cGMP can be translated into the cell's electrical response through knowledge of the gating relation for the outer segment channels. From experiments on excised patches of outer segment membrane (Section 3.5; see also Ref. [46]), the fraction of open channels can be expressed by the cooperativity expression (or Hill equation) as

$$\frac{J_{cG}}{J_{cG,\,max}} = \frac{cG^{n_{cG}}}{cG^{n_{cG}} + K_{cG}^{n_{cG}}} \tag{21}$$

Here J_{cG} is the current through the cGMP-gated channels in a region of membrane (at a fixed membrane voltage), and $J_{cG,\,max}$ is its maximum value at high concentrations of cGMP. The exponent n_{cG} is termed the Hill coefficient, and K_{cG} is the concentration of cGMP that produces a half-maximal current. In excised patches of rod membrane n_{cG} has been found to be 2–3, while K_{cG} is typically about 20 μM.

Equation (21) is also expected to govern the cGMP-activated currents of whole rod outer segments, although testing the relation in intact cells is complicated by the problem of delivering cGMP uniformly to the outer segment, and by the impracticality of space-clamping the rod membrane potential when all the cGMP-gated channels are open [135]. The most compelling results supporting the validity of Eq. (21) for whole outer segments have been obtained with the truncated rod outer segment preparation [121,136]; it bears mention that in the latter investigations n_{cG} was found to be ~2.

Recently it has been found that in intact photoreceptors K_{cG} may shift as a result of the calcium-dependent binding of calmodulin, thus contributing to light adaptation (see Section 8). This finding may also account for the fact that in different studies the values reported for n_{cG} range from about 1.8 to 3; thus, it is possible that spatial variations in K_{cG} within the outer segment might make the cooperativity of the ensemble of channels appear lower than the actual cooperativity of the individual channel [137].

The application of Eq. (21) to the intact photoreceptor is simplified by two factors. First, the concentration of cGMP is always much less than the half-activation concentration (i.e. $cG < cG_{\text{dark}} \ll K_{cG}$), so that Eq. (21) simplifies to a power relation

$$\frac{J(t)}{J_0} = \left[\frac{cG(t)}{cG_0}\right]^{n_{cG}} \tag{22}$$

where, for simplicity, we have dropped the subscript 'cG' from $J(t)$. Second (at least in rods), the current through the outer segment channels is essentially independent of transmembrane voltage over the normal operating range (see Section 3.5). Thus, it is valid to substitute the relationship for $cG(t)$ from (20) into (22) to obtain

$$F(t) = \frac{J(t)}{J_0} = \exp\left[-\tfrac{1}{2}\Phi \nu_{RE} \beta_{sub} n_{cG} (t - t_{\text{eff}})^2\right] \quad \text{for } t > t_{\text{eff}} \tag{23}$$

The term $F(t)$ has been introduced to denote the circulating current during the flash response, expressed as a fraction of the original resting steady-state level; thus $F(t)$ always begins at $F(0) = 1$.

In Eq. (23) a new delay term, t_{eff}, has been introduced, because intervening between the decline in cGMP concentration predicted by (20) and the decline in circulating current are at least three additional delay processes: (i) the time t_r for radial equilibration of cGMP concentration in the outer segment, (ii) the time t_B for equilibration of cGMP with 'fast' buffer sites in the cytoplasm, and (iii) the time t_C for gating of the channels in response to an applied change in cGMP concentration. As discussed in Section 5.2, diffusion theory predicts that t_r should be of the order of a few ms for an amphibian rod, and much shorter for a mammalian rod [134]. We have no information on the size of t_B but expect it to be very small, while t_C has been measured as 2–3 ms at room temperature [129]. On the time scale of the light response, these delays are very brief, and can simply be accumulated with the other delays discussed in Section 4. Thus, the effective delay t_{eff} in (23) may be written as the sum

$$t_{\text{eff}} = t_R + t_G + t_E + t_r + t_B + t_C \tag{24}$$

From analysis of voltage-clamped responses from salamander rods, the total delay time t_{eff} has been estimated as about 15 ms. In mammalian rods at body temperature it is no more than 2 ms [26,138,139].

Two points should now be mentioned about Eq. (23). First, this expression accounts only for current flowing through the cGMP-gated channels, and excludes current carried by the calcium exchanger (Section 7.3), as well as capacitive current (treated below). Second, in our previous work [26,27] we used the symbol F to denote J/J_{dark} (i.e., relative to the dark state), but here we are using F in a more general sense to denote current relative to the steady resting condition in any arbitrary background intensity. For the special case where the background is zero, we shall use the symbol $F_{\text{dark}}(t)$.

The form of Eq. (23) is plotted in Fig. 5C. Interestingly, the electrical response has exactly the same delayed Gaussian shape as the cGMP concentration. Thus the effect of the channel gating cooperativity is simply to increase the steepness of the response, without fundamentally changing its nature.

If the photoreceptor is not voltage-clamped, then capacitive charging of the cell membrane contributes an additional low-pass filtering step to the photocurrent, with a time constant τ_m given by the product of the cell's membrane capacitance and its leakage resistance. The size of this membrane time constant can either be estimated from the kinetics of the responses to extremely intense flashes, or can be measured directly from experiments on isolated cells in the whole-cell patch clamp configuration (e.g., [49]). Typical values (Table 1) are roughly: 10–20 ms for amphibian rods; 40–70 ms for salamander and fish cones; 1 ms for mammalian rods *in vivo*; and 2–4 ms for mammalian cones *in vivo* [48,49,52,57,64,125,139,140].

For the slow responses elicited by flashes of low or moderate intensity, τ_m can be treated as a fixed delay which can be cumulated with t_{eff}, along with any additional delays contributed by electronic or digital filtering used by the investigator. (In papers in which we have applied Eq. (23), we have used the symbol t'_{eff} to represent t_{eff} plus these additional delays [26,27,138,141].) For intense flashes, one would ideally wish to obtain a full analytical solution taking account of the individual 'chemical' delays set out in Eq. (24), but this is impractical and those delays are simply incorporated into t_{eff}. On the other hand, as the membrane time constant and the equipment delays are readily measurable, their effects can be explicitly incorporated into any analysis. Thus, the filtering effect of the membrane time constant can be calculated either by numerical convolution of Eq. (23) with an exponential decay [140], or by substitution into an analytical solution [139]. And the delay in the electronics can be compensated for by a shift on the time axis (as in Fig. 3).

5.7. *Significance of the equations: Amplification and response kinetics*

Equation (23) has great utility, both theoretically, in explaining the nature of the amplification of the phototransduction cascade and the basis for the shape of the rising phase kinetics, and also practically, in providing a simple means of fitting experimentally measured responses and extracting from them quantitative information about the cascade.

Amplification. Equation (23) embodies the manner in which the individual stages of the transduction cascade illustrated in Fig. 4A contribute to the overall amplification (or gain) of the response. Thus, we can rewrite Eq. (23) in the simpler form

$$F(t) = \exp\left[-\tfrac{1}{2}\Phi A(t - t_{\text{eff}})^2\right], \quad t > t_{\text{eff}} \tag{25}$$

(which we shall use from here on), where we define

$$A = \nu_{\text{RG}}\, c_{\text{GE}}\beta_{\text{sub}} n_{\text{cG}}, \tag{26}$$

as the 'amplification constant' of phototransduction (in s^{-2} per photoisomerization); the rate ν_{RE} has been replaced by $\nu_{\text{RG}}\, c_{\text{GE}}$, using Eq. (6).

This definition of amplification is referenced to an input of a single photoisomerization per outer segment, and the four factors on the right-hand side of

Eq. (26) allocate the overall gain to four discrete stages. The first stage of the cascade is defined as having unity gain: an isomerization activates a single R* molecule. The second stage, the enzymatic activation of G* by R*, generates a component of amplification characterized by the catalytic rate ν_{RG} (in s^{-1}). The third stage, linking G* to E* activation, contributes an amplification somewhat less than unity, characterized by the coupling efficiency c_{GE}. The fourth stage, the PDE-catalyzed hydrolysis of cGMP, contributes the second major component of amplification, characterized by β_{sub} (in s^{-1}), the rate constant of decline in free cGMP concentration elicited by a single E* subunit per volume of the outer segment cytoplasm (Eq. (11)). Finally the fifth stage, the cooperative gating of the channels by cGMP, contributes a component of amplification characterized by the Hill coefficient n_{cG}.

According to Eq. (11), the factor β_{sub} is inversely proportional to V_{cyto}, revealing the important role that the volume of the outer segment plays in amplification. Thus, the 25-fold smaller outer segment volume of mammalian rods accounts for much of the factor of 100-fold by which their amplification constant (A = 5–10 s^{-2}) exceeds that of amphibian rods (A = 0.05–0.1 s^{-2}) [27]. Because of the term $A\,t^2$ in Eq. (25), the time taken to reach a given fractional suppression of circulating current is predicted to vary inversely as $\sqrt{A}$. Hence, for a given number of photoisomerizations, the response of a mammalian rod should achieve a given fractional suppression of the cGMP-activated current about $\sqrt{100} = 10$ times faster than the response of an amphibian rod.

To further explore the implications of Eq. (25), we consider the relative merits of packaging a given quantity of rhodopsin (and its associated transduction machinery) into a single large outer segment, versus packaging it into 25 smaller outer segments. Because the total number of rhodopsin molecules is the same in the two cases, we can assume that a given intensity of incident light will elicit the same total number of photoisomerizations in the two cases. Accordingly the *average* response per rod will be the same in the two cases – but the qualification 'average' is crucial. If just a single photon were absorbed, the situation in the two scenarios would be very different. Absorption of the single photon in the large outer segment would generate a response that we can take as a reference. But absorption of a single photon in the group of 25 smaller rods would generate no response in 24 of them, together with a response in one rod that rose in $\frac{1}{5}$ of the time of the reference response in the large rod (or, at a fixed early time, to 25 times the level in the large rod). Hence Eq. (25) shows the important advantage that accrues in the 'photon counting' regime of scotopic light intensities through the use of small outer segments, such as possessed by some fish and birds, and by mammals. As the reactions occur in a smaller cytoplasmic volume, the response at a fixed time is much larger (and the time taken to reach a given level is much shorter). Provided that the retinal circuitry can handle the signals from the greater number of rods, this increased amplification is likely to provide a significant advantage in increased time resolution in the photon counting regime.

Contributions to response kinetics. The theoretical development captures the way in which the sequential steps in phototransduction contribute to the waveform of the activation phase. As summarized in Fig. 5, three of the cascade stages act as 'integrating' mechanisms: (i) the absorption of a photon is integrated to a *step* of activity, through the change in conformation of R to its active state R*; (ii) the catalytic activity of R* leads to a second stage of integration, generating a ramp of activation of G* and E*; (iii) the catalytic hydrolysis of cGMP by E*s contributes a third stage of integration, converting the E* ramp into a Gaussian time-course of cGMP concentration, which at early times approximates an inverted parabola. Thus, an impulse of activity at the input is shaped, via three stages of integration, into a parabolic decline in cGMP concentration, and in electrical current.

5.8. Comparison between experiment and theory

The practical significance of Eq. (25) lies in its general applicability to the responses of a great variety of photoreceptors, including not only single-cell measurements but also the massed extracellular potential of the electroretinogram (ERG). Thus, it provides a general tool for non-invasive and comparative assessment of the amplification of photoreceptor responses in many species, including humans.

Perhaps the main reason for the ready applicability of Eq. (25) to experimental results arises from the fact that only a single parameter, A, is required to fit (and to explain) the entire family of responses over a wide range of stimulus intensities. What this means is that all the parameters of the molecular description – the protein concentrations and diffusion coefficients, the enzyme activities, the physical dimensions of the cell, the buffering power, and the channel gating properties – collapse down into just a single parameter, A. The way in which these parameters coalesce may be appreciated by substitution of β_{sub} from Eq. (11) into Eq. (26), to obtain

$$A = \nu_{RE}\left(\frac{\frac{1}{2}k_{cat}/K_m}{N_{Av}V_{cyto}B_{cG}}\right)n_{cG} \tag{27}$$

Figure 6 compares the predictions of Eq. (25) with experimental results from a salamander rod and a mammalian rod using the suction pipette recording technique, while Fig. 7 illustrates the comparison for the *a*-wave of the ERG for both rod-driven and cone-driven responses from human and mouse retina.

For the dark-adapted salamander rod in Fig. 6A, the predictions of Eq. (25) provide a good description of the rising phase of the response, over the entire range of intensities, from about 10 to 2000 photoisomerizations per flash, for the first 500 ms or so. At later times, however, the theoretical predictions diverge from the experimental traces, as the onset of inactivation reactions renders the simple model invalid. In this experiment, the calcium buffer BAPTA had been incorporated into the cytoplasm in order to retard the decline in $[Ca^{2+}]_i$ that normally accompanies the light response [65]. In the same cell under control conditions (prior to incorporation of BAPTA) the responses began diverging from theory at even earlier times (see Fig. 7A of [26]), showing that the earliest signs of inactivation under normal conditions are calcium mediated. Similar fits with changes in $[Ca^{2+}]_i$ pre-

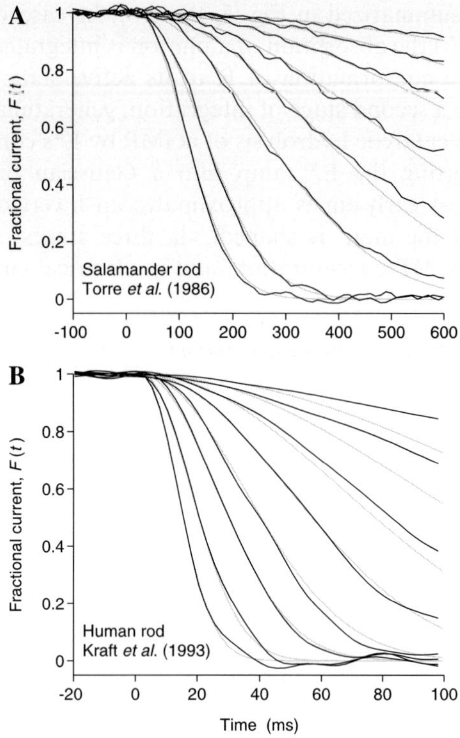

Fig. 6. Comparison of the predictions of Eq. (25) with experimentally recorded suction pipette photocurrents, for a salamander rod and a human rod. In both panels, Eq. (25) has been plotted using the calculated number of photoisomerizations, Φ, and a single value of the amplification constant and effective delay. (A) Recordings from the salamander rod illustrated in Fig. 3A, following incorporation of the calcium buffer, BAPTA; data from Fig. 2B of Ref. [65]. Dark current, −29 pA; amplification constant, $A = 0.065\ \mathrm{s}^{-2}$; delay time, $t_{\mathrm{eff}} = 20$ ms. (B) Recordings from a human rod, for flashes delivering from 34 to 3800 isomerizations; data from Fig. 4 of Ref. [53], kindly supplied by Dr. J.L. Schnapf. Dark current, −13.5 pA; amplification constant, $A = 2\ \mathrm{s}^{-2}$; delay time, $t_{\mathrm{eff}} = 2.2$ ms. In both panels the origin of time has been set to the middle of the light flash, after allowance for the delay introduced by electronic filtering.

vented have been illustrated in Figs. 10 and 11 of Ref. [120] and in Fig. 4 of Ref. [142].

Figure 6B illustrates suction pipette recordings from a human rod (from Fig. 4 of Ref. [53]). Apart from the much faster time scale (due to the higher amplification constant, A), the form of the responses is remarkably similar to that of the amphibian rod, and again the rising phase of the whole family of responses is well described by Eq. (25), out to at least 50 ms.

Figure 7 shows recordings of the a-wave of the electroretinogram (ERG) from an anesthetized mouse, under conditions that isolate rods (Fig. 7A), and from a human subject for conditions that isolate rods (Fig. 7B) or cones (Fig. 7C). There is

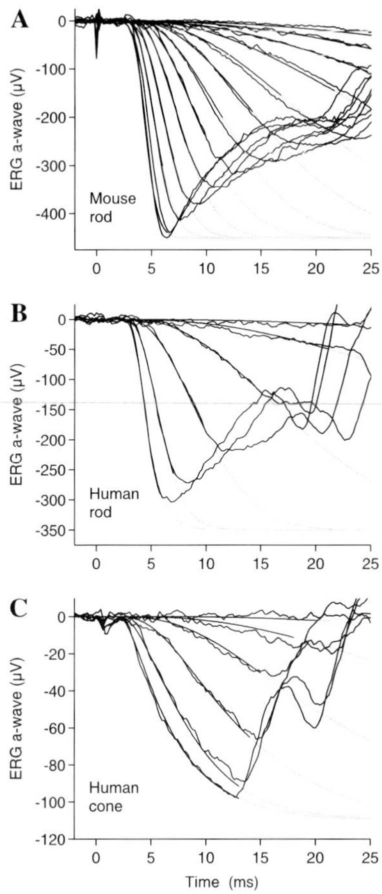

Fig. 7. Comparison of the predictions of Eq. (25) with experimentally recorded ERG *a*-waves. Prior to intrusion by the *b*-wave, the *a*-wave provides a measure of the circulating current in the photoreceptors. In each panel, Eq. (25) has been plotted using the number of isomerizations, Φ, together with constant values for the amplification constant, membrane time constant, and effective delay. (A) Mouse rod *a*-waves; data from Figs. 2 and 3 of Ref. [141]. Amplification constant, $A = 5.6\ \mathrm{s}^{-2}$; membrane time constant ignored; delay time, $t_{\mathrm{eff}} = 3.6$ ms; maximal response, $a_{\max} = -450\ \mu\mathrm{V}$. (B) Human rod *a*-waves; data from Fig. 5 of [139]. Amplification constant, $A = 4.5\ \mathrm{s}^{-2}$; membrane time constant, $\tau_{\mathrm{m}} = 1.1$ ms; delay time, $t_{\mathrm{eff}} = 2.3$ ms; maximal response, $a_{\max} = -350\ \mu\mathrm{V}$. (C) Human cone *a*-waves; data from Fig. 6 of Ref. [139]. Amplification constant, $A = 6.7\ \mathrm{s}^{-2}$ (assuming that the conversion factor from photopic Td s to photoisomerizations per red-sensitive cone is $K_{\mathrm{red}} = 200$ isomerizations per cone per Td s; see Ref. [139]); membrane time constant, $\tau_{\mathrm{m}} = 4.4$ ms; delay time, $t_{\mathrm{eff}} = 1.5$ ms; maximal response, $a_{\max} = -55\ \mu\mathrm{V}$ in both the red- and green-sensitive cones. Eq. (25) is plotted solid for the early times over which fitting was performed, and is shown as gray thereafter.

G*–E*. The critical feature in termination of activity is that RGS9–Gβ5 binds to (Gα–GTP)–PDEγ to form (at least transiently) a quaternary complex: RGS9–Gβ5–(Gα–GTP)–PDEγ. Formation of this complex may allow access of water to the GTP-binding site, which (unlike the representation in the simplified schematic in Fig. 4) is located in the interior of the Gα subunit. It is probably most realistic to view this interaction as tripartite, because the RGS9–Gβ5 complex is so tightly bound that it can effectively be regarded as a single entity [92].

The requirement for concerted action of three distinct entities is almost certainly related to the need to time the shut-off of G*–E*. The slow GTPase rate intrinsic to G* allows high efficiency of cascade activation, by ensuring that excitation does not normally decay until the G* has bound to the PDE effector enzyme. However, once the PDE has had an opportunity to function, then it is critical that the activity of the cascade is terminated rapidly, so that the photoreceptor can recover quickly and be ready to respond to further illumination. The employment of a third protein complex (the RGS9–Gβ5) to detect and quench the active state of G*–E* seems ideally suited to achieving the desired timing [173].

Phosducin. Phosducin (PD) is a 28 kDa soluble 'phospho-protein' that binds tightly to Gβγ when Gα is not attached [174–178]. Although no function has yet been established for PD in phototransduction in living rods, it bears mention for several reasons. First, it is present in mammalian rod outer segments at a concentration at least as high as the G-protein. Second, its capacity to bind to Gβγ is regulated by a cycle of phosphorylation–dephosphorylation: Gβγ only binds to phosducin when PD is unphosphorylated [95]. Third, the phosphorylation is regulated by protein kinase A (PKA), which is in turn regulated by CM [179]. Fourth, PD is a member of a family of phosducin-like phospho-proteins ('Phlops') which appear to play modulatory roles in other G-protein cascades (e.g., [180,181]).

Phosducin could function during light adaptation to lower the amplification constant of mammalian rods (Section 5.7) by effectively removing holo-G-protein from the available pool. While there is at present no experimental evidence to support this hypothesis, it remains possible that such a mechanism could be used by the rod primarily when it is in saturation, and therefore not signaling. Lowering the gain of the cascade while the rod is in the saturated, non-signaling state would conserve metabolic energy that would otherwise be spent in vain.

6.3. Regulation of cGMP synthesis: Guanylyl cyclase (GC) and its activating proteins (GCAPs)

Guanylyl cyclase. In order to maintain a cGMP concentration of the order of 1 μM in the face of hydrolysis by PDE, either in the dark or in the presence of steady light, it is necessary for synthesis of cGMP to occur. This is mediated by guanylyl cyclase (GC), an enzyme that synthesizes cGMP from GTP. Two general classes of guanylyl cyclase exist – soluble and membrane-bound ('particulate') – and the rod GC is of the latter class. The rod guanylyl cyclase appears to function as a dimer, formed by a pair of identical 110–114 kDa proteins, and it is present in the disc membrane at a concentration (surface density) of around 0.2–0.5% that of rhodopsin – i.e., from around 1 GC:500 R (Table 3) to 1 GC:200 R (K.-W. Koch, personal communication).

At least two distinct particulate guanylyl cyclases (termed GC1 and GC2) have been found in photoreceptors, but their functional differences and distinctive roles (if any) in rods and cones are not yet clear. Both have been identified in a number of species, where they have been given various names. GC1 has been localized to the outer segments of both rods and cones [182–184], and its properties measured *in vitro* appear consistent with the GC activity deduced from electrophysiological experiments on intact and truncated rods (reviewed in Ref. [81]). GC2 has been localized to the outer segments of rods [185], and appears to have properties broadly similar to, though distinct from, those of GC1. Recent work in which GC–E, the rodent homologue of GC1, has been knocked out in mice, shows that rods lacking GC1 can nonetheless generate normal circulating currents, suggesting the possibility that GC2 either accounts for basal cyclase activity, or can substitute for GC1 [186]. Interestingly, cones degenerate in the GC–E knockout mice at 4–5 weeks, suggesting that GC–E is necessary for cone function [186]. This latter result is consistent with previous evidence that homologues of GC1 are strongly expressed in the cone outer segments of some species [184].

A fundamentally important feature of photoreceptor GCs is their regulation by intracellular Ca^{2+} concentration. Although it has been shown that other particulate GCs, with weak homology to the rod GC, are regulated by extracellular signals such as ANF [187], the only established regulation of rod outer segment guanylyl cyclase activity *in situ* is by either of two calcium-binding 23 kDa 'guanylyl cyclase activating proteins' (GCAP1 and GCAP2) [188–190]. Thus, it is GCAPs that confer functional Ca^{2+}-sensitivity on guanylyl cyclase.

Guanylyl cyclase activating proteins. GCAPs are members of a large family of calcium-binding proteins, which includes CM and Rec. The mode of GCAP's action in stimulating guanylyl cyclase is illustrated schematically in Fig. 4A. In the dark, when $[Ca^{2+}]_i$ is relatively high (Table 2), a high proportion of GCAP molecules have Ca^{2+} ions bound to them, and in this form they remain soluble in the cytoplasm and do not interact with the guanylyl cyclase. But when the Ca^{2+} concentration declines during the light response, Ca^{2+} ions unbind and the calcium-free form of GCAP is able to bind to a cytoplasmic site on the GC, thereby switching on its enzymatic activity. The quantitative details of this activation will be presented in Section 7.5.

Physiological and biochemical investigations had established the logical need for, and the existence of, a calcium-dependent regulator of the rod guanylyl cyclase a decade before the first such protein, GCAP1, was positively identified [188,190]. A second and distinct GC regulator, GCAP2, was soon discovered [191,192]. Although both types of GCAP have been shown to regulate the guanylyl cyclase activity of rod outer segments, it has been established through work with recombinant GC mutants that they act at distinct cytoplasmic epitopes, with GCAP1 acting near the N-terminal and GCAP2 acting near the C-terminal [193,194]. Both GCAPs activate GC1, but the saturated activity evoked by GCAP2 is higher than that evoked by GCAP1 [193,194]. Investigations underway in several laboratories, using mutant mice with different combinations of GCs and GCAPs knocked out, should soon help to resolve the roles of these four proteins in mammalian rod phototransduction.

In cones, too, guanylyl cyclase is of course present and regulated by calcium [56,195]. In contrast with other enzymes of the transduction cascade, which are expressed as distinct isoforms in rods and cones, it seems that GC1 is expressed in both classes of cell [182,183,196], along with GCAP1 [197].

6.4. Ca^{2+} efflux: The Na^+/Ca^{2+}, K^+ exchanger (NCKX)

Virtually all tissues seem to express one or more forms of Na^+/Ca^{2+} exchanger (NCX); in particular, the NCX1 gene has been reported to be expressed ubiquitously in mammalian tissues, along with several isoforms and splice-variants [198–202]. Na^+/Ca^{2+} exchangers use the energy stored in the transmembrane Na^+ concentration gradient to extrude Ca^{2+} against its electrochemical gradient, which exceeds 1000-fold in most cells. The standard stoichiometry of operation for the Na^+/Ca^{2+} exchanger is 1 Ca^{2+} extruded in exchange for 3 Na^+ ions entering. The NCX molecule is thought to comprise 11 transmembrane segments, and contains a large hydrophilic domain between the fifth and sixth segments that appears likely to serve in ion translocation [200–202].

The exchanger in the rod (and probably cone) outer segment is of a distinct variety from the NCX exchanger, because it has an obligatory requirement for internal K^+; to highlight this distinctive requirement, we adopt the label 'NCKX'. The normal stoichiometry of the rod exchanger was established to be 1 Ca^{2+} and 1 K^+ extruded, in exchange for 4 Na^+ ions entering [203]. The functional utility of the distinctive stoichiometry was proposed to be its ability to reduce the cytoplasmic calcium concentration to a lower level than was possible with the conventional exchanger [203]. In the absence of any other Ca^{2+} influx or efflux pathways (e.g., when the cGMP-gated channels are completely closed), the NCKX on its own could theoretically set a final steady-state level of internal free calcium concentration, $[Ca^{2+}]_i$, given by the relation

$$[Ca^{2+}]_i = [Ca^{2+}]_o \frac{[Na^+]_i^4 [K^+]_o}{[Na^+]_o^4 [K^+]_i} \exp(V_m F/RT) \tag{28}$$

(analogous to the Nernst potential relation). The fourth-power dependence of $[Ca^{2+}]_i$ on the inward Na^+ concentration gradient, and the assistance provided by the outward K^+ gradient, in principle enable the rod to achieve $[Ca^{2+}]_i$ concentrations less than 1 nM in the light, as much as 500-fold lower than would be possible if the standard NCX stoichiometry applied [203]. Although it is extremely doubtful that this theoretical limit could be approached under any physiological conditions, the calculation nonetheless shows the potential utility of the NCKX mechanism. A low $[Ca^{2+}]_i$ in the light is needed to ensure that guanylyl cyclase is maximally activated, and might also serve to maximally engage other calcium-dependent inactivation processes, as discussed below.

The genes for bovine and human photoreceptor NCKX1 have been cloned [204,205], and while hydropathy plots are consistent with the 11-transmembrane segment model of the NCX family, the sequence homology between the two gene families is quite small except in the transmembrane segment region. The first functional expression of the bovine NCKX in a heterologous system could not demonstrate the K^+ sensitivity that is the hallmark of the exchanger [206], but this sensitivity has been found in a more recent heterologous expression of dolphin NCKX [207].

For further review, see Refs. [41–43,200,202,208,209].

6.5. Regulation of the cyclic GMP gated channel by CM

Calmodulin (CM), a ubiquitous calcium binding protein with four functional EF-hand motifs, is present in frog and bovine rods at a concentration of 5–10 μM [94,210,211]. Calmodulin binds four Ca^{2+} ions in a highly cooperative manner, with a $K_{\frac{1}{2}}$ of about 2.4 μM [212]. The calcium-associated form, Ca–CM, binds to the cGMP-gated channels of rods, and this binding increases the affinity of CM for Ca^{2+} by 10-fold or more [94], with the result that most of the channels have Ca–CM bound in the dark-adapted state. There are two sites for CM binding on the rod channel, on the N- and C-terminal cytoplasmic tails of the β subunit [213,214]. The binding of Ca–CM to the channels increases the K_{cG} of the channels for cGMP by about 1.5-fold, without altering the Hill coefficient or the maximal current [94,215,216].

From these properties of CM and its interaction with the cGMP-gated channels, it can be predicted that the dissociation of CM from the channels at lowered $[Ca^{2+}]_i$ will, by increasing their affinity for cGMP, serve to restore some of the circulating current that would otherwise be suppressed during exposure to steady light. In rods, though, this restoration effect is predicted to be relatively small [94,136,217]. For cone channels, a related but much more powerful mechanism exists, that utilizes a calcium-binding protein distinct from CM [218].

7. Recovery phase of the response: Predicted form of flash families

7.1. Equations for the inactivation of R and of G*–E**

*Inactivation of R**. As discussed in Section 6.1, the catalytic activity of R* is decreased by RK-mediated phosphorylation and by the subsequent binding of arrestin. The phosphorylation of a single R* is inherently a stochastic event: that is, the attachment of a single phosphate is a discrete (rather than a gradual) event, and furthermore it will occur not at a precisely repeatable time after each photoisomerization, but at a time exhibiting some randomness (to a degree that will depend on the precise nature of the mechanisms involved). From a simple model analogous to that in Section 4.2 (Eqs. (2) and (3)), the mean time to first encounter between an R* and an RK would be of the order of 10 ms in an amphibian rod, and 3–4 ms in a dark-adapted mammalian rod (assuming the density to be 1 RK per 500 rhodopsin molecules; Table 3). However, at least 70% (and perhaps more than 90%) of the RK is expected to be complexed with Rec–2Ca in the dark-adapted rod, and any encounter between R* and Rec–2Ca–RK is unlikely to be successful, since the latter does not phosphorylate R* (Section 6.1). Allowing for the amount of RK complexed with Rec–2Ca, the time to initial encounter between R* and a free RK in a dark-adapted rod may exceed 100 ms in amphibian rods and tens of milliseconds for mammalian rods. Since it is unlikely that every encounter between an R* and an RK will lead to binding and phosphorylation, it follows that the time to a 'productive' encounter could be several hundred milliseconds in amphibian rods, and several tens of milliseconds in mammalian rods.

Likewise, the binding of arrestin to R*–P would be a stochastic event, though the expected time for initial encounter would be much shorter, due to the much higher concentration of Arr. It follows that once an R*–P is produced, its capping by arrestin is likely to occur relatively rapidly.

Although it is critical to take account of the stochastic nature of R* shut-off when analyzing single-photon events, the responses to flashes delivering many photons (say 10 or more) can be treated by the much simpler 'mean' behavior. In particular, if one assumes that the stochastic termination of R* activity occurs with a rate constant that is independent of time, then the average time-course of $R^*(t)$ following a brief flash will be a simple exponential decay, characterized by an effective time constant τ_R (see Fig. 9A). This assumption clearly represents an oversimplification of the actual molecular mechanism (as discussed in Section 8), it nevertheless constitutes a productive starting point for analysis [142].

Inactivation of G–E*.* As discussed in Section 6.2, it is now apparent that under physiological conditions inactivation of G* does not occur significantly until the G* is bound to PDEγ (in the active E* complex), and in addition a second GAP factor, the RGS9–Gβ5 complex, also binds. Although the inactivation of each individual G*–E*–RGS9–Gβ5 ternary complex will again be stochastic in nature, the time-course of $E^*(t)$ can be treated by mass action principles because there will typically be at least hundreds of activated E* complexes present during most of the response to a single R*. In other words, if a population of E*s were created instantaneously at $t = 0$, then as time proceeded the average number still active would decay exponentially with a time constant τ_E.

Combining the mass action inactivation reactions for R* and G*–E*, the equation for $E^*(t)$ derived previously in Eq. (5) in the absence of inactivation can be replaced by

$$E^*(t) = \Phi \nu_{RE} U_{RE}(t - t_{RGE}) \tag{29}$$

where the function $U_{RE}(t)$ represents the convolution of a ramp with two first-order time constants τ_R and τ_E, and is defined by

$$U_{RE}(t) = \frac{\exp(-t/\tau_R) - \exp(-t/\tau_E)}{\tau_R - \tau_E}\tau_R\tau_E, \quad \text{for } t > 0 \tag{30}$$

Equation (29) is plotted in Fig. 9B. Note that as $t \to 0$, when the exponential terms admit first-order expansion, $U_{RE}(t) \to t$ (i.e., a unit ramp), so that at very early times Eq. (29) is identical to Eq. (5).

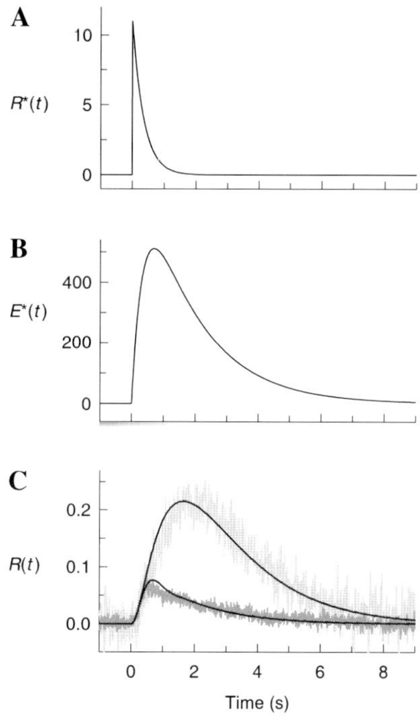

Fig. 9. Predicted kinetics of recovery for R*, E* and the electrical response. Compare these traces with the pure activation model in Fig. 5. (A) Following photoisomerization, the average number of R*s is predicted to decline exponentially, with time constant τ_R. (B) The number of activated PDEs, $E^*(t)$, as predicted by Eq. (29). The initial rising phase has exactly the same slope, $\Phi\ \nu_{RE}$, as in the case of pure activation. The shape of the function is derived from the convolution of the R* activity in A with the impulse response for E* activity, which has a time constant of τ_E. (C) Electrical response, in the cases where calcium is either clamped (upper trace) or free to change (lower trace), as predicted by numerical integration of Eqs. (16) and (31), with E* time course as in panel B (Eq. (30)), and with Eqs. (22), (32) and (33). The noisy traces are normalized responses of a salamander rod to stimuli estimated to deliver 11 photoisomerizations per flash, in calcium clamping solution (larger response) and in Ringer's solution (smaller response); data from Fig. 13 of Ref. [142]. For the theoretical traces in this figure, the parameters of activation were: $\Phi = 11$ photoisomerizations; $\nu_{RE} = 150\ s^{-1}$; $A = 0.16\ s^{-2}$ (implying $\beta_{sub} \sim 4 \times 10^{-4}\ s^{-1}$). The primary parameters of inactivation were: $\tau_R = 0.35$ s; $\tau_E = 1.7$ s; $1/\beta_{dark} = 0.9$ s; the additional parameters specifying calcium feedback are given in Ref. [142].

Limitations. As mentioned above, the assumption of first-order termination of R* activity ignores any dynamic modulation of the R* shut-off reaction. This issue will be examined in Sections 8.3.3 and 9.3, where we consider the possibility of Ca^{2+} feedback onto R* lifetime via recoverin and RK. The assumption of first-order termination also ignores any stochastic features of R* inactivation, which will clearly be important for the description of the single-photon response. The stochastic termination of R* activity may in part be governed by the calcium-recoverin feedback mechanism, and such feedback will likely further result in deviation from exponential decay. This issue will be examined in Section 9.

Another limitation may be imposed by the finite quantity of RGS9–Gβ5 and PDE (Table 3). Thus, for flashes of such intensity that amounts of G* are generated in excess of the total quantities of either of these two proteins, the inactivation time of the 'excess' G* may be delayed relative to the time, τ_E, that applies at lower flash intensities.

A third limitation is inherent in the fact that the time constants in Eqs. (29) and (30) are interchangeable. Thus, the predicted form of $E^*(t)$ remains unaltered if τ_R and τ_E are interchanged. Although the fitting of Eq. (29) to the results (see Figs. 9 and 10) provides the magnitudes of two time constants, it is not possible to determine which corresponds to τ_R and which to τ_E, without additional evidence [142].

7.2. Equations for cGMP concentration

The relation governing cGMP hydrolysis and synthesis, as expressed in Eq. (13) or (16), is completely general to the lumped case, and needs no modification in order to describe response recovery. However, several ancillary relations and parameters implicit in the solution of the differential equation may require modification from those used in the case of pure activation. One such parameter is the incremental cyclase activity, $\Delta\alpha(t)$, which was previously assumed to be negligible, but which will now require explicit formulation (Section 7.4), because of the dynamic change in $[Ca^{2+}]_i$ that accompanies the light response.

7.3. Equations for calcium fluxes and concentration

On the assumption that the only calcium fluxes into or out of the cytoplasm are carried by the cGMP-gated channels and the NCKX exchanger, and that the outer segment behaves as a well-stirred compartment, the differential equation governing the free calcium concentration can be expressed as

$$\frac{\mathrm{d}Ca}{\mathrm{d}t} = -\frac{\frac{1}{2}f_{Ca}J_{cG}(t) - J_{ex}(t)}{\mathcal{F}V_{cyto}B_{Ca}} \tag{31}$$

where Ca symbolizes $[Ca^{2+}]_i$, J_{ex} is the electrogenic exchange current, f_{Ca} is the fraction of the current, J_{cG}, through the cGMP-gated channels that is carried by Ca^{2+}, B_{Ca} is the buffering power of the cytoplasm for calcium, and $\mathcal{F}$ is Faraday's constant (the charge carried by a mole of monovalent cations) [50,70,121,136]. Note that the minus sign multiplying the equation arises from the convention that current flowing outward across the membrane is positive; thus J_{cG} and J_{ex} are normally both negative. The factor multiplying $f_{Ca}J_{cG}$ is $\frac{1}{2}$ because each Ca^{2+} ion carries two charges, while the factor multiplying J_{ex} is unity because the extrusion of each Ca^{2+} ion is accompanied by the net movement inward of a single positive charge, i.e., $4Na^{+} - (Ca^{2+} + K^{+})$.

An expression for J_{cG} was derived in Section 5.6, while J_{ex} has been found to follow thc Michaelis saturation relation [70]

$$\frac{J_{ex}(t)}{J_{ex,sat}} = \frac{Ca(t)}{Ca(t) + K_{ex}} \tag{32}$$

where $J_{ex,\ sat}$ is the maximal exchange current, and K_{ex} is the half-saturating Ca^{2+} concentration of the exchanger. For salamander rods K_{ex} has been estimated as 1.6 μM, and $J_{ex,sat}$ as about 20 pA [70].

7.4. Equations for calcium-dependent activation of guanylyl cyclase

The relationship between free calcium concentration and the activity of GC can be expressed as

$$\alpha(t) = \alpha_{min} + \frac{\alpha_{max} - \alpha_{min}}{1 + (Ca(t)/K_{cyc})^{n_{cyc}}}. \tag{33}$$

Here α_{min} is the residual component of cyclase activity at very high calcium concentrations, and $(\alpha_{max} - \alpha_{min})$ is the calcium-sensitive component of cyclase activity, while K_{cyc} is the calcium concentration for half-maximal activation, and n_{cyc} is the cooperativity coefficient. This equation has been shown to provide a good account of rod GC activity in a number of *in vitro* measurements, using $\alpha_{min} = 0$. Eleven studies of mammalian rod GC near body temperature reported values of K_{cyc} ranging from 30 to 280 nM (154 ± 95 nM, mean ± S.D.), and six of the studies provided estimates of n_{cyc} ranging from 1.7 to 3.9 (2.0 ± 0.3, mean ± S.D., after exclusion of a single outlier at 3.9) (reviewed in Ref. [81]).

A limitation of most *in vitro* studies of GC activity is that the absolute rates are not readily quantifiable in units applicable to physiology; e.g., in μM s^{-1} with respect to the cytoplasm of a single rod. However two investigations of amphibian rods deserve special mention, because they overcame these limitations. Experiments with truncated and dialyzed toad rods gave estimates of $\alpha_{max} \approx 20$–30 μM s^{-1}, $\alpha_{min} \approx 0$, $K_{cyc} = 100$ nM, and $n_{cyc} = 2$ [121], while experiments with fractured bullfrog rods gave estimates of $\alpha_{max} \approx 10$ μM s^{-1}, $\alpha_{min} = 1.9$ μM s^{-1}, $K_{cyc} = 250$ nM, and $n_{cyc} = 1.8$ [122]. Other studies using intact vertebrate rods are reviewed in Table II of Ref. [81].

7.5. Solution for the flash response in the presence of inactivation reactions

The solution for the flash response in the presence of inactivation reactions can be obtained by solving the differential equation for cGMP metabolism, Eq. (16), with the 'output' given by the CNG channel relation, Eq. (22), with the time-course of E* specified by Eqs. (29) and (30) and the PDE hydrolytic activity given by Eq. (12), and with 'calcium feedback' specified by Eqs. (31)–(33). Solutions have recently been obtained, both analytically and numerically, in the interesting (but also complicated) case where the internal calcium concentration is free to change, as well as in the more straightforward case where the calcium concentration is clamped [142].

The predicted time-course for R*, E*, and the electrical response, are presented graphically in Fig. 9.

Protein activation and recovery. Fig. 9A plots the exponential decline in R* activity predicted for the population of molecules, while Fig. 9B plots the time-course of E* activity, described by the function $U_{\mathrm{RE}}(t)$ in Eq. (30). In the language of systems theory, this equation is the convolution of two exponentially decaying functions: the R* activity (with time constant τ_{R}) and the E* activity (with lifetime τ_{E}).

It is interesting to compare the predictions for R^* and E^* in Figs. 9A and 9B with the corresponding predictions in Fig. 5, where inactivation was ignored. In the absence of inactivation, each photon generates a step increase in R*, and this increase is maintained indefinitely; in the presence of R* inactivation, the mean number of active molecules, $R^*(t)$, declines exponentially with time constant τ_{R}. In both Figs. 5A and 9B the activity of E* is predicted to begin rising linearly (i.e., as a ramp with time). The effects of inactivation are only seen at later times, and $E^*(t)$ is found to recover with a final time constant given by the slower of τ_{R} and τ_{E}.

cGMP and the electrical response. The time-course of PDE hydrolytic activity, defined by the PDE time course in Fig. 9B, then acts as the 'driving function' for the hydrolysis of cGMP, and the resulting cGMP concentration determines the electrical response. In the simpler of the two cases that we wish to consider, the cytoplasmic calcium concentration is clamped, so that no change in the cyclase activity is permitted; i.e., $\Delta\alpha(t) = 0$. In this case the predicted time-course of cGMP concentration is obtained as the convolution of the incremental PDE activity, $\Delta\beta(t)$, with the impulse response of a 'first-order filter' with time constant $1/\beta_0$. The resulting expression for $\Delta cG(t)$ therefore corresponds to a cascade of three first-order reactions, with time constants of τ_{R}, τ_{E} and $1/\beta_0$. This solution can be expressed as the weighted sum of three exponential terms with these time constants (Eq. (19) of Ref. [142]).

In the normal (unclamped) case, where the cytoplasmic calcium concentration is free to change, the equations for calcium metabolism, (31)–(33), must also be included, in order to predict the response. The analysis in this case is more complicated, and the small signal response (obtained by deriving linear approximations for the equations) is intrinsically second order [35,63,142,220]. The solution for $\Delta cG(t)$ is given by Nikonov et al. (Eq. (20) of Ref. [142]; see also Eq. (19) of Ref. [63]).

The kinetics of the electrical response predicted in the two cases (calcium clamped and calcium free to change) are shown in Fig. 9C as the smooth traces (where they are also compared with experimental responses obtained under the respective conditions; noisy traces). Importantly, the early rising phase of the predicted response is the same in the two cases, and furthermore is the same as in the absence of inactivation. Thus, as in Fig. 5C, the electrical response at early times is again predicted to follow the delayed Gaussian time-course, $R(t) \approx \frac{1}{2}\Phi A t^2$ (where the short delay t_{eff} has been ignored).

In both cases, the driving function of PDE activity, $\Delta\beta(t)$, is the same, as illustrated by the trace in Fig. 9B. In the calcium-clamped case, the cGMP concentration is simply the convolution of this trace with an exponential decay of time constant

$1/\beta_0$. In the unclamped case, $\Delta\beta(t)$ is instead convolved with a second-order function, representing the impulse response of the feedback loop: cGMP → current → Ca^{2+} → cyclase → cGMP (see Eq. (A6.8) of Ref. [142]). Thus, the differences between the two predicted curves in Fig. 9C correspond to activation of the guanylyl cyclase feedback loop. The intrinsically second-order character of the equations describing the feedback loop has been noted by other investigators [35,63,220].

In both cases, the steady-state cGMP-hydrolysis rate constant β_0 has an important role in shaping the dim-flash responses. In the calcium-clamp condition its reciprocal, $1/\beta_0$, is one of the three time constants in the cascaded reactions, and it therefore affects both the speed and amplitude of the response. In the normal Ringer's condition the role of β_0 is less obvious, because of the complexity of the relations governing the feedback to the cyclase, i.e., Eqs. (31)–(33). Nevertheless, it can be shown that β_0 retains an important role in determining the shape and amplitude of the dim-flash response, particularly in the presence of steady illumination, which causes β_0 to increase above its dark-adapted value β_{dark} (see Section 8). At very late times, the final form of recovery of the electrical response is determined by the slowest of the three time constants τ_R, τ_E and a third time constant, which can be shown to be smaller than $1/\beta_0$ (Eq. (6.8) of Ref. [142]).

Brighter flashes: non-linearity. For brighter flashes the predicted responses cannot be obtained analytically, because of the existence of non-linearities in the set of equations, and instead numerical integration is required. In Fig. 10 the responses of a salamander rod at a range of sub-saturating flash intensities are compared with the predictions of numerical integration [142].

7.6. Validity of the solutions, and limitations

Perhaps the most important caveat in the theoretical prediction of the flash response is the approximation of the decay of R* activity as a simple exponential with time constant τ_R. This issue will be considered further in Section 8.

Another concern is that the Michaelis relation in Eq. (32) might oversimplify the behavior of the exchange current, which exhibits at least two components of decay following a saturating flash [73,74,221]. This does not in fact appear to be a shortcoming. Rather it seems likely that the occurrence of distinct components in the decay of exchange current results from the release of Ca^{2+} from binding sites with different affinities, and that this in no way conflicts with Eq. (32). A distinct but related concern stems from the voltage-dependence of the exchange current [70]. Although the hyperpolarization of the rod can be neglected in computing the decline in cGMP-gated current (Eq. (22)), because of the shallowness of the current-voltage relation for the cGMP-gated channels, the corresponding approximation may not be appropriate when considering the behavior of the NCKX. However, neglect of the exchanger's voltage-dependence will cause least error in the case of dim flashes, which elicit minimal hyperpolarization.

Another concern, mentioned in the presentation of Eqs. (29) and (30), is the allocation of the two time constants to τ_R and τ_E. This issue will be examined in Section 8.3.3, after we review additional evidence relating to light adaptation.

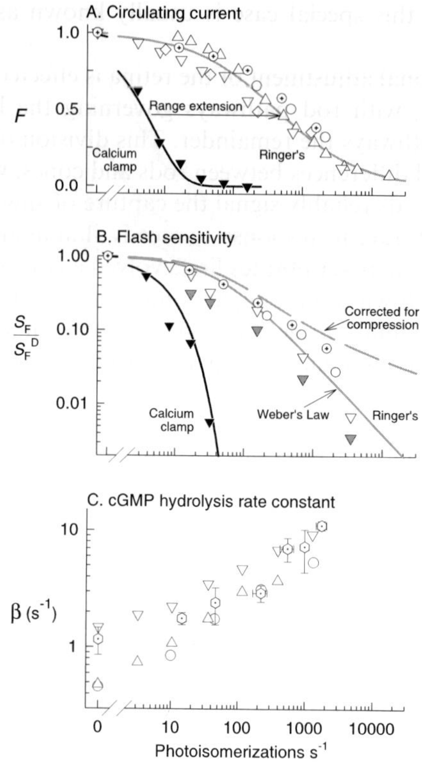

Fig. 11. Normalized circulating currents (A), flash sensitivities (B) and steady-state cGMP hydrolysis rate constants (C) of salamander rods as a function of the intensity of steady illumination (note that all panels share a common abscissa). Open symbols represent data obtained in Ringer's solution; the filled triangles represent data obtained in a solution that held $[Ca^{2+}]_i$ at its resting (dark) level, while the gray-filled symbols in panel B illustrate data obtained in solutions that held $[Ca^{2+}]_i$ at the lowered levels set during a prior exposure (in Ringer's solution) to a steady background of the given intensity. The data were taken from the following studies. In panels A, B: ◇ from Fig. 2 of Ref. [118]; ▼ and ▽, Fig. 2 of Ref. [237]; open up triangles, Fig. 9 of Ref. [136]; ○, ⊙, from Ref. [119]. In panel C: ○ from Fig. 2 of Ref. [118], △, ▽ from Figs. 5 and 7 of Ref. [238]; and ⊙ [119]. For the octagons, the vertical error bars are SDs of measurements from at least three different rods; the horizontal error bars are SDs of the intensities over which measurements were pooled. The smooth curves through the data are empirical and serve to illustrate the main trends of the data. In panel A, the curve through the calcium-clamped data has the formula $F = \exp(-I/I_0)$, with $I_0 = 7$ photoisomerizations s^{-1}, and the curve drawn through the Ringer's data in panel A has the form $F = I^n/(I^n + I_0^n)$, with $I_0 = 425$, $n = 0.7$. In panel B, the curve near the filled triangles has the same equation as that through the corresponding data in panel A, while the curve near Ringer's data is given by Weber's Law, $S_F/S_F^D = I/(I + I_0)$, with $I_0 = 40$ photoisomerizations s^{-1}; the 'corrected' curve was obtained by dividing Weber's Law by the smooth curve through the Ringer's data in panel A.

8.2. Role of calcium

One idea that has pervaded the literature on light adaptation is that the reduction in calcium concentration *causes* the decline in flash sensitivity, but this simplistic idea does not survive rigorous examination. Figure 11A illustrates that when changes in $[Ca^{2+}]_i$ are prevented, the rod is driven into saturation at intensities of about 50 photoisomerizations s^{-1}, and Fig. 11B shows that when this occurs the cell's sensitivity is drastically attenuated. Thus, in the absence of changes in calcium concentration, the cell's sensitivity is reduced enormously. In contrast, when the cell's calcium concentration is free to change, saturation does not occur at such low intensities and the cell instead desensitizes in a gradual manner with increasing light intensity. From this perspective it is more reasonable to take the view that the normal decline in $[Ca^{2+}]_i$ *rescues* the cell from saturation, and prevents its sensitivity from being annihilated at low backgrounds, thereby extending the operating range (to nearly 10,000 photoisomerizations s^{-1} in Fig. 11). In other words, the decline in $[Ca^{2+}]_i$ does not actually *cause* the desensitization – rather, it tends to do exactly the opposite.

This viewpoint also makes it clear that mechanisms that extend the cell's range of operating intensities need not necessarily reduce its sensitivity, although conversely mechanisms that directly reduce the sensitivity are almost certain to extend the operating range. With this distinction between extension of operating range and reduction of gain in mind, it is useful to analyze the individual mechanisms underlying light adaptation. Accordingly, Table 5 gathers together current hypotheses about the molecular mechanisms underlying light adaptation, and considers the explicit effect of each mechanism on (i) the operating range, and (ii) the sensitivity, of phototransduction [119,222]. In addition it specifies (iii) whether the mechanism is directly activated by the decline in $[Ca^{2+}]_i$.

8.3. Analysis of individual mechanisms underlying light adaptation

We shall now consider the nine potential mechanisms of light adaptation listed in Table 5, in the order of the proteins underlying the effects. We begin with one that does not involve any particular protein.

8.3.1. Response compression (row 1 of Table 5)

As a consequence of the partial closure of cGMP-activated channels during steady illumination, the circulating current available for modulation by an incremental flash is reduced. In order to 'correct' for this reduction, so as to uncover the state of the transduction mechanism, it turns out that one simply needs to normalize the circulating current $J(t)$ with respect to its value J_0 in the steady state prior to the flash [65].

The basis for this statement may be seen most readily in the case of calcium clamping. Section 5.5 showed that the *fractional* change in cGMP concentration should be independent of the initial cGMP level. Thus, Eq. (18) showed that for a given driving function $\Delta\beta(t)$ the normalized solution $cG(t)/cG_0$ will be independent of cG_0, or of any other aspect of the adaptational state. Hence, in the absence of any adaptational process other than output compression, the normalized current response, $r(t) = 1 - F(t)$, at early times after a flash of constant intensity should be independent of background intensity (Eq. (23)).

In Fig. 11B the correction has been applied to the curve (rather than to the raw data points). Accordingly the broken curve reveals the component of sensitivity reduction that must be accounted for through biochemical mechanisms distinct from compression. In the next subsection we instead apply the normalization to the raw responses.

8.3.2. Reduction in the intrinsic gain of R (row 2 of Table 5)*

It has been hypothesized that a component of the light-induced decrease in flash sensitivity might result from a lowering of the 'gain' with which R* produces G*s [243]; i.e. from a reduction in ν_{RG} in Eq. (4). The characteristic feature of such a reduction in R* gain (as distinct from a reduction in R* lifetime) is that it will be manifest at the very earliest times in the light response. Compelling evidence against an effect of this kind has recently been reported in salamander rods, for light adaptation that produces up to two-thirds suppression of the circulating current [119,222]. Thus, when response families are normalized to the steady circulating current (as described in the previous paragraphs), the activation phase of the response to a given flash is found to be independent of background intensity. Fig. 12 illustrates this invariance in the case of dim flashes: it shows that the early phase of the normalized dim flash response is independent of background. From this analysis it has recently been concluded that none of the factors that contribute to the amplification constant in Eq. (26) are decreased by light adaptation, and specifically that the R* 'gain' is unaltered [119]; this conclusion contrasts with the interpretations of other recent work [73,74,209,243,244].

Figure 12 additionally reveals two further important results. First, it shows that the simple 'activation only' model of Eq. (25) provides a good description of the kinetics of the onset phase of the dim-flash responses, for background intensities that suppress up to two-thirds of the dark current. Second, it reveals one manifestation of the conclusion to be presented in the next paragraph: that the lifetime of one or more of the steps in phototransduction is reduced in the presence of an adapting light.

8.3.3. Reduction in the lifetime of R (row 3 of Table 5)*

A substantial body of physiological evidence supports the conclusion that the lifetime of one or more of the disc-associated intermediates in amphibian rod phototransduction is shortened by light adaptation, and that this shortening is calcium-dependent [65,73,168,241,245,246]. In addition, there is evidence that the period of calcium-sensitivity is quite short, so that the effect disappears with a time constant of ca. 0.5 s [245,246]. The prime candidate for the intermediate whose lifetime is shortened by the decline in $[Ca^{2+}]_i$ is R* itself [36,65,120,245,246].

There are three lines of evidence supporting this identification. First, biochemical experiments have shown that the activity of RK, the enzyme that initiates the inactivation of R*, is modulated by the calcium-binding protein Rec, as indicated

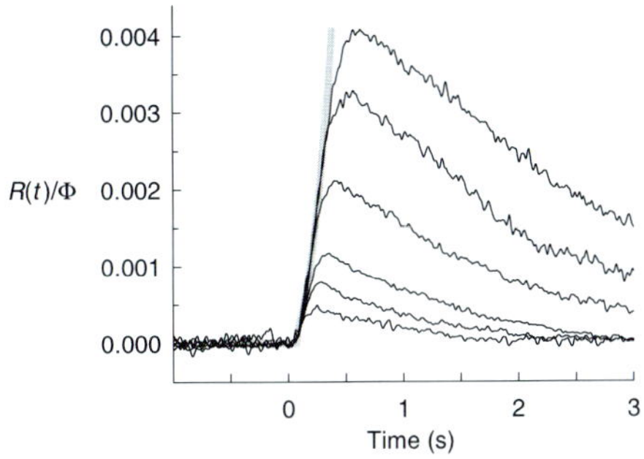

Fig. 12. Normalized dim-flash responses of a salamander rod in the presence of different backgrounds. The topmost trace was obtained in the dark, and successive traces in the presence of steady illumination producing 16, 50, 157, 500 and 1570 photoisomerizations s^{-1}. Each trace is the average of 3–5 individual responses, and was obtained from the raw response $r(t)$ by first dividing by the maximum response (i.e. the circulating current magnitude), to obtain $R(t) = r(t)/r_{max}$; this was then divided by the number of photoisomerizations produced by the flash, Φ, to obtain normalized response per photoisomerization. The corresponding circulating currents and raw sensitivities (i.e. not normalized by the circulating current) were plotted in Fig. 11 as ⊙. The smooth curve through the rising phase of the responses was computed with the pure activation model of Eq. (25), with amplification constant $A = 0.06\ s^{-2}$, and delay t_{eff} = 55 ms. The good fit of this curve to the initial portion of all the responses is consistent with the idea that the gain of the activation reactions is not changed by light adaptation. Data from Ref. [119].

schematically in Fig. 4B; see Section 6.1. Thus, there exists a mechanism by which the R* lifetime could be shortened in a calcium-dependent manner. Second, there is clear biochemical evidence that the inactivation of the G*–E* complex is *not* calcium-sensitive [248] (Section 6.2). Third, the 'dominant' or rate-limiting step in the inactivation of the disc-associated reactions *in situ* has also been shown to be insensitive to light adaptation and to changes in calcium [120,142,167]. The latter observations support the identification of the dominant time constant of inactivation with shut-off of the G*–E* complex, and consequently they indicate that the R* lifetime is 'non-dominant' (i.e., short). This conclusion fits neatly, both with the idea that R* lifetime is modulated by calcium concentration, and with experimental results, because alteration of the non-dominant lifetime will not affect the kinetics of recovery (as measured in 'Pepperberg plots' [167]), but will instead primarily appear to alter the sensitivity of the response, measured at the peak. Similarly, it will not affect the early rising phase of the response, provided that this is measured at sufficiently early times. (For a different view of the dominant inactivation time constant of amphibian rods, see Refs. [35,167], where the case is argued that R* activity in amphibian rods decays with a time constant of ca. 2 s.)

The hypothesized connection between R* lifetime and recoverin level can be formulated through the equation $\tau_R \propto 1/RK_{free}$, where RK_{free} is the concentration of RK without Rec bound to it. This molecular description, when coupled with the reaction scheme for recoverin's interaction with calcium and RK [85,249], has been shown [222] to provide a good account of the observed acceleration of the bright-

Table 5
Mechanisms of photoreceptor light adaptation

	Mechanism/hypothesis (as background increases)	Protein mediating the effect	Predicted effect on operating range	Predicted effect on 'cascade' sensitivity	Predicted effect on 'electrical' sensitivity	Ca^{2+} involvement	Identity of regulatory Ca^{2+}-binding protein	Comment	Refs.
1	Response compression (fewer channels open)	–	Reduces	No effect	Decreases	No	–	Saturation causes drastic decline in S_F	[59,234, 237,242]
2	Decrease in R* catalytic gain	R*	Extends	Decreases	Increases			Can be rejected; see text	[243,244]
3	Decrease in R* lifetime (τ_R)	RK	Extends	Decreases	Increases	Yes	Recoverin	Accelerates recovery	[36,54,65, 136,168, 241,245, 246]
4	Increase in steady-state cGMP hydrolysis rate constant (β_0)	PDE	Reduces	No effect	Decreases	No	–	Principal Ca^{2+}-independent factor	[118,142]
5	Steady increase in cGMP synthesis rate (α_0)	GC	Extends	No effect	Increases	Yes	GCAP1,2	Restores circulating current	[81,93,118, 121,136, 217,231]
6	Transient increase in cGMP synthesis rate ($\Delta\alpha(t)$)	GC	Slightly increases	No effect	Slightly decreases	Yes	GCAP1,2	Accelerates recovery	[142]
7	Decrease in K_{cG} of channels	Channel	Extends	No effect	Increases	Yes	Calmodulin	Restores circulating current	[94,213–216, 218,247]
8	Increase in Ca^{2+} buffering (B_{Ca}) of cytoplasm	Recoverin	Slightly decreases	No effect	Slightly increases	Yes	Recoverin	Results from decrease in Rec–2Ca	[119,222]
9	Pigment bleaching	Cone photopigment	Extends	Decreases	May increase or decrease	No	–	Incapacitates rods, yet underlies much cone light adaptation	[226]

Note: The column 'cascade sensitivity' refers to the sensitivity of the biochemical cascade, as PDE activity per incident photon. The column 'electrical sensitivity' refers to the overall sensitivity of the photoreceptor, in units of electrical response per incident photon, which is plotted as S_F in Fig. 11.

flash response that is elicited by a preceding step of light (the so-called 'step/flash' protocol [241]). This molecular model also explains the absence of the step/flash effect in mice lacking Rec [54].

8.3.4. Increase in β_0, the steady cGMP hydrolytic rate constant (row 4 of Table 5)
Until recently it had not been appreciated that the steady-state level of phosphodiesterase activity might play a major role in the desensitization and acceleration of the response that accompanies background illumination. In the presence of steady background illumination generating E^* activated catalytic subunits of PDE, the enzymatic activity of the PDE is defined by the hydrolytic rate constant $\beta_0 = E^*\beta_{sub} + \beta_{dark}$, the steady-state equivalent of Eq. (12). As we discussed in Section 7.5, the reciprocal of β_0 acts as one of the three time constants contributing to recovery of the dim-flash response [142]. In the case of clamped calcium, it is straightforward to show that an increase in β_0 should shorten the time-to-peak and decrease the flash sensitivity S_F. Although the analysis is more complicated when $[Ca^{2+}]_i$ is free to change, the same prediction is found to hold [142]. Furthermore, calculations and simulations employing measured values of the transduction parameters show that the increase in β_0 with steady illumination is the single most important factor in decreasing the flash sensitivity [222]. It is worth emphasizing that this effect of elevated β_0 is not caused by the decline in $[Ca^{2+}]_i$, i.e., is not a calcium-mediated mechanism.

Bath-tub analogy. This situation can be appreciated in qualitative terms by considering an analogy, which is a qualitative description of the behavior of Eq. (16), the general rate equation governing free cGMP in the outer segment. Imagine a bath-tub, with water running in from a tap (the cyclase) at a fixed rate, and with water draining out through the plug-hole (the PDE) at a rate proportional to the depth of water (the cGMP concentration). When a steady-state is reached, the depth of water will equal the rate of influx (α_0) divided by the rate constant of efflux (β_0); i.e., $cG_0 = \alpha_0/\beta_0$. Now imagine that a second plug-hole is briefly opened and then closed, causing a transient increment in the rate of efflux ($\Delta\beta(t)$). This will elicit a transient drop in water level ($\Delta cG(t)$) and eventual recovery to the original steady level.

Next, as an analogy with the application of a steady background illumination, imagine that the size of the plug-hole is enlarged (increased β_0), and in addition that the tap is turned on harder so as to increase the steady influx of water (increased α_0). If the two parameters were increased in the same ratio, then the steady-state depth of water would be unchanged. Finally, imagine that the same incremental stimulus is given as previously – the transient opening of a second plug-hole. This stimulus will now cause a smaller drop in water level than previously, together with a faster recovery; in fact the initial rate of drop in water level will be the same as previously, but recovery will occur more rapidly. In general terms, the effect of enlarging the plug-hole will be to accelerate the rate at which the water level re-equilibrates whenever it is perturbed, whereas the effect of opening up the tap will simply be to scale-up the depth of the water. Hence, if one expresses the depth of water relative to its original resting level (i.e., $cG(t)/cG_0$) the result will be independent of how rapidly (α_0) the water is flowing in from the tap (provided that this rate is constant). And one can show from analysis of Eq. (16) that the kinetic shape of the response corresponds to the convolution of the driving function ($\Delta\beta(t)$) with the exponential time constant of equilibration ($1/\beta_0$).

A final point worth drawing from the bathtub analogy concerns the condition when the bathtub is empty. Then, the effect of having different sized plug-holes completely disappears. This condition parallels that when a bright flash is given and the rod response is driven into saturation: this happens when β/α_{max} is very large, so that $cG(t)$ is driven too low to hold any channels open. The steady cGMP hydrolysis rate constant β_0 only has an effect on the rod's response kinetics when a finite level of cGMP is present.

8.3.5. Guanylyl cyclase: Steady-state increase in activity (row 5 of Table 5)
Perhaps the most thoroughly investigated molecular mechanism of photoreceptor light adaptation is the activation of guanylyl cyclase that is brought about by the decline in $[Ca^{2+}]_i$ accompanying the steady-state light response (Section 7.4). From Eqs. (14) and (22) one can readily derive the steady-state relationship

$$\frac{J_0}{J_{dark}} = \left[\frac{\alpha_0/\beta_0}{\alpha_{dark}/\beta_{dark}}\right]^{n_{cG}} = \left[\frac{\alpha_0/\alpha_{dark}}{\beta_0/\beta_{dark}}\right]^{n_{cG}} \tag{34}$$

which would apply exactly in the absence of any effect of calcium on the cGMP-gated channels. From this relationship two conclusions follow. First, in the absence of GC activation, an x-fold increase in PDE activity β_0 would cause the circulating current to decrease by $x^{n_{cG}}$; thus a quadrupling of β_0 would decrease the current to just 1/64 of its previous level, for $n_{cG} = 3$. Secondly, in the face of such an x-fold increase in PDE activity, the original circulating current could be completely restored through an x-fold increase in GC activity, α_0. Many studies have shown that, when Ca^{2+} drops, GC can be activated to at least 10-times its basal level by GCAPs (Section 7.4). In salamander rods the level of PDE activity that drives the cell into saturation corresponds roughly to a 20-fold increase over the dark level ($\beta_0 \approx 20\ s^{-1}$ [222]; see Fig. 11C). Together these findings are consistent with the view that a substantial component of range extension in salamander rods is mediated by the activation of GC, but that this mechanism of extension is exhausted beyond a ca. 10-fold activation, as discussed previously [121,217]. Finally, as indicated in Table 5, it is important to note that by rescuing the transduction mechanism from saturation, the elevation of steady cyclase activity has the effect of *increasing* the photoreceptor's sensitivity from the level that would apply in the absence of this mechanism.

8.3.6. Guanylyl cyclase: Transient increment in activity (row 6 of Table 5)
During the normal flash response, $[Ca^{2+}]_i$ declines transiently, and causes a transient increase $\Delta\alpha(t)$ in GC activity. Paradoxically, this transient activation of cyclase has an effect on sensitivity opposite to that of the steady activation, described in the previous paragraph. Thus, the transient flash-induced increment in cyclase activity accelerates the recovery of the flash response and thereby *reduces* the sensitivity

measured at the peak (cf. Fig. 9C). On the other hand, this transient mechanism has relatively little effect on the *shape* of the sensitivity *vs* background relation, because the desensitizing effect (relative to the calcium-clamped condition) is about the same magnitude, independent of background intensity. This may be seen in Fig. 11C, where the decline of sensitivity under calcium-clamped conditions (symbols) closely parallels the normal decline [237].

8.3.7. Calmodulin effect on channels (row 7 of Table 5)

A second well understood mechanism of range-extension is the calmodulin-dependent shift in the K_{cG} of the cyclic nucleotide gated channels (Section 6.5); however, the contribution of this mechanism to the overall range extension in rods is far less than the contribution of the activation of GC [217]. Indeed, for primate rods, the range extension has been modeled in terms of GC activation alone [231]. For amphibian rods, where the range extension by decline in $[Ca^{2+}]_i$ is greater, the behavior has been modeled with a combination of GC activation, a CM-dependent shift in the K_{cG} of the channels, and a third mechanism, identified as a reduction in the average number of phosphodiesterase catalytic subunits (E*) activated per photo-excited rhodopsin (R*) at higher background intensities [136,217]. A candidate molecular mechanism for this third effect is the regulation of rhodopsin kinase by Rec, as discussed previously.

In cones, it appears that the modulation of the K_{cG} of the channels is much larger than in rods [218], and this no doubt contributes to the much larger operating range of cones [56,226]. Therefore it is possible that cones may have retained more prominently a feature of the progenitor cell of photoreceptors and olfactory neurons, since olfactory receptor cells also display a very large calmodulin-dependent modulation of the $K_{\frac{1}{2}}$ of their cyclic nucleotide-gated channels [117].

8.3.8. Increased calcium buffering (row 8 of Table 5)

Calcium buffering acts to slow the transient decrease in $[Ca^{2+}]_i$ that accompanies the light response, and in doing so it retards the transient increase in GC activation discussed in Section 8.3.6, and thereby delays recovery and increases the peak response amplitude (and thus the flash sensitivity). As noted earlier (Section 6.1) it appears likely that the primary calcium buffer in rods is recoverin, with a concentration in excess of 30 μM. Analysis of the binding of Ca^{2+} to recoverin shows that, as $[Ca^{2+}]_i$ declines with increasing steady illumination, the free concentration of recoverin (Rec_{free}) will increase, and that over an intermediate range of calcium concentrations the calcium buffering power will also increase. An interesting feature of buffering by Rec is that, because of its cooperative binding of calcium, it is effective over a narrower range of calcium concentrations than a buffer with a single, non-cooperative binding site (compare Eqs. (7), (8)). We calculate that the incremental buffering power for calcium reaches its maximum when $[Ca^{2+}]_i$ drops to 100 nM, and that this peak is at least 3-fold higher than the buffering power in darkness [222]. This increase in buffering power will increase the rod's flash sensitivity relative to that which would be calculated if the buffering were independent of $[Ca^{2+}]_i$.

8.3.9. Pigment bleaching in cones (row 9 of Table 5)
Rod photoreceptors are substantially desensitized by bleached pigment, even (in the case of cells isolated from the retinal pigment epithelium) after long periods in darkness; reviewed in Refs. [250,251]. In contrast, cone photoreceptors can function almost normally with substantial fractions of their pigment in the bleached state – i.e., in the absence 11-*cis* retinal bound to their opsin (see Fig. 1B of Ref. [226]).

By avoiding saturation up to the bleaching range of intensities, cones are guaranteed to achieve Weber Law behavior at all higher levels (even though these levels may rarely be encountered). In the steady-state, the amount of pigment remaining unbleached at very high intensities is inversely proportional to the incident intensity, provided that the regeneration rate is unchanged [252]. Furthermore, the rate of photon capture is directly proportional to the amount remaining unbleached. Accordingly, the number of photoisomerizations elicited by a flash of fixed intensity superimposed on the background will be inversely proportional to background intensity, so that in the 'large bleach' regime the cone's flash sensitivity will obey Weber's Law purely through reduced quantal catch.

8.4. Summary

Some useful general conclusions can be drawn from our analysis of light adaptation, by reviewing Table 5. Increasing steady illumination is known or hypothesized (column 1) to affect the activity of proteins throughout the transduction cascade (column 2). To fully understand the mechanisms, one must consider their effects on both the overall operating range of the cell (column 3), and on its biochemical sensitivity (column 4) and on its electrical sensitivity (column 5). In many cases, the actions are caused by the decline in $[Ca^{2+}]_i$ (column 6) that accompanies steady illumination, and are mediated by specific calcium-binding proteins (column 7). However, Table 5 shows that no general rule applies to the mechanisms underlying adaptation. Indeed, perhaps the most important conclusion from the analysis is that there is no universal relationship between the effect of an adaptational mechanism on operating range and its effect on flash sensitivity. Another important conclusion is that the two dominant mechanisms in desensitization of the rod, response compression (row 1) and elevated cGMP hydrolysis rate constant (row 4), are not caused by the decline in $[Ca^{2+}]_i$ [222].

9. Single-photon responses: Implications of observed variability

We now turn to one of the most perplexing problems in the history of rod physiology: that of understanding the molecular mechanisms governing the single-photon response. This problem is no mere curiosity, since, as discussed at the beginning of Section 8, the ability of rods to generate reliable responses to single photons is a defining characteristic. We have saved this problem for last, because its analysis and discussion necessarily require prior analysis of the mechanisms of response activation and inactivation. Moreover, a notable feature of a long series (or ensemble) of single-photon responses is their similarity in shape and, as we shall discuss, the mechanisms underlying this reproducibility are likely to include those we have examined in the context of light adaptation.

9.1. Reproducibility: Ensemble behavior

A number of investigations have reported that rod responses to single-photon are highly reproducible in amplitude and shape. The reproducibility of the amplitude has been characterized by the coefficient of variation (cv = s.d./mean) of the response amplitude; in five separate studies of two species the cv of the amplitude (measured at the time-to-peak of the mean response) has been reported to be about 0.2 [34–36,38,253]. In four of these studies it was reported that the shapes of the individual responses were very similar to the shape of the mean response.

These results have been difficult to reconcile with the presumption that the inactivation of R* is likely to be triggered by a single molecular event (phosphorylation of a single residue), since this would suggest that shut-off should be inherently stochastic. The apparent contradiction has led to the suggestion that a series of undiscovered intra-molecular processes must occur, to 'smooth' the decline of activity of an individual R* [35]. However, a recent paper has presented compelling evidence that the variability in kinetics is considerably greater than previously reported, in a way that appears compatible with stochastic shut-off of R* triggered by a single molecular event [36].

A sample set of single-photon responses from a toad rod is illustrated in Fig. 13A; the traces show 50 consecutive responses to very dim flashes (delivering on average less than 1 photoisomerization per trial). At the bottom of the panel a group of 'failures' is clearly separated from a group of presumed 'singleton' events that (as a group) have peak amplitudes of a little under 2 pA; these events are in turn separated from four larger events that are presumed to represent multiple photon hits. The amplitude histogram measured at the time-to-peak of the mean response (1.9 s) is plotted in Fig. 13B, for the complete set of 350 trials delivered to the cell. This histogram closely resembles the form reported in all such studies, with a failures peak clearly evident at 0 pA, and with a broader peak (near 1.8 pA in this cell) that presumably represents singletons.

One of the main lines of evidence supporting the notion that the shape of the single-photon response is stereotypical has come from comparison of the shape of the variance, $\sigma^2(t)$, with the shape of the squared mean, $\mu^2(t)$, of the ensemble of responses [34,35,38,253]. This comparison is made in Fig. 13C. Up to well beyond the peak (until about 3 s), the shape of $\sigma^2(t)$ is indistinguishable from the shape of a suitably scaled version of $\mu^2(t)$.

However, it turns out that this finding does not provide evidence that the single-photon responses all have the same underlying shape. Although analysis of the Poisson distribution (of photon hits) indicates that if the shapes of the single-photon responses were all identical, then the ensemble variance would have the same time course as the squared ensemble mean, the converse cannot be concluded; namely, that a match between the ensemble variance and the mean squared response implies identity of shape of the individual responses. Instead, for very dim flashes, it can be

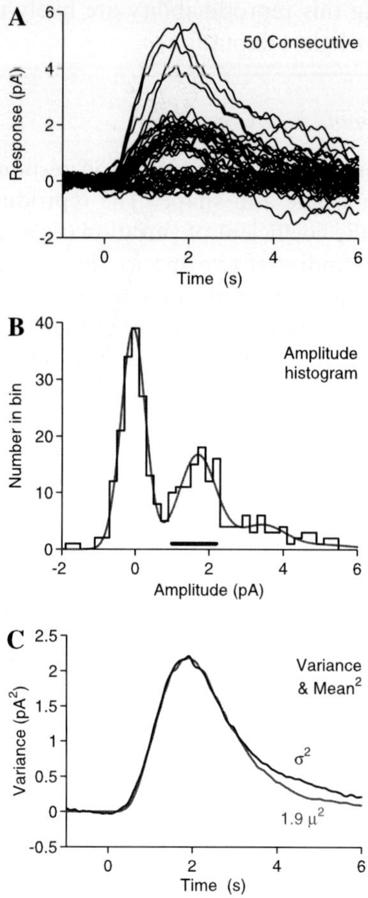

Fig. 13. Suction electrode recordings from a toad rod stimulated with a series of 350 very dim flashes, in the 'single photon' range. In each trial a brief flash was presented at time zero, delivering 0.08 photons μm^{-2}, estimated to cause about 0.6 photoisomerizations on average. (A) Sample of 50 consecutive responses from the 350 trials. Note that the responses separate fairly clearly into: failures, singletons, and multiple-photon responses. (B) Amplitude histogram for the complete set of 350 responses, determined at the time-to-peak of the average response, $t_{peak} = 1.9$ s. The curve is the 'sum of Gaussians' expression from Eq. (10) of Ref. [34], with: $\Phi = 0.65$; mean single-photon response amplitude, $a = 1.75$ pA; standard deviation of the failures, $\sigma_0 = 0.33$ pA; additional standard deviation of the singletons, $\sigma_1 = 0.38$ pA. Horizontal bar shows the range of amplitudes taken as conventional singletons. (C) Ensemble variance, $\sigma^2(t)$, for the complete set of 350 responses, compared with a scaled version of the square of the mean response, $\mu^2(t)$. Modified from Fig. 1 of Ref. [36].

shown that the ensemble variance will be dominated by the fact that the responses divide into 'failures' on the one hand and 'singletons' or 'multiples' on the other. The resulting component of the ensemble variance, which will have the time course of the mean squared response, is likely to swamp the component of variance contributed by variations between individual singleton responses.

9.2. Variability of the individual singletons

Rather than attempting to apply an ensemble test, there would be much to be gained by examining the individual responses. Although visual inspection of Fig. 13A might lead one to think that individual singleton events are broadly similar in shape, the problem is that individual variations are obscured by the process of superimposing the traces. In their recent study, Whitlock and Lamb [36] suspected that fluctuations in the amplitude of the individual responses might have been camouflaging fluctuations in kinetics in Fig. 13A. To try to avoid fluctuations in the underlying amplification, they selected (from the complete set of 85 singletons recorded in this experiment) the 40 responses that initially rose along a time-course most closely resembling the rise of the mean response. These 40 selected traces are superimposed in Fig. 14A, where they have been color-coded into five groups according to their amplitude at a later time (4 s); this color-coding has been introduced as an aid to visualization of inter-trace variability at later times. Although the individual singleton traces rise broadly along a common curve (because they have been selected to do so), they exhibit widely different behavior at later times. Thus, some traces reach peak much later than others, and in doing so they attain larger amplitudes and take considerably longer to achieve final recovery. Because of the significant level of recording noise in the individual traces, the authors averaged the colored groups of responses, to generate the traces in Fig. 14B; this panel confirms the general properties that can be seen in the raw traces in Fig. 14A.

In order to test the possibility that the recorded traces might have arisen from stochastic shut-off of R*, the authors developed a simplified model of the response kinetics that would be expected if R* activity were 'all-or-nothing' rather than graded. (To do this they used the model developed in the Section 7, with $\tau_R = \infty$, and they calculated the response to a step 'on' minus the response to a delayed step 'off'.) By least-squares fitting of this model to the raw traces, they extracted the R* lifetimes that provided the best fit for each individual singleton response, and the distribution of these extracted lifetimes is plotted in Fig. 14C. The lifetimes extracted in this way ranged from less than 0.8 s to more than 2.5 s for the illustrated cell, with a mean of 1.7 s and a coefficient of variation of 0.4 – much larger than the coefficient of variation for the kinetics of 0.2 reported previously [35].

9.3. Implications of the observed variability

The first conclusion of this analysis was that, despite the close similarity in form of the ensemble variance and squared mean responses, the individual singleton responses exhibit marked variations in kinetics. Furthermore, the time-course of the individual responses appears in no way inconsistent with the occurrence of an all-or-nothing lifetime for R*. Nevertheless the degree of variability in kinetics (with $cv \approx 0.4$) was smaller than expected on the simplest models of stochastic R* lifetime [35,254].

In an attempt to account for the extracted distribution of presumed R* lifetimes, and to account for the results of other experiments with calcium buffering, Whitlock and Lamb [36] proposed a molecular model incorporating three main features: (1)

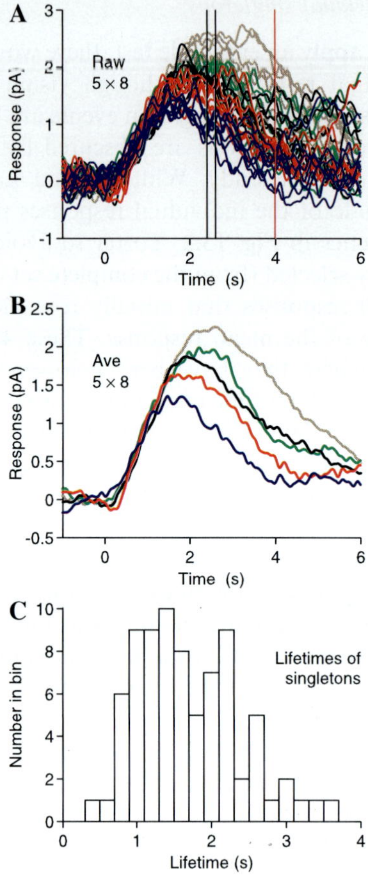

Fig. 14. Variability of the recovery kinetics of singleton responses. (A) Forty selected singleton responses. From the 350 trials in Fig. 13, the forty responses that most closely matched the early rising phase of the mean singleton response have been selected. In this particular selection, responses exhibiting spontaneous thermal isomerizations have automatically been eliminated, by the selection of only those traces that were well described by a theoretical model of the recovery phase; see Ref. [36] for details. Purely as a visual aid, the selected raw responses have been color-coded into 5 groups of 8, according to their amplitude over the indicated time window (2.5 ± 0.1 s). (B) Averages of the 5 groups of colored raw responses from A. (C) Distribution of extracted lifetimes, t_{life}, of the singletons, when a stochastic model of R* shut-off was applied to the individual singleton responses, over the time interval up to 4 s indicated by the vertical red line in A. Note that the extracted lifetimes range from less than 1 s to more than 2.5 s. Modified from Fig. 2 of Ref. [36].

all-or-nothing activity of R*; (2) feedback modulation of R* lifetime, via Ca^{2+}, through the binding of Rec–2Ca to RK; and (3) the longitudinal diffusion of cGMP and Ca^{2+} in the outer segment. The second feature is critical, because it serves to change the rate of encounters of R* with RK in a calcium-dependent manner (as described in Section 8). Initially, when $[Ca^{2+}]_i$ is high, most RKs are bound to Rec, so that very few RKs are capable of interacting with R*. As the response proceeds, and $[Ca^{2+}]_i$ declines, Rec dissociates and more and more RKs become available to interact; thus the rate constant of R* shut-off increases steadily as the response proceeds, in a manner that is cooperatively dependent on $[Ca^{2+}]_i$. The third feature, which has not be incorporated into any previous models of rod photoresponses, also appears important, because calculations and measurements suggest that the decline in $[Ca^{2+}]_i$ may be highly localized spatially [133,134]. As a result the magnitude of the change in $[Ca^{2+}]_i$ at the disc where the single R* was activated could be many-fold greater than the average change over the whole outer segment. It was estimated that, in response to a single photoisomerization, $[Ca^{2+}]_i$ at the site of photon absorption could drop to one-third of its resting value within 2 s; a change as great as this might well be sufficient to trigger a massive rise in the rate of R* shut-off.

While the proposed molecular scheme is not yet proven (and is likely to undergo refinement), it points towards a resolution of one of the most perplexing issues in the history of rod phototransduction – how the shut-off of a single R* molecule could be stochastic, yet could lead to kinetic variability ($cv \approx 0.4$) far lower than expected on models with fixed parameters.

References

1. Scott, K., Becker, A., Sun, Y., Hardy, R. and Zuker, C. (1995) Neuron **15**, 919–927.
2. Gold, G.H. (1999) Ann. Rev. Physiol. **61**, 857–871.
3. Mombaerts, P. (1999) Ann. Rev. Neurosci. **22**, 487–509.
4. Zufall, F. and Leinders-Zufall, T. (1998) Ann. N. Y. Acad. Sci. **855**, 199–204.
5. Krieger, J. and Breer, H. (1999) Science **286**, 720–723.
6. Huff, R.M. (1996) Cell. Signal. **8**, 453–459.
7. Lohse, M.J. (1996) Zeit. Kardiol. **85** Suppl 7, 1–3.
8. Strosberg, A.D. (1997) Ann. Rev. Pharmacol. **37**, 421–450.
9. Doupnik, C.A., Davidson, N., Lester, H.A. and Kofuji, P. (1997) Proc. Natl Acad. Sci. USA **94**, 10461–10466.
10. Catterall, W.A. (1997) Adv. Second Mess. Phosphoprot. Res. **31**, 159–181.
11. Ikeda, S.R. and Dunlap, K. (1999) Adv. Second Mess. Phosphoprot. Res. **33**, 131–151.
12. GCRDb G Protein Receptor Database. http://www.gcrdb.uthscsa.edu/GCR_Fam.html.
13. SwiftEMBL. http://swift.embl-heidelberg.de/7tm/.
14. Buck, L. and Axel, R. (1991) Cell **65**, 175–187.
15. Skoufos, E., Marenco, L., Healy, M.D., Singer, M.S., Nadkarni, P.M., Miller, P.L. and Shepherd, G.S. The Olfactory Receptor Database. http://ycmi.med.yale.edu/senselab/ORDB/.
16. Bourne, H., Sanders, D. and McCormick, F. (1991) Nature **349**, 117–126.
17. Sondek, J., Lambright, D., Noel, J., Hamm, H.E. and Sigler, P. (1994) Nature **372**, 276–279.
18. Lambright, D., Noel, J., Hamm, H.E. and Sigler, P. (1994) Nature **369**, 621–628.
19. Coleman, D., Berghuis, A., Lee, E., Linder, M., Gilman, A. and Sprang, S. (1994) Science **265**, 1405–1412.
20. Mixon, M., Lee, E., Coleman, D., Berghuis, A., Gilman, A. and Sprang, S. (1995) Science **270**, 954–960.
21. Wall, M., Coleman, D., Lee, E., Iniguez-Lluhi, J., Posner, B., Gilman, A. and Sprang, S. (1995) Cell **83**, 1047–1058.
22. Sondek, J., Bohm, A., Lambright, D., Hamm, H.E. and Sigler, P. (1996) Nature **379**, 369–374.

23. Sprang, S. (1997) Annu. Rev. Biochem. **66**, 639–678.
24. Hamm, H.E. (1998) J. Biol. Chem. **273**, 669–672.
25. Liri, T., Farfel, Z. and Bourne, H. (1998) Nature **394**, 35–38.
26. Lamb, T.D. and Pugh, E.N. Jr (1992) J. Physiol. **449**, 719–758.
27. Pugh, E.N. Jr and Lamb, T.D. (1993) Biochim. Biophys. Acta **1141**, 111–149.
28. Snyder, A.W. and Menzel, R. (1975) Photoreceptor Optics. Springer, Berlin.
29. Enoch, J.M. and Tobey, F.L. (1981) Vertebrate Photoreceptor Optics. Springer, Berlin.
30. Stiles, W.S. and Crawford, B.H. (1933) Proc. Roy. Soc. B **113**, 496–530.
31. Stiles, W.S. (1939) Proc. Roy. Soc. B **127**, 64–105.
32. Harosi, F.I. and Malerba, F.E. (1975) Vision Res. **15**, 379–388.
33. Liebman, P.A. (1972) in: Photochemistry of Vision, Vol. VII, ed H.J.A. Dartnall. pp. 481–528, Springer, New York.
34. Baylor, D.A., Lamb, T.D. and Yau, K.-W. (1979) J. Physiol. **288**, 613–634.
35. Rieke, F. and Baylor, D.A. (1998) Biophys. J. **75**, 1836–1857.
36. Whitlock, G.G. and Lamb, T.D. (1999) Neuron **23**, 337–351.
37. Aho, A.C., Donner, K., Helenius, S., Larsen, L.O. and Reuter, T. (1993) J. Comp. Physiol. A **172**, 671–682.
38. Baylor, D.A., Nunn, B.J. and Schnapf, J.L. (1984) J. Physiol. **357**, 575–607.
39. Chen, J., Makino, C.L., Peachey, N.S., Baylor, D.A. and Simon, M.I. (1995) Science **267**, 374–377.
40. Xu, J., Dodd, R.L., Makino, C.L., Simon, M.I., Baylor, D.A. and Chen, J. (1997) Nature **389**, 505–509.
41. Lagnado, L. and McNaughton, P.A. (1990) J. Memb. Biol. **113**, 177–191.
42. Schnetkamp, P.P. (1995) Cell Calcium **18**, 322–330.
43. Schnetkamp, P.P. (1989) Prog. Biophys. Molec. Biol. **54**, 1–29.
44. Molday, R.S. and Molday, L.L. (1993) Methods Neurosci. **15**, 131–150.
45. Finn, J.T., Grunwald, M.E. and Yau, K.-W. (1996) Ann. Rev. Physiol. **58**, 395–426.
46. Molday, R.S. and Kaupp, U.B. (2000) in: Handbook of Biological Physics, Vol. 3, eds D.G. Stavenga, W.J. DeGrip and E.N. Pugh Jr. pp. 143–181, Elsevier, Amsterdam.
47. Nakatani, K. and Yau, K.-W. (1988) J. Physiol. **395**, 695–729.
48. Baylor, D.A. and Nunn, B.J. (1986) J. Physiol. **371**, 115–145.
49. Cobbs, W.H. and Pugh, E.N. Jr (1987) J. Physiol. **394**, 529–572.
50. Miller, J.L. and Korenbrot, J.I. (1994) J. Gen. Physiol. **104**, 909–940.
51. Baylor, D.A., Lamb, T.D. and Yau, K.-W. (1979) J. Physiol. **288**, 589–611.
52. Schneeweis, D.M. and Schnapf, J.L. (1995) Science **268**, 1053–1056.
53. Kraft, T.W., Schneeweis, D.M. and Schnapf, J.L. (1993) J. Physiol. **464**, 747–765.
54. Dodd, R.L. (1998) The role of arrestin and recoverin in signal transduction by retinal rod photoreceptors. Ph.D. thesis. Stanford University.
55. Hestrin, S. and Korenbrot, J.I. (1990) J. Neurosci. **10**, 1967–1973.
56. Matthews, H.R., Fain, G.L., Murphy, R.L.W. and Lamb, T.D. (1990) J. Physiol. **420**, 447–469.
57. Perry, R.J. and McNaughton, P.A. (1991) J. Physiol. **433**, 561–587.
58. Baylor, D.A. and Hodgkin, A.L. (1973) J. Physiol. **234**, 163–198.
59. Baylor, D.A., Hodgkin, A.L. and Lamb, T.D. (1974) J. Physiol. **242**, 685–727.
60. Baylor, D.A. and Fettiplace, R. (1975) J. Physiol. **248**, 433–464.
61. von Greef, R. (1902) A Guide to Microscopic Examination of the Eye. Blackiston, Philadelphia.
62. Miller, W.H. and Snyder, A.W. (1973) Vision Res. **13**, 2185–2194.
63. Schnapf, J.L., Nunn, B.J., Meister, M. and Baylor, D.A. (1990) J. Physiol. **427**, 681–713.
64. Schneeweis, D.M. and Schnapf, J.L. (1999) J. Neurosci. **19**, 1203–1216.
65. Torre, V., Matthews, H.R. and Lamb, T.D. (1986) Proc. Natl Acad. Sci. USA **83**, 7109–7113.
66. Hodgkin, A.L. (1971) Proc. Roy. Soc. A **326**, v-xx.
67. Ratto, G.M., Payne, R., Owen, W.G. and Tsien, R.Y. (1988) J. Neurosci. **8**, 3240–3246.
68. McCarthy, S.T., Younger, J.P. and Owen, W.G. (1994) Biophys. J. **67**, 2076–2089.
69. Korenbrot, J.I. and Miller, D.L. (1989) Vision Res. **29**, 939–948.
70. Lagnado, L., Cervetto, L. and McNaughton, P.A. (1992) J. Physiol. **455**, 111–142.

71. Sampath, A.P., Matthews, H.R., Cornwall, M.C. and Fain, G.L. (1998) J. Gen. Physiol. **111**, 53–64.
72. Sampath, A.P., Matthews, H.R., Cornwall, M.C., Bandarchi, J. and Fain, G.L. (1999) J. Gen. Physiol. **113**, 267–277.
73. Gray-Keller, M.P. and Detwiler, P.B. (1994) Neuron **13**, 849–861.
74. Detwiler, P.B. and Gray-Keller, M.P. (1996) Curr. Opin. Neurobiol. **6**, 440–444.
75. Unger, V.M., Hargrave, P.A., Baldwin, J.M. and Schertler, G.F. (1997) Nature **389**, 203–206.
76. DeGrip, W.J. and Rothschild, K.J. (2000) in: Handbook of Biological Physics, Vol. 3, eds D.G. Stavenga, W.J. DeGrip and E.N. Pugh Jr. pp. 1–54, Elsevier, Amsterdam.
77. Mathies, R.A. and Lugtenburg, J. (2000) in: Handbook of Biological Physics, Vol. 3, eds D.G. Stavenga, W.J. DeGrip and E.N. Pugh Jr. pp. 55–90, Elsevier, Amsterdam.
78. Dumke, C., Arshavsky, V.Y., Calvert, P., Bownds, M. and Pugh, E.N. Jr. (1994) J. Gen. Physiol. **103**, 1071–1098.
79. Artemyev, N.O., Surendran, R., Lee, J.C. and Hamm, H.E. (1996) J. Biol. Chem. **271**, 25382–25388.
80. Shyjan, A.W., de Sauvage, F.J., Gillett, N.A., Goeddel, D.V. and Lowe, D.G. (1992) Neuron **9**, 727–737.
81. Pugh, E.N. Jr., Duda, T., Sitaramayya, A. and Sharma, R.K. (1997) Biosci. Rep. **17**, 429–473.
82. Yau, K.-W. and Baylor, D.A. (1989) Ann. Rev. Neurosci. **12**, 289–327.
83. Rispoli, G., Navangione, A. and Vellani, V. (1996) Ann. N.Y. Acad. Sci. **779**, 346–355.
84. Schwarzer, A., Kim, T.S.Y., Hagen, V., Molday, R.S. and Bauer, P.J. (1997) Biochem. **36**, 13667–13676.
85. Klenchin, V.A., Calvert, P.D. and Bownds, M.D. (1995) J. Biol. Chem. **270**, 16147–16152.
86. Granzin, J., Wilden, U., Choe, H.W., Labahn, J., Krafft, B. and Buldt, G. (1998) Nature **391**, 918–921.
87. Hirsch, J.A., Schubert, C., Gurevich, V.V. and Sigler, P.B. (1999) Cell **97**, 257–269.
88. Ames, J.B., Ishima, R., Tanaka, T., Gordon, J.I., Stryer, L. and Ikura, M. (1997) Nature **389**, 198–202.
89. Kawamura, S. (1994) Neurosci. Res. **20**, 293–298.
90. Kawamura, S., Kuwata, O., Yamada, M., Matsuda, S., Hisatomi, O. and Tokunaga, F. (1996) J. Biol. Chem. **271**, 21359–21364.
91. He, W., Cowan, C.W. and Wensel, T.G. (1998) Neuron **20**, 95–102.
92. Makino, E.R., Handy, J.W., Li, T. and Arshavsky, V.Y. (1999) Proc. Natl Acad. Sci. USA **96**, 1947–1952.
93. Polans, A., Baehr, W. and Palczewski, K. (1996) Trends Neurosci. **19**, 547–554.
94. Bauer, P.J. (1996) J. Physiol. **494**, 675–685.
95. Wilkins, J.F., Bitensky, M.W. and Willardson, B.M. (1996) J. Biol. Chem. **271**, 19232–19237.
96. Yokoyama, R. and Yokoyama, S. (2000) in: Handbook of Biological Physics, Vol. 3, eds D.G. Stavenga, W.J. DeGrip and E.N. Pugh Jr. pp. 257–296, Elsevier, Amsterdam.
97. Birge, R.R. (1990) Biochim. Biophys. Acta **1016**, 293–327.
98. Nathans, J. (1999) Neuron **24**, 299–312.
99. Ohguro, H., Fukada, Y., Takao, T., Shimonishi, Y., Yoshizawa, T. and Akino, T. (1991) EMBO J. **10**, 3669–3674.
100. Kokame, K., Fukada, Y., Yoshizawa, T., Takao, T. and Shimonishi, Y. (1992) Nature **359**, 749–752.
101. Bigay, J., Faurobert, E., Franco, M. and Chabre, M. (1994) Biochem. **33**, 14081–14090.
102. Fukada, Y., Matsuda, T., Kokame, K., Takao, T., Shimonishi, Y., Akino, T. and Yoshizawa, T. (1994) J. Biol. Chem. **269**, 5163–5170.
103. Hofmann, K.-P. (2000) in: Handbook of Biological Physics, Vol. 3, eds D.G. Stavenga, W.J. DeGrip and E.N. Pugh Jr. pp. 91–142, Elsevier, Amsterdam.
104. Hofmann, K.-P. (1986) Photobiochem. Photobiophys. **13**, 309–327.
105. Bornancin, F., Pfister, C. and Chabre, M. (1989) Euro. J. Biochem. **184**, 687–698.
106. Kahlert, M. and Hofmann, K.-P. (1991) Biophys. J. **59**, 375–386.
107. Stroop, S.D. and Beavo, J.A. (1992) Adv. Second Mess. Phosphoprot. Res. **25**, 55–71.
108. Catty, P. and Deterre, P. (1991) Euro. J. Biochem. **199**, 263–269.

109. Catty, P., Pfister, C., Bruckert, F. and Deterre, P. (1992) J. Biol. Chem. **267**, 19489–19493.
110. Lamb, T.D. (1994) Biophys. J. **67**, 1439–1454.
111. Beavo, J.A. (1995) Physiol. Rev. **75**, 725–748.
112. Fuortes, M.G.F. and Hodgkin, A.L. (1964) J. Physiol. **172**, 239–263.
113. Hagins, W.A. (1972) Ann. Rev. Biophys. Bioeng. **1**, 131–158.
114. Miller, W.H. (1981) Current Topics in Membranes and Transport **15**. Academic, New York.
115. Fesenko, E.E., Kolesnikov, S.S. and Lyubarsky, A.L. (1985) Nature **313**, 310–313.
116. Kaupp, U.B. (1995) Curr. Opin. Neurobiol. **5**, 434–442.
117. Yau, K.-W. and Chen, T.Y. (1995) in: Ligand- and Voltage-Gated Ion Channels, ed R.A. North. CRC Press, Boca Raton, FL. pp. 307–335.
118. Hodgkin, A.L. and Nunn, B.J. (1988) J. Physiol. **403**, 439–471.
119. Pugh, E.N. Jr, Nikonov, S. and Lamb, T.D. (1999) Curr. Opin. Neurobiol. **9**, 410–418.
120. Lyubarsky, A.L., Nikonov, S. and Pugh, E.N. Jr (1996) J. Gen. Physiol. **107**, 19–34.
121. Koutalos, Y., Nakatani, K., Tamura, T. and Yau, K.-W. (1995) J. Gen. Physiol. **106**, 863–890.
122. Calvert, P.D., Ho, T.W., LeFebvre, Y.M. and Arshavsky, V.Y. (1998) J. Gen. Physiol. **111**, 39–51.
123. Hofmann, K.-P. and Heck, M. (1996) Biomemb. **2A**, 141–198.
124. Baumann, C. (1978) J. Physiol. **279**, 71–80.
125. Penn, R. and Hagins, W.A. (1972) Biophys. J. **12**, 1073–1094.
126. Naqvi, K.R. (1974) Chem. Phys. Lett. **28**, 280–284.
127. Leskov, I.B., Klenchin, V.A., Handy, J.W., Whitlock, E.E., Govardovskii, V.I., Bownds, M.D., Lamb, T.D., Pugh, E.N. Jr and Arshavsky, V.Y. (2000) Neuron. **27**, 1–20.
128. Lamb, T.D. "Walk": A program for numerical simulation of activation in the G-protein cascade. http://www.physiol.cam.ac.uk/staff/Lamb/Walk.
129. Karpen, J.W., Zimmerman, A.L., Stryer, L. and Baylor, D.A. (1988) Proc. Natl Acad. Sci. USA **85**, 1287–1291.
130. Gillespie, P.G. and Beavo, J.A. (1989) Proc. Natl Acad. Sci. USA **86**, 4311–4315.
131. Pugh, E.N. Jr and Lamb, T.D. (1990) Vision Res. **30**, 1923–1948.
132. Cote, R.H. and Brunnock, M.A. (1993) J. Biol. Chem. **268**, 17190–17198.
133. Gray-Keller, M., Denk, W., Shraiman, B. and Detwiler, P.B. (1999) J. Physiol. **519**, 679–692.
134. Lamb, T.D., McNaughton, P.A. and Yau, K.-W. (1981) J. Physiol. **319**, 463–496.
135. Cameron, D.A. and Pugh, E.N. Jr (1990) J. Physiol. **430**, 419–439.
136. Koutalos, Y., Nakatani, K. and Yau, K.-W. (1995) J. Gen. Physiol. **106**, 891–921.
137. Ruiz, M.L., Brown, R.L., He, Y., Haley, T.L. and Karpen, J.W. (1999) Biochem. **38**, 10642–10648.
138. Breton, M.E., Schueller, A.W., Lamb, T.D. and Pugh, E.N. Jr (1994) Invest. Ophthalmol. Vis. Sci. **35**, 295–309.
139. Smith, N.P. and Lamb, T.D. (1997) Vision Res. **37**, 2943–2952.
140. Cideciyan, A.V. and Jacobson, S.G. (1996) Vision Res. **36**, 2609–2621.
141. Lyubarsky, A.L. and Pugh, E.N. Jr (1996) J. Neurosci. **16**, 563–571.
142. Nikonov, S., Engheta, N. and Pugh, E.N. Jr (1998) J. Gen. Physiol. **111**, 7–37.
143. Hood, D.C. and Birch, D.G. (1990) Vis. Neurosci. **31**, 2070–2081.
144. Hood, D.C. and Birch, D.G. (1993) Vision Res. **33**, 1605–1618.
145. Pugh, E.N. Jr, Falsini, B. and Lyubarksy, A.L. (1998) in: Photostasis and Related Phenomena, eds T.P. Williams and A.B. Thistle. pp. 93–128, Plenum Press, New York.
146. Robson, J.G. and Frishman, L.J. (1998) Doc. Ophthalmol. **95**, 187–215.
147. Fersht, A. (1977) Enzyme Structure and Mechanism. W.H. Freeman, San Francisco.
148. Palczewski, K. and Benovic, J.L. (1991) Trends Biochem. Sci. **16**, 387–391.
149. Palczewski, K. (1997) Euro. J. Biochem. **248**, 261–269.
150. Zhao, X., Palczewski, K. and Ohguro, H. (1995) Biophys. Chem. **56**, 183–188.
151. Hurley, J.B., Spencer, M. and Niemi, G.A. (1998) Vision Res. **38**, 1341–1352.
152. Wilden, U., Hall, S.W. and Kuhn, H. (1986) Proc. Natl Acad. Sci. USA **83**, 1174–1178.
153. Palczewski, K., Rispoli, G. and Detwiler, P.B. (1992) Neuron **8**, 117–126.
154. Ohguro, H., Van Hooser, J.P., Milam, A.H. and Palczewski, K. (1995) J. Biol. Chem. **270**, 14259–14262.

155. Chen, C.-K., Burns, M.E., Spencer, M., Niemi, G.A., Chen, J., Hurley, J.B., Baylor, D.A. and Simon, M.I. (1999) Proc. Natl Acad. Sci. USA **96**, 3718–3722.
156. Kühn, H. (1978) Biochem. **17**, 4389–4395.
157. Palczewski, K. (1994) Protein Sci. **3**, 1355–1361.
158. Gurevich, V.V. (1998) J. Biol. Chem. **273**, 15501–15506.
159. Ames, J.B., Porumb, T., Tanaka, T., Ikura, M. and Stryer, L. (1995) J. Biol. Chem. **270**, 4526–4533.
160. Zozulya, S. and Stryer, L. (1992) Proc. Natl Acad. Sci. USA **89**, 11569–11573.
161. Ames, J.B., Tanaka, T., Stryer, L. and Ikura, M. (1996) Curr. Opin. Struct. Biol. **6**, 432–438.
162. Ames, J.B., Tanaka, T., Stryer, L. and Ikura, M. (1994) Biochem. **33**, 10743–10753.
163. Kawamura, S. (1993) Nature **362**, 855–857.
164. Chen, C.-K., Inglese, J., Lefkowitz, R.J. and Hurley, J.B. (1995) J. Biol. Chem. **270**, 1–7.
165. Fung, B.K., Hurley, J.B. and Stryer, L. (1981) Proc. Natl Acad. Sci. USA **78**, 152–156.
166. Fung, B.K. and Stryer, L. (1980) Proc. Natl Acad. Sci. USA **77**, 2500–2504.
167. Pepperberg, D.R., Jin, J. and Jones, G.J. (1994) Vis. Neurosci. **11**, 53–62.
168. Murnick, J.G. and Lamb, T.D. (1996) J. Physiol. **495**, 1–13.
169. Arshavsky, V.Y. and Bownds, M.D. (1992) Nature **357**, 416–417.
170. Arshavsky, V.Y., Dumke, C.L. and Bownds, M.D. (1992) J. Biol. Chem. **267**, 24501–24507.
171. Angleson, J.K. and Wensel, T.G. (1994) J. Biol. Chem. **269**, 16290–16296.
172. Angleson, J.K. and Wensel, T.G. (1993) Neuron **11**, 939–949.
173. Arshavsky, V.Y. and Pugh, E.N. Jr (1998) Neuron **20**, 11–14.
174. Lee, R.H., Brown, B.M. and Lolley, R.N. (1984) Biochem. **23**, 1972–1977.
175. Lee, R.H., Fowler, A., McGinnis, J.F., Lolley, R.N. and Craft, C.M. (1990) J. Biol. Chem. **265**, 15867–15873.
176. Lee, R.H., Lieberman, B.S. and Lolley, R.N. (1987) Biochem. **26**, 3983–3990.
177. Lee, R.H., Ting, T.D., Lieberman, B.S., Tobias, D.E., Lolley, R.N. and Ho, Y.K. (1992) J. Biol. Chem. **267**, 25104–25112.
178. Yoshida, T., Willardson, B.M., Wilkins, J.F., Jensen, G.J., Thornton, B.D. and Bitensky, M.W. (1994) J. Biol. Chem. **269**, 24050–24057.
179. Willardson, B.M., Wilkins, J.F., Yoshida, T. and Bitensky, M.W. (1996) Proc. Natl Acad. Sci. USA **93**, 1475–1479.
180. Muller, S., Straub, A., Schroder, S., Bauer, P.H. and Lohse, M.J. (1996) J. Biol. Chem. **271**, 11781–11786.
181. Lohse, M.J., Bluml, K., Danner, S. and Krasel, C. (1996) Biochem. Soc. Trans. **24**, 975–980.
182. Liu, X., Seno, K., Nishizawa, Y., Hayashi, F., Yamazaki, A., Matsumoto, H., Wakabayashi, T. and Usukura, J. (1994) Exp. Eye Res. **59**, 761–768.
183. Hisatomi, O., Honkawa, H., Imanishi, Y., Satoh, T. and Tokunaga, F. (1999) Biochem. Biophys. Res. Comm. **255**, 216–220.
184. Cooper, N., Liu, L., Yoshida, A., Pozdnyakov, N., Margulis, A. and Sitaramayya, A. (1996) J. Mol. Neurosci. **6**, 211–222.
185. Lowe, D.G., Dizhoor, A.M., Liu, K., Gu, Q., Spencer, M., Laura, R., Lu, L. and Hurley, J.B. (1995) Proc. Natl Acad. Sci. USA **92**, 5535–5539.
186. Yang, R.-B., Robinson, S.W., Xiong, W.H., Yau, K.-W., Birch, D.G. and Garbers, D.L. (1999) J. Neurosci. **19**, 5889–5897.
187. Paul, A.K., Marala, R.B., Jaiswal, R.K. and Sharma, R.K. (1987) Science **235**, 1224–1226.
188. Gorczyca, W.A., Gray-Keller, M.P., Detwiler, P.B. and Palczewski, K. (1994) Proc. Natl Acad. Sci. USA **91**, 4014–4018.
189. Gorczyca, W.A., Polans, A.S., Surgucheva, I.G., Subbaraya, I., Baehr, W. and Palczewski, K. (1995) J. Biol. Chem. **270**, 22029–22036.
190. Palczewski, K., Subbaraya, I., Gorczyca, W.A., Helekar, B.S., Ruiz, C.C., Ohguro, H., Huang, J., Zhao, X., Crabb, J.W. and Johnson, R.S. (1994) Neuron **13**, 395–404.
191. Dizhoor, A.M., Lowe, D.G., Olshevskaya, E.V., Laura, R.P. and Hurley, J.B. (1994) Neuron **12**, 1345–1352.

192. Dizhoor, A.M., Olshevskaya, E.V., Henzel, W.J., Wong, S.C., Stults, J.T., Ankoudinova, I. and Hurley, J.B. (1995) J. Biol. Chem. **270**, 25200–25206.
193. Duda, T., Goraczniak, R.M., Pozdnyakov, N., Sitaramayya, A. and Sharma, R.K. (1998) Biochem. Biophys. Res. Comm. **242**, 118–122.
194. Krishnan, A., Goraczniak, R.M., Duda, T. and Sharma, R.K. (1998) Molec. Cell. Biochem. **178**, 251–259.
195. Matthews, H.R., Fain, G.L. and Cornwall, M.C. (1996) J. Physiol. **490**, 293–303.
196. Otto-Bruc, A., Buczylko, J., Surgucheva, I., Subbaraya, I., Rudnicka-Nawrot, M., Crabb, J.W., Arendt, A., Hargrave, P.A., Baehr, W. and Palczewski, K. (1997) Biochem. **36**, 4295–4302.
197. Howes, K., Bronson, J.D., Dang, Y.L., Li, N., Zhang, K., Ruiz, C., Helekar, B., Lee, M., Subbaraya, I., Kolb, H., Chen, J. and Baehr, W. (1998) Invest. Ophthalmol. Vis. Sci. **39**, 867–875.
198. Yu, L. and Colvin, R.A. (1997) Molec. Brain Res. **50**, 285–292.
199. Kim, I., Koh, G.Y. and Lee, C.O. (1998) Comp. Biochem. Physiol. B **119**, 157–161.
200. Reeves, J.P. (1998) J. Bioenerg. Biomemb. **30**, 151–160.
201. Cook, O., Low, W. and Rahamimoff, H. (1998) Biochim. Biophys. Acta **1371**, 40–52.
202. Blaustein, M.P. and Lederer, W.J. (1999) Physiol. Rev. **79**, 763–854.
203. Cervetto, L., Lagnado, L., Perry, R.J., Robinson, D.W. and McNaughton, P.A. (1989) Nature **337**, 740–743.
204. Reilander, H., Achilles, A., Friedel, U., Maul, G., Lottspeich, F. and Cook, N.J. (1992) EMBO J. **11**, 1689–1695.
205. Tucker, J.E., Winkfein, R.J., Cooper, C.B. and Schnetkamp, P.P. (1998) Invest. Ophthalmol. Vis. Sci. **39**, 435–440.
206. Navanglone, A., Rispoli, G., Gabellini, N. and Carafoli, E. (1997) Biophys. J. **73**, 45–51.
207. Cooper, C.B., Winkfein, R.J., Szerencsei, R.T. and Schnetkamp, P.P. (1999) Biochem. **38**, 6276–6283.
208. Perry, R.J. and McNaughton, P.A. (1991) Curr. Opin. Neurobiol. **1**, 98–104.
209. Rispoli, G. (1998) J. Photochem. Photobiol. B **44**, 1–20.
210. Kohnken, R.E., Chafouleas, J.G., Eadie, D.M., Means, A.R. and McConnell, D.G. (1981) J. Biol. Chem. **256**, 12517–12522.
211. Nagao, S., Yamazaki, A. and Bitensky, M.W. (1987) Biochem. **26**, 1659–1665.
212. Huppertz, B., Weyand, I. and Bauer, P.J. (1990) J. Biol. Chem. **265**, 9470–9475.
213. Grunwald, M.E., Yu, W.P., Yu, H.H. and Yau, K.-W. (1998) J. Biol. Chem. **273**, 9148–9157.
214. Weitz, D., Zoche, M., Müller, F., Beyermann, M., Körschen, H.G., Kaupp, U.B. and Koch, K.W. (1998) EMBO J. **17**, 2273–2284.
215. Hsu, Y.T. and Molday, R.S. (1994) J. Biol. Chem. **269**, 29765–29770.
216. Hsu, Y.T. and Molday, R.S. (1993) Nature **361**, 76–79.
217. Koutalos, Y. and Yau, K.-W. (1996) Trends Neurosci. **19**, 73–81.
218. Rebrik, T.I. and Korenbrot, J.I. (1998) J. Gen. Physiol. **112**, 537–548.
219. Hilgemann, D.W., Nicoll, D.A. and Philipson, K.D. (1991) Nature **352**, 715–718.
220. Hodgkin, A.L. (1988) in: Proceedings of the Retina Research Foundation, ed D.M.K. Lam. Portfolio Publishing Co., pp. 6–30, The Woodlands, Texas.
221. McCarthy, S.T., Younger, J.P. and Owen, W.G. (1996) J. Neurophysiol. **76**, 1991–2004.
222. Nikonov, S., Lamb, T.D. and Pugh, E.N. Jr (2000) J. Gen. Physiol., in press.
223. Rodieck, R.W. (1998) The First Steps in Seeing. Sinauer, Sunderland.
224. Shevell, S.K. (1977) Vision Res. **17**, 427–434.
225. Finkelstein, M.A. and Hood, D.C. (1981) Vision Res. **21**, 319–328.
226. Burkhardt, D.A. (1994) J. Neurosci. **14**, 1091–1105.
227. Wässle, H. (1999) in: Color Vision: From Genes to Perception, eds K.R. Gegenfurtner and L.T. Sharpe. CUP, New York.
228. Matthews, H.R. (1991) J. Physiol. **436**, 93–105.
229. Tamura, T., Nakatani, K. and Yau, K.-W. (1989) Science **245**, 755–758.
230. Nakatani, K., Tamura, T. and Yau, K.-W. (1991) J. Gen. Physiol. **97**, 413–435.
231. Tamura, T., Nakatani, K. and Yau, K.-W. (1991) J. Gen. Physiol. **98**, 95–130.

232. Bownds, M.D. and Arshavsky, V.Y. (1995) Behav. Brain Sci. **18**, 415–424.
233. Perlman, I. and Normann, R.A. (1998) Progr. Ret. Eye Res. **17**, 523–563.
234. Baylor, D.A. and Hodgkin, A.L. (1974) J. Physiol. **242**, 729–758.
235. Fain, G.L. (1976) J. Physiol. **261**, 71–101.
236. Baylor, D.A., Matthews, G.G. and Yau, K.-W. (1980) J. Physiol. **309**, 591–621.
237. Matthews, H.R., Murphy, R.L.W., Fain, G.L. and Lamb, T.D. (1988) Nature **334**, 67–69.
238. Cornwall, M.C. and Fain, G.L. (1994) J. Physiol. **480**, 261–279.
239. Nakatani, K. and Yau, K.-W. (1988) Nature **334**, 69–71.
240. Nakatani, K. and Yau, K.-W. (1989) J. Physiol. **409**, 525–548.
241. Fain, G.L., Lamb, T.D., Matthews, H.R. and Murphy, R.L.W. (1989) J. Physiol. **416**, 215–243.
242. Dowling, J.E. and Ripps, H. (1970) J. Gen. Physiol. **56**, 491–520.
243. Lagnado, L. and Baylor, D.A. (1994) Nature **367**, 273–277.
244. Jones, G.J. (1995) J. Physiol. **487**, 441–451.
245. Matthews, H.R. (1996) J. Physiol. **490**, 1–15.
246. Matthews, H.R. (1997) J. Gen. Physiol. **109**, 141–146.
247. Chen, T.Y., Illing, M., Molday, L.L., Hsu, Y.T., Yau, K.-W. and Molday, R.S. (1994) Proc. Natl Acad. Sci. USA **91**, 11757–11761.
248. Arshavsky, V.Y., Gray-Keller, M.P. and Bownds, M.D. (1991) J. Biol. Chem. **266**, 18530–18537.
249. Erickson, M.A., Lagnado, L., Zozulya, S., Neubert, T.A., Stryer, L. and Baylor, D.A. (1998) Proc. Natl Acad. Sci. USA **95**, 6474–6479.
250. Fain, G.L., Matthews, H.R. and Cornwall, M.C. (1996) Trends Neurosci. **19**, 502–507.
251. Leibrock, C.S., Reuter, T. and Lamb, T.D. (1998) Eye **12**, 511–520.
252. Hollins, M. and Alpern, M. (1983) J. Gen. Physiol. **62**, 430–447.
253. Schnapf, J.L. (1983) J. Physiol. **343**, 147–159.
254. Baylor, D.A. (1996) Proc. Natl Acad. Sci. USA **90**, 560–565.

CHAPTER 6

Comparative Molecular Biology of Visual Pigments

S. YOKOYAMA

Department of Biology,
Syracuse University

R. YOKOYAMA

Department of Physiology,
State University of New York,
Syracuse

Handbook of Biological Physics
Volume 3, edited by D.G. Stavenga, W.J. DeGrip and E.N. Pugh Jr

Contents

1. Introduction

Organisms have adapted to diverse photic environments by modifying not only their morphological characteristics but also their visual systems. A large number of examples of protective coloration, mimicry, and sexual display have been well documented [1–4]. Animal colorations are particularly important in territorial and courtship displays, often being greatly enhanced by exaggerated and stereotyped movements and postures [5]. Vision is an essential ingredient in this biological network and has profound effects on the evolution of organisms through such behaviors as mating, foraging, and predator avoidance.

In vertebrates, vision begins when photons are absorbed by photosensitive molecules, visual pigments, in the outer segments of the photoreceptor cells, called rods and cones [6]. Visual pigments are generally called rhodopsin. In vertebrates, however, visual pigments in rod cells are referred to as rhodopsin, whereas those in cone cells are often called cone pigments or color visual pigments [7]. The visual pigments, representing a group of a superfamily of G protein-coupled receptors [8,9], are made up of an integral membrane protein, called opsin, and a chromophore. The chromophore is covalently linked via a Schiff base to a specific lysine residue of the opsin [10,11]. The light sensitivity of the visual pigments is due to interactions between chromophore and opsin and each pigment is tuned to a particular wavelength of maximal absorption, called λ_{max}. The chromophore of most vertebrate and invertebrate pigments is 11-*cis*-retinal, whereas in some fish, amphibian, and reptilian species [12], it is 11-*cis*-3,4-dehydroretinal. The 11-*cis*-3,4-dehydroretinal contains an additional carbon–carbon double bond in the β-ionone ring and increases the λ_{max} value of the pigment containing it (λ_2) in comparison to that of the same pigment with 11-*cis*-retinal (λ_1) [13,14]. In many insect species the chromophore is 11-*cis*-3-hydroxyretinal [15].

Having a crucial role in the light-dependent step in visual signal transduction, visual pigments have been subjected to intense scrutiny [7,16]. It has been found that a number of structural and functional properties are shared by different visual pigments. Visual pigments have seven stretches of hydrophobic amino acids, which is remarkably similar to that of bacteriorhodopsin, a non-G protein-coupled, light-driven proton pump from *Halobacterium halobium*. Thus, a bacteriorhodopsin-like topography with seven transmembrane helices has been proposed [17] (Fig. 1). Many features of this model appear to be valid according to a general model for G protein-coupled receptors based on a large number of amino acid sequence comparisons [18] and the projection structure of bovine rhodopsin determined by cryo-electron microscopy [19–21].

Using bovine rhodopsin, functionally important residues have been identified (Fig. 2). They include the site of Schiff base linkage to the chromophore (K296) [22],

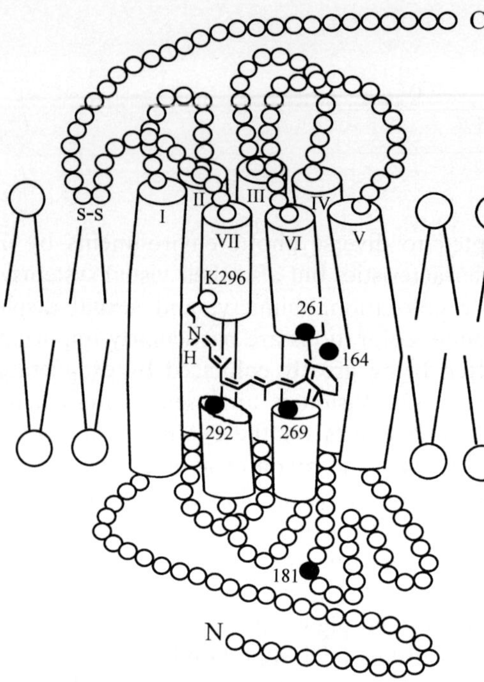

Fig. 1. Schematic model of bovine rhodopsin showing the seven transmembrane helices together with the intradiscal amino-terminus and the cytoplasmic carboxy-terminus (modified from Ref. [153]). Helices VI and VII have been cut away to demonstrate the interior placement of the chromophore. Amino acid sites are based on the human red and green pigments. The locations of five sites that are important in differentiating the spectral sensitivities of red and green pigments are shown in black circles.

N-glycosylation sites (N2 and N15) [23], palmitoylation sites (C322 and C323) [24], sites for a disulfide bond (C110 and C187) [25] and a site (E122) that together are required for the stability of metarhodopsin II [26,27], the Schiff base counterion (E113) [28–30], sites important for tranducin binding and stabilizing the inactivated state of rhodopsin (E134 and R135) [31–33], sites involved in forming the retinal binding pocket (W126, W265, and Y268) [34], sites affecting chromophore regeneration and transducin activation (A117, P267, and A292) [34], sites important for metarhodopsin I to II transition (H65, H152, and H211) [35], sites involved in both palmitoylation and phosphorylation (C140 and C185) [36], and multiple serines and threonines in the C-terminal region [37] and cytoplasmic loops I–II, III–IV, and V–VI [38–40] for potential phosphorylation. Some structural changes of the rhodopsin upon light exposure are also known [41–44]. Furthermore, most of the point mutations in the transmembrane regions identified in autosomal dominant retinitis pigmentosa (ADRP) cause partial or complete non-retinal-binding rhodopsins [45,46].

The molecular genetic analyses of color vision became feasible when Nathans and his colleagues cloned and characterized the human red, green, and blue color

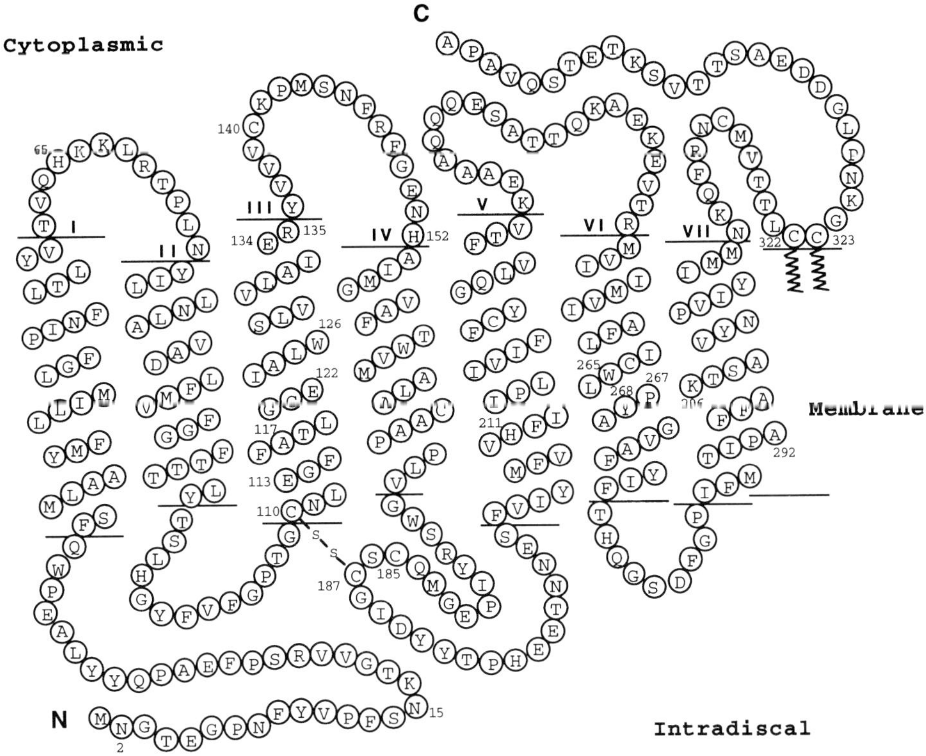

Fig. 2. Schematic representation of bovine rhodopsin (modified from Ref. [17]). Numbers next to amino acids indicate amino acid sites.

opsin genes [47]. With the rapid accumulation of nucleotide sequences of various opsin genes in vertebrate and invertebrate species, we now have a much better understanding than before on the molecular bases of color-blindness [30,47–49] and ADRP [46,50–53]. Comparative analyses of opsins have also revealed the importance of gene duplications and accumulation of mutations leading to the extant diverse opsins. The comparative analyses also identify amino acid changes that are potentially important in controlling the absorption spectra of the visual pigments [54]. Fortunately, virtually any opsin can now be easily manipulated, expressed in cultured cells, reconstituted with 11-*cis*-retinal, and the λ_{max} value of the regenerated visual pigment can be measured [26,28,30,55–60]. Thus, genetic hypotheses derived from the comparative analyses of opsin sequences can now be experimentally tested.

A central unanswered question in phototransduction is how visual pigments achieve a wide range of λ_{max} values, from the ultraviolet, at about 350 nm, to the far red, at about 630 nm. Several authors have constructed mutant pigments and tested the effects of various mutations on the shift of the λ_{max} values of the regenerated visual pigments [26,28–30,34,55]. However, virtually none of these mutants tested are found in nature, and the significance of these amino acid changes in the spectral

Table 1
Visual pigments from vertebrates[a] and invertebrates[b]

Visual pigments	λ^{c}_{max} (nm)	Reference	Notation	GenBank
Rhodopsin (RH1)				
River lamprey	P500	[158]	R. lamprey-1	M63632
Marine lamprey	P500	[159]	M. lamprey-1	U67123
American Chameleon	P491	[98]	Chameleon-1	L31503
Skate	P500	[160]	Skate-1	U81514
River European eel	P502	[161]	R. Eur. eel-1	L78007
Conger eel	P487	[162]	Conger eel-1	S82619
Marine European eel	P482	[161]	M. Eur. eel-1	L78008
Cavefish	P504	[58]	Cavefish-1	U12328
Carp	P499	[163]	Carp-1	S74449
Goldfish	P492	[96]	Goldfish-1	L11863
John Dory	P492	[164]	John Dory-1	Y14484
Squirrel fish	?		Sq. fish-1	U57536
Sand Goby	P501	[165]	Sand goby-1	X62405
Killifish	?		Killifish-1	AB001606
Guppy	P500	[166]	Guppy-1	Y11147
Mosquito fish	P505	[167]	Mosq. fish-1	Y11146
Salamander	P506	[168]	Salamander-1	U36574
African Clawed frog	P502	[169]	Clawed frog-1	L07770
Leopard frog	P502	[170]	Leop. frog-1	S49004
Bullfrog	P500	[171]	Bullfrog-1	S79840
Alligator	P499	[172]	Alligator-1	U23802
Chicken	P503	[91]	Chicken-1	D00702
Dolphin	P488	[60]	Dolphin-1	AF055456
Bovine	P500	[173]	Bovine-1	M21606
Dog	P508	[174]	Dog-1	X71380
Rat	P498	[175]	Rat-1	Z46957
Mouse	P498	[175]	Mouse-1	M55171
Chinese Hamster	?		Hamster-1	X61084
Rabbit	P502	[175]	Rabbit-1	U2168
Human	P497	[176]	Human-1	U49742
Macaque monkey	P500	[177]	Macaque-1	S76579
Rhodopsin-like (RH2)				
Killifish	?		Killifish-2	AB001603
Cavefish	?		Cavefish-2	S75251
Goldfish-Green1	P511	[96]	Goldfish-G1	L11865
Goldfish-Green2	P506	[96]	Goldfish-G2	L11866
Chicken	P508	[91]	Chicken-G	M92038
American Chameleon	P495	[98]	Chameleon-2	AF134189
Gecko	P466	[178]	Gecko-B	M92035
SWS1				
Killifish	?		Killifish-V	AB001605
Goldfish	P359	[111]	Goldfish-UV	D85863
African Clawed frog	P425	[179]	Clawed frog-V	U23463
Chicken	P415	[91]	Chicken-V	M92039
Pigeon	P393	[111]	Pigeon-near UV	AF149234
American Chameleon	P358	[98]	Chameleon-UV	AF134192

Visual pigments	λ^{c}_{max} (nm)	Reference	Notation	GenBank
Bovine	?		Bovine-B	U92557
Mouse	P359	[111]	Mouse-UV	U49720
Rat	P358	[111]	Rat-UV	U63972
Marmoset	P424	[180]	Marmoset-B	L22218
Squirrel monkey	P433	[134]	Sq. monkey-B	U53875
Human	P424	[57]	Human-B	M13295
Talapoin monkey	?		Talapoin-B	L76226
SWS2				
Killifish	?		Killifish-B	AB001602
Cavefish	P453	[102]	Cavefish-B	AF134762
Goldfish	P441	[96]	Goldfish-B	L11864
Chicken	P455	[91]	Chicken-B	M92037
American Chameleon	P437	[98]	Chameleon-B	AF133907
LWS/MWS				
Gecko	P521	[178]	Gecko-G	M92036
Cavefish-G101	P533	[102]	Cavefish-G101	M38619
Cavefish-G103	P533	[102]	Cavefish-G103	U12025
Killifish	?		Killifish-R	AB001604
Goldfish	P525	[96]	Goldfish-R	L11867
Cavefish-R	P563	[102]	Cavefish-R	M38625
African clawed frog	P611	[181]	Clawed frog-R	U90895
Chicken	P571	[91]	Chicken-R	M62903
American Chameleon	P561	[98]	Chameleon-R	U08131
Goat	P553	[182]	Goat-R	U67999
Dolphin	P524	[60]	Dolphin-G	AF055457
Rabbit	P509	[183]	Rabbit-R	AF054235
Rat	P509	[183]	Rat-G	AF054241
Mouse	P508	[59]	Mouse-G	AF011389
Human-R	P560	[57]	Human-R	M13300
Marmoset	P560	[134]	Marmoset-R	Z22218
Human-G	P530	[57]	Human-G	K03490
P				
Channel catfish-P	?		Catfish-P	AF028014
Marine lamprey	P545	[184]	M. lamprey-P	U90667
Salmon-VA	?		Salmon-VA	AF001499
American Chameleon	P482	[98]	Chameleon-P	AF134767
Chicken	P470	[121]	Chicken-P	U15762
Pigeon	?		Pigeon-P	U50598
Invertebrate opsin				
Fly (mela)-1	P480	[185,186]	Fly (mela)-1	K02315
Fly (pseudo)-1	P480	[187]	Fly (pseudo)-1	X65877
Blowfly-1	P490	[188]	Blowfly-1	M58334
Fly (mela)-2	P450	[189,190]	Fly (mela)-2	M12896
Fly (pseudo)-2	P420	[187]	Fly (pseudo)-1	X65878
Crayfish	P535	[191]	Crayfish	AF003543
Horseshoe crab	P520	[192]	Hshoe crab	L03791
Locust-2	?		Locust-2	X80072
Mantis	?		Mantis-1	X71665

Table 1 (*continuation*)

Visual pigments	λ^{c}_{max} (nm)	Reference	Notation	GenBank
Silver ant	?		Silver ant	U32501
Carpenter ant	?		Carpenter ant	U32502
Honeybee-G	P540	[193]	Honeybee-G	U26026
Fly (mela)-6	?		Fly (mela)-6	Z86118
Crab-1	P480	[194]	Crab-1	D50583
Crab-2	P480	[194]	Crab-2	D50584
Fly (mela)-3	P370	[195–197]	Fly (mela)-3	M17718
Fly (pseudo)-3	?		Fly (pseudo)-3	X65879
Fly (mela)-4	P370	[196,198]	Fly (mela)-4	M17719
Fly (pseudo)-4	?		Fly (pseudo)-4	X65880
Locust-1	?		Locust-1	X80071
Honeybee-UV	P353	[193]	Honeybee-UV	U70841
Fly (mela)-5	?		Fly (mela)-5	U80667
Scallop	?		Scallop	AB006454
Flying squid	P480	[199,200]	Flying squid	X70498
Octopus	?		Octopus	X07797
European squid	?		Eur. squid	X56788

[a] *Vertebrates*: Alligator, *Alligator mississippinensis*; American chameleon, *Anolis carolinensis*; bovine, *Bos taurus*; bullfrog, *Rana catesbeiana*; carp, *Cyprinus carpio* (estimated from λ_2 using a formula $\lambda_1 = (\lambda_2 + 263)/1.575$) [14]; cavefish, *Astyanax fasciatus*; channel catfish, *Ictalurus punctatus*; Chicken, *Gallus gallus*; Chinese hamster, *Cricetulus griseus*, clawed toad, *Xenopus laevis*; conger eel, *Conger conger*; dog, *Canis familiaris*; dolphin, *Tursiops truncatus*; gecko, *Gekko gekko*; goat, *Capra hircus*; goldfish, *Carassius auratus*; guppy, *Poecilia reticulata*; human, *Homo sapiens*; John Dory, *Zeus faber*; killifish, *Oryzias latipes*; macaque, *Macaca fascicularis*; marine European eel, *Anguilla anguilla*; marine lamprey, *Lamptera marinus*; marmoset, *Callithrix jacchus*; mosquito fish, *Gambusia affinis*; mouse, *Mus musculus*; Northern leopard frog, *Rana pipiens*; pigeon, *Columba livia*; rabbit, *Oryctolagus cuniculus*; rat, *Rattus norvegicus*; river European eel, *Anguilla anguilla*; river lamprey, *Lamptera japonica*; salamander, *Ambystoma tigrinum*; salmon, *Salmo salaar*; sand Goby, *Pomatoschistus minutus*; skate, *Raja erinacea*; squirrel fish, *Neoniphon sammara*; squirrel monkey, *Saimiri boliviensis*; and Talapoin monkey, *Miopithecus talapoin*.
[b] *Invertebrates*: Blowfly, *Calliphora vicina*; carpenter ant, *Cataglyphis bombycina*; crab, *Hemigrapsus sanguineus*; crayfish, *Procambarus milleri*; European squid, *Loligo forbesi*; fly (mela), *Drosophila melanogaster*; fly (pseudo), *Drosophila pseudoobscura*; flying squid, *Todarodes pacificus*; honeybee, *Apis mellifera*; horseshoe crab, *Limulus polyphemus*; locust, *Schistocerca gregaria*; mantis, *Sphodromantis* sp.; octopus, *Paroctopus defleini*; scallop, *Patinopecten yessoensis*; silver ant, *Camponotus abdominalis*.
[c] The λ_{max} values of regenerated visual pigments using cultured cells are underlined; "?" – λ_{max} unknown.

Chameleon-UV, Rat-G, and Chameleon-P genes span more than 19 kb in length from start to stop codons. On the other hand, the corresponding lengths of the intron-less Cavefish-1 (and other fish RH1 genes [86]), Fly (mela)-3, and Fly (pseudo)-3 genes are only 1059, 1152, and 1149 bp, respectively. Furthermore, the lengths of the paralogous genes in the same species can vary substantially (Fig. 3).

Comparing the American chameleon and Fly (mela) opsin genes (Fig. 4), the vertebrate opsin genes classified as RH1, RH2, SWS1, SWS2, and P-opsin groups (Table 1) [54] contain 5 exons, while those in the LWS/MWS group [54] contain 6 exons. The introns 1, 2, 3, and 4 of the RH1, RH2, SWS1, and SWS2 genes and introns 2, 3, 4, and 5 of the LWS/MWS genes interrupt their coding sequences at

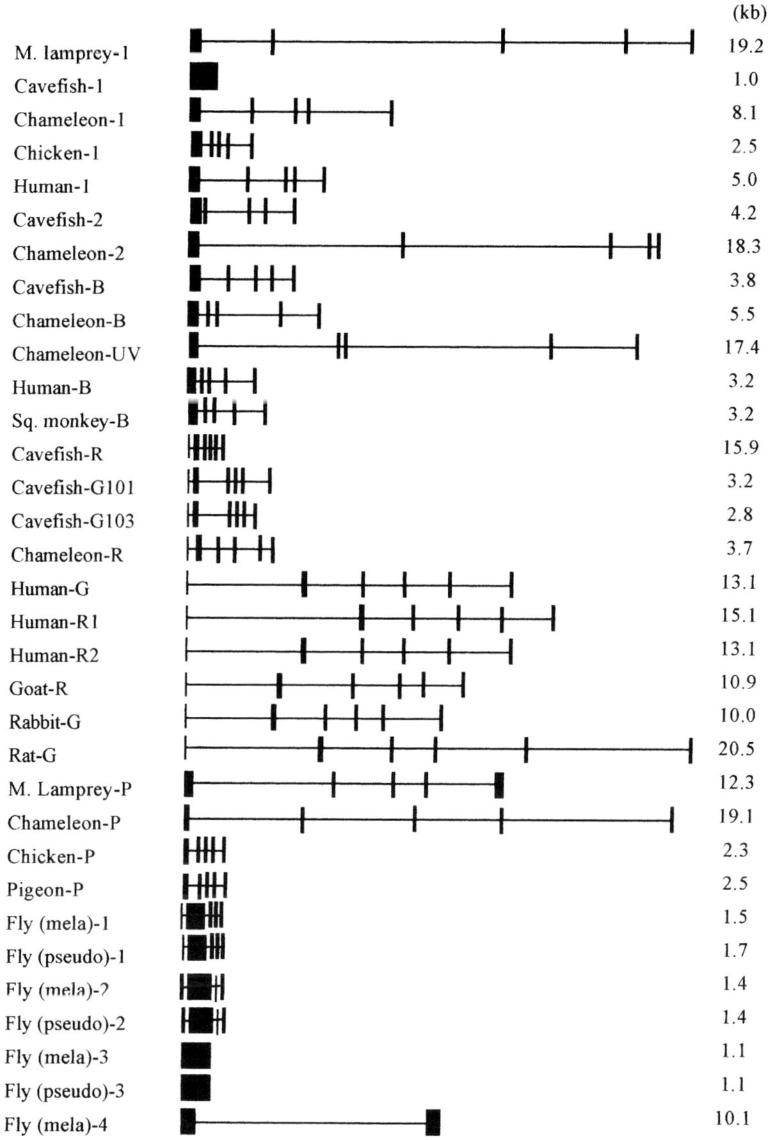

Fig. 3. Structures of opsin genes, where exons and introns are represented by black boxes and lines, respectively. Human-R1 and Human-R2 genes have intron 1-length polymorphism, one of them being 2 kb longer than the other [142].

exactly the same corresponding sites. The position of the second intron of the P-opsin gene is displaced 14 nucleotides toward the 3′ direction compared to those of the equivalent intron of the retinal opsin genes [87–89]. Except for the intronless RH1 fish opsin genes characterized, all vertebrates have completely conserved intron

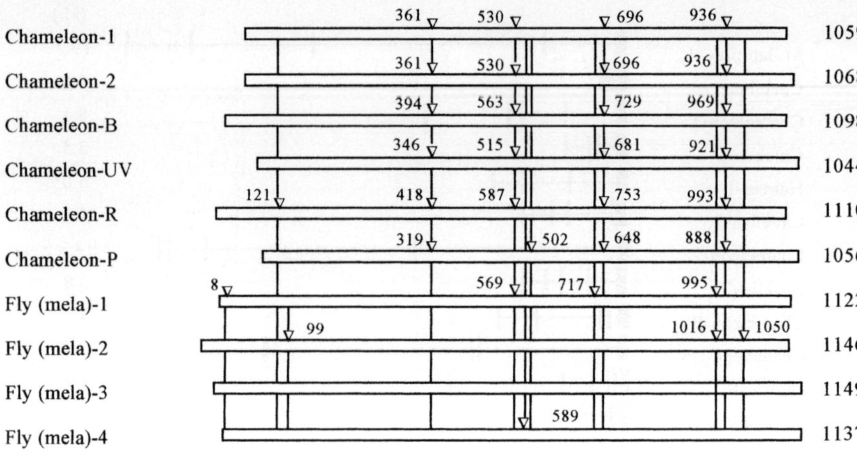

Fig. 4. Positions of introns, indicated by open triangles, in the opsin genes. The numbers indicate the position where introns start.

positions within each opsin group. Since the M. lamprey-1 gene contains four introns at the same position of the corresponding sites of other RH1 genes [90], the ancestral fish species must have lost all introns at an early stage of fish evolution. The number and position of introns in invertebrate opsin genes are more variable than those of the corresponding genes of vertebrates. The Fly (mela)-1, -2, -3, and -4 genes contain 4, 3, 0, and 1 introns, respectively (Fig. 3). The positions of these introns are not conserved among the four groups of invertebrate opsin genes, with the exception of one intron position between Fly (mela)-1 and -2 genes (Fig. 4). When the invertebrate and vertebrate opsin genes are compared, only the intron 2 of Fly (mela)-1 gene is found to match the intron 2 positions of RH1, RH2, SWS1, and SWS2 genes and the intron 3 positions of LWS/MWS genes in vertebrate (Fig. 4).

5. Evolution of opsins

The 105 opsins deduced from DNA and cDNA sequences can be classified into vertebrate and invertebrate opsin groups (Fig. 5). The vertebrate opsins can be classified into RH1, RH2, SWS1, SWS2, LWS/MWS, and P groups (Fig. 5). The RH1, RH2, LWS/MWS, SWS1, and SWS2 groups correspond to Rh, M2, L, S, and M1 groups in Okano et al.'s classification [91]. Since the divergence of the six opsin groups predates that of different vertebrate lineages (Fig. 5) [6,54], the ancestors of all vertebrates must have possessed all of the six types of opsin genes. The invertebrate opsins can be classified into four major groups: (1) Rh1/2/6 group; (2) crab opsins; (3) Rh3/4/5 group; (4) mollusc group (Fig. 5). Since at least one intron position has been conserved between vertebrate and invertebrate opsin genes, the intron loss must have occurred after the divergence of vertebrate and invertebrate species. The phylogenetic relationships in Fig. 5 clearly show that the loss of introns

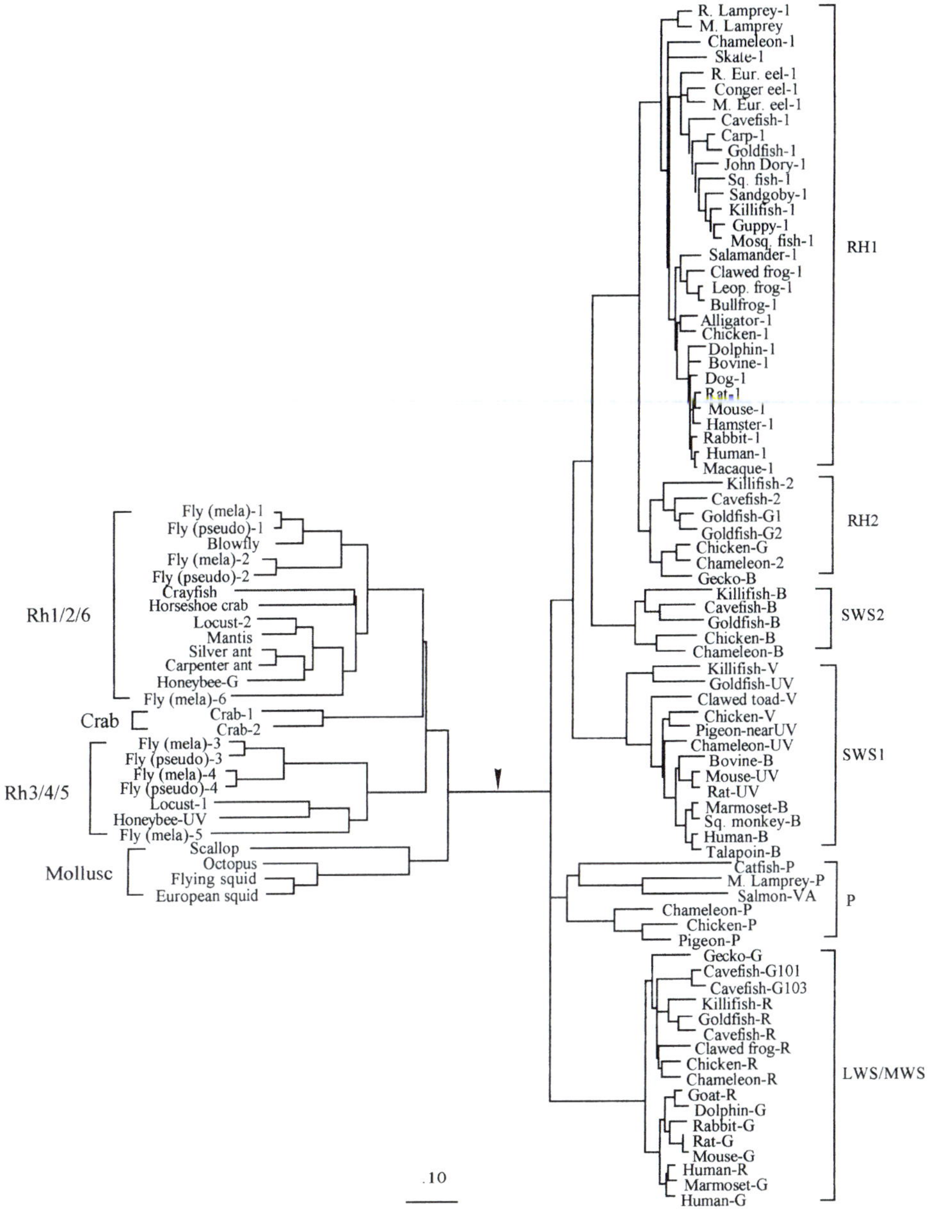

Fig. 5. Phylogenetic tree for the vertebrate and invertebrate pigments based on the numbers of amino acid replacements per site. The root of the tree, shown by the triangle, is located by using olfactory proteins in rat (GenBank M64376) and human (X64994). The bar at the bottom indicates evolutionary distance measured as the number of amino acid replacements per site.

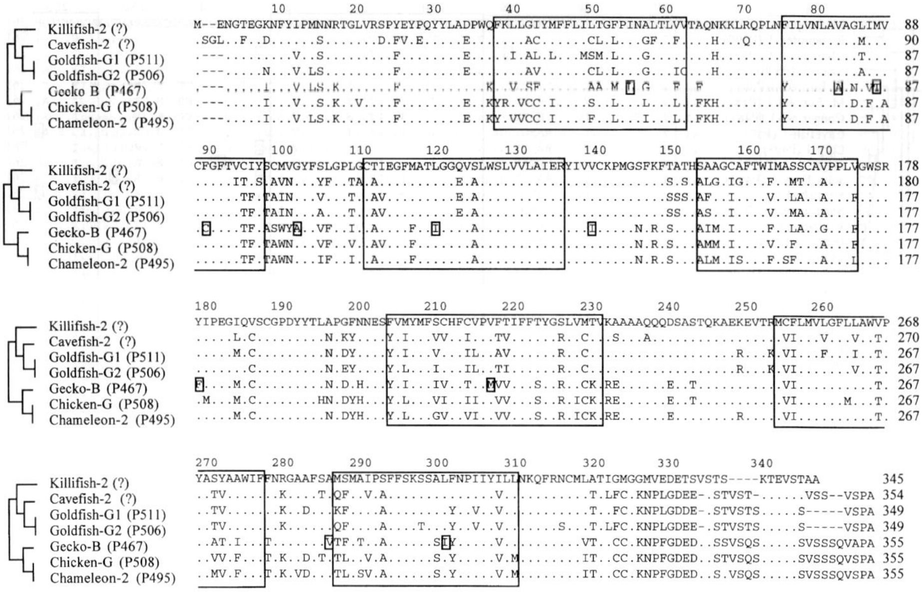

Fig. 8. Amino acid sequences of RH2 pigments. Amino acid numbering at the top corresponds to those of Bovine-1 pigment.

UV vision is available to many fish, bird, amphibian, reptilian, and mammalian species [103]. The petals of bird-pollinated flowers have substantial UV reflectance, which may provide attractive targets to birds with UV vision. Similarly, the scales of fish and the feathers of birds often reflect UV. In these species, UV vision is important in enhancing the visibility of their body coloration patterns [104,105]. The recent finding of dewlap UV reflection of five species of Puerto Rican anoline species provides another example of possible adaptive change. Three species with UV-bright dewlaps live in microhabitats that are often exposed to direct sunlight, while the two species with no UV reflection live in the understory of closed-canopy forest, where little UV light is available [106]. Thus, studies on the relationship between UV vision of these species and availability of UV light in their habitats will shed light on the adaptation and development of UV vision.

It was previously reported that the UV opsin cDNA clone was isolated from zebrafish [107]. This UV opsin was shown to be evolutionarily most closely related to the Goldfish-1 opsin [6,54,61,92,108]. Robinson et al. [107] suggested that only one amino acid replacement W126K was responsible for the λ_{max}-shift from 500 to 360 nm in the zebrafish UV pigment. However, subsequent work showed the supposedly UV opsin is in fact expressed in rods rather than in cones [109] and is not the real UV opsin of zebrafish [110]. In fact, the K126 reported by Robinson et al. [107] appears to be in error. Thus, all currently known violet and UV opsins belong to the SWS1 cluster (Fig. 5).

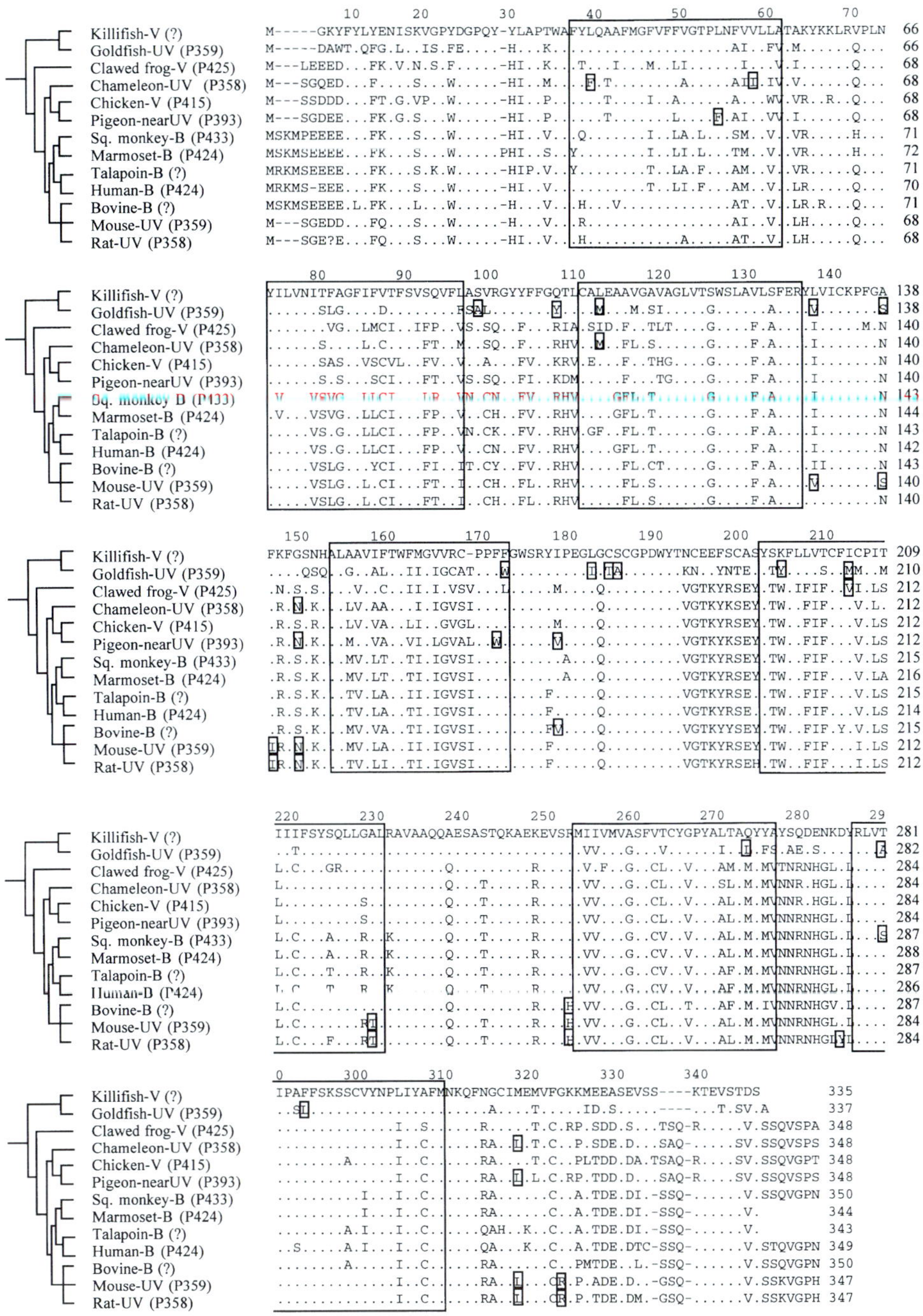

Fig. 9. Amino acid sequences of SWS1 pigments. Amino acid numbering at the top corresponds to those of Bovine-1 pigment.

To understand the molecular basis of UV vision, it is essential to construct the UV pigments and analyze their mutants using cultured cells. We succeeded for the first time to regenerate the American chameleon UV pigments in cultured COS cells, and we evaluated its λ_{max} value to be 358 nm [98]. More recently, the visual pigments of mouse, rat, goldfish, and pigeon with the respective λ_{max} values of 359, 358, 359, and 393 nm have also been regenerated (Fig. 10) [111]. The construction and functional assays of the UV and near-UV pigments from a wide range of vertebrate species allows us to study the molecular bases of UV vision, and specifically the effects of amino acid polymorphism of opsins on the λ_{max} value. In Fig. 9, 28 UV or near-UV pigment-specific amino acid replacements at highly conserved sites can be identified. Among these, amino acid replacements L112M, I137V, S149N, A229T, R252H, and M318L are shared by at least two different pigments. In particular, both S149N and M318L occurred in Chameleon-UV, Pigeon-near UV, Mouse-UV, and Rat-UV pigments. There is no common amino acid replacement to all UV pigments, strongly suggesting that no single mechanism is responsible for the development of UV pigments. Curiously, despite the different molecular bases of UV sensitivity and a wide range of available UV light at 40–400 nm [112], many UV pigments are tuned to detect a surprisingly narrow range of wavelengths at around 360 nm (Table 1).

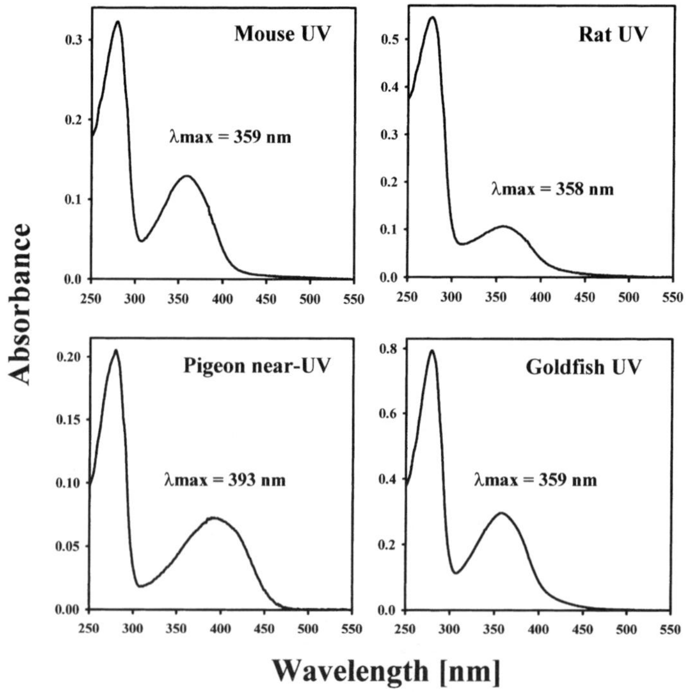

Fig. 10. Absorption spectra of Mouse-UV, Rat-UV, Pigeon-near UV, and Goldfish-UV pigments. Spectra were set to 0 at 550 nm (after Ref. [111]).

5.4. LWS/MWS group

All green and red pigments in the LWS/MWS group are responsible for achieving λ_{max} values of 508–533 and 553–611 nm, respectively (Table 1). As already noted, Chicken-G, Goldfish-G1, and Goldfish-G2 pigments of the RH2 group also mediate green sensitivity. One LWS and two RH2 opsin cDNA clones have been isolated from goldfish [96], whereas the relatively closely related cavefish has one RH2, two MWS, and one LWS genes. Why do goldfish and cavefish use their RH2 and LWS/MWS genes, respectively, for their green vision? It appears that gene duplication of the ancestral green and red opsin genes precedes the divergence of the two fish species (Fig. 5), suggesting that goldfish should also have an additional gene which is orthologous to that encoding the cavefish green opsin (Cavefish-G101 and Cavefish-G103) [113]. This argument is also supported by Southern analysis [113].

In addition, when the goldfish red opsin, which is orthologous to the cavefish red opsin, was expressed in cultured cells and reconstituted with 11-*cis*-retinal, the resulting pigment had a λ_{max} value of 525 nm rather than the expected value of 560 nm [96]. Thus, the real goldfish red opsin has not yet been identified. If the second LWS/MWS gene implicated by Southern analysis turns out to be the real LWS/MWS opsin gene, then the actual functions of Goldfish-G1, Goldfish-G2, and Goldfish-R pigments in goldfish color vision need to be re-evaluated.

5.5. P(pineal opsin) group

Virtually all eukaryotes and some prokaryotes are known to express daily rhythms in their behavior, physiology, and biochemistry [114]. These rhythms are synchronized by environmental cycles of light and temperature. The pineal glands of birds, reptiles and fish [115], the hypothalamic suprachiasmatic nucleus of mammals [116], and the retina of amphibians [117] and mammals [118] are known to reveal circadian oscillations. Dissociated pineal cell cultures reveal three major components: (1) a photosensitive input pathway; (2) a circadian oscillator that generates the rhythm; (3) an output pathway that results in the synthesis of melatonin [119].

Recently, the pineal gland-specific opsin (pinopsin or P-opsin) genes have been isolated from marine lamprey [120], chicken [87,121], pigeon [88], and American chameleon [89]. The phylogenetic analysis suggests that Salmon-VA and Channel catfish-P pigments (Table 1) also belong to this group, and furthermore that these P-opsins and the retinal opsins are evolutionarily related (Fig. 5). How commonly is the P-opsin gene found among vertebrates? Southern hybridization analysis detects the P-opsin gene in marine lamprey, birds, and reptiles, but not in frog (*Xenopus laevis*), man, and bovine, implying that it has disappeared from the latter species during vertebrate evolution [89]. Then, what types of opsins are used in the pineal glands of teleosts, amphibians, and even in some mammals? Perhaps these species use retinal opsins and/or entirely new, yet unknown, pineal gland-specific opsins for pineal phototransduction.

Lizards have an additional non-visual photosensitive organ, the parietal eye, that is suspected to enhance the detection of dawn and dusk [122]. RT-PCR analysis of American chameleon shows that: (1) all RH1, RH2, SWS1, SWS2, and LWS/MWS opsins, but not the P-opsin, are expressed in the retina; (2) the expression of the P-opsin and much lower levels of SWS1, SWS2, and LWS/MWS opsin expression are detected in the pineal gland; (3) SWS1 opsin and much less SWS2 and LWS/MWS opsins and P-opsin are detected in the parietal eye [89]. In chicken, both P-opsin and LWS/MWS opsin are expressed in the pineal gland [121]. The significance of these expression patterns of different visual opsins in non-visual photosensitive organs remains to be elucidated.

5.6. Paralogous opsins with similar functions

Without using colored oil-droplets or 11-*cis*-3,4-dehydroretinal, blue and red/green vision of vertebrates are typically mediated by SWS and LWS/MWS pigments, respectively. However, the RH2 pigments are used for blue vision (gecko) and green vision (chicken and possibly goldfish). Why were paralogous opsins used for blue and green vision? Although the chicken has six major groups of opsins, it lacks the MWS opsin so that it probably had to use the RH2 opsin for green vision. Why did not gecko, like other vertebrates, use either SWS1 or SWS2 pigments for its blue vision? It turns out that the nocturnal gecko has only rods in its retina [6]. Since RH2 is more closely related to rod-specific RH1 than to cone-specific SWS1 and SWS2 (Fig. 5), the RH2 might have become adapted to the pure-rod retina more easily than the SWS opsins. At present, it is not known if gecko still possess SWS opsins or not.

6. Spectral tuning

6.1. Theory

A protonated retinylidene Schiff base alone in solution absorbs light maximally at 440 nm [123]. Any shift from this value, called the 'opsin shift', may come from interactions between the opsin and its retinylidene chromophore. Thus, Kropf and Hubbard [124] suggested an electrostatic interaction between the retinylidene Schiff base and charged amino acid sites in the opsin (see also Ref. [125]). To test this 'electrostatic interaction' model, several authors modified amino acids of bovine rhodopsin by site-directed mutagenesis [28–30,55]. Eight amino acid changes out of about 30 such mutations caused a shift of more than 5 nm in the λ_{max} [54,61,92]. Among these, E113Q causes the most dramatic effect, shifting λ_{max} from 500 to 380 nm [28–30,126]. How realistic are these mutagenesis experiments with respect to spectral tuning? In fact, no species is known to have used E113Q for an actual hypsochromic shift in λ_{max}. So far, virtually none of these 'electrostatic' mutants has been found in nature [54,61,92].

How can we conduct more realistic mutagenesis experiments? In principle, the answer is simple. Since some mutations have allowed organisms to adapt to different photic environments by natural selection, our task is to identify such mutants that are in fact responsible for the shift in λ_{max}. Obviously, evolutionary genetic methods will provide one of the most effective ways in solving this problem. In turn, this approach will elucidate the molecular mechanisms involved in the spectral tuning. Molecular mechanisms involved in the natural selection of visual pigments must be studied basically in two steps. First, we need to identify amino acid changes of visual pigments that may change their λ_{max} values. Once potentially important amino acid changes are identified, their effects on the shifts in λ_{max} values have to be determined. To establish the correlation between certain amino acid changes in the visual pigments and the directions of the λ_{max}-shifts, we first construct the evolutionary tree for different visual pigments and infer not only the times and directions of their λ_{max}-shifts in the evolutionary tree but also amino acid sequences of ancestral organisms [127]. Then, we can determine the directions of spectral shifts induced by potentially important amino acid changes. As we already witnessed, we may consider particularly the highly conserved sites, because evolutionary conservation often implies functional importance in this process [7,54,61,65]. Perhaps surprisingly, the amino acid changes indicated by the 'evolutionary models' usually do not overlap with those in the 'electrostatic interaction models'. They hence provide an entirely new set of amino acid substitutions of possible functional importance [61]. This observation is in fact fully reasonable, because evolution often depends on conservative amino acid substitutions. Without comparative analyses, mutagenesis experiments tend to rely on amino acid changes with drastic physicochemical differences and they are therefore often biologically unrealistic.

So far, considering mutations based on evolutionary models and on electrostatic interaction models together with actual opsin polymorphism among ADRP patients, a total of about 160 bovine rhodopsin mutants has been constructed and the λ_{max} values of the mutants have been evaluated (Fig. 11). Among these, 28 mutants show λ_{max} shifts of more than 5 nm, of which E113D, E122Q, H211C, F261Y, W265Y, A269T, and A292S are found in nature and I123R, A124R, S127R, and A164I are found among ADRP patients; the remaining 17 mutants have not yet been found in natural populations.

We have already encountered some applications of the evolutionary model in our discussion of polymorphic RH1, RH2, and UV pigments. Among the RH1 pigments, Chameleon-1, M. Eur. eel-1, Conger eel-1, Goldfish-1, John Dory-1, and Dolphin-1 pigments have ~10 nm hypsochromic-shifted λ_{max} values, whereas the Dog-1 pigment has ~10 nm bathochromic-shifted λ_{max} value (Table 1). For these pigments, a total of 18 different types of amino acid replacements can be found at highly conserved sites (Fig. 7): S22N, M155I, F159C, N199H, E232A, and T319M for Chameleon-1 pigment; A292S and C322F for M. Eur. eel-1 pigment; F159L, A292S, and C322F for Conger eel-1 pigment; V81I, A82S, Q237H, and Q238E for Goldfish-1 pigment; A168S, E232D, T251S, and Y274F for John Dory-1 pigment; and F159L, A292S, and K325R for Dolphin-1 pigment. Of the latter mutations, A292S is known to cause about a 10 nm hypsochromic shift in the λ_{max} value of Bovine-1 pigment (Fig. 11). Thus, the hypsochromic-shifted λ_{max} values of M. Eur. eel-1, Conger eel-1, and Dolphin-1 pigments can be explained by this single amino acid replacement. The phylogenetic tree in Fig. 7 suggests that A292S occurred in the common ancestor of the two eel pigments and in the ancestor of Dolphin pigment independently from each other. The molecular mechanisms for the

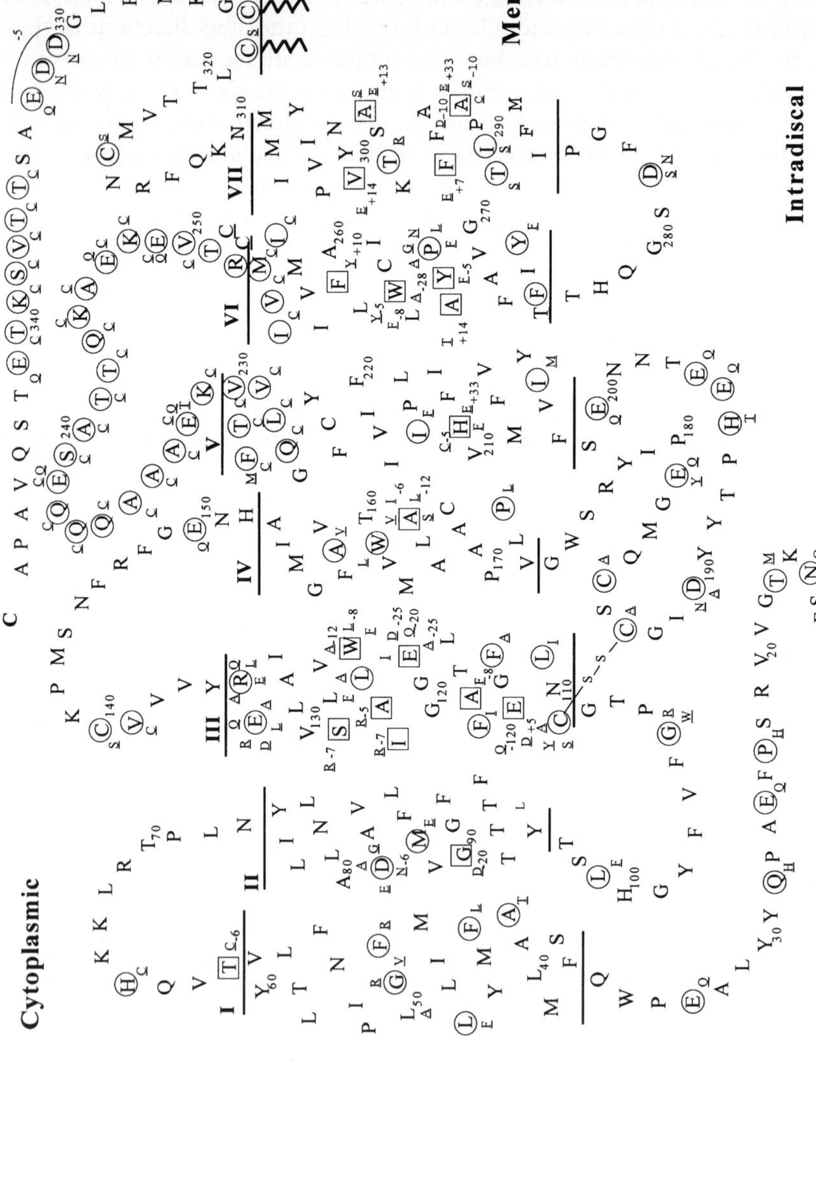

Fig. 11. Caption opposite.

hypsochromic shift in the λ_{max} values of Chameleon-1, Goldfish-1, and John Dory-1 pigments still need to be elucidated. As for the bathochromic shift in the λ_{max} of Dog-1 pigment, two amino acid replacements, V198I and T213A, are good candidates for shifting the λ_{max} value toward longer wavelengths, because amino acids at these sites are highly conserved among the mammalian pigments (Fig. 7). Similarly, a total of 25 potentially important amino acids that may cause the hypsochromic shifts of the LWS/MWS pigments can be identified (Fig. 12). Among these, T11, E21, L62, I127, I132, S180, Y197, F277, A285, and S308 are shared by different pigments. As we will see in the following, amino acids at the last five sites are shown to be important in distinguishing the red and green pigments [56,59,62,128,129]. The effects of these amino acid changes on λ_{max}-shifts have to be studied by constructing appropriate opsin mutants in cultured cells.

The evolutionary model can make false predictions either due to sequencing errors or due to insufficient data (e.g., see Ref. [54]). These negative results demonstrate that, in order to identify important amino acid changes that are responsible for λ_{max}-shifts, we need a larger number of visual pigments with variable λ_{max} values and their opsin sequences. To make the prediction based on the 'evolutionary model' more accurate, it is desirable to have additional biophysical analyses. The tentative tertiary structure of visual pigments and experimental analyses on visual pigment activation have been used to elucidate functionally important processes during activation of rhodopsin and other G protein-coupled receptors [130]. The

---

Opposite: Fig. 11. Amino acid changes and λ_{max}-shifts. Underlined amino acids indicate mutant forms, where red- and blue-shifts in λ_{max} are shown by positive and negative values, respectively. Squares and circles indicate λ_{max}-shifts with more than or less than 5 nm, respectively. Numbers next to amino acids indicate amino acid sites. N2Q [154], G3C [154], T4K [52,154], E5Q [30], N15Q [154], T17M [52,154], P23H [45,52], E25Q [30], Q28H [52], E33Q [30], F45L [52], G51V [46,52], G51R [52], G51A [46], T62C [41], H65C [41,43,44], D83A [29], G83E [29], D83G [55], D83N [29,30,34,55,60], M86E [55], G90D [52], G106R [52], G106W [52], C110A [27], C110S [36], C110Y [52], E113Q [28–30], E113D [28], F115A [34], A117F [34], E122I [55], E122D [28,29,34], E122Q [28–30,34,55], E122A [29,34], I123R, A124R, L125A, L125F, S127R (all [53]), W126F, W126L, W126A (all [34]), E134A [28], E134R [28], E134Q [28–30,55,155,156], E134D [28,155], E134L [55], R135Q [28,155,156], R135E [28,55,155], R135L [55], R135A [155], V139C [42], C140S [44,156], E150Q [30], W161L [34], A164V [46], A164S [56], A164I [46], A164L [46], P171L [52], E181Q [30], C185A [27], C187A [27], D190N [30], D190A [45], E196Q [30], E197Q [30], E201Q [29,30], H211C [55], H211F [55], H211E [26], Q225C, L226C, V227C, F228C, T229C, V230C, K231C, E232C, A233C, A234C, A235C, Q236C, Q237C, Q238C, E239C, S240C, A241C, T242C, T243C, Q244C, K245C, A246C, E247C, K248C, E249C, V250C, T251C, R252C, M253C, V254C, I255C, and I256C (all in [157]), E232Q [30], E239Q [30,155], E247Q [30], E249Q [30], F261Y [56], W265Y, W265F, W265A, P267G, P267A, P267S (all [34]), P267L [46], Y268F [34], A269T [56], Y274F [34], D282N [30], A292E [26], A292D [34], A292S [59,60], A292C [59,60], F293E [26], T297R [46], A299S [60], A299E [26], V300E [26], C316S [36,156], C322S [36,41,43], C323S [36,41,43], T335C, T336C, V337C, S338C, K339C, and T340C (all in [44]), E341Q [30], a triple mutant D330N/D331N/E332N [30], and A42T, L46F, L99F, L112I, F116I, A158I, E181Y, H195T, I214F, F228M, E232T, F276T, D282S, T289S, and I290S [S. Yokoyama, G. Yu, and F.B. Radlwimmer, unpublished results].

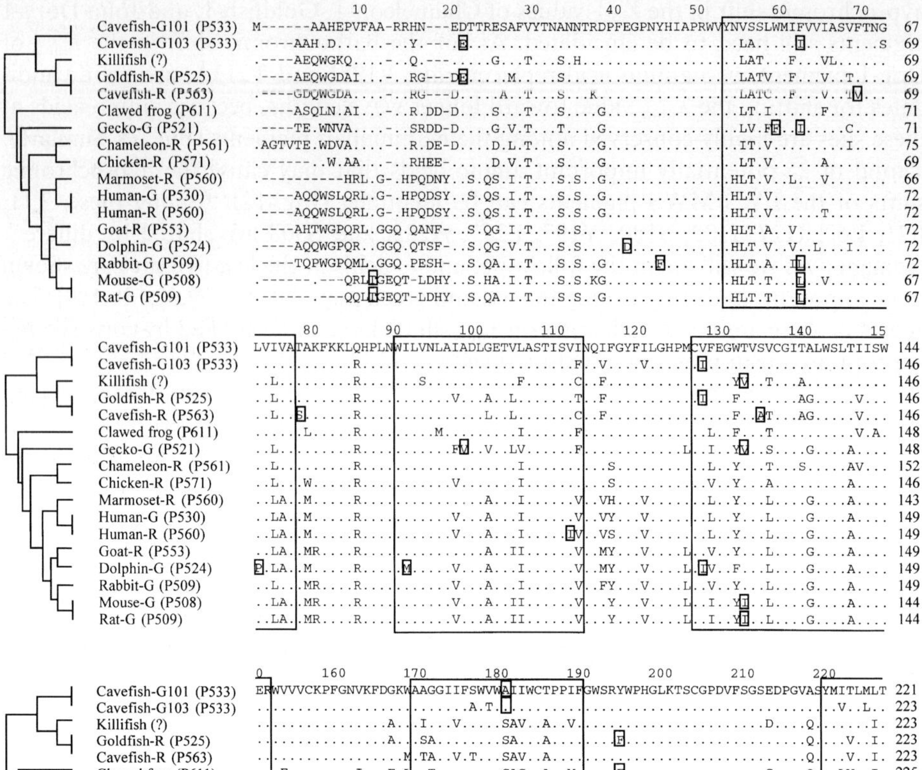

Fig. 12. Amino acid sequences of LWS/MWS pigments. Amino acid numbering at the top corresponds to those of Human-R and Human-G pigments.

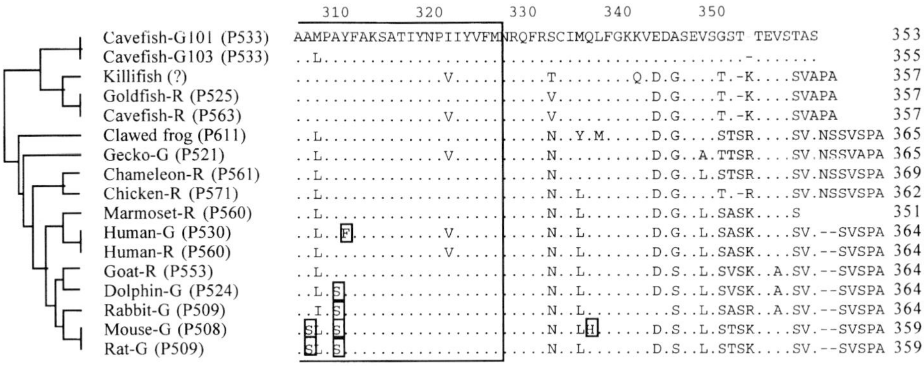

Fig. 12. Caption opposite.

ultimate goal of the 'evolutionary model' approach in understanding the mechanisms of spectral tuning is to accurately infer the magnitudes of λ_{max}-shift caused by certain amino acid replacements. For this purpose, it is important to develop biophysical theories for the absorption spectrum of a visual pigment based on its tertiary structure (e.g. [131]). These analyses will significantly enhance the power of predicting potentially important amino acid changes.

6.2. *'Three-sites' rule of red/green color vision*

A survey of color vision of different organisms shows that only a limited number of species in the bony fishes, birds, reptiles, and primates have trichromatic color vision, while many vertebrates are color-blind (e.g. [6,132–134]). For example, most mammals have dichromatic color vision, having blue pigments with either green or red pigments [134]. The trichromatic color vision of higher primates is believed to have evolved to facilitate the detection of yellow and red fruits against dappled foliage [135].

The method based on the evolutionary model was first applied to the red/green color vision in vertebrates [128]. In order to elucidate the molecular mechanisms underlying the spectral tuning of the green and red pigments, Cavefish-R, Cavefish-G101, and Cavefish-G103 opsin genes were cloned and sequenced [128,136]. The amino acids deduced from these DNA sequences were compared to those from Human-1, Human-B, Human-G, and Human-R opsin genes [128]. The results suggested that (1) the green and red opsins of the two species are derived by independent gene duplications (nodes 'a' and 'b' in Fig. 13) and (2) the red opsins in both humans and fish evolved independently from the green opsin by three identical amino acid changes from AFA (A180, F277, and A285) to SYT (at the corresponding positions; Fig. 13) [128]. Furthermore, it has been predicted that "the amino acid substitution at this residue (180) may not have been as important as those at residues 277 and 285 in the development of the red visual pigment [128]". Thus, the 'evolutionary model' method sometimes can detect even very subtle

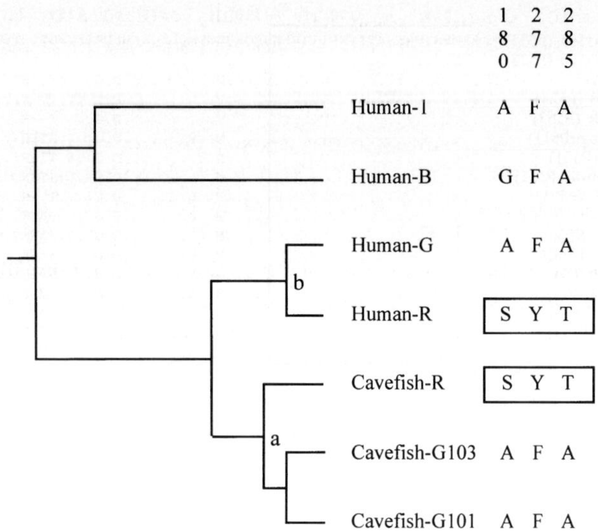

Fig. 13. Phylogenetic tree and amino acid replacements, indicated by boxes, of the red and green pigments of cavefish and human.

differences. By characterizing eight red and green opsin sequences from primates, Neitz et al. [129] also concluded that the spectral difference between red and green vision was due to the difference between AFA of the green pigment and SYT of the red pigments.

The amino acid sites 180, 277, and 285 correspond to residues 164, 261, and 269 in Bovine-1 pigment. They are located near the chromophore (Fig. 1) [7,54,92], suggesting the possible interaction between these amino acids and the chromophore. The three corresponding amino acid changes A164S, F261Y, and A269T in Bovine-1 pigment increased λ_{max} values by 2, 10, and 14 nm, respectively, showing that these changes could explain the majority of the 30 nm difference between MWS and LWS vision [56]. Qualitatively, the same conclusion has been reached by constructing Human-R and Human-G mutants, except that the entire 30 nm of λ_{max}-shift also requires minor contributions from Y116S, T230I, S233A and F309Y [62]. An amino acid change A180S in Human-R pigment is also known to shift λ_{max} by 4 nm toward longer wavelengths [63,64].

6.3. 'Five-sites' rule of red/green color vision

Amino acid differences at 180, 277, and 285, particularly those at the latter two sites, cause the major difference between red and green color sensitivity in a diverse array of vertebrate species. However, it is also known that a Goldfish-R pigment with SYT at the three critical sites is green-sensitive and has a λ_{max} value of 525 nm [96]. The inconsistency of this SYT-containing pigment with the green-sensitivity has remained unnoticed. Recently, however, it has been shown that, despite having AYT at the three critical sites (Fig. 12), Mouse-G pigment has a λ_{max} value of 508 nm and that this hypsochromic shift in λ_{max} can be fully explained by two amino acid changes: H197Y and A308S [59]. The amino acid sites 197 and 308 of Mouse-G pigment correspond to sites 181 and 292 of Bovine-1 pigment, respectively (Fig. 1). Fig. 12 demonstrates that amino acid changes H197Y and A308S and the hypsochromic shifts in λ_{max} values of the visual pigments are highly correlated.

The amino acid changes S180A, H197Y, Y277F, T285A, and A308S are known to reduce the λ_{max} values of Human-R pigment by approximately 7, 28, 10, 16, and 18 nm, respectively [137]. The effects of these amino acid substitutions on the λ_{max} value are considered to be more or less additive [59,62,137]. To test the generality of the five-sites rule, we have determined the partial amino acid sequences of red and green opsins from cat (*Felis catus*), dog (*Canis familiaris*), deer (*Odocoileus virginianus*), guinea pig (*Cavia porcellus*), and gray squirrel (*Sciurus carolinensis*) [137]. One way to study the evolutionary processes of color vision is to infer the amino acid sequences of opsins of ancestral organisms using the present-day sequences. Thus, these and the currently available red and green opsin sequences (Table 1) are also subjected to the inference of the amino acid sequences of opsins of mammalian ancestors.

The analysis shows that the mammalian ancestor appears to have had only one green–red hybrid gene in its genome and that most extant mammalian green and red pigments must have evolved from the green–red hybrid pigments by directed amino acid substitutions [137]. Hominoids, Old World monkeys, and some New World monkeys have distinct blue, green, and red genes [134,138]. These green and red genes might have been derived either by gene duplication of the hybrid gene, followed by the directed nucleotide substitutions or by the formation of the green and red pigment alleles, followed by the duplication of the two loci by unequal recombination [139,140]. From the existing data, we cannot reject either one of these possibilities [127]. Even in these species, however, directed amino acid substitutions must have played an important role during the evolution of the green and red pigments.

If we apply the five-sites rule directly to the inferred ancestral amino acid sequences in Fig. 14, the mammalian ancestor at node ‘a’ appears to have had a green pigment with a λ_{max} of ~530 nm. This ancestral green pigment has been transmitted directly to squirrel and the pigment of the ancestral murine species at node ‘i’ has subsequently achieved a further hypsochromic shift by two amino acid substitutions, S180A and A308S (Fig. 14). The ancestor of primates, rabbit, horse, cat, dog, goat, and deer achieved their red sensitivity at node ‘c’ primarily by one amino acid change, Y197H. The extant green pigments of human and deer seem to have evolved from the ancestral red pigment at nodes ‘e’ and ‘g’, respectively, by two independent amino acid substitutions, Y277F and T285A. Finally, the extant green pigment of rabbit appears to have evolved from the ancestral red pigment at node ‘d’ by H197Y and A308S. Since the molecular basis of red and green color vision is not firmly established yet (see the following), the details of this evolutionary scenario may require further revision. However, both sites 197 and 308 of the green and red pigments of hominoids, Old World and New World monkeys are monomorphic and consist of H197 and A308. The five-sites rule is therefore reduced to the three-sites

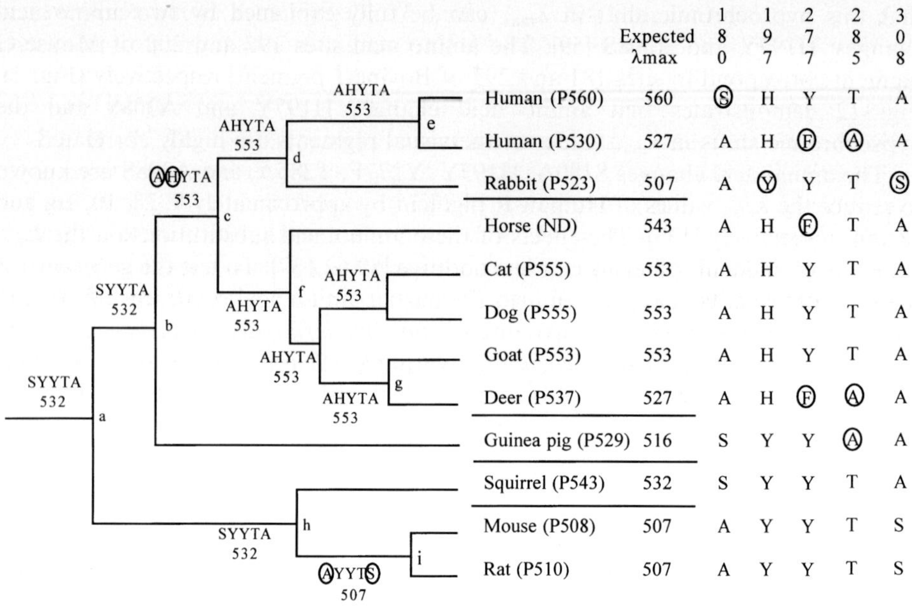

Fig. 14. Phylogenetic tree topologies of the mammalian red and green pigments and ancestral amino acids inferred [137].

rule. Thus, the primate ancestor appears to have had the red pigment first, and the green pigment evolved from the red pigment [127].

6.4. Green vision of rabbit, deer, guinea pig, and squirrel

How general is the five-sites rule as the molecular basis of red and green color vision in mammals? The slightly hypsochromic-shifted λ_{max} values of the red pigments of goat, cat, and dog are accurately predicted by S180A of the five-sites rule (Fig. 14). Both rat and mouse pigments with λ_{max} values of ~510 nm are also consistent with the five-sites rule. Amino acid replacements Y277F and T285A in the deer pigment predict a hypsochromic-shifted λ_{max}, but the observed value is 10 nm higher than the expected value (Fig. 14). Similarly, the observed λ_{max} values for the green pigments of rabbit, guinea pig, and squirrel are consistently ~10 nm higher than the corresponding expected values. Thus, the spectral absorption of the green pigments in rabbit, deer, guinea pig, and squirrel cannot be adequately explained by the 'five-sites' rule.

Two causes for the discrepancies can be considered. First, the effects of different amino acids at the five sites on the λ_{max} values may not be additive after all. So far, amino acid interactions have been studied considering either sites 180, 277, and 285 or sites 197 and 308 separately. Such interactions might have been important in causing the higher observed λ_{max} values, which needs to be tested experimentally. Second, there may exist currently unknown amino acids at sites other than 180, 197,

277, 285, and 308 that shift the λ_{max} values. To identify such amino acids, we can again follow the evolutionary model of determining potentially important amino acids that may shift the λ_{max} values. For example, we can identify a total of 17 amino acid replacements in the four species: I115F, L244M, A248T, A298S, and L306I for rabbit, A101V, V133A, and K157R for deer, I61T, F62I, V68I, Q241H, and A298S for guinea pig, and F62I, L135V, V162M, S211Y, and V216M for squirrel (Fig. 15). Among these changes, F62I is common to both guinea pig and squirrel, while A298S is common to rabbit and guinea pig. With the exception of these two, all amino acid changes are unique to a particular species. Thus, the hypsochromic shifts in λ_{max} of the green pigments of the four species appear to have different molecular bases.

6.5. *Red vision of goldfish*

As mentioned earlier, Goldfish-R (Goldfish-R′) pigment with the red pigment-specific amino acids SHYTA at the five critical sites (Fig. 12) is known to have a λ_{max} value of 525 nm [96], which is ~35 nm lower than the expected value. The direction of this λ_{max}-shift is opposite compared to those of the green pigments of rabbit, deer, squirrel, and guinea pig. In this case, we need to search for new amino acid changes that are responsible for the hypsochromic shift of Goldfish-R′ pigment. At present, the real goldfish red opsin gene has not been found. Both phylogenetic and Southern analyses strongly suggest that goldfish has an additional green opsin gene [113]. Does this gene encode the real red opsin? Or, as suggested by Johnson et al. [96], does an allelic form of the Goldfish-R′ opsin gene encode the real red pigment? In either case, once the real Goldfish-R opsin gene is cloned, comparison of Goldfish-R and Goldfish-R′ pigments, together with the mutagenesis analyses using cultured cells will reveal the molecular mechanisms that cause hypsochromic shifts in the λ_{max} of the latter pigment. The significant hypsochromic shift in λ_{max} value of Goldfish-R′ pigment and the unexpectedly higher λ_{max} value of the green pigment of rabbit, deer, guinea pig, and squirrel occurred independently from each other. Accordingly, the sites that account for the different absorption spectra of the Goldfish-R pigment and the green pigment of the four mammalian species can be very different. Thus, the detailed molecular bases for the green and red pigments still remain to be clarified.

6.6. *LWS/MWS opsin genes on X chromosomes*

Men with trichromatic color vision are initially considered to have typically one red opsin gene and one, two, or more green opsin genes [47,75,141,142]. The proximity and sequence homology among the head-to-tail array of genes appear to provide a suitable condition for unequal homologous recombination events. The intragenic and intergenic recombinations produce various types of hybrid (or fusion) genes and a variable number of genes, respectively. Intragenic recombination between red and green opsin genes produce both 5′ green–red 3′ and 5′ red–green 3′ hybrid genes [47,48,75,143]. The addition of the 5′ green–red 3′ gene to an otherwise normal gene array, containing wildtype red and green opsin genes, causes deuteranomaly, the

```
                     50        60        70        80        90       100       110
Human (P560)        APRWVYHLTSVWMIFVVTASVFTNGLVLAATMKFKKLRHPLNWILVNLAVADLAETVIASTISIVNQVS
Human (P530)        .................I.............................................V....Y
Rabbit (P523)       ..........A...L..I..........V...R..............................V...FY
Horse (ND)          .......V..A......I..............R.......................I......V...IY
Goat (P553)         ..........A..V...I..............R................I......I......V...MY
Deer (P537)         ..........A..V...I..............R................I...V..I......V...MY
Cat (P555)          .......V..A......I......................................I......V...IY
Dog (P555)          ..........A......I..........V...R................I......I......V...IY
Guinea pig (P529)   ..........A..TI..I..I.......V...R..............................V....Y
Squirrel (P543)     .......I..T...I..I..........V....................I.............V...LY
Rat (P508)          ..........T...L..I..............R.......................I......V...IY
Mouse (P510)        ..........T...L..V..............R.......................I......V...IY

                     120       130       140       150       160       170       180
Human (P560)        GYFVLGHPMCVLEGYTVSLCGITGLWSLAIISWERWLVVCKPFGNVRFDAKLAIVGIAFSWIWSAVWTA
Human (P530)        ....................................M..........................A.....
Rabbit (P523)       ........L..V..........................................A........A.....
Horse (ND)          ...........V........................M................VA........A.....
Goat (P553)         ........L..V........................M................TA........A.....
Deer (P537)         ........L..V....A...................M...R.............A........A.....
Cat (P555)          ......................................................A........A.....
Dog (P555)          ........L..V..........................................A........A.....
Guinea pig (P529)   ........L..V.............................................V...V.......
Squirrel (P543)     ........L..V......V..........................M.......................
Rat (P508)          ........L..I...I.....................................T...V...V.A.....
Mouse (P510)        ........L..I.........................................T...V...V.A.I...

                      190       200       210       220       230       240       250
Human (P560)        PPIFGWSRYWPHGLKTSCGPDVFSGSSYPGVQSYMIVLMVTCCIIPLAIIMLCYLQVWLAIRAVAKQQK
Human (P530)        ............................................T..S..V.................
Rabbit (P523)       ...........Y.............T.........M...........SV.V.......M...T.....
Horse (ND)          .........................Q.............I.......SV.V.................
Goat (P553)         .......................................I...F...SV.I.................
Deer (P537)         .......................................I...F...SV.V.................
Cat (P555)          .......................................I.......SV.V.................
Dog (P555)          ......G................................T....N..SV.I.................
Guinea pig (P529)   ...........Y.............T.........M........T..S..V....H............
Squirrel (P543)     ...........Y.............Y....M....M...........S..I.................
Rat (P508)          ...........Y.............T.........M........F..S..V.................
Mouse (P510)        ...........Y.............T.........M........F..S..V.................

                       260       270       280       290       300       310       320
Human (P560)        ESESTQKAEKEVTRMVVVMIFAYCVCWGPYTFFACFAAANPGYAFHPLMAALPAYFAKSATIYNPV
Human (P530)        ...................VL.F.F.....A............P..........F...........
Rabbit (P523)       ...................V....L............T.H...S....V..I.S...........I
Horse (ND)          ................M..V..F.L..............H........V................I
Goat (P553)         ................M.......L..............H........V................I
Deer (P537)         ................M.....F.L.....A........H........V................I
Cat (P555)          ................M......................H........V................I
Dog (P555)          ...................VL...L..............H........V................I
Guinea pig (P529)   ...................VL...L.....A......T.....S....V................I
Squirrel (P543)     ...................V....L............T..........V................I
Rat (P508)          ...................V....L............T.H........V.S..S...........I
Mouse (P510)        ...................V....L..............H........V.S..S...........I
```

Fig. 15. Amino acid sequences of the red and green pigments in mammals. Amino acid numbering at the top corresponds to those of Human-R and Human-G pigments.

most common inherited color vision defect [47]. Interestingly, most deuteranomalous men have wildtype green gene(s), that do not contribute to correct the defective color vision [49]. This particular example may give the impression that the hybrid genes always lead to color blindness. However, some males with normal trichromatic color vision have a 5′ green–red 3′ hybrid gene instead of the wildtype red opsin gene [144]. This observation shows that, in humans, as long as the exon 5 of the gene encodes Y277 and T285, red sensitivity is restored. More generally, the absorption spectra of the green and red pigments in humans are determined mainly by the amino acids at sites 277 and 285, that are encoded by exon 5 of the red and green opsin genes [49,128,144,145]. Thus, in humans, we may define any genes, including hybrid genes, that encode Y277 and T285 as red opsin genes, while those encoding F277 and A285 as green opsin genes. Then, many human X chromosomes have more than one, sometimes up to four, red opsin genes [144,145].

In order to attain red and green sensitivity, most animals use both red and green opsin genes. However, many New World monkeys have only one X-linked opsin gene locus with multiple alleles [129,135,138,146]. Jacobs et al. [138] showed that one genus of New World monkeys (*Alouatta*; howler monkey) has two X-linked opsin genes that are probably due to a recent gene duplication confined to this genus. In squirrel monkey (*Saimiri sciureus*), the visual pigments based on three alleles have λ_{max} of 532, 547, and 561 nm and consist of amino acids AFA, AFT, and SYT at 180, 277, and 285, respectively [129]. In this visual system, all males and homozygous females at this locus are dichromatic. However, if heterozygous females happen to have both pigments with λ_{max} of 532 and 561 nm, then they attain full-fledged trichromatic color vision [132,146].

7. Rhodopsin mutants and retinal diseases

Protein polymorphisms of visual pigments are of considerable interest, not only because they give us insight into the molecular mechanisms of spectral tuning and other functions of visual pigments, but also because they are found in patients with retinal diseases. Retinitis pigmentosa (RP) is a group of inherited disorders, causing a progressive loss of retinal function, including retinal degeneration and night blindness [147]. Molecular genetic studies of RP patients have revealed a wide spectrum of mutations in Human-1 pigments that result in autosomal dominant RP (ADRP), which accounts for about 10–20% of RP cases [148,149].

The ADRP mutants may be distinguished into three classes according to their phenotypes. Class I mutants closely resemble Human-1 opsin expression in cultured cells, fold correctly to bind to 11-*cis*-retinal, and are transported to the cell surface [50–53]. Class II mutants are defective in folding, do not bind to 11-*cis*-retinal, and stay in the cytoplasmic reticulum [52] (they correspond to Class IIa mutants in Refs. [50], [51]). Class III mutants bind to 11-*cis*-retinal poorly and remain in the cytoplasmic reticulum [52] (these correspond to Class IIb mutants in Refs. [50], [51]). The rhodopsin mutants classified in this way are shown in Fig. 16. The three classes of ADRP mutants are not always clearly distinguished and need to be refined in the

Contents

1. Introduction

Visual pigments (VPs) constitute the interface between the photophysical processes of light absorption by matter and light-induced physiological reactions of living organisms. According to this function, VPs have to meet a number of functional and structural constraints. In order to ensure an appropriate behavior of animals also under conditions of low light intensity, VPs have been optimized with respect to a high quantum efficiency for the conversion of light energy into a physiological signal. For the same purpose, the construction of the eyes has also been improved with respect to their optical parameters in order to maximally sample the incident light. Amongst other parameters of functional optimization, this strategy becomes obvious from a very tight packing of visual cells – preferentially in the optical axis of the eyes. The environmental adaptation of animals can be recognized in many cases in the morphological optimization of eyes and the subsequent neuronal connections. This ensures a broad variety of structural and functional flexibility, providing the animals with optimal sensitivity, maximal contrast or excellent spectral discrimination. Besides their occurrence in eyes, equipped with focussing lenses for image formation, VPs are also found in less structured light-sensitive regions of lower eukaryotes, and are identified in unicellular organisms and prokaryotes where they are localized in specialized arrays or even randomly distributed, and serve as photosensory and phototactic pigments [1,2].

VPs consist of a protein moiety and a covalently bound chromophoric group (Fig. 1). In all cases identified so far, the protein component is an intrinsic membrane protein, and the chromophore is a derivative of the vitamin-A aldehyde. Besides the above mentioned high photon capture probability, VPs have also to ensure a high thermal stability, i.e., a low dark activity. This feature is essential for a correct function of the VP molecule at very low light intensity where thermal noise might compete with rare photoisomerization events and cause false visual sensations [3–5]. Furthermore, the basic structural principle of VPs has to allow tuning of the spectral selectivity of a VP over a wide spectral range on the basis of very few chromophore structures. Finally, VPs have to be regenerated very efficiently in order to guarantee a permanently high probability of light absorption.

Spectral and functional properties of a large number of insect VPs are compiled in a number of review and handbook articles [6–15]. Accordingly, in this chapter special emphasis will be given to more recent findings, and relation to formerly published results on VP structure and function will only been drawn in selected cases. Although this article is dedicated to the VPs of invertebrates, reference to VPs of vertebrates will be made in cases of structure/function relationship.

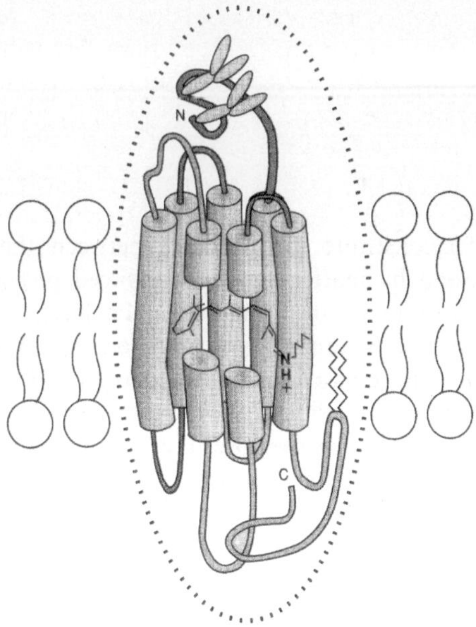

Fig. 1. Principal structure of VPs. VPs consist of a highly hydrophobic polypeptide chain which is embedded in specialized membrane structures of the visual cell. The polypeptide folds into seven membrane-spanning α-helices, which are connected by hydrophilic loops. The interior of the helix bundle forms a binding cavity for the visual chromophore, 11-*cis* retinal, which is covalently attached to the protein via a protonated Schiff base to a lysine residue from the seventh helix (modified from Ref. [396]).

2. Visual cell morphology

The broad optical and morphological variety of eyes which have been developed will not be discussed here in detail (for a more comprehensive description see Refs. [11,16,17]). One important aspect for the analysis and functional description of insect VPs is the principal structural difference between vertebrate and invertebrate visual cells. Whereas the photoreceptor cells of vertebrates are of ciliary origin, those from the invertebrates derive in most cases from microvillar structures. During the evolution of vertebrate photoreceptor cells, specialized cellular compartments (the so-called outer segments, OS) were generated. These OS accumulate as a result of excessive membrane biosynthesis at the distal end of the visual cell. The newly synthesized membranes which already contain VP molecules in high concentration, fold into layers which eventually seal at their bending edges, and form stacks of membrane discs enveloped by an outer cell membrane. These OS contain a relatively small volume of cytosol, but consist predominantly of phospholipid bilayer structures. Since the ciliary connection between the outer segments and the cellular body is very narrow and fragile, it can be easily disrupted upon the application of external mechanical stress. This represents an easy way of preparation of the outer segments which are obtained in large quantities and high purity. In such preparations, the visual pigment comprises the dominating protein and has thus become a preferred object for biochemical and spectroscopical studies [18–21]. As a consequence, there exists a much better knowledge of structure and function of vertebrate in contrast to the invertebrate photoreceptors [12].

As outlined above, the experimental problems for the preparation of invertebrate and in particular insect VPs are due to the morphological structure of their visual cells (for reviews on invertebrate eye morphology, see Refs. [6–8,10,11]). The preferential building principle of invertebrate eyes are the compound eyes, which are composed of a large number of ommatidia. In the case of the insects, each ommatidium harbors generally eight (in some cases nine) photoreceptor ("retinula") cells [14,22,23], which carry the light-sensitive molecules embedded in a membraneous rim, extending nearly along the entire length of the photoreceptor cell. Inspection of the structure of this hydrophobic part of the photoreceptor cell reveals a similar principle of construction as found for the vertebrates: evaginations of the plasma membrane (microvilli) enlarge the part of the photoreceptor cell destinated for VP packing, despite the fact that these membraneous compartments are not disconnected from the cell body of the photoreceptor cell as is the case for the vertebrates. The photoreceptor cells are arranged in a bundle (the so-called ommatidium) with the microvilli arranged towards the center of this cylindrical structure where in some cases the microvilli from several visual cells even interlace into a rhabdom. Accordingly, the rhabdom is an area of high VP concentration and utmost spectral sensitivity.

Taking together the smaller amounts of VP per compound eye and the more complex morphology of the photoreceptor cells, this makes the preparation of insect VPs remarkably difficult compared to that of the vertebrate VPs. The formation of the morphological structure of the membraneous microvillar part of the visual cell is tightly coupled to the biosynthesis of the VP [24–26]. Interestingly, the determining parameter for the development of the rhabdom structure is the presence of the 11-*cis* (but not the all-*trans*) isomer of the VP chromophore [25,27,28], which has to be provided in sufficient amounts. It could even been shown from rescue experiments (placing the gene encoding the major VP, *rh1*, under the control of a heat-shock inducible promoter) in rhodopsin-deficient fruitfly mutants (*ninaE*, *nina* = neither inactivation nor afterpotential) that an exact timing for the expression of the VP is required for a correct microvillar development [29].

3. General structure of VPs

The fact that the VPs are intrinsic membrane chromoproteins allows a very tight packing of proteins within the membrane without fatally altering the membrane properties. A similar arrangement would probably not be possible by ordering proteins in the cytosol, since it may increase the viscosity and thereby slow down diffusion-controlled processes. The embedding of proteins in membranes furthermore allows special arrangements of reduced rotational flexibility, which might be

concur with the dependency of retinoid provision. In this case, the control animals and those raised on the carotenoid-deprived diet showed nearly equal amounts of the mRNAs encoding the VPs Manop1 and Manop2 [82].

Great effort has been given to the changes in rhabdom size and pigment variation in crabs [123–129] and Limulus [130]. Besides changes in the optical properties aiming at higher photon capture probability by, e.g., movements of the screening pigment in proximal direction and alternatively exposing the rhabdom more towards the distal part of the compound eye, also morphological changes of the rhabdom structure due to protein biosynthesis or degradation are known for many invertebrates. The crabs undergo very prominent variations of the rhabdom structure. Changes of the rhabdom size are clearly correlated with the variation of light intensity, i.e., following the day/night rhythm [131]. As an example, the rhabdom volume in the grapsid crab *Hemigrapsus sanguineus* during the night is estimated to be ca. eight times larger than during the day [123]. In this crab, the degradation pathway has been identified in great detail, and might serve as a paradigm for rhabdom (and VP) degradation [124,125]. Initiation of the rhabdom degradation process could recently be coupled to the appearance of diacylglycerol, which is a component of the visual transduction pathway. The rhabdom breakdown takes place via pinocytosis, and also by shedding the tips or even groups of microvilli into the extracellular medium [132,133]. This cell debris is then phagocytosed by further processes.

6. Visual pigment sequence analysis

Molecular biology technology has allowed to identify a broad variety of visual pigment sequences. Originally by employment of the cDNA encoding bovine opsin, the cDNA for the dominating VP, Rh1, of *Drosophila melanogaster* (which is expressed in the photoreceptor cells R1-R6) had been characterized [134,135]. This sequence in turn allowed the identification of several other *D. melanogaster* VP sequences [136–139]. Making use of the consensus motifs from computerized sequence alignments, a number of other invertebrate VPs have been characterized during recent years with respect to their primary structure (see below).

6.1. Molecular biology technology

For quite a long time, the sole possibility to at least partially characterize insect visual pigments was by employment of microspectrophotometry or the recording of action spectra by electrophysiological methods. These techniques, however, yielded information only on the spectral properties (absorption maximum, the ratio of the pigment and metapigment extinction coefficient, and to some extent on the pigment → metapigment conversion process), and are prone to artifacts caused by screening pigments or electrical coupling. With the establishment of molecular biology technology, primary structure information (sequence analysis) has also become feasible, and quite recently, these methods have opened the door for a heterologous production of insect VPs.

6.1.1. Identification strategy for VP sequences

The technique to isolate genes using PCR-based methods still relies on RACE (rapid amplification of cDNA ends) [140]. This technique is based on the preparation of cDNA with a modified oligo-dT synthesis primer which consists of a polythymidine tract (e.g., T_{17}) at its 3′ end, but being flanked with a specially designed sequence of 30–40 nucleotides. In 3′ RACE, an in gene-located sense-encoding gene specific primer, known from the previously identified opsin gene fragment, is used with a primer which is complementary to the flanking sequence of the cDNA synthesis primer. The fragment which arises will have sequence overlap with the opsin gene sequence which is already known and will extend to the terminus of the open reading frame and then through the 3′ untranslated region (UTR), which should have a recognizable polyadenylation signal prior to the poly-A sequence.

The characterization of the 5′ end sequence is still a formidable task, since for the application of the RACE technique it is necessary to provide a sequence motif which can be exploited by PCR. Generally, the first-strand cDNA is extended by incubation with terminal deoxynucleotidyl transferase (TdT) and one type of nucleotide triphosphates which will add a string of nucleotides to its 3′ end (homopolymer tailing), corresponding to the 5′ end of the gene. This material is used in a PCR reaction with an antisense gene specific primer and a primer which is complementary to the added nucleotide tract. The sequence of the product which arises will overlap with the known sequence and should have a clearly recognisable initiation codon which specifies the start of the gene. An alternative method using inverse PCR to obtain the 5′ end of genes relies on the use of two gene specific primers with double strand cDNA which has been self-ligated to form circular molecules [141]. The sense-coding primer is based on a sequence slightly upstream of the polyadenylation signal and the antisense encoding primer is based on penultimate sequences available in the 5′ region of the gene. The resulting PCR product will have two regions of overlap with the previously known sequence and will also encode the remainder of the N-terminal region and 5′ UTR.

The sequence information from gene fragments encoding an opsin is invaluable, because it can serve as a probe in library screening, or as a basis for the synthesis of gene specific primers which can be used in additional PCR reactions to obtain the 5′ and 3′ regions of the opsin-encoding gene. The resources required for successful PCR are minimal and in the primary stages only require first strand synthesized cDNA prepared from small amounts of tissue without the need to extensively purify the mRNA.

6.1.2. Dipteran VPs

The investigation of gene organization and sequence of the encoded insect VPs started with the fruitfly *Drosophila melanogaster*, where many mutations and genotypes causing visual impairment had been characterized [142–144]. The first two reports of the gene sequence encoding the most abundant VP, Rh1, located in the six peripheral visual cells R1-6, were published simultaneously [134,135]. Genomic clones were isolated by screening a *Drosophila* genomic library under stringent conditions with a bovine opsin cDNA probe followed by isolation of cDNAs encoding the VP.

Subsequent to the identification of Rh1, which is a VP of 373 amino acids, three other VP sequences were identified (Rh2, Rh3, and Rh4 with 381, 383 and 378 amino acids, respectively) [136–139] and allocated to specific photoreceptor cell types in *Drosophila* by using gene-specific probes for *in situ* hybridization or screening cDNA prepared from various *Drosophila* mutants. Using these techniques, Rh2 had been primarily assigned to the R8 photoreceptor cell from *sev*- and *ora*-derived mRNA (*sev* = sevenless, *ora* = outer rhabdomeres absent) [136]. This assignment was revised after comparison with results from other groups who reported preferential *in situ* hybridization of an Rh2 probe to the ocelli [145], and from "targeted misexpression" of Rh2 [146]. For that latter approach, the Rh2-encoding DNA was cloned under the control of the *rh1* promoter and expressed in outer photoreceptor cells R1-6 of the *ninaE* mutant (*nina* = neither inactivation nor afterpotential), now allowing microspectrophotometric investigation of the gene product which showed similarity to the known spectral sensitivity of the ocelli and not to that from the R8 photoreceptor cell.

Rh3 and Rh4 have been allocated to subclasses of R7 photoreceptor cell type. According to P-element mediated expression and microspectrophotometric investigations [147], these VPs showed maximal sensitivity in the blue/near UV range. Opsin-encoding gene(s) in R8 have remained elusive for quite a while. Quite recently, two new VP sequences, Rh5 [148] and Rh6 [149] from *D. melanogaster* were reported which are expressed in the R8 photoreceptor cells. Rh5 was identified by RT-PCR of mRNA from a *Drosophila* mutant (*ninaE*), which in addition was lacking the R7 visual cells (*vis*), using degenerated primer sequences designed from conserved cDNA stretches of the known Rh1–Rh4 sequences [148]. The choice of the mutant flies was advantageous, since the majority of the other pigments (Rh1–Rh4) was absent. This new sequence was placed into P-element vectors to generate transgenic flies. Although no action spectrum for these flies was measured, ERG recordings demonstrated that when the Rh5 protein is expressed in the R1–R6 cells it can induce the phototransduction process there. Rh5 was cloned nearly simultaneously by Papatsenko et al. [150] who also reported detailed promotor analysis and found a strict correlation between the expression of Rh5 and Rh3. Rh6 was originally found as a homologous cDNA from screening a *Calliphora* cDNA which revealed a clone with less than the expected inter-species similarity [149]. This relatively low similarity led to the assumption that a new VP type had been identified. This assumption was proven correct by using this cDNA to screen a *Drosophila* cDNA which yielded the Rh6 open reading frame (ORF). As found for the Rh3- and Rh4-type VPs which are expressed in two subsets of the R7 visual cells, these new VPs were also localized both in R8-type visual cells, again making a selective function probable. Interestingly, the expression of Rh5 is strictly coordinated with the expression of Rh3 [148], shedding light onto rhabdomere development. In fact, the developmental correlation appears to be even stronger: if R7 cells are removed (e.g., in the *Drosophila* mutants *sevenless*, *boss* or *sina*) and thus the control of development is absent, the expression of Rh6 is strongly upregulated on the expense of Rh5 [151]. Apparently, Rh6 is the default VP in R8 cells, and the equilibrium between Rh5 and Rh6 is dependent on the controlling function of Rh3 on the biosynthesis of Rh5. However, whereas Rh3 and Rh4 both show features reminiscent to blue/UV-absorbing pigments, and accordingly, also segregate on the same limb of a phylogenetic tree (see below, and Fig. 12), the two VPs found in R8 cells exhibit clear differences in their sequences and deduced spectral properties. Whereas Rh5 (382 amino acids) shows maximal similarity to Rh3 (50%), Rh6 (369 amino acids) is mostly related to Rh1 and Rh2 (53% and 51% identity, respectively). This would place Rh6 into the phylogenetic proximity of long-wavelength (LW) absorbing pigments. Taking into account that the R8 cells exist as two subsets of cells with different spectral absorptions (R8y and R8p with maximal sensitivity around 530 and 460 nm, respectively [14]), the Rh6 VP can be proposed to be responsible for the LW sensitivity of the R8y cells. This spectral assignment was confirmed and specified by ectopic expression of Rh5 and Rh6 in the *Drosophila* mutant *ninaE* [404]. The transgenic flies showed maximal sensitivity at 437 (Rh5) and 508 nm (Rh6), quite in agreement to intracellular recordings of *Calliphora* and *Musca*.

A number of VP sequences from other *Drosophila* species have been identified in the meantime (see Table 1), all making use of the high homology between the various *Drosophila* strains.

Besides the VP sequences from *Drosophila* there is only one dipteran species for which a VP DNA sequence was identified. The abundant pigment (371 amino acids) from photoreceptor cells R1-6 from the blowfly (*Calliphora vicina*) was identified by screening a cDNA library with a *Drosophila* Rh1 selective probe which led to the isolation of a cDNA of 1113 nucleotides [42]. The close relationship of Rh1 between *Drosophila* and *Calliphora* is reflected by the high degree of homology between the two sequences whose polypeptide chains are 86% identical and 98% similar.

6.1.3. Other insect VP sequences

A survey for VP sequences in insects from different classes, which in some cases could be phylogenetically fairly distant from the well-characterized Diptera required a more sophisticated strategy and primer design. The still high degree of similarity in some parts of the VPs (of vertebrate and invertebrate origin) allows the preparation of degenerate oligonucleotide primers for use in RT-PCR with an mRNA preparation from eye-cup tissue. The most effective primary PCR reaction is to use a number of gene-specific primers from these conserved regions of the receptor gene in various combinations which yield several PCR products of expected sizes. In most cases, when new VP sequences were characterized, insect species were selected for which the number and the estimated absorption range of different VPs could be assumed from spectroscopic studies.

6.1.4. Mantid opsin

Following chronologically after the identification of the *Drosophila* VP sequences, the VP from the mantid *Sphodromantis* sp. was the first non-dipteran VP which was identified [152]. Wavelength-dependent behavioral experiments performed with the

```
                   290       300       310       320       330       340       350
Limpo520  270 CRLAKVAMMTVGLWFMAWTPYLIISWAGVFSSGTRLTPLATIWGSVFAKANSCYNPIVYGISHPRYK-AA
Limpo530  270 CRLAKVAMMTVGLWFMAWTPYLIIAWAGVFSSGTRLTPLATIWGSVFAKANSCYNPIVYGISHPRYK-AA
HemsaRh1  277 IRIAKTALVNVSLWFICWTPYAAITIQGLLGNAEGITPLLTTLPALLAKSCSCYNPFVYAISHPKFR-LA
HemsaRh2  277 IRIAKTALVNVSLWFICWTPYALISLKGVMGDTSGITPLVSTLPALLAKSCSCYNPFVYAISHPKYR-LA
Camsc     241 CRLAKIAMTTVALWFIAWTPYLLINWVGMFAR-SYLSPVYTIWGYVFAKANAVYNPIVYAIS--------
Camhu     241 CRLAKNAMTTVALWFIAWTPCLLINWVGMFAH-SYLSPVYTIWGYVFAKANAVYNPIVYAIS--------
Camlu     241 CRLAKIAMTTVALWFIAWTPYLLINWVGMFAR-SYLSPVYTIWGYVFAKANAVYNPIVYAIS--------
Camma     240 CRLAKIAMTTVALWFIAWTPYLLINWVGMFAR-SYLSPVYTIWGYVFAKANAVYNPIVYAIS--------
Orcau     241 CRLAKIAMTTVALWFIAWTPYLLINWVGMFAR-SYLSPVYTIWGYVFAKANAVYNPIVYAIS--------
Orcvi     241 CRLAKIAMTTVALWFIAWTPYLLINWVGMFAR-SYLSPVYTIWGYVFAKANAVYNPIVYAIS--------
Procl     274 CRLAKIAMTTVALWFIAWTPYLLINWVGMFAR-SYLSPVYTIWGYVFAKANAVYNPIVYAISHPKYR-AA
Promi     241 CRLAKIAMTTVALWFIAWTPYLLINWVGMFAR-SYLSPVYTIWGYVFAKANAVYNPIVYAIS--------
Proor     238 CRLAKIAMTTVALWFIAWTPYLLINWVGMFAR-SYLSPVYTIWGYVFAKANAVYNPIVYAIS--------
Prose     241 CRLAKIAMTTVALWFIAWTPCLLINWVGMFAR-SYLSPVYTIWGYVFAKANAVYNPIVYAIS--------
Allsu     250 MKLAKISIVIVTQSLLSWSPYAIVALLAQFGPIEWVTPYAAQLPVMFAKASAIHNPMIYSVSHPKFR-EA
Lolfo     257 MKLAKISIVIVTQFLLSWSPYAVVALLAQFGPIEWVTPYAAQLPVMFAKASAIHNPMIYSVSHPKFR-ER
Octdo     258 MKLAKISMVIITQFMLSWSPYAIIALLAQFGPAEWVTPYAAELPVLFAKASAIHNPIVYSVSHPKFR-EA
Sepof     257 MKLAKISIVIVTQFLLSWSPYAVVALLAQFGPIEWVTPYAAQLPVMFAKASAIHNPLIYSVSHPKFR-EA
Todpa     257 MRLAKISIVIVSQFLLSWSPYAVVALLAQFGPLEWVTPYAAQLPVMFAKASAIHNPMIYSVSHPKFR-EA
Patye1    274 LRISKIAMTVTCLFIISWSPYAIIALIAQFGPAHWITPLVSELPMMLAKSSSMHNPVVYALSHPKFR-KA
Patye2    236 QRVTFISFLMMAAFMVAWTPYAIMSALAIGSF--NVENSFAALPTLFAKASCAYNPFIYAFTNANFRDTV

                   360       370       380       390       400       410       420
Limpo520  339 LYQRFPS--LACGSGESGSDVKSEASA-TTTMEEKPKIPEA-----------------------------
Limpo530  339 LYQRFPS--LACGSGESGSDVKSEASA-TMTMEEKPKSPEA-----------------------------
HemsaRh1  346 ITQHLPW--FCVHEKDPNDVEENQSSN-TQTQEKS-----------------------------------
HemsaRh2  346 ITQHLPW--FCVHETETKSNDDSQSNS-TVAQDKA-----------------------------------
Camsc         ----------------------------------------------------------------------
Camhu         ----------------------------------------------------------------------
Camlu         ----------------------------------------------------------------------
Camma         ----------------------------------------------------------------------
Orcau         ----------------------------------------------------------------------
Orcvi         ----------------------------------------------------------------------
Procl     342 MEKKLPC--LSC---KTESDDVSESAS-TTTSSAEEKAESA-----------------------------
Promi         ----------------------------------------------------------------------
Proor         ----------------------------------------------------------------------
Prose         ----------------------------------------------------------------------
Allsu     319 IASNFPWILTCCQYDEKEIEDDKDAEA-EIPAAEQS-GGESVDAAQMKEMMAMM-----QKMQAQQQQQP
Lolfo     326 IASNFPWILTCCQYDEKEIEDDKDAEA-EIPAGEQS-GGETADAAQMKEMMAMM-----QKMQAQQQQQP
Octdo     327 IQTTFPWLLTCCQFDEKECEDANDAEE-EVVASERG--GESRDAAQMKEMMAMM-----QKMQAQQAAYQ
Sepof     326 IAENFPWIITCCQFDEKEVEDDKDAET-EIPATEQS-GGESADAAQMKEMMAMM-----QKMQQQQAAYP
Todpa     326 ISQTFPWVLTCCQFDDKETEDDKDAET-EIPAGESSDAAPSADAAQMKEMMAMM-----QKMQQQQAAYP
Patye1    343 LYQRVPWLFCCCKPKEKADFRTSVCSKRSVTRTESVNSDVSSVISNLSDSTTTLGLTSEGATRANRETSF
Patye2    304 VEIMAPWTTRRVGVSTLPWPQVTYYPRRRTSAVNTTDIEFPDDNIFIVNSSVNG-----PTVKREKIVQR

                   430       440       450       460       470       480       490
Limpo520      ----------------------------------------------------------------------
Limpo530      ----------------------------------------------------------------------
HemsaRh1      ----------------------------------------------------------------------
HemsaRh2      ----------------------------------------------------------------------
Camsc         ----------------------------------------------------------------------
Camhu         ----------------------------------------------------------------------
Camlu         ----------------------------------------------------------------------
Camma         ----------------------------------------------------------------------
Orcau         ----------------------------------------------------------------------
Orcvi         ----------------------------------------------------------------------
Procl         ----------------------------------------------------------------------
Promi         ----------------------------------------------------------------------
Proor         ----------------------------------------------------------------------
Prose         ----------------------------------------------------------------------
Allsu     382 ---AYPPQG-YPP-QGYPPPPPQG-YP-PQGYPP--QGYPP-------QGYPP----PPQGPPPQGPPPQ
Lolfo     389 ---AYPPQG-YPP-QGYPPPPPQG-YP-PQGYPP--QGYPP-------QGYPP----PPQGPPPQGPPPQ
Octdo     389 P--PPPPQG-YPP-QGYPP---QGAYPPPQGYPP--QGYPP-------QGYPPQG-YPPQGAPPQVEAPQ
Sepof     389 PQGAYPPQGGYPP-QGYPPPPAQGGYP-PQGYPPPPQGYPPA------QGYPPQGYPPPQGAPPQGAPPQ
Todpa     390 ----P--QGYAPPPQGYPP---QG-YP-PQGYPP--QGYPP-------QGYPP----PPQGAPPQGAPP-
Patye1    413 RRSVSIIKGDEDP-CTHPDTFLLAYKEVEVGNLFDMTDDQNRRDSNLHSLYIPTRVQHRPTTQSLGTTPG
Patye2    369 NPINVRLGIKIEP--RDSRAATENTFTADFSVI-------------------------------------

                   500       510       520       530       540       550       560
Limpo520      ------------------
Limpo530      ------------------
HemsaRh1      ------------------
HemsaRh2      ------------------
Camsc         ------------------
Camhu         ------------------
Camlu         ------------------
Camma         ------------------
Orcau         ------------------
Orcvi         ------------------
Procl         ------------------
Promi         ------------------
Proor         ------------------
Prose         ------------------
Allsu     432 ----AAPPQGVD------
Lolfo     439 ----AAPPQGVDNQAYQA
Octdo     442 ----GAPPQGVDNQAYQA
Sepof     451 ----AAPPQGVDNQAYQA
Todpa     435 ----AAPPQGVDNQAYQA
Patye1    482 GVYIVDNGQRVNGLTFNS
Patye2        ------------------
```

Fig. 4b. *(continued)*

helices (V and VII) the hydrophobicity can be extended quite long, making them ca. 30 and 27 residues in length. The helix–loop transitions are more precisely determined for the insect VPs than for the other non-insect invertebrate VPs (Fig. 4b). It remains to be seen on a three-dimensional protein structure whether these assignments rcmain correct. Two of the interconnecting loops are remarkably short (ca. 12 amino acids between TM I and II, and between 10 and 11 amino acids for the loop between TM VI and VII), and probably do not fold into a secondary structure motif. The first connecting loop (i1) exhibits a remarkably high degree of similarity in all invertebrate sequences. As outlined above (see Section 5), this part of the protein is involved in correct transport towards and insertion into the membrane [43].

7. Structurally and functionally important protein domains

All following discussions have to be considered in comparison to the VPs of vertebrates and other related receptors due to the extensive information known there and the lack of structural information for the invertebrate VPs. Quite recently, the crystal structure of bovine rhodopsin has been determined at a resolution of 2.8 Å [406]. This work now allows for the first time a detailed view of the arrangement of the seven helices, and of the interactions between the retinal chromophore and the surrounding protein.

7.1. Primary and secondary structure of proteins

As mentioned above, the VPs from vertebrates are not included in the discussion on structure and function. Also, another class of retinal-binding proteins, the retinochromes from insects, up to now identified in the blowfly [115,188] and in the honeybee [85,189] will not be discussed in detail, since there is nearly no sequence information available. Furthermore, the insect retinochromes are claimed to be water-soluble proteins, which constitute a different class of proteins than the VPs and the retinochromes of the cephalopods [190]. In one case of squids (*Todarodes pacificus*), the retinochrome encoding the cDNA gene has been cloned and its primary sequence determined [172]. The gene product is an intrinsic membrane protein of 301 amino acids with seven transmembrane helices. However, a sequence comparison of the VP and the retinochrome from *T. pacificus* reveals many differences between both proteins. It should be kept in mind that the retinochromes fulfil a quite different physiological role, i.e., VP regeneration – and also mediate a different photoisomerization of the covalently bound retinal (all-*trans* to 11-*cis*; for a detailed discussion of the function of retinochrome see Section 8.3.3).

The majority of the invertebrate VP sequences consists of 369–385 amino acids, corresponding to molecular weights of slightly larger than 41 kDa. The VPs of the cephalopods are lacking a stretch of ca. 15 amino acids at the beginning of the polypeptide chain (the N-terminal part consists of ca. 37 amino acids up to the start of the first transmembrane helix, compared to ca. 50 amino acids for *D. melanogaster* Rh1) and carry an insertion of ca. six amino acids at the start of the C-terminal tail (given as the distance between the exit from TM VII to the vicinal cysteine motif). They also show a highly repetitive motif (PPQGY) with only very little variation, which is found in 10–11 copies at the very end of the C-terminus. The five listed VPs of cephalopods consist of 439 (*A. subulata*), 448 (*T. pacificus*), 452 (*L. forbesi*), 455 (*O. dofleini*), and 464 amino acids (*S. officinalis*), corresponding to a MW of about 50 kDa. Some of the VP sequences (Fig. 4a,b) appear to be significantly shorter than the general size (e.g., the freshwater crayfish-10 species – and two Rh1 sequences from *Drosophila simulans* and *D. virilis*). However, these truncated sequences are due to an only partial sequence elucidation using PCR primers for only the core portion of the VP.

Secondary structure alignment reveals the common structural motif of seven α-helical transmembrane domains for all VP proteins. The α-helical region spanning the membrane varies from ca. 21–28 amino acids, calculated for a membrane thickness of ca. 50 Å, and taking into consideration the overall hydrophobic character of the amino acids. In some cases charged or polar amino acids can be located at positions which are highly conserved and most probably indicate the beginning or end of a transmembrane segment, e.g., an Asp-Arg unit at the cytosolic side of TM III (position 167/168, Fig. 4a, and 154/155, Fig. 4b, respectively) and an arginine residue at the extracytosolic (N-terminal) side of TM IV (position 211), except for the Rh2 pigments of *D. melanogaster* (Ala) and *D. pseudoobscura* (Ser). In the alignment of the non-insect sequences (Fig. 4b), this site is not similarly well-conserved (position 198). Another motif, His–Glu–Lys (H–E–K, positions 265–267 in Fig. 4a, and 252–254 in Fig. 4b), can be identified in all invertebrate VPs at the C-terminal side of TM V. This tripeptide is known to be involved in G-protein binding (see below, signal transduction), and thus should be found at the cytosolic side of the membrane. Similarly, the entrance to TM VI (cytosolic side) is always indicated by a strictly conserved lysine (positions 298, 285, respectively). The appearance of positively charged amino acids (Lys, Arg) at the boundaries between cytosolic and transmembrane domains may indicate potential ionic interactions to the phosphate headgroups of the lipid bilayer, which anchor this region of the protein at the membrane surface.

Within the transmembrane hydrophobic stretches several hydrophilic amino acids can be identified, including dissociable and polar residues. Their presence can be correlated to functional tasks such as supporting the sub-ps photoisomerization of the chromophore, and wavelength tuning via the interaction of strategically placed amino acids with the π-electron system of the chromophore (see Chapters 1 and 2).

7.2. Structural motifs common to all invertebrate VPs

The most significant structural motif of 7TM receptors is the presence of two cysteine residues, located at the extracellular side of the membrane, which allow the formation of a Cys–Cys disulfide bridge for structural stabilization. One of these cysteines is located close to the N-terminal entrance into the third transmembrane helix (position 143 for the insects, Fig. 4a, and 130 for the other invertebrate sequences, Fig. 4b). Its putative reaction partner is usually found ca. 10 residues from the exit position of the fourth helix (position 221 for the insects and 208 for the other sequences, respectively). Formation of this disulfide bridge is indispensable for protein stability, as could be demonstrated by site-directed mutagenesis of either one of these two positions in bovine rhodopsin [45,191], resulting in a remarkable loss of chromophore binding capability.

A number of highly or even fully conserved proline residues are found (at positions 47, 48, 62, 66, 103, 140, 214, 247, 313, 331, 349, 357, and 368) in the insect and in the other sequences (at 47, 48, 52, 89, 111, 192, 201, 234, 300, 318, 336, and 344). Interestingly, an accumulation of five prolines within the first one hundred amino acids is found in all sequences. It may be assumed that this arrangement might serve as possible targets for a prolyl-isomerase activity which supports the correct folding of the nascent polypeptides chain, and thereby plays an important role for the translation control. This suggestion gains credence from a characterization of a prolyl isomerase in *Drosophila*, which upon misfunction causes the *ninaA* phenotype [48] (for a discussion of the role of prolines as intrahelical residues that cause conformational flexibility, see below).

Also common to all VPs are clusters of serines and threonines at the C-terminal end of the protein serving as targets for kinases (see Section 9 and Chapters 3, 4 and 9). The clustering is extreme in the Rh4 of *D. virilis* (between positions 381 and 390) and also in the UV-sensitive pigment of the honeybee *Apis mellifera* (positions 378–387), where eight out of 10 amino acids are serines or threonines. This motif is less pronounced in the *Limulus*, crayfish, crab, the cephalopod and the scallop sequences (Fig. 4b), indicating a different mode of VP activity regulation for these invertebrates.

7.3. *Extracytosolic N-terminal portion*

Though highly divergent, the N-terminal portion of the invertebrate VPs, spanning the first 62 amino acids prior to the entrance of the first transmembrane section, show several conserved amino acid positions. The gathering of up to five prolines within the first 45 amino acids has already been discussed in the preceding section. Furthermore, a conserved motif can be identified around position 60 with a fully conserved Trp; W_{58} for the insects and W_{43} for the other invertebrates – except for one of the scallop sequences (Patye2): H(Y)–W–Y(X)–X–Y/F–P (symbols separated by "/" indicate similar occurrence of structurally or functionally related amino acids; symbols in brackets refer to rare exchanges; X indicates amino acids with no obvious structural or functional properties). Possibly all these residues are essential for the function of the VPs, because positions of many of the above mentioned prolines are also conserved in the vertebrate VPs and in muscarinic acetylcholine receptors [192]. Support for the proposed importance of these amino acids is gained also from mutations of some of these positions in the vertebrate VPs, causing genotypes of retinitis pigmentosa when expressed in transgenic mice. The resulting mutated pigments are very unstable and are badly transported into the photoreceptor cell membrane [192,193]. Apparently the N-terminal portion of the VPs is important for correct transport and membrane insertion, although no obvious signal sequence motif can be identified.

7.4. Cytosolic loop between TMs I and II, and extracytosolic loops

Strong conservation can be recognized in the cytosolic loop (i1) between TMs I and II for all (nearly 60) sequences, though no functional role has been ascribed to this region of the protein. For bovine rhodopsin, an involvement of i1 in G-protein binding/activation was originally assumed similarly as for the loops i2, i3, and i4 (i4 is formed in some VPs by palmitoylation of a cysteine residue that immobilizes a portion of the C-terminus) [44]. This suggestion had to be revised from competition experiments employing synthetic polypeptides with sequences common to all four cytosolic loops [194]. These experiments have clearly revealed a competition for the i2-, i3- and C-terminal derived peptides (i4), whereas the peptide corresponding to the i1 loop hardly interfered with transducin binding. The conserved motif in this loop, from positions 96 through 105, can be identified as T/S(A/G)–T/S(A)–K–S(G/N)–L_{100}(I)–R/K–T(S)–P(A/S)–A/S–N. The high degree of conservation together with the prevalence of hydroxylated (S,T) and positively charged residues (K,R), and the proline at position 104 indicate a structurally important role for this peptide. Since this loop is the first cytosolic loop, this part of the protein might be involved in the transport of the nascent polypeptide chain towards the membrane, and in anchoring of the polypeptide on the cytosolic side of the membrane to ensure the correct folding of the polypeptide, putatively involving also proline isomerization [13]. Mutation experiments at this loop have proven this suggestion correct: changing the residues L_{100} or N_{105} in *D. melanogaster* Rh1 (L_{81} and N_{86}, respectively, in the numbering of the *Drosophila* sequence) blocks the nascent polypeptide chain in the glycosylated state, from which it is rapidly degraded with severe damage of the complete visual cell structure [43]. The contribution of proline for the protein stability can be inferred from *Drosophila ninaA* mutants. These flies synthesize the protein, but due to the lack of a prolyl-isomerase (cyclophilin) the polypeptide is not functional [46,48]. This hypothesis gains additional credence from the observation that also in the vertebrate VPs the characteristic features of this domain are highly conserved, and that mutations in this region are responsible for one genotype of retinitis pigmentosa in man [195].

No functional role has yet been identified for the extracytosolic loops, although a high degree of homology is readily seen, in particular for the loop e2 between TMs IV and V. Though speculative, this region of the protein could be involved in structural stability similar to the first cytosolic loop between TMs I and II. This argument is supported by the fact that a highly conserved motif can be detected: R(A,S)–Y/F–V–P–E–G–N(Y/F,I/V)–L–T–S, T(G,A,N)–C–X–X–D–Y(F)–L(M,F), (positions 211–226), which includes the tripeptide PEG with an acidic residue (E_{215}), and one of the cysteine residues (position 221) involved in disulfide bridge formation. Inspection of the insect sequences alignment reveals a basic residue close to the entrance into TM V (R_{233}) which might form a dipole with the acidic residue E_{215} and add stability to the protein structure. In some blue-absorbing pigments, R_{233} is exchanged to a lysine, in only one pigment – Manse3 of *M. sexta* – a glutamate is at that position, and in another pigment – Apimc from *A. mellifera cerana* – an alanine is found.

The divergent evolution of UV/blue absorbing insect pigments (Rh3/Rh4-type) from the Rh1/Rh2-type VPs can be traced to some of the residues in this loop. The Rh1/2 photoreceptors carry at position 212 a tyrosine, at position 217 an asparagine and at position 223 either isoleucine, threonine, or valine, whereas at the corresponding positions in the SW-absorbing species a phenylalanine (F_{212}), a tyrosine (Y_{217}) and another phenylalanine (F_{223}) are located.

In both alignments loop VI/VII is very short, containing ca. 10 amino acids which have no obvious similarity, so it probably functions only to fix TM VI close to VII.

7.5. *Transmembrane sections*

The diverging evolution of LW- and SW absorbing VPs becomes also obvious in the TM domains (see boxed areas in TM V, VI, and VII). The most remarkable feature of TM I is found at the C-terminal exit between positions 86–94 (72–80 in Fig. 4b), with the sequence G–N–G(F/A/C)–X–V(T)–X–W/Y–X–F(M,Y), with X = non-specific unpolar amino acid (I, L, M, V). The exchanges G74F and V76T (Fig. 4b) are found exclusively in the crayfish pigments, and the replacement of $F_{94/80}$ by either methionine or tyrosine is found only in *Limulus* (F80M) or in the SW-absorbing pigment of the locust – Schgr2, respectively. It should be kept in mind that glycine may change the helix structure.

In TM II three amino acid positions will be highlighted: an aspartate at position 115 (present in all invertebrate VPs except *A. mellifera cerana* (Asn); for a suggested function of this amino acid, see below), a methionine at position 119 (found in all insect VPs except the Rh4- and Rh5-type pigments), and a proline residue at position 124 (in some cases shifted by one position, i.e., 123 or 125). Concerning the non-insect invertebrate sequences, the aspartate and the proline are found in all sequences, the methionine is found in all except the VPs of the cephalopods (D_{101}, M_{104}, and P_{111}). In TM II of all SW-absorbing pigments, except the Rh5/Rh6 pigments of *D. melanogaster*, charged residues can be identified in TM II that are supposed to be components of wavelength regulation in these UV/blue absorbing pigments (see below, Section 8.6.2).

TM III exhibits a fairly strong sequence variation. The conserved motifs are one of the cysteine residues ($C_{143/130}$, respectively) of the disulfide bridge and two fully conserved glycine residues at positions 150 and 154 (Fig. 4a), and 137 and 141, (Fig. 4b), respectively.

The TM IV exhibits even stronger sequence divergence than TM III. It might be even more important that portions of the transmembrane section in TM IV can be proposed to interact with the chromophore. In all insect VPs three tryptophans are present (positions 194, 200, and 209; 177, 182, and 188 in Fig. 4b, respectively), except for some SW-absorbing pigments, where these positions are occupied by one of the other aromatic amino acids, tyrosine or phenylalanine. The occurrence of aromatic amino acids usually is indicative for a close proximity with the chromophore, allowing electronic interactions in sandwich-like structures, as found for the chromophore–protein interaction in bacteriorhodopsin [196–198]. The intimate contact between residues from TM IV to the polyene chain of the chromophore is discussed in Section 8.6.2. The counterparts of the above-mentioned charged/polar amino acids in TM II, selectively in the SW absorbing pigments, can also be identified in TM IV (see below).

The most interesting features of TM V are a fully conserved proline at position 247 (234 in Fig. 4b), and the clustering of between nine and 11 aromatic amino acid residues within the 29 amino acids of this transmembrane segment, which can be assumed to interact with the chromophore.

TM VI (and the preceding connecting loop) show a regular pattern of variation related to the phylogenetic position of pigments, either being part of the Rh1/Rh2 class or belonging to the SW-absorbing pigments, e.g., Rh3/Rh4 (see arrangement of boxes indicating conserved residues). This statement becomes even more solid from the newly included fragments of the butterfly *Papilio glaucus* (PapglRh1-6). Interestingly also for the non-insect VP alignment (Fig. 4b), a separation into classes becomes obvious in TM VI. The cephalopod sequences and those from the scallop segregate together, being distinct from the crayfish, crab and horseshoe crab sequences. A conserved motif of four amino acids acids W–T(S)–P–Y (positions 311 through 314, 298–301 in Fig. 4b) is preceded by another conserved dipeptide–L–W (L–F in the SW-absorbing pigments). This dipeptide is absent in all five cephalopod and in the two scallop sequences. A glycine residue at position 321 (308 for Fig. 4b), which in the cephalopod and scallop sequences is at position 311/310, may terminate the helix.

TM VII again shows a clear grouping into Rh1/Rh2-type or SW-absorbing pigments, and also in the non-insect sequence alignment (Fig. 4b) one finds a segregation into groups (*Limulus* and crayfish on the one hand and crab, cephalopod and scallop on the other). The C-terminal half of TM VII which carries the chromophore binding lysine residue (position 342, 329 in Fig. 4a,b, respectively), still exhibiting class-related changes of the amino acid sequences, is clearly more conserved than the N-terminal half. The conserved sequence (F–A–K...) starts at position -2 with respect to the chromophore attachment site, Lys342/329. Fully conserved in all sequences are two proline residues: P_{349} and P_{357} in the insect sequences and P_{336} and P_{344} in the non-insect alignment (Fig. 4b). The crayfish sequences, except the matching sequence from *P. clarkii* "Procl", were truncated, due to the experimental approach, before reaching this second proline. An interesting sequential variation can be identified exclusively in the case of the SW-absorbing pigments: at position +6 with respect to the chromophore-binding lysine (position 348 in Fig. 4a) one finds a negatively charged amino acid, which in all cases is an aspartate, except for the Rh4 from *D. pseudoobscura*, where a glutamate is found. This charged amino acid is also absent in the non-insect VP sequences, indicating an important functional role for this amino acid (discussed in Section 8.6.2). Within this part of the protein, one finds the beginning of one of the most strongly conserved stretches of amino acids. This allowed the design of oligonucleotide primers for use in PCR based cloning strategies (see Section 6.1.1). The motif P–I(F,W)–V–Y–G(A,S)–I(V)–S(N)–H–P–K(R)–Y(F)–R(K) (positions 349 through 360, and 336 through 347, Fig. 4b) exhibits only few changes.

7.6. *Glycosylation sites*

The post-translational N-linked glycosylation has been found to be essential for anchoring the nascent polypeptide after passing through the hydrophobic membrane bilayer, and for a correct folding. It should be kept in mind that in contrast to the permanent glycosylation of vertebrate rhodopsin, the insect VPs (shown for *Calliphora* and *Drosophila*) are only transiently glycosylated, such that the carbohydrate moiety can be identified in the newly synthesized polypeptide and is removed from the mature protein [42]. The importance of these sites has been demonstrated from a mutation in Rh1 of *Drosophila*, N20I, (pos. 39 in Fig. 4a) where a target site for glycosylation has been removed. This mutant suffers from accelerated degradation of rhabdomere structure compared to the WT [39].

The sequence analysis reveals two glycosylation sites (consensus motif Asn-X-Ser/Thr), one located in the N-terminal part, present in all sequences, and a second one on the extracellular side of the membrane in loop e2 between TM IV and V, present only in some of the VPs. Although the sequence alignments in Fig. 4a,b are not quite exact for the extreme N-terminal parts, and place the residue for glycosylation at different positions, they can be readily identified. The location of the glycosylation motifs follows the segregation of the VPs into several subgroups. VPs of Rh1 type carry the Asn residue 15–20 amino acids apart from the N-terminal methionine (position 39 in Fig. 4a); interestingly, Rh6 from *D. melanogaster* also falls into this group. In the SW-absorbing VPs, e.g., Rh3/4 from *D. melanogaster*, the UV-sensitive pigments from *A. mellifera* or those from both ants, the glycosylation site is already one of the first three to eight amino acids. Only the LW-absorbing pigments show a second glycosylation motif in the loop between TM IV and TM V (position 217). However, it has not yet been shown that this site is of similar importance as the N-terminally located one.

Due to the fragmentary sequences of the crustaceans (Fig. 4b), the presence of the first, N-terminal glycosylation site cannot precisely be assigned. However, the two sequences of the crab *H. sanguineus* both carry the putative glycosylation site at position three, whereas in the other sequences (*Limulus*, *P. clarkii*, cephalopods, and sequence #1 of the scallop) the asparagine is located at a distance of at least eight residues from the N-terminal methionine. The divergent structure of the crab sequences, already alluded to, becomes again obvious in the domain of the putative second glycosylation site (e2 loop between TM IV and TM V, see also the discussion in the previous paragraph on this site): all sequences exhibit a glycosylation motif in this part of the protein (position 204 or 207 in Fig. 4b), except those from the crab (and the scallop which, as outlined already exhibit an overall higher degree of dissimilarity).

7.7. *Chromophore attachment site, counterion situation and suggested protein–chromophore interactions*

Common to all VP protein sequences is the location of the chromophore attachment site, a lysine residue in the center of the seventh transmembrane helix. Resonance

Raman measurements on squid rhodopsin unambiguously demonstrated the protonated state of the Schiff base [199,200]. The same structure can be inferred for the insect VPs, although the only resonance Raman measurements of an insect VP were not fully conclusive [201]. Interestingly, the N-terminal halves of TM VII of the insect and also of the other invertebrate pigments show a strong variation of the regular pattern, which allows assignment of the VPs to either the Rh1-type or to the SW-absorbing type. The sequences from the horseshoe crab and the crayfish follow the pattern of the Rh1-type pigments such that they exhibit a Trp-Gly motif at the identical position as the LW-absorbing pigments of the insects (positions 323–324 in Fig. 4b). This is at a distance of six residues (in N-terminal direction) apart from the chromophore attachment site. The Trp is followed at a distance of four amino acids (corresponding to one helical turn) by a phenylalanine, thus adding to the aromatic character of this part of the seventh helix. In contrast, the sequences from crab, cephalopods and scallop all show a Leu-Pro motif, identical in position and quite similar to the motif of the SW-absorbing insect pigments that show Ile-Pro.

The necessity of a counterion in the invertebrate VPs is of no doubt. However, the structural description remains unresolved. Whereas in the vertebrate pigment a glutamate residue (E_{113} in bovine rhodopsin, dissociated in the parent state of the receptor) has been assigned to supply the counterion on the basis of site-directed mutagenesis experiments [202], such a dissociable amino acid cannot be identified at a corresponding position (position 146 in Fig. 4a, 133 in Fig. 4b, respectively) in the N-terminal part of TM III of any invertebrate VP. Until recombinant VPs are available, the presence of counterions in invertebrates can only be inferred. In fact, since E_{113} of bovine rhodopsin has been also characterized as a catalyst for the hydrolysis of the Schiff base [202,203], such a residue would not be anticipated to be found in the invertebrate VPs, because they form thermostable metastates. At the corresponding position of E_{113}, all invertebrate VPs carry an aromatic amino acid, preferentially a tyrosine, with the SW-absorbing pigments of Rh3/4-type having a phenylalanine at that position. Whether this exchange (Y $\rightarrow$ F) contributes to UV sensitivity, or is a spectrally ineffective substitution, remains speculative. The Rh1-type insect pigments, and those from *Limulus* and the crab *Hemigrapsus*, carry a charged residue (D/E_{144}, D/E_{131} in Fig. 4b) in direct proximity to Cys143/130 (which is one partner of the disulphide bridge), but this negatively charged amino acid is not strictly conserved (Lys in the cephalopods) and is putatively located at the membrane boundary or even within the cytosolic loop. Only one charged (or dissociable) residue, which is completely conserved in all invertebrate VPs, can be unambiguosly identified within the hydrophobic domain. This residue, D_{115} (D_{101} in Fig. 4b) in TM II, is found at an identical position also in many vertebrate visual pigment sequences, and even in many other 7TM receptors [204], indicating a more general role for the stability of the protein, possibly through the formation of hydrogen bonds.

For the invertebrates, a more complex arrangement of counterions may exist involving more than one residue. An arrangement could be imagined similar to that existing in the halobacterial retinal-binding protein bacteriorhodopsin (BR) [197,198,205,206], where several amino acids and a water molecule contribute to the stabilization of the Schiff base [207,208].

A large number of aromatic amino acids is found in the TM regions of all VP sequences, which are believed to interact with the chromophore. Again, this assumption is substantiated from the known three-dimensional structure of bacteriorhodopsin, where several tryptophan, tyrosine and phenylalanine residues are associated with the chromophore and dictate its exact placement within the protein as well as its spectroscopic properties [196–198]. An impression of the binding site structure for a retinal-bearing protein is obtained from the protein structure of bacteriorhodopsin, which is available at high resolution (Fig. 5). The retinal chromophore in BR is enveloped by four tryptophan and a tyrosine residue (W86, W138, W182, Y185, W189), and additional polar and charged amino acids (T89, T90, D115, S141) increase the influence of the protein on the chromophore. Ser141 has been identified as an important residue for the absorption properties of the retinal chromophore: its hydroxy group is located close to the chromophore ring site and interacts with the π-electron system of the chromophore in a way that the alternation of double bond/single bond character of the polyene chain is intensified (for a discussion of the electronic structure of the protonated retinal Schiff base see Section 8.6). Interestingly, a proline is also found in close proximity to the retinal chromophore in BR (Pro186). Via its capability to adopt the *cis* or the *trans* configuration the chromophore might add flexibility to this part of the protein after its photoisomerization. A first view on the binding site of rhodopsin is now possible due to the recently published crystal structure [406].

An interesting pattern emerges when the LW- and the SW-absorbing pigments are compared, which most probably is involved in the wavelength tuning. The mechanisms for adjusting the spectral absorption are different between invertebrate and vertebrate pigments (see also Chapter 2), and will be discussed in greater detail below (see 8.6.2).

Recently, a gradual adaptation of the protein environment to either retinal or 3-OH retinal as chromophore has been suggested on the basis of a limited number of

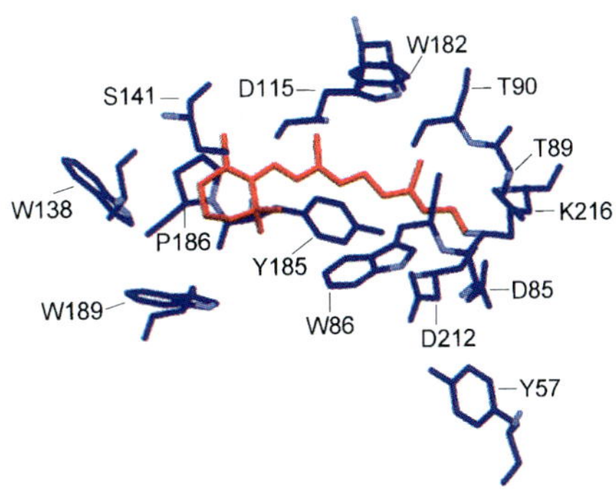

Fig. 5. Arrangement of the retinal chromophore (in red) and surrounding amino acids (in blue) in the halobacterial membrane protein bacteriorhodopsin (BR, adapted from Ref. [198]). Besides several polar and charged amino acids, specifically the aromatic residues are depicted. Please note that the chromophore in BR adopts the all-*trans* geometry.

Rh1-type pigments exhibit a lysine at position 279 and a glutamic acid at position 285, and a motif of negatively/positively charged amino acids (Glu/Asp–Lys) at positions 289/290. In the Rh3/4-type insect VPs, all three sites are either converted into an oppositely charged amino acid or are replaced by a non-charged amino acid. Inspection of the sequence alignment for the non-insect VPs reveals again the particular position of the cephalopods compared to the other sequences in this alignment (Fig. 4b), e.g., the positively charged Arg257 and also the directly adjacent negatively charged Asp/Glu all are replaced by alanines. Another change of polarity is found for position 267 (S/T converts to E in the cephalopods). The negative charge contributed by E_{272} and E_{273} is either converted or neutralized in the cephalopods (K_{272}, A_{273}).

When compared to the i3 loop of bovine rhodopsin, which consists of 24 amino acids, highest similarity is found in the C-terminal section with the amino acids: **S**–**S**–E–D–**A**–**D**–**K**–S–A–E–G–**K** (sequence of *Calliphora*, position 283–295, the alignment generates a deletion at position 287) vs **T**–**T**–Q–K–**A**–**E**–**K**–E–V–T–**R** (position 242–252 of bovine rhodopsin; conserved and semi-conserved residues shown in bold. No similarity to this motif can be identified either in the crayfish or in the cephalopod VPs. The N-terminal half of this loop is more diverse between the invertebrate VPs and bovine rhodopsin, exhibiting two insertions of six and four amino acids. In particular, these insertions generate sequences which render them similar to the N-terminal portion of the α-subunit of bovine transducin [234]. Since this sequence motif cannot be detected in the vertebrate VPs, it was assumed that for the invertebrates a different activation process for the G-protein might take place, probably involving a competition between part of photoreceptor molecule itself and the N-terminal portion of the attached G-protein. On the basis of much more sequences available, this suggestion deserves additional attention. Fig. 6 collects the sequences of the i3 loop for a number of invertebrate VPs and of the N-terminal part of the G-protein α-subunits of *Drosophila* and the squid *Todarodes*. For comparison,

	1 10 20 30
1. *D. melanogaster*, Rh1	H**EK**AM**RE**Q**AKKMN**VKS---**LR**SSEDA--EK-**SAE**G**KLA**-**K**
2. *D. melanogaster*, Rh3	H**EK**AL**RD**Q**AKKMN**VES---**LR**SNVDKNKE--**TAEIRIA**-**K**
3. *D. melanogaster*, α-SU	S**E**E**A**-**KE**Q-**KRIN**QEIEKQ**LR**R-----------**ELKLL**--
4. (bov.) transducin, α-SU	E**EK**HS**RE**LE**KK**--------**LK**--EDA-------**E**-**K**D**A**-**R**
5. *L. polyphemus*	H**EK**Q**LRE**Q**AKKMNV**AS---**LR**A--NADQQKQ**SAE**C**RLA**-**K**
6. *P. clarkii*	H**EK**G**MRD**Q**AKKM**G**IK**S---**LR**---NEEAQKT**SAE**C**RLA**-**K**
7. *T. pacificus*	H**EK**E**M**AAM**AKRLNAK**E---**LR**K---AQAGA-N**AEMRLA**-**K**
8. *T. pacificus*, α-SU	EA**KE**Q----**KRIN**QEIEKQ**LR**RDKRDARR----**ELKLL**--

Fig. 6. Sequence comparison for the i3 loop of several invertebrate VPs and the N-terminal portion of the α-subunit of the respective G-proteins (from top: 1, 2, VPs of *D. melanogaster*, Rh1 and Rh3, respectively; 3, 4, *D. melanogaster* and bovine G-protein (transducin); 5, 6, 7, VPs of *L. polyphemus*, *P. clarkii*, and *T. pacificus*, and 8, G-protein of *T. pacificus*. Amino acids which are identical or show conservative exchanges, such as Asp/Glu, Lys/Arg, or Ala/Val/Ile/Leu/Met, are given in bold.

the corresponding part of the bovine G-protein transducin has also been included (the insect sequences correspond to positions 265–298 in Fig. 4a, the non-insect VPs correspond to 252–285; the G-protein sequences relate to positions 6–37 for the *Drosophila* sequence, and to positions 8–37 for the squid sequence). An alignment of this highly charged protein stretches is fairly difficult and remains somewhat arbitrary. For example, the squid G-protein sequence (#8) shows a nearly exact threefold repeat of a tripeptide (**RRDKRDARRE**, positions 21 through 34, mind the gaps). This gives several possibilities for alignment. Inspection of the sequence alignment reveals that the above-mentioned similarity between the bovine transducin sequence and the i3 loop of *Drosophila* VP is to some extent incidental, in particular when the genuine fly G-protein sequence is included: the charged motif EK which is present in all VP sequences and in bovine transducin (positions 2 and 3 in Fig. 6) is present neither in the fly nor in the squid G-protein. Also, the tripeptide EDA (positions 24–26) is identical only in the *Drosophila* Rh1 and the bovine transducin sequences. There remain three sites of significant sequence similarity (if conservative changes, e.g., E/D, Q/N, R/K, I/L/V/M/A, are allowed): the tetrapeptide at positions 10–13 always reads K–K/R–M/I/L–N (except for *P. clarkii* which carries a G at the last position), the two amino acids at positions 20, 21 are always L–R, and the pentapeptide from positions 34 to 38 shows a high degree of similarity: E–G/I/L/C/M–K/R–L/I–A/L. An additional site is found in all but the cephalopod VP sequences: positions 6–8 are always R/K–E/D–Q; interestingly, the corresponding stretch in the squid VP reads A–A–M. Although the indications are vague, one might speculate that the preceding tripeptide in the squid VP (positions 3–5, K–E–M) is the equivalent to this conserved part.

Interestingly, the N-terminal portion of the i3 loop is highly conserved between the insects and the other invertebrates, whereas, as mentioned above, the C-terminal half of this loop is more similar between the vertebrate and the insect VPs. Probably, this part of the protein reflects the diverging evolution of vertebrates, insects, the cephalopods and the other invertebrates listed here.

In the i4 loop of some invertebrate VPs, a vicinal cysteine couple (position 369/370, Fig. 4a) is located within the C-terminal end of the protein. In bovine rhodopsin the cysteines are known to be palmitylated [44]. It was shown for the vertebrate pigment that this covalent attachment anchors the C-terminal sequence in the lipid bilayer and forms a fourth cytosolic loop (i4) which is involved in G-protein binding/activation [194]. The presence of this cysteine couple in the invertebrate VPs varies. Two vicinal cysteines are present in the Rh1 pigment of *Drosophila* and *Calliphora*, in all cephalopod VPs, and in the first sequence of the scallop. The VP of the crayfish *P. clarkii* exhibits an insertion of two amino acids (C–L–S–C) at the corresponding positions. Interestingly, such an arrangement again could allow the two cysteine residues to point into the same direction, if one assumes that this sequence motif is part of a short α-helix. In *Limulus* a further modification with S–L–A–C can be identified. This modified motif is also present in Lo-1 and in the mantid VP. In the insect sequences, the Cys–Cys couple is preceded directly by, or is in close proximity to a fully conserved proline (except for the two blue absorbing pigments from the honeybees, and the Lo-2 pigment of the locust which carry a lysine). In the blue/UV absorbing VPs, *Drosophila* Rh3/4 and Lo-2, cysteines are absent from this region. Also, the red and green absorbing VPs of the vertebrates lack the cysteines at that position, whereas some blue-absorbing vertebrate pigments have either one or two cysteines [235], see also Chapter 6. Though Cys → Ser mutants in bovine rhodopsin still can induce transducin activation, the change alters signal transduction, presumably due to a modified interaction with the G-protein [236].

The primary step in signal transduction within the photoreceptor molecule includes the transfer of information from the photoisomerized chromophore to amino acids from the chromophore binding site and further on to the cytosolic loops, where a conformational change is triggered. Some histidine residues, due to their pK value allowing protonation changes at physiological pH, and proline residues, due to their capability to undergo *cis–trans* isomerization, have attracted specific attention. In bovine rhodopsin a histidine (H_{211}) has been identified to take part in the protonation changes of the chromophore during meta I – meta II equilibrium formation [237], see also Chapter 3. The invertebrate VPs do not carry a histidine at the corresponding position, i.e., the extracytosolic half of TM V. This indicates that a catalytic participation of a histidine in a change of the protonation state of the metapigment is of no relevance for signal transduction. Remarkably, four conserved proline residues can be considered as participants in the signal transduction within the protein, namely in TM IV, V, VI and VII, positions 204 (205), 247, 313 and 349 (Fig. 4a) or positions 192, 234, 300, and 336 (Fig. 4b), respectively.

In all insect, *Limulus* and crayfish VPs there are up to twelve hydroxylated amino acids (S/T) clustered in the C-terminus. These residues serve as targets for a kinase which is believed to switch off the active state of the vertebrate VP ([99,226]; for an overview see also Ref. [98]). In *Calliphora*, however, phosphorylation of these sites is preceded by arrestin binding. Whereas the kinase reaction rate is relatively slow, the binding of arrestin is a rapid process and could deactivate the metastate by occupying the transducin binding site [93,94].

7.9. *C-terminus of the cephalopods*

The C-terminal tail of the cephalopod VPs is much longer due to the insertion of up to 11 copies of a pentapeptide motif: P–P–Q–G–Y, beginning around position 430 (Fig. 4b). This repetitive motif is not found in any other VP characterized so far. A number of investigations were dedicated to this peculiar structure which has some similarity to other, functionally not related membrane proteins (for an overview see Ref. [238]). This sequence motif interferes with the mobility of the protein in the microvillar membrane as shown by proteolytic digestion which changed the rotational mobility and altered the arrangement of the VP molecules in clusters [239,240]. In contrast to the other invertebrate VPs, those from the cephalopods exhibit only three to five serine or threonine residues in the C-terminus, but exhibit a large number of negatively charged residues within the first fifty residues of the C-terminal domain, before the ongoing of the repeated motif. It has thus been proposed that these negatively charged amino acids might be involved in the adaptation/deactivation of the cephalopod VPs by a yet unknown mechanism [168]. Detailed analysis, however, has revealed that also in the cephalopods phosphorylation of Ser/Thr residues in the C-terminus is regulating deactivation (see below).

8. Spectral properties of invertebrate VPs and stability of the metapigments

Extensive information has been collected for the spectral position of invertebrate rhodopsins and their metapigment state (see also Chapter 10; for a list of comprehensive reviews and overviews see references given in the introduction). It has clearly been demonstrated that the majority of the insects have developed di-, tri- or, as in the case of the butterflies, polychromatic vision. Polychromaticity has been demonstrated also for some crustaceans (e.g., the mantid shrimps [241–245]) that have developed a structurally and functionally highly complex polychromatic visual system, not only in the visible spectral range, but also in the ultraviolet region [405]. The compound eyes of these mantid shrimps are also able to detect polarized light. Yet, many crayfish species appear to have only monochromatic vision, and also many cephalopod species are considered to be monochromatic. Despite putatively being monochromatic, the eyes of squids are excellently designed for the detection of polarized light, which gives additional visual information in an environment of very low light intensity (see Refs. [31,238] and references cited therein). However, as has been alluded to already, the firefly squid *Watasenia scintillans* and some other, related squid species have presumably developed a trichromatic visual system [69,246].

In particular, the explanation of the UV-sensitivity of many VPs is challenging, since in some cases the absorbance maximum is shorter than that of the free retinal molecule in organic solvents, indicating a sophisticated mechanism by which the protein surrounding induces a hypsochromic shift of the absorption of the visual chromophore. The dominating type of VP, however, absorbs maximally in the blue/green part of the spectrum. As in the vertebrate pigments, absorption of light causes photoisomerization of the 11-*cis* into the all-*trans* isomer of the retinal chromophore on the (sub)picosecond timescale, followed by a series of conformational changes of chromophore and protein in the nanoseconds to seconds time range. The latter, thermally driven motions are accompanied by changes in the absorption properties of the formed intermediates and thus can be studied spectroscopically. Due to the problems to prepare VP from insect eyes and the yet unsuccessful attempts to express recombinant VPs in sufficient amounts for spectral and kinetic analysis, detailed information on the intermediates of the phototransformation for the insect VPs is still sparse, although quite detailed information could be obtained for VPs from cephalopods, due to their large retinae (see Section 8.3).

In contrast to the vertebrate pigments, which bleach irreversibly, the VPs from invertebrates form a metapigment which is thermally stable and which can be reconverted into the parent state upon irradiation with light of appropriate wavelength. Accordingly, many invertebrate VPs function as a light-driven biological switch with strong photochromic properties. When the absorption maximum of the pigment is below 500 nm that of the corresponding metapigment is shifted bathochromically, but when the peak wavelength of the pigment is above 500 nm the shift is hypsochromic; this observation is not yet explained [22,23].

Photoequilibrium studies performed for the *Calliphora vicina* mutant *chalky* revealed that for Rh1 the ratio of the peak extinction coefficients (metapigment: pigment) is ca. 1.8. This observation also holds true for many other insect species, as has been seen for the recombinant *Drosophila* and locust rhodopsins (see below). Assuming an absorption coefficient for the blowfly pigment of ca. 30–40.000 M^{-1} cm^{-1}, a coefficient for the metapigment of ca. 54–72.000 M^{-1} cm^{-1} would result. The upper limit for this value appears unusually high, making a lower value for the pigment more probable. Pigment and metapigment states also differ by their fluorescence properties. Whereas the parent state has a negligibly small fluorescence quantum efficiency (a value of lower than 10^{-3} was determined for bovine rhodopsin [247]), the metapigments show a fairly strong fluorescence. This capability of strong light emission appears to be a general feature of the metapigments, especially for the invertebrate VPs. For the blowfly, a Stokes shift of ca. 80 nm was determined (λ_{max} of metapigment = 580 nm, λ_{em} = 660 nm [248]).

It was shown for the vertebrate pigment that the counterion (Glu113 in rhodopsin) is involved in the hydrolysis of the Schiff base, and that the *de novo* synthesis of 11-*cis* retinal is accomplished by a sophisticated enzymatic machinery in the vertebrate photoreceptor cells (see Refs. [89,90] and references cited therein). As mentioned above, no charged amino acid is present at the corresponding position in the invertebrates, which either carry a tyrosine or a phenylalanine. Yet, invertebrate pigments also can be bleached by permanent irradiation, albeit at a relatively slow rate (see above, Section 5).

8.1. Insect VPs

8.1.1. Methods for the functional analysis of insect VPs

A variety of methods have been developed for the functional analysis of insect VPs. However, due to the failure of heterologous expression, many applications were restricted to the study of living animals, dissected eye cup tissue or preparations of the VP by treatment with detergent, but nearly all investigations suffer from low amounts of material, making the development of highly sophisticated detection methods necessary. Yet, in general, none of these techniques makes a detailed kinetic analysis of the pigment → metapigment conversion possible, nor does it allow variation of the type of chromophore or the protein primary structure. Exchange of the chromophore and application of vibrational spectroscopy – Fourier-transform infrared and resonance Raman – so far has been successfully employed only for cephalopod VP (see Sections 8.3 and 8.8), and in only one case a resonance Raman study was reported for an insect VP [201].

8.1.2. Recording of electroretinograms (ERGs)

Electrophysiology has been extensively applied for spectral and functional characterization of insect, *Limulus* and crayfish VPs. Since ERG recordings reflect changes of the membrane potential of the various visual cells, this method also gives information on the visual transduction pathway [14,15,249]. Particularly useful cases are *Drosophila* mutants [144,220], or transgenic flies which have expressed a non-*Drosophila* VP (see below). From measurements recorded in a wavelength-dependent manner, "action spectra" are obtained. If, however, from these action spectra information on the pure absorption spectra of a given VP should be extracted, one has to take into account the different neuronal processing, the signal amplification processes and possible cross-talk (electric coupling [250]) between the various photoreceptor cells. Furthermore, possibly interfering components in the compound eyes may cause partial shading and thereby shift the spectral sensitivity. Such spectral shading can be accomplished by screening pigments, or by other VPs due to the tiered arrangement of visual cells (cf. the situation of R8-type cells in the retina of Diptera, which are obscured by the distally located R7-type visual cells [14,250–252]). Also, carotenoid compounds which are embedded in the cell tissue can induce shading and thus cause a shift of the wavelength of maximal sensitivity. Since the screening pigments are the most critical problem for electrophysiology and measurements of absorption spectra (see Section 8.1.3), mostly mutant flies devoid of screening pigments have been employed (specifically the mutant *chalky* of *C. vicina* and the *white* mutant of *D. melanogaster*).

The disturbing effect of screening pigments in analytical methods such as electrophysiology of course is an endogenous property of their physiological function where they are involved in polychromatic vision. Many insects have optimized the wavelength selectivity of their VPs by partial shading by a screening pigment. An example where polychromatic vision is accomplished by screening pigments is the butterfly *Papilio xuthus* [78,253]. An ultraviolet-absorbing pigment, presumably 3-OH retinol, dramatically narrows the spectral absorption of a violet-absorbing VP. Simultaneously, it covers the *cis* peak of a green-absorbing pigment which would cause spectral absorption in the wavelength range around 350 nm [78]. Also for the long-wavelength absorbing VPs, red-transmitting screening pigments shift the peak sensitivity bathochromically and narrow the absorption band [253]; see Chapter 10.

8.1.3. Microspectrophotometry

Very elegant spectroscopic techniques have been developed for the study of light-induced absorption changes during the pigment/metapigment conversion in living animals or intact eye cup preparations (see below). The microspectrophotometric technique makes use of transmittance and/or reflectance measurement of light applied onto a single compound eye, arranged in the optical axis [17,22,248,254]. This technique has successfully been applied for the spectral identification of VPs expressed in transgenic flies (see Section 8.2.2).

8.1.4. Eye cup preparation

Pigment → metapigment conversion can also be followed by preparing a single eye cup in a regular spectrophotometer. Specifically, these measurements have been performed with material from the *chalky* mutant (lacking screening pigments) of the blowfly *Calliphora vicina*. With such a preparation, the photoreversible conversion between both thermally stable pigment forms can be followed over several hours, and allows, e.g., the effect of a carotenoid-deprived diet for VP formation (Fig. 7). The thermal stability of the metapigment and the photoreversibility between both pigment forms allows to determine a ratio of ca. 1.8:1 for the peak absorption coefficient of the metapigment relative to that of the parent state. Based on the knowledge of the parent state spectrum, the ratio of the extinction coefficients and the possibility to generate photoequilibria with various content of both pigment forms in dependence of the applied wavelength (see traces 2 and 3 in Fig. 7), the relative content of the parent state in a photoequilibrium can be calculated for any irradiation wavelength (S-shaped curve in Fig. 7, relating to the ordinate at the right hand side,% rhodopsin). Unfortunately, since there exists no P-element mediated transformation for blowfly, this experimental method cannot be applied for a routine characterization of recombinant VPs.

8.1.5. Detergent extraction

Up to now, successful preparation of insect visual pigment for *in vitro* studies has only been achieved by detergent treatment of compound eye tissue. This disrupts the membrane structure of the visual cell and results in relatively stable, tube-like preparations of the resealed microvilli membranes which contain still photoreversibly active VPs in solution [27]. This procedure suffers from a low yield of functional material, however. Only few examples of this methodology have yet been reported, since, besides the low yield, large amounts of screening pigments also are extracted together with the VP. This seriously obstructs the measurement of VP absorption spectra. Accordingly, detergent solubilization is best performed with the blowfly mutant *chalky* or the fruitfly mutant *white*, which lack screening pigment in their compound eyes. One of the earliest detergent extractions using digitonin was performed on the *Drosophila* mutant *norp* A^{P12} [255]. A reduced amount of visual pigment (compared to WT flies) was obtained, which however showed normal photochemistry. It was thus concluded that a more complex defect in the visual transduction pathway causes this mutant phenotype. The same treatment with digitonin or CHAPS of the moth *Manduca sexta*, which possesses larger eyes than blowfly, allowed cleaner preparations. This yielded three VP in soluble form [162] with absorbance maxima at 520, 450 and 357 nm. The absorbance peak wavelengths of the corresponding metapigments are 485, 485 and 470 nm, respectively. The relative abundance of the pigments in *Manduca* eyes has been estimated as P520:P450:P357 = 100:25:8.

8.2. Recombinant approaches

For a detailed understanding of the function of VPs and the underlying chromophore–protein interactions, the establishment of heterologous expression, giving access to changes of the protein sequence and allowing variation of the chromophore structure, is of utmost importance. This has clearly been demonstrated for

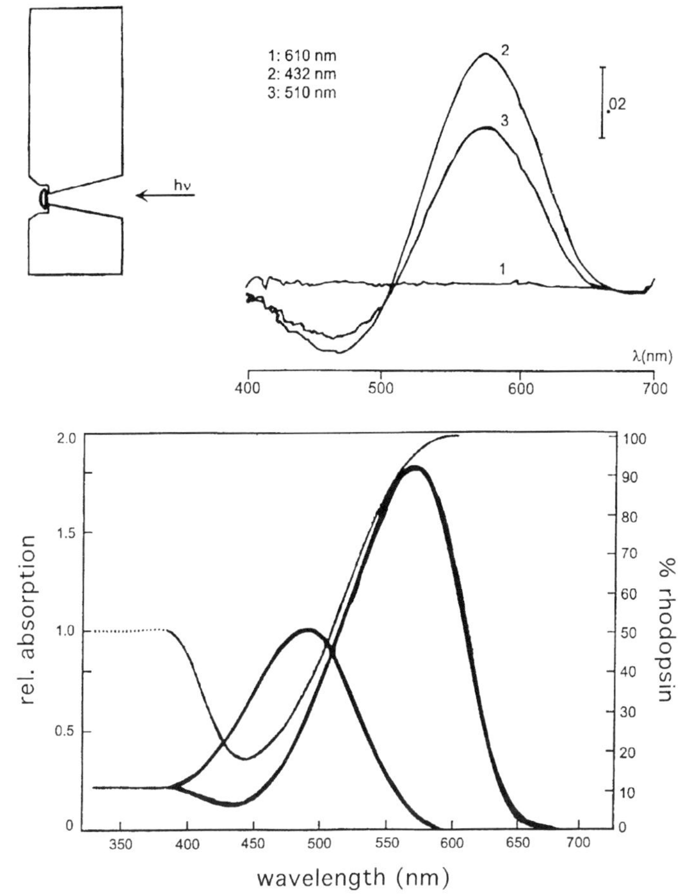

Fig. 7. (Top) Experimental set-up and absorption difference spectra (from pigment → metapigment conversion) recorded on a single eye cup preparation of the blowfly *Calliphora vicina* (mutant *chalky*). Left: A single eye cup is placed on the rear side of a cuvette-sized (1 × 1 cm) metal block equipped with a cylindrical hole allowing the measuring light beam (hν) to pass through. The sample chamber can be sealed with thin quartz plates on both sides after applying some Ringer solution in order to prevent desiccation of the sample. Right: difference absorption spectra obtained from a single eye cup preparation. Baseline: the complete VP content of the eye cup was converted into the parent state by irradiation with long-wavelength light ($\lambda \geqslant 610$ nm), and a baseline spectrum recorded (trace 1). The sample was then irradiated with light of 432 nm (yielding maximal amount of metapigment), until photoequilibrium was obtained, and a new spectrum was recorded (spectrum 2). Exhaustive irradiation at $\lambda = 510$ nm (isosbestic point of both spectra) yielded the spectrum 3. Between the recording of spectra 2 and 3 a shift in the baseline occurred, since in a difference spectrum the isosbestic point has to meet the baseline. Yet the spectra are shown uncorrected. Bottom: from the spectrum of the parent state ($\lambda_{max} = 495$ nm, obtained by long wavelength irradiation), and any difference spectrum, the absorption spectrum of the metapigment ($\lambda_{max} =$ ca. 570 nm, relative in intensity to the parent state spectrum) can be generated (left ordinate). The right ordinate indicates the relative amount (in percent) of pigment in a photoequilibrium in dependence of the applied wavelength (modified from Ref. [92]).

vertebrate opsin [256,257], but for the invertebrate VPs it still represents a formidable task. Two principal approaches can be imagined:

(i) The expression of foreign DNA in photoreceptor-deficient fruitflies by P-element mediated transformation (i.e., injection of foreign DNA into the posterior pole of *Drosophila* embryos). This method is restricted to fruitfly, since no comparable transformation system is available for any other invertebrate species. Furthermore, this approach is strongly limited with respect to providing sufficient amounts of material for biochemical or spectroscopic studies.

(ii) The heterologous expression of invertebrate VPs in various hosts (preferentially in eukaryotic, e.g., mammalian or insect cell lines), as is already established for the expression of the vertebrate VPs. This latter technique, however, has yet not been successfully applied.

8.2.1. Expression in transgenic flies

A number of *Drosophila* strains can be transformed with foreign DNA by means of the P-element technology [258,259]. In brief, this method is based on the activity of a transposase, which allows stable insertion of the foreign DNA into the genome of the fly. The DNA encoding the transposase is located on a helper plasmid, whereas the plasmid used for cloning the foreign DNA is furnished with the recognition sequences required for insertion. The multiple cloning site of these vectors directly succeeds the promoter of Rh1-type VP of *Drosophila*, thereby placing the foreign DNA under the control of the Rh1 promoter. Transformation is performed by injection of the P-element DNA into the posterior pole of *Drosophila* eggs at a precisely determined time of development. Besides the DNA subjected to transformation, usually a phenotypical marker is also located on the same plasmid, allowing identification of transformants by simple inspection of the offspring. After hatching, the G_1 generation of flies has to be selected for having received the marker, and has to be crossed until a homozygous population is obtained. For the P-element mediated transformation, a *Drosophila* double mutant is used which lacks the outer photoreceptor cells R1-R6 (*ninaE*), and simultaneously shows the *white* phenotype (absence of screening pigments; for a description of mutants being employed for electrophysiology/microspectrophotometry see overview in Ref. [144]). Since it has been found that a rescue of this genotype by transformation with VP-encoding DNA also re-establishes the degenerated rhabdomeric structure, flies with a nearly normal eye structure result, which can be analyzed by either microspectrophotometry or electrophysiology.

8.2.2. Microspectrophotometry of transgenic flies

Microspectrophotometry of transgenic flies has first been performed with the *ninaE Drosphila* mutant that was rescued by its genuine VP, Rh1 [146,147]. The cloning construct for this experiment is described as P[Rh1 + Rh1] (P-element mediated transformation of [Rh1 promoter + cDNA encoding a VP, here Rh1]). By this method, also Rh2 [146], and Rh3 and Rh4 [147] were expressed, the latter two either separately or simultaneously in the same cell. The *ninaE*; P[Rh1 + Rh2] flies showed a blue shifted absorbance compared to Rh1 in microspectrophotometric experiments with a maximum around 420 nm and a metapigment peak at ca. 520 nm. Also, the transgenic flies expressing Rh2 produced a sensitizing pigment similar to that reported from measurements of the ocelli [260], with the position of the central peak of the fine-structured absorbance band being similar to recordings from normal ocelli ($\lambda_{max} \cong 350$ nm; for a description of the sensitizing pigment in Diptera, see below). This capability of forming a sensitizing pigment could be intrinsic to the photoreceptor molecule itself or, alternatively, could be provided by the host cell as being triggered by the expression of a photoreceptor molecule. The resulting absorption maximum, however, favors the former explanation. The expression of Rh3 and/or Rh4 led to the assignment of Rh3 to the pigment of R7p cells ($\lambda_{max} = 345$ nm), and revealed clear differences to the VP of R7y cells ($\lambda_{max} - 425$ nm), which predominantly functions through its sensitizing pigment with a maximal sensitivity around 350 nm [251]. Though the spectral contributions of Rh4 to the sensitivity of R7 could be demonstrated [147], a mechanism for a possible interaction between Rh3 and Rh4 VP is still elusive. The microspectrophotometric analysis of Rh5 and Rh6 expressing *Drosophila* matched excellently with intracellular recordings [404]. The ectopic expression of these two VPs also revealed a mosaic-like distribution in the photoreceptor cells.

Whereas originally employed only for *Drosophila*-derived VP-genes, the generation and spectral characterization of transgenic flies has now been performed for quite a number of VP-encoding DNAs, making it a general procedure for describing spectral properties of insect VPs. Correct assignment of the absorption maxima has recently also been achieved for the VPs from the honeybee (*Apis melifera*). The spectral analysis of living animals revealed the presence of three pigments with spectral absorption in the green, blue and UV region of the spectrum, respectively [261,262]. The first sequence which was obtained from cloning cDNAs was ascribed to the green absorbing pigment, by sequence alignment and also by analysis of the relative amount of pigments [263], since it corresponded to the most abundant mRNA. The second ORF was proposed to encode a UV-sensitive pigment [159]. A detailed re-analysis revealed again two sequences, however, none of them matched with the formerly published LW-pigment. Besides the already reported sequence, which originally had been claimed to be a UV-sensitive pigment, a second sequence was found which showed even higher homology to sequences of UV-pigments (sequences of UV-sensitive pigments had not been available at the time when the first honeybee sequence was reported). This paradox could be solved by microspectrophotometry of transgenic *Drosophila* which expressed both new honeybee sequences. The spectral analysis revealed that the latter assignment was correct, such that the sequence, originally assigned to a UV pigment, encodes a blue-absorbing VP, whereas the second of the latter identified sequences encodes the UV-sensitive pigment [160]. Thereby, the approach of determining the spectral sensitivity from sequence comparison in combination with microspectrophotometry had demonstrated its usefulness for the characterization of insect VPs. Besides the honeybee VPs, VPs from ants (Smith, W.C., pers. commun.), and from the locust *Schistocerca gregaria* [264] have been expressed and spectrally characterized in transgenic fruitflies.

shifted with respect to both pigment forms (*Todarodes pacificus*: 480, 488, and 495 nm, values are given in the order P, M, and Re, respectively; *Sepia esculenta*: 486, 495, and 508 nm; *Sepiella japonica*: 500, 500, and 522 nm). If a strong bathochromic shift occurs upon formation of the metapigment, the absorption of the retinochrome is found in between both pigment forms (*Octopus vulgaris*: 475, 503, and 490 nm; *Octopus ocellatus*: 477, 512, and 490 nm).

The photochemistry of the retinochromes and the intermediates formed during this reaction sequence could be identified in detail. Very rapid processes (in the sub-ps time range) form a prelumi retinochrome ($\lambda_{max} = 460$ nm). This state already contains the isomerized 11-*cis* retinal chromophore and decays within ca. 100 ps into the lumi-intermediate ($\lambda_{max} = 475$ nm). The photoisomerization could also be followed by FTIR spectroscopy at low temperature [291]. From this intermediate, the metaretinochrome ($\lambda_{max} = 470$ nm) is formed in a relatively slow, biphasic process with time constants of 80 and 290 ms at ambient temperature. In the dark, metaretinochrome slowly converts back to the parent state of retinochrome. It is assumed that the reconversion is accompanied by a thermal conversion of the 11-*cis* isomer of retinal, first to the 13-*cis* and then to the all-*trans* state [289] – in case the 11-*cis* retinal is not released from the binding site via interaction with the retinal-binding protein (see below). Due to the possibility to obtain the chromophore-free retinochrome apoprotein, a quite detailed view of the binding site and of the chromophore-protein interactions could be obtained by utilizing structurally and electronically modified chromophores for pigment reconstitution (summarized in Ref. [290]). It was concluded that the apoprotein adopts the chromophore in its 6-s-*cis* configuration. Particularly detailed information was obtained with 14-substituted retinals. The use of 14-methyl retinal slowed down the process of pigment formation, indicating a significant steric hindrance for this chromophore derivative. No pigment at all was formed when 13-demethy-14-methyl retinal was used for assembly [292]. The use of 14-halogenated retinals (14-fluoro, 14-chloro) caused a strong effect on the position of the absorption band. When 14-fluoro retinal was used, the unprotonated Schiff base was mainly obtained. This was not the case for the 10- and 12-fluoro retinals. The 14-chloro derivative showed a much more moderate effect on the deprotonation of the Schiff base. This indicates a very precise activity of the counterion in retinochrome [293].

The capability of the retinochromes to produce the 11-*cis* isomer of retinal (and of retinal derivatives) has been utilized in preparative procedures [292,294]: in cases where 11-*cis* isomers of chemically synthesized retinal derivatives, designed for reconstitution studies, were difficult to generate due to their chemical instability or the low quantum yield by irradiation in organic solvents, the incorporation of the all-*trans* forms of these compounds into aporetinochrome, followed by long-wavelength irradiation, led to the formation of the 11-*cis* isomers in a very smooth manner.

In both, insects and cephalopods, in addition to retinochromes, retinal-binding and retinal-transporting proteins also have been identified. These might serve as a shuttle between the VPs and the retinochromes. The retinal-binding protein, identified in the insect visual system, apparently binds the retinal molecule covalently via a Schiff base, as was deduced from a cyanogen borohydride treatment [86]. In the squid visual system, a retinal binding protein (RALBP), interacts with the retinochrome for the purpose of retinal exchange as could elegantly be shown by the use of A_2-(3,4-dehydro) retinal [295,296]. *In vitro* experiments further elucidated the chromophore shuttling in the cephalopod visual system: the 11-*cis* isomer of retinal, formed by retinochrome upon irradiation, is transferred to RALBP when the retinochrome converts to its meta state. A second experiment demonstrated that RALBP loaded with 11-*cis* retinal provides the chromophore for visual pigment formation when the VP apoprotein is added. Detailed analysis of squid RALBP revealed that it exhibits similarities to the retinol-binding proteins of vertebrates. Similar to these, RALBP also exhibits hydrophobic domains which most probably are found in the interior part of the protein, whereas the domains exposed to the hydrophilic environment show a large number of hydrophilic residues. The cephalopod protein most probably adopts to a large extent a β-sheet secondary structure, as its counterpart from the vertebrates [297].

8.4. Other invertebrate VPs

A number of other invertebrate VPs have been characterized with respect to their spectral properties. Since most of them are reviewed elsewhere [6–8], only few more recently characterized cases shall be highlighted here.

A comprehensive study has been presented for the VPs of crabs covering some 25 different species [242,244,245]. Crabs are monochromatic with absorption maxima of their VPs between 473 and 515 nm. The λ_{max} value of the metapigment is similar (between 473 and 510 nm), such that in some cases the absorption spectra of both pigment forms strongly overlap. The light-induced formation of the metapigment then only becomes apparent from an intensity change of the absorption band. In other cases the spectrum of the metapigment is clearly discernible. The ratio of the absorption coefficients in crabs is smaller than the ca. 1:1.8 (P:M) of insects. In the crab VPs, this ratio is sometimes smaller than 1.0 (as small as 1:0.85, P:M), but mostly is around 1:1.1–1.2, with few exceptions where it rises to 1:1.4 [242].

The VP of the mantid shrimps have gained remarkable interest, since the animals have developed a most sophisticated visual system [241,298]. Already Exner [299] noticed the remarkable optical capabilities of these animals. These animals are "formidable predators" [300] equipped with a pair of forward appendages which catch prey in an extremely fast movement. Mantid shrimps have developed compound eyes which show a remarkable concave morphology and a spatial differentiation such that a large part of the eye can screen the visual horizon. The mid-band of the compound eyes comprises maximal sensitivity and covers a very broad spectral range. The broad spectral sensitivity is accomplished by an astonishingly large number of VP: at least eight different pigments with absorbance maxima from 400 to 540 nm have been identified. The photoreceptor cells are arranged in tiers, and probably a system of filtering pigments contributes to the spectral variation. Some regions of the compound eye were optimized for detection of polarized light. The visual properties of these animals have been extensively studied and are reviewed in Refs. [244,245].

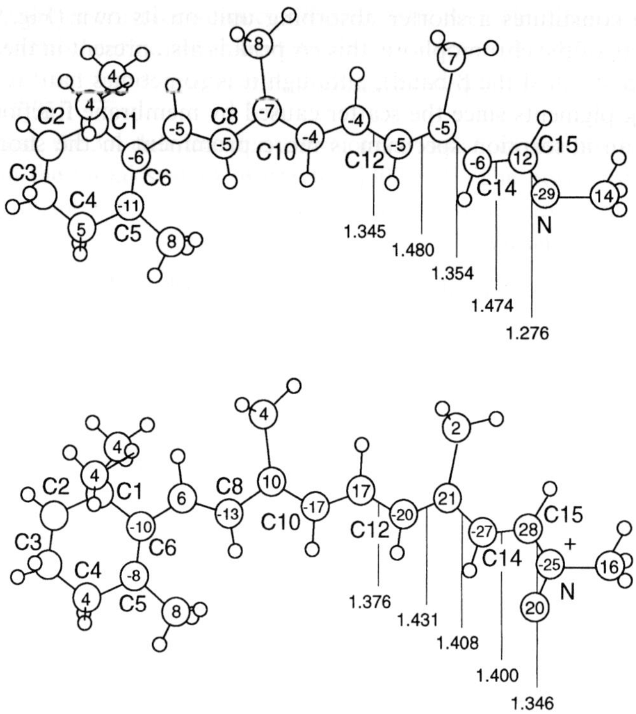

Fig. 10. Quantum chemical calculation of charge densities and bond lengths of all-*trans* retinal N-methyl Schiff base in unprotonated (top) and in the protonated state (bottom). The changes in electron charge are given in percent of one elementary charge, and the bond lengths in Å. Picture modified from Ref. [72].

Calculations for the model compound, all-*trans* retinal, showed that this effect is strongest for the region C_{13}—C_{14}—C_{15}, leading to nearly equal lengths for these two bonds; the values of 135.4 and 147.4 pm in the unprotonated compound turn into 140.8 and 140.0 pm, respectively, upon protonation. It can be assumed that this effect might be extended further into the polyene chain, and becomes important also for the 11-*cis* isomer, when the influences of precisely placed counterions are taken into account. Coupled to this variation in bond order, the protonation also induces a polarization, i.e., a redistribution of electron density, such that even-numbered carbon atoms of the polyene chain gain, and the odd ones lose negative charge. Again for the model compound, this effect is most pronounced on the C13–C14 double bond. The change in electron density upon protonation for C13 is −0.26 e, i.e., 26% of an elementary charge, and for C14 0.21 e. The effect is also observable for the C11–C12 double bond. It is strongly attenuated for the C9–C10 and the C7–C8 double bonds (Fig. 10).

In contrast to the retinal itself, the absorption maximum of a PSB can be remarkably altered by the influence of polar or charged residues in its close proximity. This has been proposed as a mechanism for the fine-tuning of the absorption maxima of VPs irrespective of the type of chromophore being incorporated. The effect of external charges on the absorption properties could convincingly be demonstrated by the introduction of polar substituents along the polyene chain, mimicking the influence of amino acid residues on a protein bound retinal Schiff base [73,320–322]. The combined approaches of quantum chemistry and the synthesis of model compounds enabled Honig and Nakanishi and coworkers to develop a model able to account for the wavelength variation. In this "external-two-point-charge-model" [71] they proposed an interaction between the positively charged Schiff base and a negative amino acid that could act as a counterion. Alternatively, the negative charge could also be provided by an ensemble of amino acids grouped around the Schiff base. In addition, a dipole of amino acid residues was proposed to be located in proximity to the cyclohexene ring of the retinal chromophore (Fig. 9). Also, the contribution of the protein environment has been taken into account. For a given VP, the energy difference between the absorption peak wavelength of the retinal PSB in organic solvents and that of the pigment (values given in cm^{-1}), reflecting the electronic influence of the protein environment on the absorption properties of the chromophore, was defined as the "opsin shift".

Considering a possible effect of the structural variation of the various visual chomophores on the absorption maximum of VPs, it becomes evident that the wavelength regulation in VPs is only marginally affected by the chromophore type, since the peak wavelength of retinal-(PSB)-based VPs are distributed over a broad spectral range. The obvious case of an influence of the chromophore structure is seen upon the exchange of retinal by its 3,4-didehydro derivative, which in some animals can be induced by environmental conditions. Since no additional structural constraints accompany this exchange, this selection can add up to 30 nm – due to the additional double bond in the cyclohexene ring – to the position of the absorption peak wavelength [51]. No change in the electronic properties and consequently no change of the absorption properties will result from an exchange of retinal vs 3-OH retinal, since the hydroxy substituent is at a homoallylic position, not being expected to interact with the double bond system. Only in an artificial case (incorporation of 3-OH retinal into bovine opsin [323]) this exchange has led to a hypsochromic shift of the absorption peak of the new pigment. However, this result can readily be explained on the basis of a steric hindrance between the hydroxy group and the protein cavity of bovine opsin since the binding site has been optimized during evolution for unsubstituted retinal. The rationale of a steric hindrance is supported from the incorporation of other ring-substituted retinals for which also a spectral shift to shorter wavelength results [324]. The fairly recent identification of 4-OH retinal as a functional chromophore in the blue shifted absorbing VP of a cephalopod (*Watasenia scintillans*, $\lambda_{max} = 471$ nm), however, may well be ascribed to the electronic interaction between the π-electron system and the hydroxy substituent in allylic position [54]. This suggestion, however, has to remain speculative, since no knowledge of the protein contribution is available, and no chromophore exchange has yet been reported.

In general, spectral selectivity is controlled in a concerted manner by amino acids of the protein which are in close proximity to the chromophore, as was demonstrated by an extensive site directed mutagenesis study on the human red and green VPs [325]. These two VPs vary by only 15 amino acids and exhibit an absorbance difference of ca. 30 nm. Out of the 15 varied amino acids seven were identified by SDM to be of utmost relevance for the conversion from one pigment type to another.

For a more general explanation of wavelength regulation in VPs, computer simulations in conjunction with the preparation of polyene model compounds have yielded at least two different mechanisms which can account for absorption in the 450–600 nm wavelength range. Wavelength regulation – according to these combined theoretical/chemical approaches – can be accomplished by (i) an intimate interaction between the positively charged Schiff base and an amino acid residue in close, but variable distance, and (ii) a weaker interaction putatively provided by a dipole, generated by two polar or charged amino acids which should be located in the proximity of the cyclohexene ring. These interactions have also been implemented in the "external two point charge model" of Honig and Nakanishi. An example for the stronger of the two regulating mechanisms, (i), was found in bovine rhodopsin by identifying a simple Schiff base/counterion arrangement (Glu113/Lys296) rhodopsin [202]. On the other hand, a complex counterion structure was identified in the halobacterial retinal-protein BR from high resolution crystal structure analysis [197,198].

Great effort has been invested into a theory explaining the electronic properties and the mechanism of wavelength tuning of retinal proteins (bacteriorhodopsin and VPs) [71,317,326–329]. Localized effects from the protein have been proposed from studies using either isotope labeled or structurally modified chromophore derivatives [330,331], which yielded the "external two point charge model" [71].

8.6.1. Visual pigments with spectral sensitivity in the short-wavelength region

The tuning mechanism for the chromophore absorption by the protein environment of the UV-absorbing pigments represent a severe challenge, since in some cases the maximal absorption of the rhodopsin is even at shorter wavelengths than that of the free chromophore, e.g., the UV-sensitive pigments of the honeybee or the owlfly (*Ascalaphus macaronius*). Since in general the absorption of retinal experiences a bathochromic shift upon formation of a protonated Schiff base, the mechanisms which give rise to the hypsochromically shifted absorption maxima are not well understood. This complex situation becomes even more paradoxical when the resonance Raman measurements for the UV pigment of the owlfly are taken into consideration, as it is claimed that the Schiff base in this VP might indeed be protonated [201]. It remains to be seen from recombinant approaches and from model-building/crystal-structure analysis which effects of the surrounding protein cavity on the chromophore accomplish such a strongly blue-shifted absorption maximum.

8.6.2. Blue/UV-sensitivity based on retinal chromophores

The VPs with spectral absorption in the SW spectral range comprise a limb of their own in the phylogenetic tree irrespective of the type of chromophore ("upper" part of limb V in Fig. 12). Any attempt to explain the mechanisms for wavelength regulation – and in particular in the UV region of the spectrum – remain speculative due to the lack of a three-dimensional structure of the VPs, and has to rely on the study of model compounds, theoretical calculations and sequence alignment. A first view on the protein interior of a VP is now possible from the structure of bovine rhodopsin [406]. Interestingly, the UV-sensitive pigments of many insects have been optimized for another function: in the dorsal rim of the honeybee they are specialized for the detection of polarized light, and in fact bees use this kind of visual detection for localizing the position of the sun under conditions of an overcast sky [32,332].

The sequence alignment of the VPs (Fig. 4a,b) reveals some remarkable motifs, which might be involved in the wavelength regulation of the UV/violet-absorbing pigments, since they are located only in this type of pigment. An inspection of the sequences of all pigments gathered on the upper part of limb V (Fig. 12) yields the presence of three conserved charged amino acids at identical or precisely shifted positions, that are absent in all other VPs (Fig. 11). On the one hand, in TM VII an aspartate or glutamate is found seven residues (or two helical turns) away from the chromophore binding lysine residue in the direction towards the C-terminus (position 348 in Fig. 4a). In all other invertebrate VPs, also in those in Fig. 4b, an asparagine (serine for *M. sexta*, #2) is present at that position. This position is at very close distance to the PSB and allows a strong compensation of the positive charge of the protonated Schiff base. It thus makes this residue a likely candidate for a direct counterion. In addition, two oppositely charged residues are found on the N-terminal side of TMs II and IV (positions 121 and 207/210, respectively, depending on the type of pigment), which can be arranged close to each other to allow formation of an ion pair or dipole, depending on the state of dissociation. It may be argued that interacting amino acids should be located in adjacent protein regions, in the case of the 7TM receptors in two subsequent transmembrane helices, and not in TM II and TM IV. An inspection of the three-dimensional structure of frog and bovine rhodopsin [333,406], however, reveals that on the N-terminal side of this VP, the helices II and IV approach each other on the expense of the space for helix III which is expelled out of the helix bundle, to a certain extent. On the C-terminal side,

	TM II (helix B)	TM IV (helix D)	TM VII (helix G)
Locusta (blue)	..MLL-**E**MPM..	..PLL**R**IWS..	..SCI**D**PWV..
Dros (blue)	..MMV-**K**TPI..	..CYT**E**TWG..	..ACI**D**PFV..
Dros. (UV)	..MCL-**K**API..	..PLTQFW**D**..	..ACI**D**PFV..
Locusta (green)	..MMFTMSAP..	..PLF-GWN..	..AVFNPIV..
Mantis (green)	..MMLSMSPP..	..PLF-GWN..	..AVYNPIV..
Dros (green)	..IMITNTPM..	..PAF-GWS..	..ACYNPIV..
Calliphora (green)	..IMITNPMM..	..PVF-GWS..	..ACYNPIV..

Fig. 11. Sequence comparison between putatively SW- (blue/UV) and LW- (green) absorbing VPs. Shown are TMs II, IV and VII. Only in the case of the SW-absorbing pigments charged amino acids are found (see text).

all helices follow each other quite regularly (TM I–TM VII). In case such an arrangement of the helices exists also in the invertebrate pigments, it would readily allow the proposed interaction between the two charged amino acids.

The Rh3 and Rh4 sequences from *Drosophila*, the sequence #2 of *M. sexta*, the UV-sensitive pigment of the honeybee and of both ants carry a lysine (position 121) in TM II, whereas the blue-absorbing pigments of honeybee, #3 of *M. sexta*, and the blue-absorbing pigment of *S. gregaria* show a negatively charged glutamate at that identical position. Rh5 carries an asparagine instead, whereas in Rh6 one finds an uncharged residue (Thr121; but note that Rh6 is assumed to represent a LW-absorbing pigment). Amino acids of opposite charge to those found in TM II are identified in TM IV (position 207 or 210, respectively), which could constitute a partner in a salt bridge. In most of the discussed pigments this amino acid is located at position 207 (Glu in *Drosophila* Rh3, Arg or Lys in all others, except for the two ant pigments which carry a Gln or a His, respectively). For the Rh4 pigments of *Drosophila*, exhibiting a very hypsochromic shifted absorbance maximum ($\lambda_{max} = 345$ nm), the negatively charged amino acid is found one helical turn away (Asp210). The location of the charged amino acids in the sequences is shown schematically for several VP sequences in Fig. 11. The additive effect from all three amino acids could result in an increase in electron density and compensate the electron deficit caused by the positive charge of the Schiff base, thus explaining the hypsochromic absorption. The arrangement of the three charged amino acids that constitute a closely located counterion and a dipole in the SW-absorbing VPs shows intriguing similarity to the arrangement of amino acids, which were proposed in the "external-two-point-charge-model". In fact, this finding represents a real "experimental" proof for this model. It should be mentioned that these amino acid residues are not present in the UV/blue-absorbing VPs of vertebrates [235,334–336].

8.6.3. The sensitizing pigment of dipterans

Spectral analysis and electrophysiological experiments have shown a strong, fine structured, three-peaked sensitivity spectra around 350 nm in the eyes of several dipteran species [252,260,337]. According to its spectral structure this could not be caused by an aldehyde-based visual pigment. The extraction of retinoid compounds from the eyes of flies showed, besides the presence of the aldehyde 3-hydroxy retinal, large amounts of the corresponding alcohol (3-OH retinol [53]). Since retinoids bound to proteins show spectral characteristics similar to those determined for the sensitivity of the compound eyes in the ultraviolet, it was proposed that the SW absorption is based on retinol (or a retinol derivative) which is located closely to or even being embedded in the VP molecule itself. Accordingly, in addition to the aldehyde-based UV-sensitive VPs, a mechanism making use of an auxiliary chromophore has evolved in some Diptera enhancing the spectral absorption in the UV region without the demand to develop an additional, new type of chromoprotein. This advantageous visual property is based on the presence of 3-OH *retinol*, functioning as a "sensitizing pigment" [76]. For a discussion on evolutionary aspects and a detailed description of this light absorption mechanism see Refs. [61,77]; only the most salient features of the sensitizing pigment are summarized here.

The basis for an understanding of the function of the sensitizing pigment came from experiments which showed that the sensitivity for UV decreased proportionally with the sensitivity of the pigment in the visible range. This indicated a radiationless, Förster-like energy transfer from the UV- (retinol) to the blue–green-absorbing (retinal) chromophore. The high efficiency of energy transfer suggests that the retinol and the retinal of the VP are maximally 25 Å apart [77]. Originally, the all-*trans* isomer of 3-OH retinol, which has the highest extinction coefficient of all alcohol isomers (and thus could guarantee maximal energy transfer), was proposed as a candidate for the isomeric state of the sensitizing pigment. More recent experiments based on external supplements of various retinoid derivatives on retinoid-deprived flies or flies raised on various diets, followed by HPLC analysis of the retinol content, gave preference for the 13-*cis* isomer of 3-OH retinol as the sensitizing pigment chromophore [338].

The sensitizing pigment has emerged very "late" in the evolution of the insects, probably due to the fact that the molecular changes of the protein moiety, resulting from the acquisition of 3-OH retinal made this novel absorption principle possible. The employment of the hydroxylated retinal chromophore can thus be considered as an "evolutionary prerequisite" to also engage the hydroxylated retinol, since it allows the non-covalent fixation of the alcohol through the hydroxy groups at *both* ends of the molecule (for comparison see the changed properties of a retinol based sensitizing pigment described in Ref. [337]).

8.7. Kinetics of light-induced absorption changes in invertebrate VPs

8.7.1. Insect VPs

The formidable difficulties in insect VP sample preparation have obstructed a detailed kinetic analysis of the pigment → metapigment interconversion. Only one group (Stavenga and coworkers) has provided kinetic data so far, performed on living animals. Analysis of the pigment-to-metapigment conversion in the blowfly *Calliphora vicina* has revealed a rapid (<40 ns) formation of a red-shifted species named bathorhodopsin [339]. This intermediate decays with a time constant of 700 ns (determined at ambient temperature) into a lumirhodopsin, from which the metapigment is formed with a time constant of ca. 80 μs. The thermal stability of the metapigment allowed also the study of the reconversion into the pigment parent state, after exciting the metapigment. This process again started with the rapid formation of an intermediate. This species converted with a time constant of 4 μs into an intermediate which exhibits spectral features already very similar to the parent form of the pigment itself. However, the full formation of the pigment state requires another 13 ms [339,340]. The light-induced decay of the metapigment has also been followed by fluorescence spectroscopy, making use of the strong fluorescence of the metapigments [248]; see Chapter 10.

8.7.2. Cephalopod and other VPs

Time-resolved measurements of absorption changes with squid rhodopsin revealed extremely fast kinetics for the *cis* → *trans* photoisomerization of ca. 400 fs [341].

Limulus sequences in relative proximity to the mantid and the green absorbing pigment (Lo1) of the locust *S. gregaria* [13,155]. Based on the larger number of sequences available now, this arrangement has to be revised. Interestingly, already in the former study, the blue/UV-absorbing pigments from e.g. *Drosophila* (Rh3, Rh4) segregated on the same limb as the SW-absorbing pigment from the locust *Schistocerca* (Schgr2) (cf. limb V in this study, Fig. 12). This placement, which results irrespective of the chromophore type, was suggested to result from the functional constraint of constructing a SW-absorbing receptor [13]. Obviously UV-wavelength detection requires a strong variation in the entire protein sequence in comparison to the green/blue-absorbing pigments. The detailed discussion of sequence motifs (see above) also demonstrates a divergent evolution for the LW- and the SW-absorbing receptors. This assumption is now firmly based on 38 insect VP sequences.

The number of complete and partial sequences has grown remarkably during the last few years, making a phylogenetic arrangement of the invertebrates now much more meaningful. However, within the insects, still the majority of sequences (14 out of 38) belongs to various *Drosophila* species. On the basis of 59 sequences presently available in the databases, a phylogenetic tree was constructed (Fig. 12). For tree formation, the "rootless tree" option was chosen.

The formation of groups of similar VPs, either according to their relation to orders or classes, or to special properties is readily seen. As can be expected from their different primary structure, e.g., the extended C-terminal tail, the cephalopod sequences are all found on one limb of the tree (limb IV), but interestingly the sequence of the scallop (*Patinopecten yessoensis*, "Patye") is found in relative proximity to them. The structure of this limb is still very plain and does not show any differentiation or selective branching. Clearly, a larger number of sequences will show more detailed features of the cephalopods and will allow the assignment of sequences to emerging subgroups of similarity.

Similarly, the crayfish sequences constitute a limb of their own (limb III). Although most of the crayfish sequences are fragmentary, some detailed structure in the arrangement of the respective sequences emerges. To this group of pigments, the two pigments from *Limulus* show the highest degree of similarity, as was already outlined on the basis of molecular similarities (see above).

The two VPs of the crab *H. sanguineus* show extremely low similarity to the gross of the VPs, and consequently constitute a limb of their own. Further analysis and sequences of animals related to crabs will shed light onto a possible phylogenetic relation to other invertebrates.

Inspection of the parts of the phylogenetic tree where the insect VPs segregate (limbs I, II, and V), reveals a clear group formation according to the type of pigments. The relatively large group of the SW-absorbing pigments found in limb V (16 sequences) shows the arrangement into several subclasses. The *Drosophila* Rh3/4-type pigments, the UV-sensitive pigments from both ants, the honeybee, one butterfly pigment (PapglRh5) and pigment #2 from *M. sexta*, all segregate together on one limb in section V. The blue-absorbing pigments of the ants, the Rh6 pigment of *P. glaucus*, the pigment #3 from *M. sexta*, together with Rh5 from *D. melanogaster* constitute the second limb in section V of the phylogenetic tree. In both subsections,

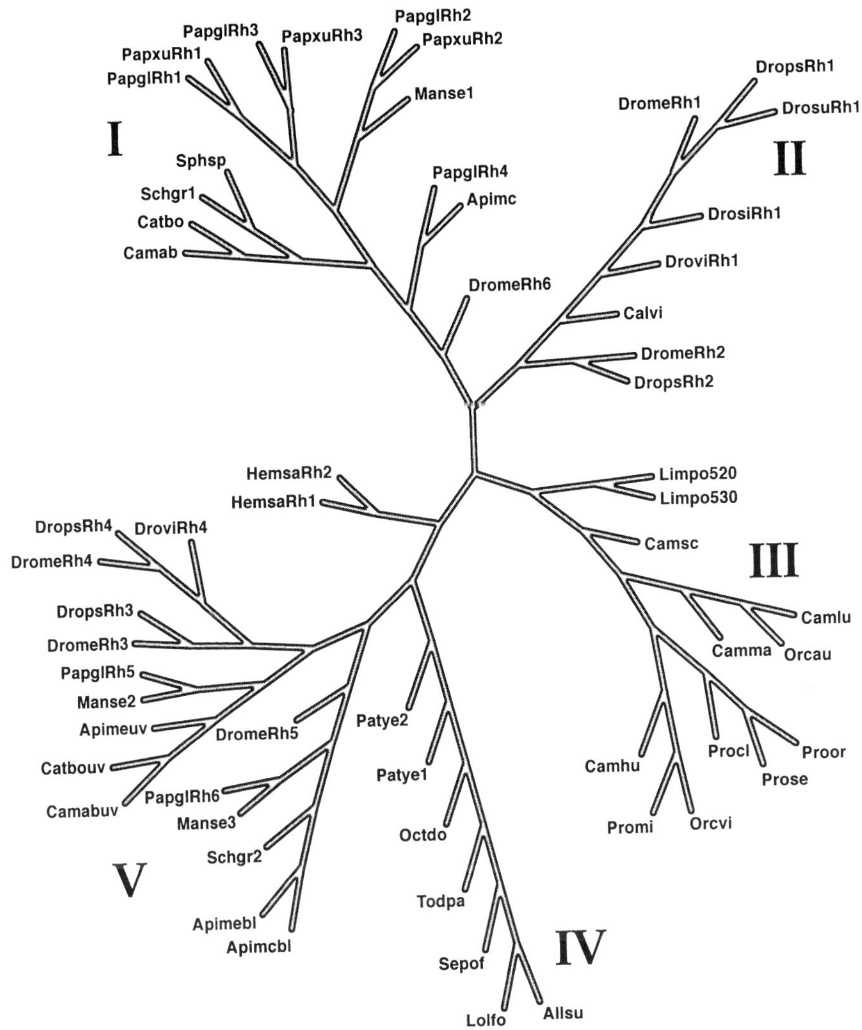

Fig. 12. Rootless phylogenetic tree of all currently available invertebrate VP sequences. The tree is based on the sequence alignments (Fig. 4a,b) and was generated on the basis of 59 invertebrate sequences by using the software package of Ref. [403]. Clearly identifiable limbs of the phylogenetic tree are labeled with Roman numbers in a clockwise manner starting with the top left limb of the retinal-based LW-absorbing pigments. For abbreviations see caption to Table 1.

the pigments *P. glaucus* #5 and *M. sexta* #2, and also *P. glaucus* #6 and *M. sexta* #3 show a higher sequence similarity to each other than to the rest of this particular subgroup. The finding that DromeRh6 (on limb I) is fairly distant from this group of pigments reflects its spectral absorption as a LW-absorbing VP (present in the R8y subclass of the R8 visual cells, λ_{max} around 530 nm [14]). The retinal-based pigments from limb I, which in all cases probably represent the most abundant VP type (the Rh1-type pigments of both ants, "Camab", "Catbo", the honeybee, "Apime", the mantid, "Sphsp", and the pigment Lo-1 from the locust, "Schgr"), constitute one limb of the tree demonstrating the close relationship between these pigments, and the relative distance from the Rh1 type pigments of the dipterans that utilize 3-OH retinal (branch II). One can conclude that the mantis, the locust and both ants show a higher degree to each other than to the rest of the sequences in limb I. As can be expected, the most abundant VPs of the Diptera, the Rh1-type of all *Drosophila* species and of *Calliphora*, exhibit more homology to each other than to other *Drosophila* VPs, and comprise their own group (branch II), only being accompanied by the ocelli pigments (Rh2-type VPs of *D. melanogaster* and *D. pseudoobscura*). The phylogenetic tree proves the variant nature of SW-absorbing pigments vs green light sensitivity (Rh1-type VPs), in placing Rh3 and Rh4 from *Drosophila* (with absorbance maxima at ca. 345 and 375 nm, respectively) together with the SW-absorbing pigment from *Schistocerca* (Schgr2) onto the same limb. A rationale for this arrangement, which includes VPs from retinal and 3-OH retinal employing insects, was already suggested (see above, and Ref. [155]) as being a consequence of the constraint to shift the sensitivity of the VP into the short wavelength region.

11. Conclusions

The knowledge of structural and functional properties of invertebrate and in particular of insect VPs has considerably grown during the last few years. A converging picture emerged also for the mechanism and the involved components of signal transduction. Whereas for quite a while any analysis was restricted to spectroscopic and electrophysiological investigations, performed on living animals or various more or less crude photoreceptor preparations, the advent of molecular biology technology has significantly contributed to our current state of knowledge. For the particular application of primary structure elucidation, quite valuable information could be derived, even in cases when the study was restricted to a plain sequence alignment (due to the lack of protein preparation protocols and the availability of purified material). As outlined in this chapter, attempts to perform more detailed functional analysis are extremely important, but still are impeded by the difficulties to provide VP in sufficient amount and purity. Thus, compared to the very detailed understanding of the structure, function and the process of signal transduction of the vertebrate VPs, and partially also for the cephalopod pigments, the lack of information in insect vision is still evident. The wealth of results obtained for bovine rhodopsin and for the cephalopod pigments compiled by time-resolved vibrational and absorption spectroscopy in conjunction with molecular biology technology is clearly unquestionable. It is thus of utmost importance to establish a routine preparation method of invertebrate photoreceptors – preferentially on the basis of recombinant technology, which will allow the convenient production of sufficient amounts of material for spectroscopic studies which then will open the door towards a more detailed understanding of insect vision.

Abbreviations

3-, 4-OH, 3-, 4-hydroxy
ε_{max}, molar extinction coefficient
λ_{max}, wavelength of maximal absorption
λ_{em}, wavelength of maximal emission
bl, blue(-absorbing)
bp, base pairs
BR, bacteriorhodopsin
cDNA, complementary DNA
cGMP, cyclic-guanosine mono phosphate
CNBr, cyanogen bromide
EOP, extraocular photoreceptor
FTIR, Fourier-transform infrared (spectroscopy)
IP_3, inositol trisphosphate
kDa, kilo-Dalton
LW, long wavelength
nina = neither inactivation nor afterpotential
ORF, open reading frame
OS, outer segments
PCR, polymerase chain reaction
PDA, prolonged depolarizing afterpotential
PDZ, post-synaptic density protein, disc-large, zo-1
PDE, phosphodiesterase
PITP, phosphoinositol transport protein
PKC, phosphokinase C
PLC, phospholipase C
PSB, protonated Schiff base
RACE, rapid amplification of C-terminal ends (by PCR)
RT-PCR, reverse transcription polymerase chain reaction
SDM, site-directed mutagenesis
SFV, Semliki Forest Virus
SW, short wavelength
TdT, terminal deoxynucleotidyl transferase
TM, transmembrane
trp = transient receptor potential (not to be confused with the amino acid trp = tryptophan)
UV, UV(-absorbing)
VP, visual pigment

One- and three-letter code of amino acids

Alanine, Ala, A
Arginine, Arg, R
Aspartate, Asp, D
Asparagine, Asn, N

Cysteine, Cys, C
Glutamate, Glu, E
Glutamin, Gln, Q
Glycine, Gly, G
Histidine, His, H
Isoleucine, Ile, I
Leucine, Leu, L
Lysine, Lys, K
Methionine, Met, M
Phenylalanine, Phe, F
Proline, Pro, P
Serine, Ser, S
Threonine, Thr, T
Tryptophan, Trp, W
Tyrosine, Tyr, Y
Valine, Val, V

Acknowledgements

This chapter would not exist without the support of Dr. K. Worm, University of Essen, Germany, who generated the sequence alignments and the phylogenetic tree. I am also indebted to Armin Engels for critical reading of the manuscript.

References

1. Kreimer, G. (1994) Intl. Rev. Cytol. **148**, 229–310.
2. Hoff, W.D., Jung, K.-H. and Spudich, J.L. (1997) Annual Rev. Biophys. Biomolec. Struct. **26**, 223–258.
3. Hecht, S., Shlaer, S. and Pirenne, M.H. (1942) J. Gen. Physiol. **25**, 819–840.
4. Barlow, H.B., ed. (1977) in: Vertebrate Photoreception. pp. 337–358, Academic Press, London.
5. Aho, A.-C., Donner, K., Hydén, C., Larson, L.O. and Reuter, T. (1988) Nature **334**, 348–350.
6. Autrum, H., ed. (1979) Handbook of Sensory Physiology, Vol. VII/6 A, Springer, Berlin.
7. Autrum, H., ed. (1981) Handbook of Sensory Physiology Vol. VII/6 B, Springer, Berlin.
8. Autrum, H., ed. (1981) Handbook of Sensory Physiology Vol. VII/6 C, Springer, Berlin.
9. Langer, H., ed. (1998) Biochemistry and Physiology of Visual Pigments. Springer, Heidelberg.
10. Prosser, C.L., ed. (1991) Neural and Integrative Animal Physiology. Wiley-Liss, New York.
11. Holmes, M.G., ed. (1991) Photoreceptor Evolution and Function. Academic Press, London.
12. Hargrave, P.A., ed. (1993) Methods in Neurosciences. Vol. 15, Photoreceptor Cells. Academic Press, San Diego.
13. Gärtner, W. and Towner, P. (1995) Photochem. Photobiol. **62**, 1–16.
14. Hardie, R.C. (1985) in: Progress in Sensory Physiology, Vol. 5, ed D. Ottoson. pp. 1–79, Springer, Berlin.
15. Ranganathan, R., Malicki, D.M. and Zuker, C.S. (1995) Annu. Rev. Neurosci. **18**, 283–317.
16. Goldsmith, T.H. (1990) Quart. Rev. Biol. **65**, 281–322.
17. Stavenga, D.G. and Wunderer, H.J. (1999) in: Atlas of Arthropod Sensory Receptors, eds E. Eguchi and Y. Tominaga. pp. 1–12, Springer, Berlin.
18. Hofmann, K.P. (1986) Photobiochem. Photobiophys. 309–327.
19. Lewis, J.W. and Kliger, D.S. (1992) J. Bioenerg. Biomembr. **24**, 201–210.

20. Hargrave, P.A. and McDowell, J.H. (1992) FASEB J. **6**, 2323–2331.
21. Nathans, J. (1992) Biochemistry **31**, 4923–4931.
22. Stavenga, D.G. (1992) Trends Neurosci. **15**, 213–218.
23. Stavenga, D.G. (1995) Progr. Retinal Eye Res. **15**, 231–259.
24. Schwemer, J. and Henning, U. (1984) Cell Tissue Res. **236**, 293–303.
25. Shukolyukov, S.A. and Denisova, N.A. (1992) Insect Biochem. Mol. Biol. **22**, 925–935.
26. Kumar, J.P. and Ready, D.F. (1995) Development **121**, 4359–4370.
27. Paulsen, R. and Schwemer, J. (1983) Eur. J. Biochem. **137**, 609–614.
28. Huber, A., Wolfrum, U. and Paulsen, R. (1994) Eur. J. Cell Biol. **63**, 219–229.
29. Kumar, J.P., Bowman, J., O'Tousa, J.E. and Ready, D.F. (1997) Developmental Biol. **188**, 43–47.
30. Laughlin, S.B., Menzel, R. and Snyder, A.W. (1975) in: Photoreceptor Optics, eds A.W. Snyder and R. Menzel. pp. 237–259, Springer, Berlin.
31. Saibil, H.R. (1990) in: Squid as Experimental Animals, eds W. Adelman, D. Gilbert and J. Arnold. pp. 371–397, Plenum, New York.
32. Rossel, S. (1989) in: Facets of Vision, eds D.G. Stavenga and R.C. Hardie. pp. 298–316, Springer, Berlin.
33. Strader, C.D., Fong, T.M., Tota, M.R., Underwood, D. and Dixon, R.A.F. (1994) Annu. Rev. Biochem. **63**, 101–132.
34. Gudermann, T., Nürnberg, B. and Schultz, G. (1994) J. Molec. Med. **73**, 51–63.
35. Nürnberg, B., Gudermann, T. and Schultz, G. (1995) J. Molec. Med. **73**, 123–132.
36. Molday, R.S. (1998) Invest. Ophthalmol. Visual Sci. **39**, 2493–2513.
37. Pepe, I.M. (1999) J. Photochem. Photobiol., B: Biol. **48**, 1–10.
38. Beck-Sickinger, A.G. (1996) Drug Discovery Today **1**, 502–513.
39. O'Tousa, J.E. (1992) Visual Neurosci. **8**, 385–390.
40. Katanosaka, K., Tokunaga, F., Kawamura, S. and Ozaki, K. (1998) FEBS Lett. **424**, 149–154.
41. Fukuda, M.N., Papermaster, D.P. and Hargrave, P.A. (1979) J. Biol. Chem. **254**, 8201–8207.
42. Huber, A., Smith, D.P., Zuker, C.S. and Paulsen, R. (1990) J. Biol. Chem. **265**, 17906–17910.
43. Bentrop, J., Schwab, K., Pak, W.L. and Paulsen, R. (1997) EMBO J. **16**, 1600–1609.
44. Ovchinnikov, Y.A., Abdulaev, N.G. and Bogachuk, A.S. (1988) FEBS Lett. **230**, 1–5.
45. Karnik, S.S., Sakmar, T.P., Chen, H.-B. and Khorana, H.G. (1988) Proc. Natl. Acad. Sci. USA **85**, 8459–8463.
46. Shieh, B.-H., Stamnes, M.A., Seavello, S., Harris, G.L. and Zuker, C.S. (1989) Nature **338**, 67–70.
47. Ondek, B., Hardy, R.W., Baker, E.K., Stamnes, M.A., Shieh, B.-H. and Zuker, C.S. (1992) J. Biol. Chem. **267**, 16460–16466.
48. Stamnes, M.A., Rutherford, S.L. and Zuker, C.S. (1992) Trends Cell Biol. **2**, 272–276.
49. Baker, E.K., Colley, N.J. and Zuker, C.S. (1994) EMBO J. **13**, 4886–4895.
50. Deininger, W., Kröger, P., Hegemann, U., Lottspeich, F. and Hegemann, P. (1995) EMBO J. **14**, 5849–5858.
51. Wald, G. (1936) J. Gen. Physiol. **22**, 391–415.
52. Vogt, K. (1983) Z. Naturforsch. **38c**, 329–333.
53. Vogt, K. and Kirschfeld, K. (1984) Naturwissenschaften **71**, 211–212.
54. Matsui, S., Seidou, M., Uchiyama, I., Sekiya, N., Hiraki, K., Yoshihara, K. and Kito, Y. (1988) Biochim. Biophys. Acta **966**, 370–374.
55. Hegemann, P., Gärtner, W. and Uhl, R. (1991) Biophys. J. **60**, 1477–1489.
56. Derguini, F., Mazur, P., Nakanishi, K., Starace, D.M., Saranak, J. and Foster, K.W. (1991) Photochem. Photobiol. **54**, 1017–1021.
57. Haupts, U., Eisfeld, W., Stockburger, M. and Oesterhelt, D. (1994) FEBS Lett. **356**, 25–29.
58. Harosi, F.I. (1994) Vision Res. **34**, 1359–1367.
59. Smith, W.C. and Goldsmith, T.H. (1990) J. Mol. Evol. **30**, 72–84.
60. Seki, T., Fujishita, S., Ito, M., Matsuoka, N. and Tsukida, K. (1987) Exp. Biol. **47**, 95–103.
61. Vogt, K. (1987) Photobiochem. Photobiophys. (Suppl.) 273–296.
62. Vogt, K. (1984) Verh. Dtsch. Zool. Ges. **77**.

63. Gleadall, I.G., Hariyama, T. and Tsukahara, Y. (1989) J. Insect Physiol. **35**, 787–795.
64. Hariyama, T., Terakita, A., Sakayori, M., Katsukura, Y., Ozaki, K. and Tsukahara, Y. (1998) J. Comp. Physiol. A **183**, 165–170.
65. Seki, T., Fujishita, S. and Obana, S. (1989) Exp. Biol. **48**, 65–75.
66. Plangger, A. (1989) Visuelle Chromophore im Libellenauge. Diplomarbeit Universität Freiburg iBr. Germany.
67. Seki, T. and Vogt, K. (1998) Comp. Biochem. Physiol. **119B**, 53–64.
68. Seki, T., Isono, K., Ozaki, K., Tsukahara, Y., Shibata-Katsuta, Y. and Ito, M. (1998) Eur. J. Biochem. **257**, 522–527.
69. Kito, Y., Narita, K., Seidou, M., Michinomae, M., Yoshihara, K., Partridge, J.C. and Herring, P.J. (1992) in: Structures and Functions of Retinal Proteins, ed J.-L. Rigaud. pp. 411–414, John Libbey, Montrouge.
70. Matsui, S., Seidou, M., Horiuchi, S., Uchiyama, I. and Kito, Y. (1988) J. Gen. Physiol. **92**, 55–66.
71. Honig, B.H., Dinur, U., Nakanishi, K., Balogh-Nair, V., Gawinowicz, M.A., Arnaboldi, M. and Motto, M.G. (1979) J. Am. Chem. Soc. **101**, 7084–7086.
72. Tavan, P., Schulten, K. and Oesterhelt, D. (1985) Biophys. J. **47**, 415–430.
73. Baasov, T. and Sheves, M. (1985) J. Am. Chem. Soc. **107**, 7524–7533.
74. Sheves, M., Kohne, B. and Mazur, Y. (1983) J. Chem. Soc., Chem. Comm. 1232–1234.
75. Grossjean, M.F. and Tavan, P. (1988) J. Chem. Phys. **88**, 4884–4896.
76. Kirschfeld, K., Franceschini, N. and Minke, B. (1977) Nature (London) **269**, 386–390.
77. Vogt, K. and Kirschfeld, K. (1983) Biophys. Struct. Mech. **9**, 319–328.
78. Arikawa, K., Mizuno, S., Scholten, D.G.W., Kinoshita, M., Seki, T., Kitamoto, J. and Stavenga, D.G. (1999) Vision Res. **39**, 1–8.
79. Isono, K., Tanimura, T., Oda, Y. and Tsukahara, Y. (1988) J. Gen. Physiol. **92**, 587–600.
80. Hamdorf, K., Paulsen, R. and Schwemer, J. (1973) in: Biochemistry and Physiology of Visual Pigments, ed H. Langer. pp. 155–166, Springer, Berlin.
81. Ozaki, K., Nagatani, H., Ozaki, M. and Tokunaga, F. (1993) Neuron **10**, 1113–1119.
82. Chase, M.R., Bennett, R.R. and White, R.H. (1996) Vis. Neurosci. **13**, 353–358.
83. Hara, R., Hara, T., Tokunaga, F. and Yoshizawa, T. (1981) Photochem. Photobiol. **33**, 883–891.
84. Hara, T., Hara, R., Hara-Nishimura, I., Nishimura, M., Terakita, A. and Ozaki, K. (1992) in: Structures and Functions of Retinal Proteins, ed J.-L. Rigaud. pp. 255–258, Libbey, Montrouge.
85. Smith, W.C. and Goldsmith, T.H. (1991) Vis. Neurosci. **7**, 237–249.
86. Pepe, I.M., Schwemer, J. and Paulsen, R. (1982) Vision Res. **22**, 775–781.
87. Pepe, I.M. and Cugnoli, C. (1993) J. Photochem. Photobiol. B: Biology **17**, 90.
88. Pepe, I.M., Cugnoli, C., Keen, J.N. and Findlay, J.B.C. (1992) in: Structures and Functions of Retinal Proteins, ed J.-L. Rigaud. pp. 267–270, Libbey, Montrouge.
89. Rando, R.R. (1990) Angew. Chemie – Intl. Ed. Engl. **29**, 461–480.
90. Rando, R.R. (1996) Chem. Biol. **3**, 255–262.
91. Autrum, H. (1981) in: Handbook of Sensory Physiology, VII/6C, ed H. Autrum. pp. 1–91, Springer, Berlin.
92. Schwemer, J. (1989) in: Facets of Vision, eds D.G. Stavenga and R.C. Hardie. pp. 112–133, Springer, Berlin.
93. Bentrop, J., Plangger, A. and Paulsen, R. (1993) Eur. J. Biochem. **216**, 67–73.
94. Plangger, A., Malicki, D., Whitney, M. and Paulsen, R. (1994) J. Biol. Chem. **269**, 26969–26975.
95. Kiselev, A. and Subramaniam, S. (1994) Science **266**, 1369–1373.
96. Kiselev, A. and Subramaniam, S. (1997) Biochemistry **36**, 2188–2196.
97. Kiselev, A. and Subramaniam, S. (1996) Biochemistry **35**, 1848–1855.
98. Scott, K. and Zuker, C. (1997) Trends Biochem. Sci. **22**, 350–354.
99. Dolph, P.J., Ranganathan, R., Colley, N.J., Hardy, R.W., Socolich, M. and Zuker, C.S. (1993) Science **260**, 1910–1916.
100. Tsunoda, S., Sierralta, J., Sun, Y.M., Bodner, R., Suzuki, E. and Becker, A. (1997) Nature **388**, 243–249.
101. Scott, K. and Zuker, C.S. (1998) Nature **395**, 805–808.

102. Chang, J.T., Milligan, S., Li, Y., Chew, C.E., Wiggs, J., Copeland, N.G., Jenkins, N.A., Campochiaro, P.A., Hyde, D.R. and Zack, D.J. (1997) J. Neurosci. **17**, 5881–5890.
103. Gehring, W.J. (1998) in: New Neuroethology on the Move, eds N. Elsner and R. Wehner. pp. 45–53, de Gruyter, Stuttgart.
104. Sheng, G.J., Thouvenot, E., Schmucker, D., Wilson, D.S. and Desplan, C. (1997) Genes Develop. **11**, 1122–1131.
105. Quiring, R., Walldorf, U., Kloter, U. and Gehring, W.J. (1994) Science **265**, 785–789.
106. Halder, G., Callaerts, P. and Gehring, W.J. (1995) Science **267**, 1788–1792.
107. Tomarev, S.I., Callaerts, P., Kos, L., Zinovieva, R., Halder, G., Gehring, W.J. and Piatigorsky, J. (1997) Proc. Natl. Acad. Sci. USA **94**, 2421–2426.
108. Hafen, E., Basler, K., Edstroem, J.E. and Rubin, G.M. (1987) Science **236**, 58–63.
109. Simon, M.A., Carthew, R.W., Fortini, M.E., Gaul, U., Mardon, G. and Rubin, G.M. (1992) Cold Spring Harb. Sym. Quant. Biol. **57**, 375–380.
110. Serikaku, M.A. and O'Tousa, J.E. (1994) Genetics **138**, 1137–1150.
111. Kurada, P., Tonini, T.D., Serikaku, M.A., Piccini, J.P. and O'Tousa, J.E. (1998) Vis. Neurosci. **15**, 693–700.
112. Nuoffer, C. and Balch, W.E. (1994) Annu. Rev. Biochem. **63**, 949–990.
113. Novick, P. and Zerial, M. (1997) Curr. Opinion Cell Biol. **9**, 496–504.
114. Shetty, K.M., Kurada, P. and O'Tousa, J.E. (1998) J. Biol. Chem. **273**, 20425–20430.
115. Schwemer, J. (1993) in: Sensory Systems of Arthropods, ed K. Wiese. pp. 54–68, Birkhäuser, Basel.
116. Schwemer, J. (1984) J. Comp. Physiol. A **154**, 535–547.
117. Schwemer, J. (1986) in: The Molecular Mechanism of Photoreception, ed H. Stieve. pp. 303–326, Springer, Berlin.
118. Cugnoli, C., Fioravanti, R., Golisano, O. and Pepe, I.M. (1992) Struct. Funct. Retl. Prot. **221**, 271–274.
119. Pepe, I.M. and Cugnoli, C. (1998) Photochem. Photobiol. **68**, 257–259.
120. Smith, W.C., Friedman, M.A. and Goldsmith, T.H. (2000) Vis. Neurosci. **8**, 329–336.
121. Bernard, G.D. (1983) Science **219**, 69–71.
122. Bernard, G.D. (1983) Biophys. Struct. Mech. **9**, 277–286.
123. Eguchi, E. (1999) in: Atlas of Arthropod Sensory Receptors, eds E. Eguchi and Y. Tominaga. pp. 87–96, Springer, Berlin.
124. Eguchi, E. and Waterman, T.H. (1967) Z. Zellforsch. **79**, 209–229.
125. Eguchi, E. and Waterman, T.H. (1976) Cell Tissue Res. **169**, 419–434.
126. Blest, A.D. (1988) Adv. Insect Physiol. **20**, 1–53.
127. Blest, A.D. and Stowe, S. (1997) J. Comp. Physiol. A **180**, 347–355.
128. Blest, A.D., Stowe, S. and Delaney, A. (1994) J. Comp. Physiol. A **174**, 747–754.
129. Cronin, T.W. and Goldsmith, T.H. (1984) J. Gen. Physiol. **84**, 63–81.
130. Dorlöchter, M. and Stieve, H. (1997) Progr. Neurobiol. **53**, 451–515.
131. Arikawa, K., Morikawa, Y., Suzuki, T. and Eguchi, E. (1988) Experientia **44**, 219–220.
132. Itaya, S.K. (1976) Cell Tissue Res. **166**, 265–273.
133. Williams, D.S. and Blest, A.D. (1980) Cell Tissue Res. **205**, 423–438.
134. O'Tousa, J.E., Baehr, W., Martin, R.L., Hirsh, J., Pak, W.L. and Applebury, M.L. (1985) Cell **40**, 839–850.
135. Zuker, C.S., Cowman, A.F. and Rubin, G.M. (1985) Cell **40**, 851–858.
136. Cowman, A.F., Zuker, C.S. and Rubin, G.M. (1986) Cell **44**, 705–710.
137. Zuker, C.S., Montell, C., Jones, K., Laverty, T. and Rubin, G.M. (1987) J. Neurosci. **7**, 1550–1557.
138. Montell, C., Jones, K., Zuker, C.S. and Rubin, G.M. (1987) J. Neurosci. **7**, 1558–1566.
139. Fryxell, K.J. and Meyerowitz, E.M. (1987) EMBO J. **6**, 443–451.
140. Frohman, M.A., Dush, M.K. and Martin, G.R. (1988) Proc. Natl. Acad. Sci. USA **85**, 8998–9002.
141. Towner, P. and Gärtner, W. (1992) Nucl. Acids Res. **20**, 4669–4670.
142. Leonard, D.S., Bowman, V.D., Ready, D.F. and Pak, W.L. (1992) J. Neurobiol. **23**, 605–626.
143. Pak, W.L. and O'Tousa, J.E. (1988) Photochem. Photobiol. **47**, 877–882.
144. Pak, W.L. (1995) Invest. Ophthalmol. Visual Sci. **36**, 2340–2357.

145. Pollock, J.A. and Benzer, S. (1988) Nature **333**, 779–782.
146. Feiler, R., Harris, W.A., Kirschfeld, K., Wehrhahn, C. and Zuker, C.S. (1988) Nature **333**, 737–741.
147. Feiler, R., Bjornson, R., Kirschfeld, K., Mismer, D., Rubin, G.M., Smith, D.P., Socolich, M. and Zuker, C.S. (1992) J. Neurosci. **12**, 3862–3868.
148. Chou, W.-H., Hall, K.J., Wilson, D.B., Wideman, C.L., Townson, S.M., Chadwell, L.V. and Britt, S.G. (1996) Neuron **17**, 1101–1115.
149. Huber, A., Schurmans, S., Bentrop, J., Groell, C., Wolfrum, J. and Paulsen, R. (1997) FEBS Lett. **406**, 6–10.
150. Papatsenko, D., Sheng, G.J. and Desplan, C. (1997) Development **124**, 1665–1673.
151. Chou, W.-H., Huber, A., Bentrop, J., Schulz, S., Schwab, K., Chadwell, L.V., Paulsen, R. and Britt, S.G. (1999) Development **126**, 607–616.
152. Towner, P. and Gärtner, W. (1994) Gene **143**, 227–231.
153. Rossel, S. (1979) J. Comp. Physiol. A **131**, 95–112.
154. Bennett, R.R., Tunstall, J. and Horridge, G.A. (1967) Z. vergl. Physiol. **55**, 195–206.
155. Towner, P., Harris, P., Wolstenholme, A.J., Hill, C., Worm, K. and Gärtner, W. (1997) Vision Res. **37**, 495–503.
156. Popp, M.P., Grisshammer, R., Hargrave, P.A. and Smith, W.C. (1996) Invert. Neurosci. **1**, 323–329.
157. Smith, W.C., Ayers, D.A., Popp, M.P. and Hargrave, P.A. (1997) Invert. Neurosci. **3**, 49–56.
158. Menzel, R. and Müller, U. (1996) Annu. Rev. Neurosci. **19**, 379–404.
159. Bellingham, J., Wilkie, S.E., Morris, A.G., Bowmaker, J.K. and Hunt, D.M. (1997) Eur. J. Biochem. **243**, 775–781.
160. Townson, S.M., Chang, B.S.W., Salcedo, E., Chadwell, L.V., Pierce, N.E. and Britt, S.G. (1998) J. Neurosci. **18**, 2412–2422.
161. Chang, B.S.W., Ayers, D.M., Smith, W.C. and Pierce, N.E. (1996) Gene **173**, 215–219.
162. Bennett, R.R. and Brown, P.K. (1985) Vision Res. **25**, 1771–1781.
163. Chase, M.R., Bennett, R.R. and White, R.H. (1997) J. Exp. Biol. **200**, 2469–2478.
164. Kitamoto, J., Sakamoto, K., Ozaki, K., Mishina, Y. and Arikawa, K. (1998) J. Exp. Biol. **201**, 1255–1261.
165. Briscoe, A.D. (1998) Naturwissenschaften **85**, 33–35.
166. Arikawa, K., Suyama, D. and Fujii, T. (1997) J. Comp. Physiol. A **180**, 295–299.
167. Ovchinnikov, Y.A., Abdulaev, N.G., Zolotarev, A.S., Artamonov, I.D., Bespalov, I.A., Dergachev, A.E. and Tsuda, M. (1988) FEBS Lett. **232**, 69–72.
168. Hall, M.D., Hoon, M.A., Ryba, N.J.P., Pottinger, J.D.D., Keen, J.N., Saibil, H.R. and Findlay, J.B.C. (1991) Biochem. J. **274**, 35–40.
169. Morris, A.G., Bowmaker, J.K. and Hunt, D.M. (1993) Proc. Royal Soc. Lond. B **254**, 233–240.
170. Bellingham, J., Morris, A.G. and Hunt, D.M. (1998) J. Exp. Biol. **201**, 2299–2306.
171. Hara-Nishimura, I., Kondo, M., Nishimura, M., Hara, R. and Hara, T. (1993) FEBS Lett. **317**, 5–11.
172. Hara-Nishimura, I., Matsumoto, T., Mori, H., Nishimura, M., Hara, R. and Hara, T. (1990) FEBS Lett. **271**, 106–110.
173. Hara-Nishimura, I., Kondo, M., Nishimura, M., Hara, R. and Hara, T. (1993) FEBS Lett. **335**, 94–98.
174. Smith, W.C., Price, D.A., Greenberg, R.M. and Battelle, B.-A. (1993) Proc. Natl. Acad. Sci. USA **90**, 6150–6154.
175. Hariyama, T., Ozaki, K., Tokunaga, F. and Tsukahara, Y. (1993) FEBS Lett. **315**, 287–292.
176. Crandall, K.A. and Hillis, D.M. (1997) Nature **387**, 667–668.
177. Crandall, K.A. and Cronin, T.W. (1997) J. Molec. Evolution **45**, 524–534.
178. Sakamoto, K., Hisatomi, O., Tokunaga, F. and Eguchi, E. (1996) J. Exp. Biol. **199**, 441–450.
179. Kojima, D., Terakita, A., Ishikawa, T., Tsukahara, Y., Maeda, A. and Shichida, Y. (1997) J. Biol. Chem. **272**, 22979–22982.
180. Burr, A.H. (1985) Photochem. Photobiol. **41**, 577–582.
181. Boom, J., Lobo, K., Smith, M.J. and Burr, A.H. (1988) Worm Breeder's Gazette **10**, 52.
182. Mismer, D. and Rubin, G.M. (1987) Genetics **116**, 565–578.

183. Mismer, D., Michael, W.M., Laverty, T.R. and Rubin, G.R. (1988) Genetics **120**, 173–180.
184. Mismer, D. and Rubin, G.M. (1989) Genetics **121**, 77–87.
185. Fortini, M.E. and Rubin, G.M. (1990) Genes Develop. **4**, 444–463.
186. Briscoe, A.D. (1999) Gene **230**, 101–109.
187. Thompson, J.D., Higgins, D.G. and Gibson, T.J. (1994) Nucl. Acids Res. **22**, 4680.
188. Schwemer, J. (1983) Biophys. Struct. Mech. **9**, 287–298.
189. Cugnoli, C., Mantovani, R., Fioravanti, R. and Pepe, I.M. (1989) FEBS Lett. **257**, 63–67.
190. Hara, T. and Hara, R. (1972) in: Handbook of Sensory Physiology. Vol. VII/1, ed H.J.A. Dartnall. pp. 720–746, Springer, Berlin.
191. Kaushal, S. and Khorana, H.G. (1994) Biochemistry **33**, 6121–6128.
192. Dryja, T.P., McGee, T.L., Reichel, E., Hahn, L.B., Cowley, G.S., Yandell, D.W., Sandberg, M.A. and Berson, E.L. (1990) Nature **343**, 364–366.
193. Roof, D.J., Adamian, M. and Hayes, A. (1994) Invest. Ophthalmol. Visual Sci. **35**, 4049–4062.
194. König, B., Arendt, A., McDowell, J.H., Kahlert, M., Hargrave, P.A. and Hofmann, K.P. (1989) Proc. Natl. Acad. Sci. USA **86**, 6878–6882.
195. Berson, E.L. (1993) Invest. Ophthalmol. Visual Sci. **34**, 1659–1676.
196. Henderson, R., Baldwin, J.M., Ceska, T.A., Zemlin, F., Beckmann, E. and Downing, K.H. (1990) J. Mol. Biol. **213**, 899–929.
197. Luecke, H., Richter, H.T. and Lanyi, J.K. (1998) Science **280**, 1934–1937.
198. Essen, L.-O., Siegert, R., Lehmann, W.D. and Oesterhelt, D. (1998) Proc. Natl. Acad. Sci. USA **95**, 11673–11678.
199. Kitagawa, T. and Tsuda, M. (1980) Biochim. Biophys. Acta **624**, 211–217.
200. Pande, C., Pande, A., Yue, K.T., Callender, R.H., Ebrey, T.G. and Tsuda, M. (1987) Biochemistry **26**, 4941–4947.
201. Pande, C., Deng, H., Rath, P., Callender, R.H. and Schwemer, J. (1987) Biochemistry **26**, 7426–7430.
202. Sakmar, T.P., Franke, R.R. and Khorana, H.G. (1989) Proc. Natl. Acad. Sci. USA **86**, 8309–8313.
203. Robinson, P.R., Cohen, G.B., Zhukovsky, E.A. and Oprian, D.D. (1992) Neuron **9**, 719–725.
204. Fryxell, K.J. and Meyerowitz, E.M. (1991) J. Mol. Evol. **33**, 367–378.
205. Trissl, H.-W. (1990) Photochem. Photobiol. **51**, 793–818.
206. Mathies, R.A., Lin, S.W., Ames, J.B. and Pollard, W.T. (1991) Annu. Rev. Biophys. Biophys. Chem. **20**, 491–518.
207. Stockburger, M. and Hildebrandt, P. (1984) Biochemistry **23**, 5539–5548.
208. Fischer, W.B., Sonar, S., Marti, T., Khorana, H.G. and Rothschild, K.J. (1994) Biochemistry **33**, 12757–12762.
209. Han, M., Lin, S.W., Smith, S.O. and Sakmar, T.P. (1996) J. Biol. Chem. **271**, 32330–32336.
210. Jäger, S., Han, M., Lewis, J.W., Szundi, I., Sakmar, T.P. and Kliger, D.S. (1997) Biochemistry **36**, 11804–11810.
211. Han, M., Groesbeck, M., Sakmar, T.P. and Smith, S.O. (1998) Proc. Natl. Acad. Sci. USA **94**, 13442–13447.
212. Han, M., Lin, S.W., Minkova, M., Smith, S.O. and Sakmar, T.P. (1996) J. Biol. Chem. **271**, 32337–32342.
213. Blumenfeld, A., Erusalimsky, J., Heichal, O., Selinger, Z. and Minke, B. (1985) Proc. Natl. Acad. Sci. USA **82**, 7116–7120.
214. Johnson, E.C., Robinson, P.R. and Lisman, J.E. (1986) Nature **324**, 468–470.
215. Yarfitz, S., Niemi, G.A., McConnell, J.L., Fitch, C.L. and Hurley, J.B. (1991) Neuron **7**, 429–438.
216. Garen-Fazio, S., Neer, E.J. and Schmidt, C.J. (1991) J. Comp. Neurol. **309**, 17–26.
217. Terakita, A., Hariyama, T., Tsukahara, Y., Katsukura, Y. and Tashiro, H. (1993) FEBS Lett. **330**, 197–200.
218. Bacigalupo, J., Bautista, D.M., Brink, D.L., Hetzer, J.F. and Oday, P.M. (1995) J. Neurosci. **15**, 7196–7200.
219. Bhatia, J., Davies, A., Gaudoin, J.B. and Saibil, H.R. (1996) J. Photochem. Photobiol., B: Biol. **35**, 19–23.

220. Hardie, R.C. and Minke, B. (1995) Cell Calcium **18**, 256–274.
221. Lee, Y.J., Dobbs, M.B., Verardi, M.L. and Hyde, D.R. (1990) Neuron **5**, 889–898.
222. Lee, Y.J., Shah, S., Suzuki, E., Zars, T., Oday, P.M. and Hyde, D.R. (1994) Neuron **13**, 1143–1157.
223. Munger, S.D., Schremser-Berlin, J.L., Brink, C.M. and Battelle, B.A. (1996) Invert. Neurosci. **2**, 175–182.
224. Ryba, N.J.P., Pottinger, J.D.D., Keen, J.N. and Findlay, J.B.C. (1991) Biochem. J. **273**, 225–228.
225. Mitchell, J., Gutierrez, J. and Northup, J.K. (1995) J. Biol. Chem. **270**, 854–859.
226. Vinos, J., Jalink, K., Hardy, R.W., Britt, S.G. and Zuker, C.S. (1997) Science **277**, 687–690.
227. Devary, O., Heichal, O., Blumenfeld, A., Cassel, D., Suss, E., Barash, S., Rubinstein, C.T., Minke, B. and Selinger, Z. (1987) Proc. Natl. Acad. Sci. USA **84**, 6939–6943.
228. Baumann, O. and Walz, B. (1989) Cell Tissue Res. **255**, 511–522.
229. Hardie, R.C. (1991) Proc. Royal Soc. Lond. **245**, 203–210.
230. Tanaka, J.C., Eccleston, J.F. and Furman, R.E. (1989) Biochemistry **28**, 2776–2784.
231. Nagy, K., ed. (1998) J. Photochem. Photobiol., B: Biol. **35**, 1–137.
232. Probst, W.C., Snyder, L.A., Schuster, D.I., Brosius, J. and Sealon, S.C. (1992) DNA Cell Biol. **11**, 1–20.
233. Weitz, C.J. and Nathans, J. (1993) Biochemistry **32**, 14176–14182.
234. Baehr, W. and Applebury, M.L. (1986) Trends Neurosci. **9**, 198–203.
235. Yokoyama, S. (1994) Molec. Biol. Evolution **11**, 32–39.
236. Karnik, S.S., Ridge, K.D., Bhattacharya, S. and Khorana, H.G. (1993) Proc. Natl. Acad. Sci. USA **90**, 40–44.
237. Weitz, C.J. and Nathans, J. (1992) Neuron **8**, 465–472.
238. Saibil, H.R., Langmack, K.A., Venien-Bryan, C. and Wilkinson, J.R. (1994) in: Cephalopod Neurobiology, eds N.J. Abott and R. Williamson. pp. 479–489, Oxford Univ. Press, Oxford.
239. Ryba, N.J., Hoon, M.A., Findlay, J.B., Saibil, H.R., Wilkinson, J.R., Heimburg, T. and Marsh, D. (1993) Biochemistry **32**, 3298–3305.
240. Ovchinnikov, Y.A., Lipkin, V.M., Shuvaeva, T.M., Bogachuk, A.P. and Shemyakin, V.V. (1985) FEBS Lett. **179**, 107–110.
241. Marshall, N.J. (1988) Nature **333**, 557–560.
242. Cronin, T.W. and Forward, R. (1988) J. Comp. Physiol. A **162**, 463–478.
243. Forward, R., Cronin, T.W. and Douglass, J.K. (1988) J. Comp. Physiol. A **162**, 479–490.
244. Marshall, N.J., Land, M.F., King, C.A. and Cronin, T.W. (1991) Phil. Trans. Royal Soc. Lond. B **334**, 33–56.
245. Marshall, N.J., Land, M.F., King, C.A. and Cronin, T.W. (1991) Phil. Trans. Royal Soc. Lond. B **334**, 57–84.
246. Seidou, M., Sugahara, M., Uchiyama, H., Hamanaka, T., Michinomae, M., Yoshihara, K. and Kito, Y. (1990) J. Comp. Physiol. **166**, 769–773.
247. Doukas, A.G., Junnarkar, M.R., Alfano, R.R., Callender, R.H., Kakitani, T. and Honig, B.H. (1984) Proc. Natl. Acad. Sci. USA **81**, 4790–4794.
248. Kruizinga, B. and Stavenga, D.G. (1990) Photochem. Photobiol. **51**, 197–201.
249. Hillman, P., Hochstein, S. and Minke, B. (1983) Physiol. Rev. **63**, 668–722.
250. Laughlin, S.B. (1976) J. Comp. Physiol. **111**, 221–247.
251. Kirschfeld, K., Hardie, R.C., Lenz, G. and Vogt, K. (1988) J. Comp. Physiol. A **162**, 421–433.
252. Hardie, R.C., Vogt, K. and Rudolph, A. (1989) J. Insect Physiol. **35**, 423–431.
253. Arikawa, K., Scholten, D.G.W., Kinoshita, M. and Stavenga, D.G. (1999) Zool. Sci. **16**, 17–24.
254. Stavenga, D.G. (1983) Biophys. Struct. Mech. **9**, 309–317.
255. Ostroy, S.E., Wilson, M. and Pak, W.L. (1974) Biochem. Biophys. Research Comm. **59**, 960–966.
256. Oprian, D.D., Molday, R.S., Kaufman, R.J. and Khorana, H.G. (1987) Proc. Natl. Acad. Sci. USA **84**, 8874–8878.
257. DeCaluwé, L.L.J., Vanoostrum, J., Janssen, J.J.M. and DeGrip, W.J. (1993) Meth. Neurosci. **15**, 307–321.
258. Spradling, A.C. and Rubin, G.M. (1982) Science **218**, 341–347.
259. Rubin, G.M. and Spradling, A.C. (1982) Science **218**, 348–353.

260. Kirschfeld, K., Feiler, R. and Vogt, K. (1988) J. Comp. Physiol. A **163**, 421–423.
261. Menzel, R. and Blakers, M. (1976) J. Comp. Physiol. A **108**, 11–33.
262. Menzel, R. (1979) in: Handbook of Sensory Physiology Vol. VII/6 A, ed. H. Autrum. pp. 503–580, Springer, Berlin.
263. Chang, B., Ayers, D.M., Smith, W.C. and Pierce, N. (1996) Gene **173**, 215–219.
264. Engels, A., Reichert, H., Gehring, W.J. and Gärtner, W. (1999) Eur. J. Biochem. **267**, 1917–1922.
265. Franke, R.R., Sakmar, T.P., Oprian, D.D. and Khorana, H.G. (1988) J. Biol. Chem. **263**, 2119–2122.
266. Harada, Y., Senda, T., Sakamoto, T., Takamoto, K. and Ishibashi, T. (1994) J. Biochem. **115**, 66–75.
267. Janssen, J.J.M., van den Ven, W.J.M., van Groningen-Luyben, W.A.H.M., Roosien, J., Vlak, J.M. and DeGrip, W.J. (1988) Molec. Biol. Rep. **13**, 65–71.
268. Lundström, K., Mills, A., Allet, E., Ceszkowski, K., Agudo, G., Chollet, A. and Liljestrom, P. (1995) J. Rec. Signal Transd. Res. **15**, 23–32.
269. Schmid, E.L., Tairi, A.-P., Hovius, R. and Vogel, H. (1998) Anal. Chem. **70**, 1331–1338.
270. Hölscher, K., Lundström, K. and Gärtner, W. (2000) Biochemistry (submitted).
271. Tsunoda, S., Sierralta, J. and Zuker, C.S. (1998) Curr. Opinion Genet. Development **8**, 419–422.
272. Huber, A., Sander, P., Gobert, A., Bähner, M., Hermann, R. and Paulsen, R. (1996) EMBO J. **15**, 7036–7045.
273. Tsuda, M., Tsuda, T., Terayama, Y., Fukada, Y., Akino, T., Yamanaka, G., Stryer, L., Katada, T., Ui, M. and Ebrey, T.G. (1986) FEBS Lett. **198**, 5–10.
274. Seki, T., Hara, R. and Hara, T. (1980) Photochem. Photobiol. **32**, 469–479.
275. Hubbard, R. and St. George, R.C.C. (1958) J. Gen. Physiol. **41**, 501–528.
276. Waterman, T.H. (1981) in: Handbook of Sensory Physiology, VII/6B, ed H. Autrum. pp. 281–496, Springer, Berlin
277. Saibil, H.R. and Hewat, E.A. (1987) J. Cell Biol. **105**, 19–28.
278. Koutalos, Y., Ebrey, T.G., Gilson, H.R. and Honig, B. (1990) Biophys. J. **58**, 493–501.
279. Dixon, S.F. and Cooper, A. (1987) Photochem. Photobiol. **46**, 115–119.
280. Dartnall, H.J.A., Goodeve, C.F. and Lythgoe, R.J. (1936) Proc. Royal Soc. A. **156**, 158–170.
281. Cronin, T.W. and Goldsmith, T.H. (1982) Photochem. Photobiol. **36**, 447–454.
282. Cooper, A., Dixon, S.F. and Tsuda, M. (1986) Eur. Biophys. J. **13**, 195–201.
283. Koutalos, Y., Ebrey, T.G., Tsuda, M., Odashima, K., Lien, T., Park, M.H., Shimizu, N., Derguini, F., Nakanishi, K., Gilson, H.R. and Honig, B.H. (1989) Biochemistry **28**, 2732–2739.
284. Arnaboldi, M., Motto, M.G., Tsujimoto, K., Balogh-Nair, V. and Nakanishi, K. (1979) J. Am. Chem. Soc. **101**, 7082–7084.
285. Iwasa, T., Colmenares, L.U., Hirata, K., Arime, Y., Nakagawa, M. and Kikkawa, S. (1998) J. Phys. Chem. A **102**, 5602–5610.
286. Hara, T., Hara, R., Kishigami, A., Koshida, Y., Horiuchi, S. and Raj, U. (1995) Zool. Sci. **12**, 195–201.
287. Saunders, W.B. and Landman, N.H. (1987) Nautilus. The Biology and Paleobiology of a Living Fossil. Plenum, New York.
288. Hara, T. and Hara, R. (1976) J. Gen. Physiol. **67**, 791–805.
289. Ozaki, K., Hara, R., Hara, T. and Kakitani, T. (1983) Biophys. J. **44**, 127–137.
290. Tokunaga, F. and Yoshihara, K. (1995) Israel J. Chem. **35**, 357–364.
291. Sekiya, N., Kishigami, A., Naoki, H., Chang, C.W., Yoshihara, K., Hara, R., Hara, T. and Tokunaga, F. (1991) FEBS Lett. **280**, 107–111.
292. Tsujimoto, K., Ohashi, M., Hara, R. and Ozaki, K. (1985) Chem. Lett. 1373–1376.
293. Sekiya, N. and Yoshihara, K. (1995) Biophys. Chem. **56**, 31–39.
294. Tsujimoto, K., Ohmura, K., Ohashi, M., Hara, R., Hara, T., Ozaki, K. and Watanabe, M. (1986) Chem. Lett. 2005–2008.
295. Ozaki, K., Terakita, A., Hara, R. and Hara, T. (1987) Vision Res. **27**, 1057–1070.
296. Terakita, A., Hara, R. and Hara, T. (1989) Vision Res. **29**, 639–652.
297. Ozaki, K., Terakita, A., Ozaki, M., Hara, R., Hara, T., Hara-Nishimura, I., Mori, H. and Nishimura, M. (1994) J. Biol. Chem. **269**, 3838–3845.

298. Cronin, T.W. and Marshall, N.J. (1989) Nature **339**, 137–140.
299. Exner, S. (1891) Die Physiologie der facettirten Augen von Krebsen und Insecten. Deuticke, Leipzig.
300. Hardie, R.C. (1988) Nature **333**, 499–500.
301. Donner, K.O., Langer, H., Lindstrom, M. and Schlecht, P. (1994) Behavioral Physiology **174**, 451–459.
302. Zeiger, J. and Goldsmith, T.H. (1989) Vision Res. **29**, 519–527.
303. Zeiger, J. and Goldsmith, T.H. (1993) Visual Neurosci. **10**, 193–202.
304. Fernandez, H. (1965) A Survey of the Visual Pigments of Decapod Crustacea of South Florida. PhD thesis. Univ. of Miami, Coral Gables.
305. Wald, G. (1967) Nature **215**, 1131–1133.
306. Zeiger, J. and Goldsmith, T.H. (1994) Vision Res. **34**, 2679–2688.
307. Hubbard, R. and Wald, G. (1960) Nature **186**, 212–214.
308. Dartnall, H.J.A. (1953) Brit. Med. Bull. **65**, 24–30.
309. Lipetz, L.E. and Cronin, T.W. (1988) Vision Res. **28**, 1083–1093.
310. Partridge, J.C. and DeGrip, W.J. (1991) Vision Res. **31**, 619–630.
311. Stavenga, D.G., Smits, R.P. and Hoenders, B.J. (1993) Vision Res. **33**, 1011–1017.
312. Gribakin, F.G., Alekseyev, Y.N. and Ukhanov, K.Y. (1996) J. Photochem. Photobiol., B: Biol. **35**, 13–18.
313. Tan, Q., Lou, J.H., Borhan, B., Karnaukhova, E., Berova, N. and Nakanishi, K. (1997) Angew. Chemie – Intl. Ed. Engl. **36**, 2089–2093.
314. Buss, V., Kolster, K., Terstegen, F. and Vahrenhorst, R. (1998) Angew. Chemie – Intl. Ed. Engl. **37**, 1893–1895.
315. Hemley, R. and Kohler, B.E. (1977) Biophys. J. **20**, 377–382.
316. Isler, O. (1971) Carotenoids. Birkhäuser, Basel.
317. Greenberg, A.D., Honig, B.H. and Ebrey, T.G. (1975) Nature **257**, 823–824.
318. Honig, B.H., Ebrey, T.G., Callender, R.H., Dinur, U. and Ottolenghi, M. (1979) Proc. Natl. Acad. Sci. USA **76**, 2503–2507.
319. Han, M. and Smith, S.O. (1995) Biochemistry **34**, 1425–1432.
320. Baasov, T. and Sheves, M. (1986) Biochemistry **25**, 5249–5258.
321. Derguini, F., Dunn, D., Eisenstein, L., Nakanishi, K., Odashima, K., Rao, V.J., Sastry, L. and Termini, J. (1986) Pure Appl. Chem. **58**, 719–724.
322. Albeck, A., Friedman, N., Sheves, M. and Ottolenghi, M. (1989) Biophys. J. **56**, 1259–1265.
323. Gärtner, W., Ullrich, D. and Vogt, K. (1991) Photochem. Photobiol. **54**, 1047–1055.
324. Nakanishi, K. and Crouch, R. (1995) Israel J. Chemistry **35**, 253–272.
325. Asenjo, A.B., Rim, J. and Oprian, D.D. (1994) Neuron **12**, 1131–1138.
326. Salem, L. and Bruckmann, P. (1975) Nature **258**, 526–528.
327. Kakitani, T., Kakitani, H., Honig, B.H. and Nakanishi, K. (1983) J. Am. Chem. Soc. **105**, 648–650.
328. Kakitani, H., Kakitani, T., Rodman, H. and Honig, B.H. (1985) Photochem. Photobiol. **41**, 471–479.
329. Birge, R.R. (1990) Annu. Rev. Phys. Chem. **41**, 683–733.
330. Harbison, G.S., Smith, S.O., Pardoen, J.A., Courtin, J.M.L., Lugtenburg, J., Herzfeld, J., Mathies, R.A. and Griffin, R.G. (1985) Biochemistry **24**, 6955–6962.
331. Palings, I., Pardoen, J.A., van den Berg, E.M.M., Winkel, C., Lugtenburg, J. and Mathies, R.A. (1987) Biochemistry **26**, 2544–2556.
332. Rossel, S. (1987) Naturwissenschaften **74**, 53–62.
333. Unger, V.M., Hargrave, P.A., Baldwin, J.M. and Schertler, G.F.X. (1997) Nature **389**, 203–206.
334. Robinson, J., Schmitt, E.A., Harosi, F.I., Reece, R.J. and Dowling, J.E. (1993) Proc. Natl. Acad. Sci. USA **90**, 6009–6012.
335. Yoshizawa, T. (1994) Biophys. Chem. **50**, 17–24.
336. Goldsmith, T.H. (1994) Vision Res. **34**, 1479–1487.
337. Kirschfeld, K. and Vogt, K. (1986) Vision Res. **26**, 1771–1777.

338. Hamdorf, K., Hochstrate, P., Höglund, G., Moser, M., Sperber, S. and Schlecht, P. (1992) J. Comp. Physiol. A **171**, 601–615.
339. Kruizinga, B., Kammann, R.L. and Stavenga, D.G. (1983) Biophys. Struct. Mech. **9**, 299–307.
340. Roebroek, J.G.H., Gagné, S. and Stavenga, D.G. (1989) J. Comp. Physiol. A **165**, 75–81.
341. Taiji, M., Bryl, K., Nakagawa, M., Tsuda, M. and Kobayashi, T. (1992) Photochem. Photobiol. **56**, 1003–1011.
342. Schoenlein, R.W., Peteanu, L.A., Mathies, R.A. and Shank, C.V. (1991) Science **254**, 412–415.
343. Ohtani, H., Kobayashi, T., Tsuda, M. and Ebrey, T.G. (1988) Biophys. J. **53**, 17–24.
344. Tsuda, M. (1979) Biochim. Biophys. Acta **545**, 537–546.
345. Kobayashi, T. and Nagakura, S. (1982) Methods Enzymol. **81**, 368–373.
346. Naito, T., Nashima-Hayama, K., Ohtsu, K. and Kito, Y. (1981) Vision Res. **21**, 935–941.
347. Nakagawa, M., Kikkawa, S., Iwasa, T. and Tsuda, M. (1997) Biophys. J. **72**, 2320–2328.
348. Siebert, F. (1990) in: Photochromism; Molecules and Systems, eds H. Dürr and H. Bouas-Laurent. pp. 756–792, Elsevier, Amsterdam.
349. Ottolenghi, M. and Sheves, M., eds (1995) Israel J. Chem. **35**, 193–513.
350. Bagley, K.A., Eisenstein, L., Ebrey, T.G. and Tsuda, M. (1989) Biochemistry **28**, 3366–3373.
351. Huang, L., Deng, H., Weng, G., Koutalos, Y., Ebrey, T., Groesbeek, M., Lugtenburg, J., Tsuda, M. and Callender, R.H. (1996) Biochemistry **35**, 8504–8510.
352. Huang, L., Deng, H., Koutalos, Y., Ebrey, T.G., Groesbeek, M., Lugtenburg, J., Tsuda, M. and Callender, R.H. (1997) Photochem. Photobiol. **66**, 747–754.
353. Deng, H., Manor, D., Weng, G., Rath, P., Koutalos, Y., Ebrey, T., Gebhard, R., Lugtenburg, J., Tsuda, M. and Callender, R.H. (1991) Biochemistry **30**, 4495–4502.
354. Nishimura, S., Kandori, H., Nakagawa, M., Tsuda, M. and Maeda, A. (1997) Biochemistry **36**, 864–870.
355. Masuda, S., Morita, E.H., Tasumi, M., Iwasa, T. and Tsuda, M. (1993) FEBS Lett. **317**, 223–227.
356. Masuda, S., Morita, E.H., Tasumi, M., Iwasa, T. and Tsuda, M. (1993) J. Molec. Struct. **297**, 29–34.
357. Paulsen, R., Bentrop, J., Bauernschmitt, H.G., Böcking, D. and Peters, K. (1987) Photobiochem. Photobiophys. (Suppl.) 261–272.
358. Tsuda, M. (1987) Photochem. Photobiol. **45**, 915–931.
359. Rayer, B., Naynert, M. and Stieve, H. (1990) J. Photochem. Photobiol., B: Biology **7**, 107–148.
360. Pak, W.L. and Shortridge, R.D. (1991) Photochem. Photobiol. **53**, 871–875.
361. Smith, D.P., Stamnes, M.A. and Zuker, C.S. (1991) Annu. Rev. Cell Biol. **7**, 161–190.
362. Ranganathan, R., Harris, W.A. and Zuker, C.S. (1991) Trends Neurosci. **14**, 486–493.
363. Zuker, C.S. (1996) Proc. Natl. Acad. Sci. USA **93**, 571–576.
364. Finn, J.T., Solessio, E.C. and Yau, K.W. (1997) Nature **385**, 815–819.
365. Gorman, A.L.F. and McReynolds, J.S. (1969) Science **165**, 309–310.
366. Han, M., Groesbeck, M., Smith, S.O. and Sakmar, T.P. (1997) Biochemistry **37**, 538–545.
367. Deer, J.L., Hurley, J.B. and Yarfitz, S.L. (1995) J. Biol. Chem. **270**, 12623–12628.
368. Cho, K.O., Hunt, C.A. and Kennedy, M.B. (1992) Neuron **9**, 929–942.
369. Huber, A., Sander, P. and Paulsen, R. (1996) J. Biol. Chem. **271**, 11710–11717.
370. Huber, A., Sander, P., Bähner, M. and Paulsen, R. (1998) FEBS Lett. **425**, 317–322.
371. Chyb, S., Raghu, P. and Hardie, R.C. (1999) Nature **397**, 255–259.
372. Bloomquist, B.T., Shortridge, R.D., Schneuwly, S., Perdew, M., Montell, C., Steller, H., Rubin, G. and Pak, W.L. (1988) Cell **54**, 723–733.
373. Bacigalupo, J., Johnson, E.C., Vergara, C. and Lisman, J.E. (1991) Proc. Natl. Acad. Sci. USA **88**, 7938–7942.
374. Lisman, J.E., Robinson, P.R. and Johnson, E.C. (1986) Photochem. Photobiol. **43**, 57S.
375. Gorman, A.L.F. and McReynolds, J.S. (1974) Science **185**, 620–621.
376. McReynolds, J.S. and Gorman, A.L.F. (1974) Science **183**, 658–659.
377. Kikkawa, S., Tominaga, K., Nakagawa, M., Iwasa, T. and Tsuda, M. (1996) Biochemistry **35**, 15857–15864.
378. Trowell, S.C. (1988) FEBS Lett. **238**, 281–284.

379. Carne, A., Conway, S., Keen, J.N., Liu, S.-H., McGregor, R.A., Monk, P.D., Lott, J.S., Pottinger, J.D.D., Ryba, N., Sinclair, A. and Findlay, J.B.C. (1992) in: Structures and Functions of Retinal Proteins, ed J.-L. Rigaud. pp. 75–78, Libbey, Montrouge.
380. Saibil, H.R. and Michel-Villaz, M. (1984) Proc. Natl. Acad. Sci. USA **81**, 5111–5115.
381. Stieve, H. and Lumme, G. (1989) FEBS Lett. **253**, 6–10.
382. Terakita, A., Yamashita, T., Tachibanaki, S. and Shichida, Y. (1998) FEBS Lett. **439**, 110–114.
383. Robinson, P.R., Wood, S.F., Szuts, E.Z., Fein, A. and Hamm, H.E. (1990) Biochem. J. **272**, 79–85.
384. Pottinger, J.D.D., Ryba, N.J.P., Keen, J.N. and Findlay, J.B.C. (1991) Biochem. J. **279**, 323–326.
385. Nobes, C., Baverstock, J. and Saibil, H. (1992) Biochem. J. **287**, 545–548.
386. Schraermeyer, U., Stieve, H. and Rack, M. (1995) Tissue Cell **27**, 317–322.
387. Ryba, N.J.P., Findlay, J.B.C. and Reid, J.D. (1993) Biochem. J. **292**, 333–341.
388. Lott, J.S., Ryba, N.J., Pottinger, J.D., Keen, J.N., Carne, A. and Findlay, J.B. (1992) FEBS Lett. **312**, 241–244.
389. Baer, K.M. and Saibil, H.R. (1988) J. Biol. Chem. **263**, 17–20.
390. Wood, S.F., Szuts, E.Z. and Fein, A. (1989) J. Biol. Chem. **264**, 12970–12976.
391. Mitchell, J. and Mayeenuddin, L.H. (1998) Biochemistry **37**, 9064–9072.
392. Monk, P.D., Carne, A., Liu, S.-H., Ford, J.W., Keen, J.N. and Findlay, J.B.C. (1996) J. Neurochem. **67**, 2227–2235.
393. Tsuda, M., Tsuda, T. and Hirata, H. (1989) FEBS Lett. **257**, 38–40.
394. Tsuda, M., Hirata, H. and Tsuda, T. (1992) Photochem. Photobiol. **56**, 1167–1172.
395. Ohguro, H., Yoshida, N., Shindou, H., Crabb, J.W., Palczewski, K. and Tsuda, M. (1998) Photochem. Photobiol. **68**, 824–828.
396. Hargrave, P.A. and McDowell, J.H. (1992) Intl. Rev. Cytol. **137B**, 49–97.
397. Arcà, B., Lombardo, F., Capurro, M., della Torre, A., Dimopoulos, G., James, A.A. and Coluzzi, M. (1999) Proc. Natl. Acad. Sci. USA **96**, 1516–1521.
398. Mardulyn, P. and Cameron, S.A. (1999) Mol. Phylogenet. Evol. **12**, 282–294.
399. Carulli, J.P. and Hartl, D.L. (1992) Genetics **132**, 193–204.
400. Carulli, J.P., Chen, D.-M., Stark, W.S. and Hartl, D.L. (1994) J. Mol. Evol. **38**, 250–262.
401. Zeng, L.-W., Comeron, J.M., Chen, B. and Kreitman, M. (1998) Genetica **102–103**, 369–382.
402. Neufeld, T.P., Carthew, R.W. and Rubin, G.M. (1991) Proc. Natl. Acad. Sci. USA **88**, 10203–10207.
403. Felsenstein, J. (1989) Cladistics **5**, 164–166.
404. Salcedo, E., Huber, A., Henrich, S., Chadwell, L.V., Chou, W.-H., Paulsen, R. and Britt, S.G. (1999) J. Neurosci. **19**, 10716–10726.
405. Marshall, N.J. and Oberwinkler, J. (1999) Nature **401**, 873–874.
406. Palczewski, K., Kumasaka, T., Hori, T., Behnke, C.A., Motoshima, H., Fox, B.A., LeTrong, I., Teller, D.C., Okada, T., Stenkamp, R.E., Yamamoto, M. and Miyano, M. (2000) Science **289**, 739–745.

CHAPTER 8

Phototransduction Mechanisms in Microvillar and Ciliary Photoreceptors of Invertebrates

E. NASI and M. DEL PILAR GOMEZ

Department of Physiology,
Boston University

R. PAYNE

Department of Biology,
University of Maryland,
College Park

Handbook of Biological Physics
Volume 3, edited by D.G. Stavenga, W.J. DeGrip and E.N. Pugh Jr

Contents

2. Invertebrate photoreceptors are highly evolved detectors of light

2.1. A high quantum efficiency has been achieved

The random nature of quantal absorption by rhodopsin results in fluctuations of the rate at which rhodopsin molecules are photoisomerized during steady illumination. This "photon shot noise" is a significant limit to signal detection [3]. Because quantal absorptions obey Poisson statistics, the standard deviation of the rate of photoisomerizations is equal to the square root of the mean rate. Increasing the mean rate of photoisomerizations therefore reduces the relative amplitude of photon shot noise. Thus even photoreceptors that function in diurnal environments, where there is plenty of light to generate a signal, can benefit from high quantum efficiency. High quantum efficiency requires the maintenance of a high density of photoreceptive membrane, which may be metabolically expensive. However, it allows the detection of small fluctuations in photon flux originating from details in the environment. In locust photoreceptors, single photoisomerizations of rhodopsin have been shown to generate responses with a probability of greater than 0.6 [4], an identical performance to that of toad rod photoreceptors.

2.2. Temporal resolution is adjusted to an animal's behavior

High temporal resolution requires both a rapid response onset and termination. It might be thought that all photoreceptors should respond as rapidly as possible, so as to pass the maximal amount of information. However, the speed of response actually varies from species-to-species. The response time course is therefore probably the result of 2 compromise – there must be a penalty associated with a rapid response [5]. One penalty is a loss of amplification. In the model of Fig. 2, amplification at each stage is proportional to the time for which each intermediate in the visual cascade remains active. Higher amplification, appropriate for dark-adapted photoreceptors or nocturnal species, therefore requires longer response times. In bright light, when the need for amplification is less pressing, the response of the visual cascade can be faster, but the electrical time constant of the photoreceptor's plasma membrane then emerges as a limiting factor. Since specific membrane capacitance is fixed, a reduction in membrane resistance must occur if the time constant is to be reduced, which increases the ion flux required through the light activated channels to generate a signal. Maintaining intracellular ion concentrations in the face of this increased flux requires increased metabolic expenditure. As a result, only photoreceptors functioning in fast moving, diurnal animals respond rapidly while those of nocturnal or slowly moving animals display slower responses. Photoreceptor performance is therefore tuned to behavior and environment (Fig. 3) [1,5–8].

2.3. Phototransduction modulates large membrane currents that are matched to photoreceptor size

The maximal light-activated currents observed in invertebrate photoreceptors are comparatively large and indicate the presence of 10^4–10^6 light-activated ion

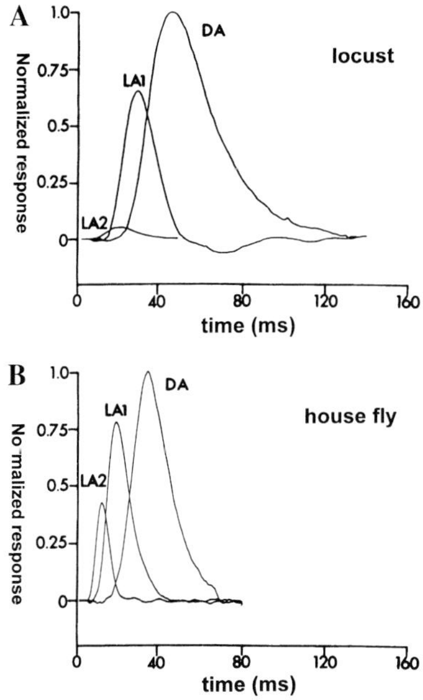

Fig. 3. The speed and amplitude of the response of insect photoreceptors depend upon the behavior of the species and the state of light adaptation. Responses to dim flashes in the photoreceptors of the compound eye of the slow-flying locust *Locusta migratoria* (A) are slower than those of the rapid-flying house fly *Musca domestica* (B) In addition, in both species, increasing levels of background illumination (LA1, LA2) speed the responses and reduce their amplitude compared to the dark-adapted (DA) responses (for details see [1]).

channels. The magnitude of the light-activated current depends on photoreceptor size, which can vary greatly between species. The giant photoreceptors of the rudimentary ventral eye of the horseshoe crab, *Limulus polyphemus*, have a cell body volume of hundreds of pl, while that of photoreceptors in the compound eye of the fruit fly, *Drosophila melanogaster*, is only approximately 1 pl. This difference in size must be related to the requirement for high spatial resolution and dense photoreceptor packing in compound eyes but not in rudimentary eye spots that do not form images. Photoreceptor size is not likely to affect the initial stages of phototransduction, since the reactants are probably confined to the immediate vicinity of a photoisomerization. However, large photoreceptors have much smaller electrical input resistance. Production of membrane potential changes in large cells therefore requires a greater light-activated current. Accordingly, the maximal light-activated current of a *Drosophila* photoreceptor is 10–20 nA (Fig. 8D), while that of a *Limulus* ventral photoreceptor may approach 1 μA [9]. The increased amplification in large cells could arise from either an increase in the number of light-activated channels or an increase in the effective conductance of each channel.

The maximal light-activated current of even a small invertebrate photoreceptor is typically much larger than the maximal current that flows through ion channels in the outer segments of vertebrate rods and cones. The latter current, which is modulated by light-induced closure of the channels, is 10–100 pA [10,11]. Thus the dynamic range of current flow available to an invertebrate photoreceptor is much greater than that available to a vertebrate photoreceptor.

2.4. The physical localization of transduction creates a linear response

A major consideration in understanding the kinetics of the light response is the localization of the visual cascade [12–14]. Phototransduction begins at the site of absorption of a light quantum. Given the limited diffusion of molecules within cells and the presence of structures like microvilli, which seem to be designed to limit the spread of intermediates to less than a micrometer, the reactions of the visual cascade are certainly highly localized. One consequence of this localization is that concentrations of intermediates may be very high and the affinity of receptors for those intermediates can be correspondingly reduced. A 200 μM increase in concentration can be achieved by the release of only one hundred molecules within the internal volume of a 0.7-μm-long squid photoreceptor microvillus [15]. It therefore may be difficult to artificially introduce into photoreceptors appropriate concentrations of chemical agents that mimic or block the intermediates of the visual cascade. In addition, the intermediates released following photoisomerizations within the same cell are effectively isolated from one another so that even if the mechanisms generating quantum bumps are highly non-linear, involving substantial feedback or co-operative reactions, linear summation of bumps will be observed up to high light intensities. This linearity aids the performance of the cell as a quantum counter. However, it also makes it extremely difficult to test whether or not transduction obeys the linear principles exemplified by models of the type illustrated in Fig. 2.

3. Structure and electrical response of microvillar photoreceptors

3.1. Distribution and anatomy of microvillar photoreceptors

Microvillar photoreceptors have been described in arthropods, mollusks, annelids and flatworms, forming the basis of simple and compound eyes as well as isolated "eye spots" having no optical resolving power. In compound eyes, photoreceptors are grouped into a complex multicellular array, an ommatidium, behind each facet of the eye (Fig. 4A). Central to the performance of each photoreceptor is the rhabdomere (Fig. 4B,C) – an array of photoreceptive microvilli positioned so as to maximally absorb light entering the eye. Each microvillus is a cylindrical outgrowth of the plasma membrane, 50–80 nm in diameter and 0.5–10 μm in length. The membrane of a typical microvillus contains a thousand or more particles, which are presumed to be mostly rhodopsin molecules [16–19]. A photoreceptor might possess 10^5 microvilli, resulting in a total rhodopsin content of $\sim 10^8$ molecules [20,21], comparable to that of vertebrate retinal photoreceptors.

Electron micrographs often show an axial filament within the microvillus and side arms which seem to link the filament to the microvillar plasma membrane

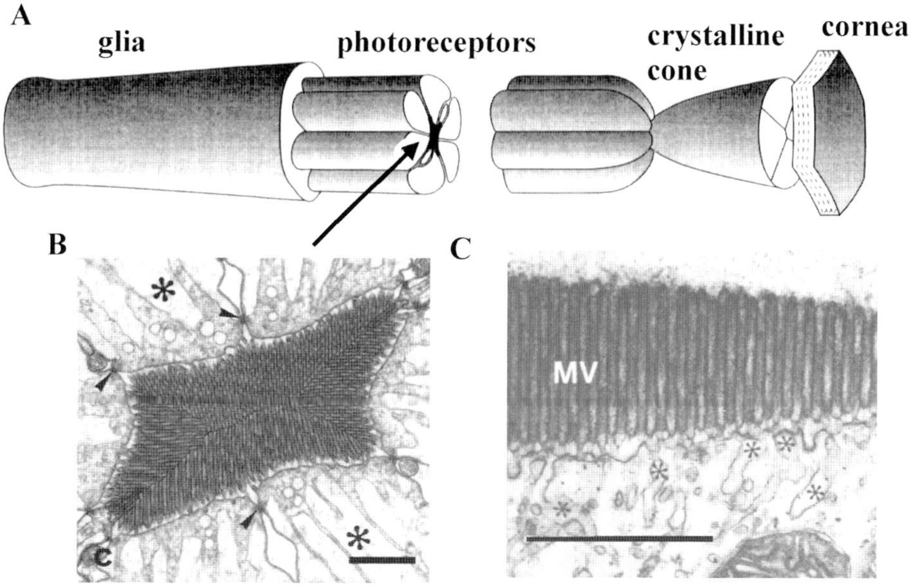

Fig. 4. (A) Diagram of ommatidium within the compound eye of a honeybee. Light enters at the cornea, travels through the crystalline cone cells and is focused onto the microvillar membrane of the central rhabdom, formed from the contributions (rhabdomeres) of several photoreceptor cells. (B) Electron micrograph of a cross-section through the rhabdom of a honeybee showing the central core of microvilli surrounded by a vacuole-like network of submicrovillar cisternae (asterisks). Arrowheads show belt desmosomes that separate the microvillar membrane from the non-microvillar membrane. (C) Electron micrograph of a longitudinal section through the rhabdom of a blowfly, *Calliphora*, showing the central core of microvilli (MV) overlying a scattered network of tubular submicrovillar cisternae (asterisks). Scale bars are 1 μm (after [45,362]).

[22–24]. The axial filament contains a bundle of actin filaments with the + end directed towards the tip of the microvillus [23,25–28]. The actin filaments appear to extend through the bottom of the microvillus into the cytoplasm, via fenestrations in the sub-rhabdomeral cisternae of smooth endoplasmic reticulum [29,30]. In addition to actin, alpha-spectrin may also associate with the sub-rhabdomeral cisternae of smooth endoplasmic reticulum to create an F-actin/spectrin cytoskeletal system [31]. The actin-binding proteins, alpha-actinin, vinculin and filamin have also been immunolocalized to the rhabdom in octopus retinae [32]. The actin cytoskeleton is responsible for the presence of unconventional myosin III in the rhabdom of *Drosophila* [33] and *Limulus* lateral eye photoreceptors [34].

In addition to the actin cytoskeleton, a novel structural element has been discovered in *Drosophila* and blowfly (*Calliphora*) photoreceptors, where the INAD protein acts as multivalent linker between membrane proteins, such as the TRP and TRPL channels and other elements of the transduction cascade, including phospholipase C (PLC), protein kinase C (PKC) and calmodulin [35–37]. In

rounding a given photoisomerization. If one assumes a linear cascade, as in Fig. 2, then variability at comparatively early stages in the cascade, when relatively few intermediate molecules have been created, is going to have the most impact on variability in the final size of the quantum bump. In *Limulus* ventral photoreceptors, it has been proposed that variability in the time taken to turn off the interaction of rhodopsin with GTP-binding proteins underlies the distribution of bump amplitudes [67,68]. However, mutant *Drosophila* photoreceptors which are deficient in calmodulin can generate a train of bumps in response to a single photoisomerization which vary much less in amplitude than do normal bumps [69]. The amplification available to a given rhodopsin in *Drosophila* may therefore be fixed, with variation in the amounts of intermediates available in different microvillar regions accounting for local variation in quantum event amplitude. It remains to be seen, however, to what extent the cycling of continuously active rhodopsin in this situation mimics normal activation of the cascade, since rapid biochemical feedback in the transduction cascade could conceivably result in stereotyped oscillatory responses.

Spontaneous events resembling bumps are often observed [70–73]. These can arise at any stage of the visual cascade due to thermal activation of the intermediates. The size of spontaneous events depends on how far upstream they are initiated in the visual cascade. Spontaneous events that mimic single photon signals in size might be expected to arise from thermal isomerizations of rhodopsin [74]. Some photoreceptors are capable of extremely low rates of these spontaneous quantum bumps. Dark-adapted locust photoreceptors, for example, maintain a spontaneous quantum bump rate of only 10 per hour [73]. But in the photoreceptors of other species, such as those of *Limulus* lateral eyes, event rates can be much higher and may show a circadian rhythm, being appropriately lower at night [74], when spontaneous events may limit the detection of dim objects. Thermal isomerization of rhodopsin may not be the only source of spontaneous discrete events. In *Limulus* ventral eye photoreceptors, the majority of spontaneous events are smaller than single photon signals, more similar to those induced by the GTP-binding protein activator, GTP-γS (see below), consistent with their arising from spontaneous activation of individual GTP-binding proteins.

Spontaneous events arising from random thermal activation of intermediates in the visual cascade and the stochastic variation in the response to individual quanta are two major causes of internal "transducer noise". In a study of locust photoreceptors, Laughlin and Lillywhite [73] found transducer noise to be as significant as photon shot noise in limiting the detection of dim flashes.

3.4. Light adaptation adjusts amplification to match the level of background illumination

While the amplification that creates quantum bumps is crucial for night vision, it turns into a liability at higher mean light intensities, because of saturation of the response. The maximum membrane potential change that light can evoke is 60–70 mV, set by the driving force for ion movements through the light-activated conductance. As the mean light intensity rises, light adaptation must progressively reduce the sensitivity of the photoreceptor, so as to avoid response saturation and take advantage of the improvement in signal-to-noise ratio resulting from the higher photon flux. This reduction in absolute sensitivity maintains a constant sensitivity to image contrast as background illumination increases [76,77]. During prolonged illumination, therefore, the photocurrent generated by microvillar photoreceptors declines within a few hundred ms and the sensitivity to further stimulation is reduced – creating a transition from "peak response" to "plateau response" (Fig. 5). Sensitivity to flashes can be reduced by several orders of magnitude by bright continuous background illumination and is accompanied by a reduction in the duration and latency of the response (Fig. 3) [78,79]. A reduction in the latency, duration and amplitude of quantum bumps underlies these changes [80–83]. Upon return to darkness, the sensitivity and time course of the response typically recover over several minutes.

Adaptation can be described kinetically as a negative feedback process, initiated by products of the visual cascade, which regulate the rates of earlier reactions and reduce their gain [2,84–87]. In the model of Fig. 2, a reduction in gain could be brought about by either a reduced catalytic efficiency of one or more of the intermediates in the cascade (a reduction in the rate constants α_n) or by an acceleration of the removal of intermediates (an increase in rate constants β_n). The latter will result in a reduction in response duration, as is usually observed during light adaptation. If the action of feedback to accelerate the rate of removal of intermediates is sufficiently rapid, then the time course of deactivation of the response to even a dim flash might be governed by the same processes that mediate adaptation.

4. Properties of the light-activated conductance of microvillar photoreceptors

4.1. The light-activated conductance is permeable to sodium and potassium

The gross features of the light response, namely the increase in membrane conductance leading to a depolarizing receptor potential, appear to be shared more or less universally by different invertebrate microvillar photoreceptors. However, paralleling the wide variety of morphological organization found across species, the effector mechanisms also appear to display significant diversity at a more detailed level. Considering that only a handful of model systems have been utilized for experimental studies, one should anticipate that the range of variations amongst the light-activated conductances could turn out to be rather extensive.

Compared to the intense efforts aimed at elucidating mechanisms that gate the light-activated conductance, the process of ion permeation through light-activated channels has received relatively little attention. The seminal work was conducted in *Limulus* ventral photoreceptors by Millecchia and Mauro [52], who first determined that the reversal potential of the photocurrent under voltage clamp is a few millivolts positive of 0. This value was independent of light intensity, and coincided with the reversal potential of spontaneous quantum bumps. Replacement of external Na^+ depressed the photoresponse, displaced the reversal voltage in the negative direction, and reduced the chord conductance. These observations were extended by Brown and Mote [88], who systematically manipulated extracellular Na^+, employing a variety of ionic and non-ionic substitutes, and observed a pronounced effect on the photocurrent reversal potential, approaching 55 mV/decade near the upper end of the Na^+ concentration range. The inward current that underlies the depolarizing receptor potential therefore appeared to be primarily carried by sodium ions, an inference that was subsequently generalized to several other species. Brown et al. [89] studied the ionic bases of the light-induced current in barnacle (*Balanus*) photoreceptors, and also measured a similar, if somewhat more positive, reversal potential (+25 mV), and a pronounced dependency of this value on the concentration of external sodium. Similar conclusions were reached in studies of other invertebrates, including marine mollusks such as *Hermissenda* [90], squid [91], *Strombus* [92] and the file clam *Lima* [93], prompting the generalization that light-dependent channels of invertebrate microvillar photoreceptors are sodium-selective. Several observations, however, indicate that additional ions must be significantly permeant. In the first place, when the extracellular sodium ion concentration is altered, the photocurrent reversal potential (V_{rev}) changes substantially less than the amount predicted by the Nernst equation, especially in the lower range of Na^+ concentrations. Furthermore, V_{rev} is several tens of millivolts more negative than the typical range expected for a pure Na^+ conductance; the existence of such a discrepancy is unambiguously revealed by directly determining intracellular ionic activities with ion-selective electrodes, as was done in *Balanus* photoreceptors [94] or by controlling the cytosolic ionic composition via intracellular dialysis [93]. As a consequence, some other ion(s) with an equilibrium potential (E) more negative than E_{Na} must be implicated. In *Limulus* ventral photoreceptors, Brown and Mote [88] determined that V_{rev} is insensitive to large changes in extracellular chloride, implying that light-dependent channels are cationic. Potassium is therefore a likely permeant species, and indeed the same authors reported a small shift in V_{rev} with changes in extracellular potassium ion concentration ($[K^+]_o$) under conditions of reduced extracellular sodium. Holt and Brown [95] had also previously observed a light-induced increase in $^{42}K^+$ efflux in whole ventral eyes of *Limulus*. The increase in K^+ permeability during the photocurrent was subsequently also measured in *Lima* [93]; the estimated ratio P_{Na}/P_K was in the vicinity of 1.8.

4.2. *Significant species differences exist in the permeability of the light-activated conductance to Ca^{2+}*

Permeation of sodium and potassium does not completely account for the ionic fluxes occurring during the light response, as several observations do not fit the prediction of a mixed Na^+/K^+ conductance: in *Lima*, for example, even in the absence of extracellular Na^+ the dependency of the photocurrent reversal voltage on $[K^+]_o$ is sub-Nernstian, and V_{rev} consistently remains more positive than E_K. In addition, in several species it has been observed that at physiological membrane potentials replacement of external Na^+ leaves a residual light-evoked current that is inwardly directed, or conversely, a depolarizing receptor potential; this occurs even if a non-ionic replacement such as sucrose is used, to rule out partial substitution effects [88,92,93]. By necessity, some additional ion(s) with a positive equilibrium potential must be implicated. Obvious candidates are Ca^{2+} (E_{Ca} > 120 mV) and/or perhaps Mg^{2+} (E_{Mg} > 30 mV). A role for divalent cations in the generation of the photocurrent was not recognized for several years, because the original measurements in *Limulus* indicated that removal of Ca^{2+} from the normal extracellular solution had little or no effect on V_{rev} [52,88]. By contrast, the response amplitude and chord conductance were substantially enhanced, prompting the suggestion that Ca^{2+} could perhaps function as a blocker of the light-sensitive channels. These results diverge from those obtained in *Balanus* photoreceptors, where V_{rev} shifts significantly upon manipulating the extracellular Ca^{2+} concentration ($[Ca^{2+}]_o$) [89]. Similarly, in *Drosophila* the reversal of the light-induced current also exhibits a pronounced dependency on external Ca^{2+}, implying that the channels carrying at least part of the photocurrent are more permeable to Ca^{2+} than Na^+. The estimated selectivity ratio ranges from 25:1 to 40:1 [96,97].

The possibility that light-dependent channels may be permeable to Ca^{2+} is of considerable importance because of its well-known involvement in the modulation of light sensitivity and its key participation in the excitatory process as well (see Sections 5 and 6). Oddly, no photocurrent could be demonstrated in an extracellular medium containing Ca^{2+} as the sole putative permeant species [97], a result that was attributed to build-up of cytosolic Ca^{2+} and desensitization of the light response. In an attempt to separate permeation from photoresponse inhibition, this issue was re-examined in isolated molluskan photoreceptors, adopting several strategies designed to minimize impairments to cytosolic Ca^{2+} homeostasis and loss of responsiveness that would otherwise result from the required ionic manipulations [93]: (i) Internal dialysis with Na-free solutions, to prevent reverse operation of the Na/Ca exchanger. (ii) Rapid solution changes, temporally limiting exposure to potentially detrimental ionic conditions. (iii) Single-channel recording, exposing only the cell-attached patch of membrane to the test solutions. With these precautions, an inward whole-cell photocurrent could indeed be measured with Ca^{2+} as the only extracellular charge carrier; the magnitude of this current and its reversal potential exhibited a clear dependency on $[Ca^{2+}]_o$, especially if intracellular permeant cations were also removed. In all cases, however, the current carried by Ca^{2+} was <5% of that measured with normal $[Na]_o$. Unitary light-activated currents were reduced in a similar way when the pipette contained only divalent cations, again indicating a substantial selectivity for Na^+ over Ca^{2+}. One must conclude that the degree of Ca^{2+} influx through light-activated channels is subject to significant interspecies differences. In this respect, it is noteworthy that a marked rise in $[Ca^{2+}]_i$ accompanies the light response, as initially established by Brown and Blinks [98]; however, the extent to which Ca^{2+} can be released from internal stores varies greatly across species, in a manner seemingly complementary to Ca^{2+} influx (see Sections 5.1, below): in *Limulus* ventral photoreceptors, where a massive light-induced internal release is observed, permeation of Ca^{2+} through light-dependent channels appears to be minute; by contrast, *Balanus* and *Drosophila* lack a conspicuous release of Ca^{2+} from intracellular stores, but display a much greater influx [98–100].

4.3. *Brief channel openings underlie the light-activated conductance*

The elementary properties of the light-activated conductance have been a key issue since the early stages of development of the field. Like in other systems, initial assessments necessarily had to resort to an indirect approach such as noise analysis, which is necessarily subject to various assumptions. Using fluctuation analysis, Wong [101] estimated the unitary conductance of light-dependent channels in *Limulus* ventral photoreceptors at 18 pS, with a mean open time of 19 ms.

While estimates derived from fluctuation analysis can provide an approximate view of the basic characteristics of elementary events underlying the photocurrent, the importance of obtaining a direct measurement of the unitary currents cannot be overstated. Dissection of light-dependent mechanisms solely relying on macroscopic measurements is problematic, due to the lack of ionic selectivity of the channels and the unavailability of pharmacological blockers. Single-channel recording largely overcomes these limitations and may provide, in principle, a tool to assess directly the action of putative modulators/activators of the current.

A major difficulty for the application of patch-electrode recording to microvillar photoreceptors is the inaccessibility of the light-sensitive membrane, either because it is buried in the inner core of the tightly packed ommatidial cluster (e.g. in compound eyes) or it is covered by a layer of glia. A breakthrough resulted when Stern et al. [102] succeeded in painstakingly removing the glial cells that encase individual photoreceptors in the ventral nerve of *Limulus*, thus exposing the plasma membrane; although the microvilli could not be directly visualized in a living preparation under a light microscope, the rhabdomeral lobe was readily identified functionally by its high light-sensitivity, using a microspot of light. This denuded-cell preparation was exploited by Bacigalupo and Lisman [103], to carry out on-cell patch clamp recording. Establishment of giga-ohm seals was facilitated by prior mild sonication of the cells, presumed to smooth out the convoluted geometry of the membrane surface. In subsequent studies, however, omission of this procedure apparently did not compromise seal formation (e.g. [104]). In patches of microvillar membrane, channels whose activity was triggered by illumination were found, with a single-channel conductance ranging from 35 to 50 pS; the lack of responsiveness to depolarizing potentials applied in the dark argues against the possibility that the activity was just a secondary consequence of the depolarizing receptor potential.

Identification of the sampled channels as the light-activated conductance requires establishing a close correspondence between the properties of the unitary light-induced currents and those of the macroscopic photoresponse. Bacigalupo and Lisman [103] demonstrated that light-evoked single-channel activity was graded with light intensity, was reduced by light adaptation, and the reversal potential agreed with that of the macroscopic photocurrent. The most salient discrepancy concerned response latency, which typically ranged from a few hundred milliseconds to several seconds (compared to tens of milliseconds for the whole-cell photocurrent). This anomaly most likely stems from membrane deformation by the applied suction, resulting in a bleb of membrane being drawn into the electrode [105–107]. This geometrical distortion may impose a long lag on the activation of ionic channels by a diffusible internal messenger. Recordings of light-activated single-channel currents in *Limulus* were subsequently also obtained by Nagy and Stieve [108,109], who reported that the largest unitary conductance was 30 pS; the lower value in part reflects the use of a lower temperature during recording (14°C vs room temperature).

Besides *Limulus*, the only other species in which light-dependent single channels have been recorded systematically in microvillar cells is the scallop, *Pecten irradians* [110]. Large numbers of isolated, intact photoreceptors are readily obtained by enzymatic dissociation of the retina; the plasma membrane is exposed, thus requiring no mechanical stripping of glia. Single-channel currents, specifically activated by light, can be recorded in cell-attached patches and are almost exclusively confined to the microvillar region, where their density is sufficiently high to account for the macroscopic light evoked current. Rarely a patch fails to respond to light stimulation, and frequently too many channels are present to permit resolution of unitary currents (Fig. 6). If one considers the amplitude of the whole-cell photocurrent (several nA at the peak) and the size of the phototransducing region, then on the basis of the measured unitary currents and fractional open time (2 pA and 0.1 ns, respectively) one would estimate as many as 15–20 channels in the small membrane area spanned by the electrode tip, in good agreement with observations.

Light-dependent channel openings are very brief, on average 1 ms or less at 20–22°C; these rapid kinetics are not the consequence of flicker block by extracellular divalent cations, as mean open times are not lengthened by removal of Ca^{2+} and Mg^{2+} from the pipette solution. The mean open time duration does not change substantially with stimulus intensity, consistent with a gating scheme involving a positive internal transmitter [111,112]. The envelope of the channel openings resembles the time-course of the macroscopic response, and the latency to the first opening is comparable to that of the photocurrent, provided that care is taken to maintain low levels of suction (< 2–3 cm H_2O) in order to avoid distortion of the membrane. Unitary currents are inward at resting potential, and have a reversal voltage similar to that of the macroscopic light response. Voltage modulates the activity of light-sensitive channels by increasing the opening rate and also by lengthening the mean open times as the patch is depolarized. The predominant

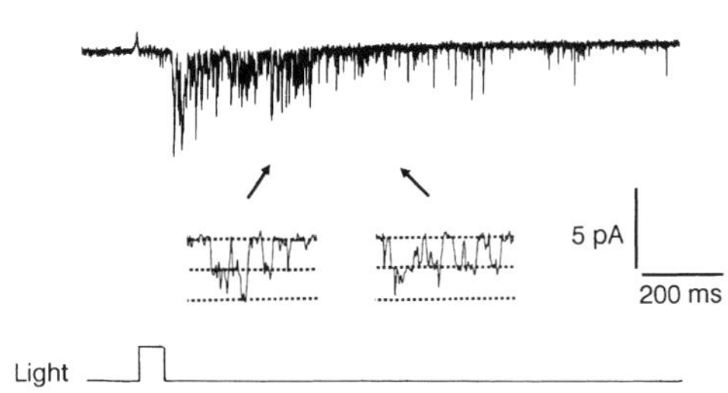

Fig. 6. Examples of light-dependent channel activity recorded from a cell-attached patch pipette placed on the microvillar lobe of a scallop (*Pecten*) photoreceptor following a flash of light. Insets show channel activity at an amplified timescale (for details see [110]).

unitary conductance is approximately 48 pS, rather similar to that reported in *Limulus* by Bacigalupo and Lisman [103].

Channel activation is graded with stimulus intensity; the dynamic range is comparable to that of the whole-cell response, and extends to extremely low levels of illumination, which only elicit quantum bumps in the macroscopic photocurrent. Under such conditions, channel openings are clustered in brief bursts, as one would expect, although considerably less frequent than the total rate of discrete waves produced by the cell. The discrepancy may be explained by the localization of single-photon responses to the region near the isomerized rhodopsin molecule [13], so that only those bumps originating in the vicinity of the patch electrode may be recorded. The possibility of extending the measurements of unitary currents to threshold light levels has an interesting implication. Bacigalupo and Lisman [103] observed that the open times of light-sensitive channels in *Limulus* were brief, and concluded that the rate-limiting step in the relaxation kinetics of quantum bumps must be the time-course of the internal transmitter activity, rather than the rate of channel closure. Such an inference was criticized by Dirnberger et al. [113] because bright stimuli were employed in those measurements, and light adaptation is known to accelerate the time-course of discrete waves [81]. The observation of millisecond open time durations with extremely dim illumination in *Pecten* is incompatible with the notion that the rate of channel closure accounts for the relaxation of discrete waves.

4.4. Multiple light-dependent conductances are present in some photoreceptors

Up to this point, an implicit assumption has been that light stimulation only affects one membrane conductance in microvillar photoreceptors. This supposition was justified not only by parsimony, but also, because Millecchia and Mauro [52] reported that the reversal potentials measured at various times during the photocurrent coincided. However, early clues suggestive of a more complex situation appeared when Lisman and Brown [115] reported that, using prolonged steps of light, they were unable to find a holding voltage at which no current was elicited. Such a result suggested that more than one ionic mechanism was altered by light: the first one, which they called the "fast process", displayed rapid activation and deactivation kinetics, a reversal potential around +20 mV and a positive slope resistance at all voltages tested. The second one, with slower kinetics, smaller amplitude and negative slope resistance throughout most of the voltage range, was labeled "slow process". Further studies [116], showed that the "slow process" was not directly triggered by light. Rather, it reflected the reduction by photostimulation of a sustained potassium current sensitive to tetraethylammonium ions (TEA) and activated by depolarization. Subsequently, it was discovered that such a modulatory effect was mediated by light-induced changes in cytosolic Ca^{2+} concentration [117]. It was proposed that this mechanism serves to stabilize the plateau of the photocurrent during repetitive stimulation, compensating for the progressive decrease in the light-activated conductance. Although this phenomenon does not entail an additional class of channels genuinely gated by light stimulation, it was the first clue to point to the possibility of multiple effects of light stimulation on ion conductances.

Subsequent observations obtained in various species indicated that light could indeed control more than one membrane conductance not gated by voltage changes. For example, the complex kinetics of the light response in *Hermissenda* photoreceptors had been initially attributed to synaptic interactions [118,119]; however, Detwiler [90] later demonstrated that axotomy fails to eliminate the multiphasic receptor potential, and characterized three discernible components of the photoresponse, with different kinetics and light sensitivity. In the presence of dim lights, cells responded with a small depolarization (G1); stronger lights elicited a second depolarizing tail (G3), and with the brightest lights the two depolarizing humps were separated by a hyperpolarization (G2). Evidence that G2 and G3 are not simply the result of membrane potential changes during the first component was obtained by mimicking G1 with depolarizing currents applied in the dark, and showing that no additional active components were elicited.

In *Limulus* photoreceptors the receptor potential (i.e. "fast process" in the nomenclature of Lisman and Brown [115]) frequently reveals complex kinetics, with a distinct notch or inflection point; this feature prompted a subdivision into two components, C1 and C2, both independent of voltage stimulation [120,121]. The appearance of these distinct depolarizing "waves", is influenced by the state of light adaptation of the cells. C1 is less prone to desensitization and becomes the predominant component when light flashes are delivered at high frequency of repetition or in the presence of an intense background light. By optimizing stimulation protocols in order to make either component dominant, their respective I–V relations could be measured, revealing a considerably greater steepness for C2, but no systematic differences in their reversal potentials. The possibility of separate light-dependent conductances was entertained, but the inability to distinguish the various kinetic components of the photoresponse on the basis of conductance properties initially prompted Maaz et al. [121] to lean towards the notion that C1 and C2 may result from a single class of light-dependent channels subject to different control mechanisms for gating and adaptation. A similar model had also been considered earlier by Lisman and Strong [122].

More compelling evidence supporting the existence of more than one light-dependent ionic conductance was obtained in isolated microvillar photoreceptors from *Lima scabra* (Fig. 7) [123,124]. Under current-clamp these cells produce a complex photoresponse reminiscent of that of *Hermissenda* [90] and *Strombus* [92]. This consists of an initial depolarizing phase (the main component under dark-adapted conditions) and, with higher stimulus intensities, a second more sluggish depolarizing wave preceded by a hyperpolarizing "dip". Recording under whole-cell clamp to prevent confounding by voltage-dependent mechanisms [124] showed that near the resting potential a single transient inward current was elicited by a flash, provided that the intensity was low or moderate. By contrast, with brighter lights an additional wave of inward current appeared after an inflection point. The late component of the photocurrent in *Lima* was less susceptible to light adaptation by repetitive stimulation, reminiscent of the differential desensitization of C1 and C2 in *Limulus* [121]. Possible contamination by Ca^{2+}-dependent K^+ channels present in these cells [125] was ruled out by showing that the complex

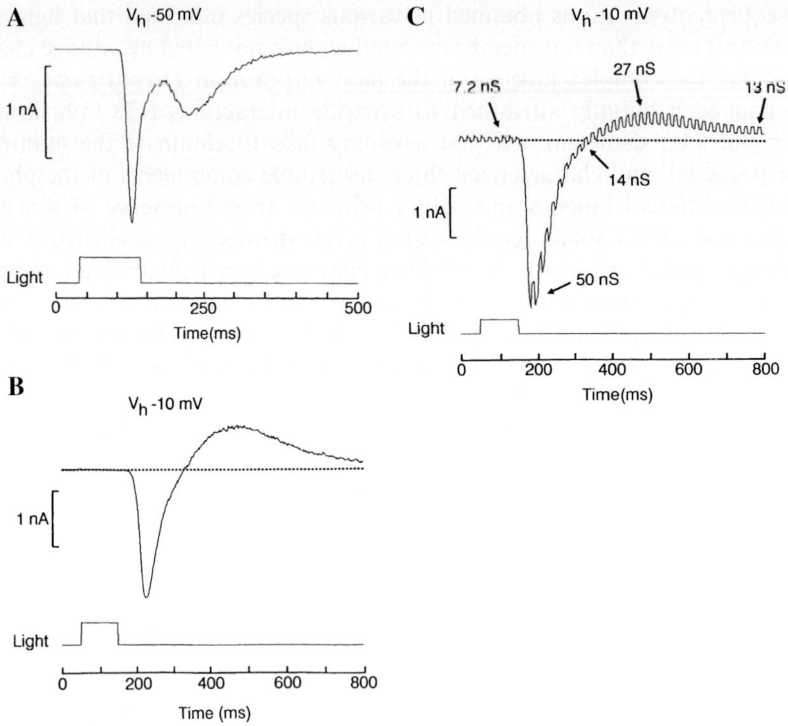

Fig. 7. Complex time-course of the light-dependent current recorded in file clam (*Lima*) photoreceptors at holding potentials of −50 (A) and −10 mV (B), the latter trace showing a biphasic response indicating the lack of a unique reversal potential for the photocurrent. Superimposed voltage perturbations (C) show that both phases of the photocurrent recorded at −10 mV are associated with an increase in membrane conductance.

time-course of the photocurrent was retained after complete pharmacological blockage of outward currents by internal Cs or by TEA. Each of the two light-evoked waves of inward current was accompanied by a distinct increase in membrane conductance, and was differentially affected by changes in the holding voltage. In particular, measurements of the reversal potential of the light response further supported the presence of separate channel populations. With dim flashes, which only elicited the first transient, the photocurrent had a single reversal potential, slightly positive of 0 mV. In contrast, bright light stimulation evoked responses that over a certain range of holding voltages had a clear biphasic time course (Fig. 7), characterized by a second wave with a more negative reversal potential. This indicates that the ionic selectivity for the two components must be different.

Multiple reversal potentials of the light response were subsequently shown in a detailed voltage-clamp study re-examining the various kinetic components of the photocurrent in *Limulus* ventral photoreceptors [126]. The small first component,

C1, was the first to recover after light adaptation. The second component, C2, needed a longer time to recover after an adapting illumination and dominated the peak response to flashes of moderate intensity. Applying a bright stimulus to a dark-adapted cell resulted in a third component, C3, late in the response. Fig. 17B provides an example of these components and their differing sensitivity to pharmacological intervention. Demonstrating the reversal potentials of the components required applying small perturbations of the holding voltage, at 1 mV increments, in the vicinity of +10 mV. In this way, a clearly multiphasic photocurrent was revealed, with C1 displaying a more positive reversal voltage than C2, which in turn inverted before C3. The closeness of the individual reversal voltages probably accounts for previous failure to observe this phenomenon. Similarly, in *Drosophila* photoreceptors, where a unique reversal potential had initially been reported [96], subsequent measurements in low extracellular Ca^{2+} unequivocally demonstrated separate components, with distinct reversal potentials and a biphasic light-induced current at certain holding voltages [97].

The coexistence of different light-dependent conductances in the plasma membrane of microvillar photoreceptors should also manifest itself in single-channel recordings. The presence of an additional class of smaller unitary light-dependent currents has been detected since the first report by Bacigalupo and Lisman [103] in *Limulus* ventral photoreceptors, but these relatively infrequent events were not analyzed further. Subsequently, Nagy and Stieve [108] described up to three different sizes of unitary currents activated by light in the same preparation, with unitary conductances clustering around 6, 10, and 29 pS (at 14°C). On the basis of their different activation latencies, the authors concluded that these reflected separate classes of ion channel proteins, and suggested that distinct biochemical pathways and internal messengers may control their gating [109]. Johnson et al. [127] later confirmed that in *Limulus* at least two classes of events were present, and determined that their respective unitary conductances were around 15–18 and 43 pS (room temperature). However, their analysis favored a single class of ion channel proteins with subconductance states. They proposed that a transition between different "modes" of gating, from the large to the small conductance state, may be responsible at least in part for the decay of the photocurrent during the late phase of light adaptation.

In *Pecten* microvillar photoreceptors, two sizes of light-evoked unitary currents were also observed in cell-attached patches [110], with single-channel conductances of 48 and 18 pS, respectively. These appeared to occur in an apparently independent manner, but the sample sizes and the presence of more than one channel per patch precluded a statistical analysis of possible modal shifts. Instead, the relative incidence of large and small events was compared in dark-adapted vs background-adapted state (after matching the effective intensity of the test light). The results, however, failed to reveal any systematic trend that may indicate an adaptation-related switch in gating mode.

One must point out a limitation shared by all the studies reported so far, which hinders a more rigorous examination: in cell-attached recordings the photoreceptors are not held under voltage clamp, so that the receptor potential produces two unfortunate consequences: (i) the driving force transiently changes, altering the amplitude of channel events and rendering classification of unitary current sizes difficult; (ii) voltage changes *per se* may foster the appearance of one or another type of opening during different phases of the response. Under these circumstances, only the activity recorded either during the plateau phase of the response or after membrane repolarization can be analyzed with confidence, but these events may not be representative of the mechanisms underlying the early transient phase of the light response.

In summary, not only do light-dependent conductances in invertebrate photoreceptors of the microvillar type exhibit a significant diversity, but also accumulated evidence strongly supports the notion that different types of channels can contribute to the light response in a given cell. This makes phototransduction in these organisms a more complicated process than initially thought; the proposition that the various light-dependent ionic mechanisms may be under the control of parallel enzymatic pathways triggered by light is one of the ideas that has emerged.

Opposite: Fig. 8. (A) Representation of light-induced Ca^{2+} elevation within a *Limulus* ventral photoreceptor cell. The cell was filled through a micropipette with Oregon Green 5N, a fluorescent Ca^{2+}-indicator dye, and exposed to a step of 488 nm light from a laser which scanned along the line drawn over the image of the cell, passing through the light-sensitive rhabdomeral (R-)lobe, the arhabdomeral (A-)lobe and the axon. As the laser scanned along the line every 2 ms, the fluorescence of the dye was sampled confocally. Following the onset of laser scanning, the Oregon Green-5N fluorescence within the R-lobe (inset graph) increased with time after a delay of a few tens of ms (bottom left-hand corner of graph), indicating a delayed rise in intracellular Ca^{2+} concentration due to the activation of phototransduction by the laser light. As scanning continued, the fluorescence increase spread from the edge of the R-lobe (left-hand side of graph) towards the A-lobe (right-hand side of graph) and Ca^{2+} diffused from release sites beneath the plasma membrane into the rest of the cell (for details see [137]). (B) Magnitude of Ca^{2+} elevation recorded during illumination of another *Limulus* ventral photoreceptor filled with the Ca^{2+}-indicator dye Calcium Green 5N and illuminated by a 488 nm laser. A stationary laser spot was placed as close as possible to the microvillar plasma membrane of the R-lobe of the photoreceptor and dye fluorescence was measured confocally at the same spot (for details see [137]). (C) Time-course of rising phase of fluorescence (dots), indicating elevation of intracellular Ca^{2+} concentration, and photocurrent (solid line) in a *Limulus* ventral photoreceptor filled with the Ca^{2+}-indicator dye fluo-3 during illumination by a 488 nm laser at a spot close to the microvillar membrane, showing detection of Ca^{2+} increase (open arrow) before the photocurrent (filled arrow). Contrary to the usual convention, inward current is shown as an upward deflection of the trace so as to facilitate the comparison with the fluorescence trace (for details see [137]). (D) Ca^{2+} elevation estimated from the fluorescence of indo-1 during bright UV illumination of an isolated *Drosophila* photoreceptor cell (for details see [128]). (E) Photocurrent (I_m) and (F) light-induced Ca^{2+} increase measured from the microvillar lobe of an isolated *Lima* photoreceptor loaded with Calcium Green 5N and voltage clamped at −50 mV (for details see [133]).

5. Biochemical events underlying excitation and adaptation in microvillar photoreceptors

5.1. An elevation of intracellular free Ca^{2+} ion concentration ($[Ca^{2+}]_i$) accompanies the light response

The intracellular free Ca^{2+} ion concentration ($[Ca^{2+}]_i$) of microvillar photoreceptors in darkness is less than 1 μM. Estimates of 160 nM have been made in *Drosophila* [128], 300 nM in *Balanus* [129] and 400–700 nM in *Limulus* ventral photoreceptors [130,131]. By contrast, resting $[Mg^{2+}]_i$ levels of approximately 2 mM have been reported in *Limulus* ventral photoreceptors and *Balanus* photoreceptors [131,132]. The light response of all microvillar photoreceptors so far examined is accompanied by an elevation of $[Ca^{2+}]_i$ (Fig. 8). Substantial release of Ca^{2+} from internal Ca^{2+} stores has been demonstrated in the *Limulus* ventral eye and *Lima* photoreceptors [98,132,133], while in the photoreceptors of *Balanus* and

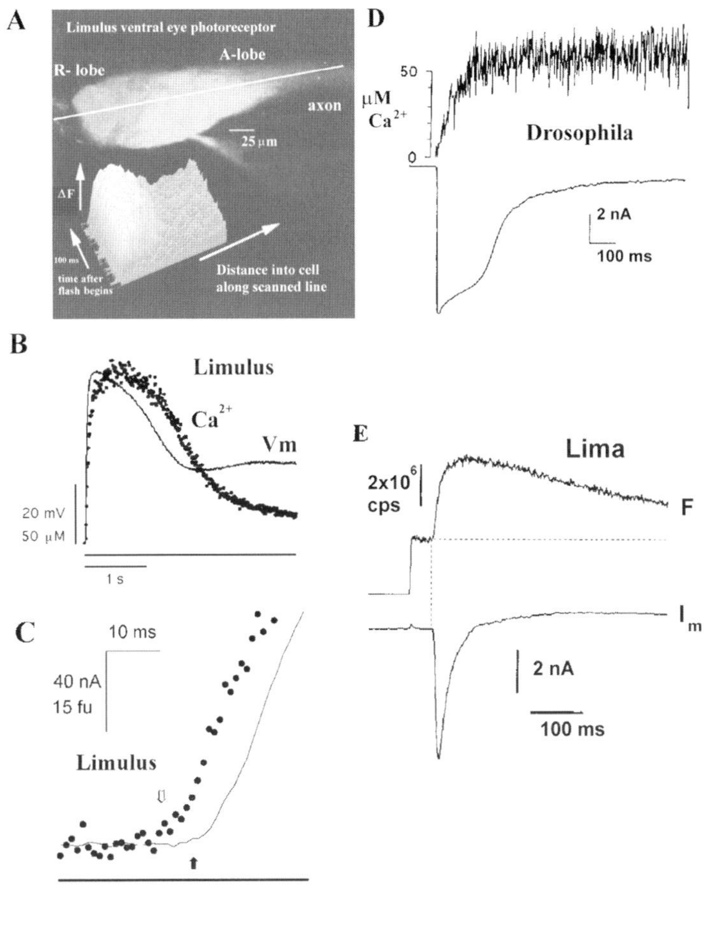

Drosophila, extracellular Ca^{2+} enters mostly through ion channels in the plasma membrane [98–100]. In the photoreceptors of the honeybee [134,135] a combination of both release and influx has been demonstrated by measurement of extracellular and intracellular ion concentrations. The source of the released Ca^{2+} is most likely the SMC. Electron-probe microanalysis of the SMC of bee photoreceptors has demonstrated a release of Ca^{2+} upon illumination [136].

In *Drosophila*, *Calliphora* and *Limulus* ventral photoreceptors, where the elevation of $[Ca^{2+}]_i$ has been measured within a few micrometers of the microvillar membrane, the initial elevation of $[Ca^{2+}]_i$ following a saturating flash rises to tens or even hundreds of micromolar within less than a second (Fig. 8) [137–139]. In fly photoreceptors where Ca^{2+} is a major carrier of the photocurrent, this rapid rise is not surprising and is presumably even greater at the mouths of the light-sensitive channels. However, even in *Limulus* ventral photoreceptors, where Ca^{2+} is primarily released from intracellular stores, the rise is just as spectacular [140,141]. Using confocal imaging and fluorescent Ca^{2+} indicators, light-induced elevation of $[Ca^{2+}]_i$ within a few μm of the microvillar membrane in *Limulus* ventral photoreceptors rises in response to saturating light flashes at an initial rate of 1–2 mM/s [137], reaching a peak of $\sim$150 μM within 500 ms following a latent period of only 14–40 ms. At least 600 Ca^{2+} ions appear to be released per effective photon [131]. These high Ca^{2+} concentrations are thought to be attainable only in the neighbourhood of Ca^{2+} channels, indicating a close proximity of the Ca^{2+} release sites to the measurement spots underneath the microvillar membrane.

Ca^{2+} elevations in all species are graded in amplitude with light intensity, and in *Limulus* ventral and *Calliphora* photoreceptors are clearly subject to adaptation, in so far as prior illumination results in a reduction in sensitivity of the elevation of $[Ca^{2+}]_i$ to light (Fig. 9) [139]. This adaptation also results in a decline in $[Ca^{2+}]_i$ during sustained bright illumination (Fig. 8B) [98,131,139,141].

5.2. *Light adaptation is mediated by the light-induced elevation of* $[Ca^{2+}]_i$

The light-induced elevation of $[Ca^{2+}]_i$ appears to function as the mediator of the feedback that is responsible for reducing the sensitivity and latency of the visual cascade during prolonged bright illumination [142–144]. Injection of Ca^{2+} ions into the cytosol of *Limulus* ventral photoreceptors diminishes the sensitivity and latency of the light response (Fig. 10A), mimicking the effect of an adapting light [144,145]. Injection of Ca^{2+} chelators such as EGTA or BAPTA blocks the decline of the light-induced current that results from light adaptation during prolonged illumination [142] and slows the light response (Fig. 10B). Since the process that determines the latency of quantum bumps is independent of that which determines their amplitude (see above), it is probable that Ca^{2+} acts at separate sites to reduce the response latency and sensitivity. Reduction of the extracellular calcium concentration greatly increases the latency of quantum bumps in dark-adapted *Limulus* ventral photoreceptors, without appreciable effect on bump amplitude, probably as a result of reduced $[Ca^{2+}]_i$ [66,146].

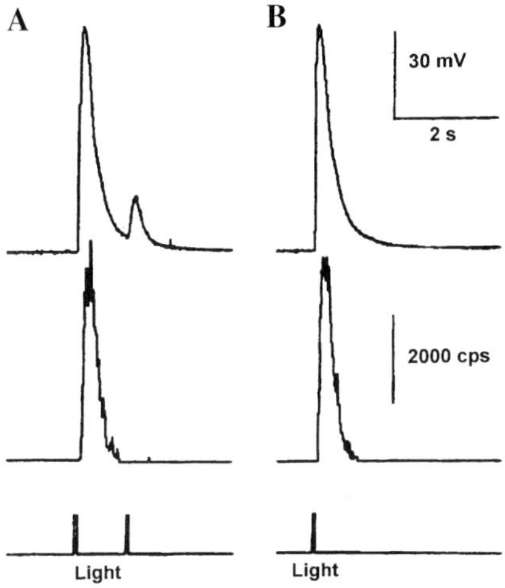

Fig. 9. Adaptation of light-induced depolarization and Ca^{2+} release in *Limulus* ventral photoreceptors. (A) Membrane potential (top trace) and luminescence of the Ca^{2+}-sensitive photoprotein aequorin (middle trace; indicating intracellular Ca^{2+} concentration) following two consecutive light flashes (bottom trace). Adaptation greatly reduces the depolarization and Ca^{2+} elevation produced in response to the second flash. (B) Recovery from adaptation of light-induced depolarization (top trace) and Ca^{2+} elevation (middle trace), recorded 100 s after the traces in (A) (for experimental details see [243]).

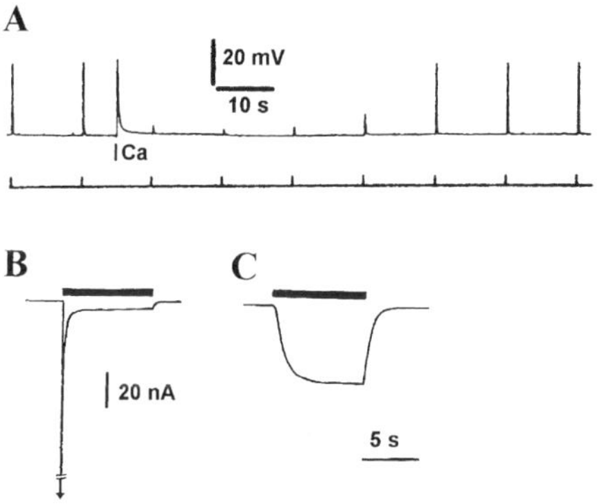

Fig. 10. (A) Adaptation of light response resulting from injection of Ca^{2+} into a *Limulus* ventral photoreceptor. Flashes of light (bottom trace) were delivered to ventral photoreceptors while recording membrane potential (upper trace) through a micropipette that also contained 2 mM calcium aspartate. At the vertical bar, pressure was briefly applied to the micropipette, expelling Ca^{2+}-containing solution into the rhabomeral lobe of the photoreceptor. The injection of Ca^{2+} resulted in a transient depolarization followed by a profound, reversible desensitization of the responses to subsequent light flashes (after [234]). (B,C) Blockade of adaptation resulting from injection of the Ca^{2+}-chelator BAPTA into a *Limulus* ventral photoreceptor. (B) Photocurrent recorded in response to a flash of light prior to injection of EGTA shows a typical large initial transient response, which saturates the recording apparatus (break in trace) followed by rapid adaptation to a plateau of a few nA. (C) Pressure injection of 20 mM CaBAPTA: 20 mM K_2BAPTA abolishes the initial transient, leaving a response that rises slowly to a larger final without adaptation.

5.3. *A light-activated phosphoinositide (PI) pathway operates in microvillar photoreceptors*

The phosphoinositide (PI) pathway is a mechanism for releasing intracellular messengers upon the activation of a receptor protein, using inositol phospholipids as a substrate. Because activation of the PI pathway elevates $[Ca^{2+}]_i$ in most cells, it is an attractive candidate for the visual cascade in microvillar photoreceptors. A GTP-binding protein mediates the response to the activated receptor (Fig. 11). In the case of photoreceptors, the receptor protein is rhodopsin (Rh). Activated Rh catalyzes the exchange of GTP for GDP bound to the alpha subunit of a heterotrimeric GTP-binding protein of the G_q sub-family, which in turn activates phospholipase C (PLC). PLC cleaves phosphatidylinositol (4,5) bisphosphate (PIP_2), a minor membrane phospholipid, into a lipid messenger, diacylglycerol (DAG), and the water-soluble messenger, D-myo-inositol 1,4,5 trisphosphate ($InsP_3$) [147]. The major known target of DAG is protein kinase C (PKC) [148]. DAG also can be metabolized into fatty acids, which might serve as downstream messengers. The best-characterized target of $InsP_3$ is a Ca^{2+} channel, the $InsP_3$ receptor protein ($InsP_3R$), located in the membrane of endoplasmic reticulum [149]. $InsP_3$ therefore releases Ca^{2+} from intracellular stores. $InsP_3$ has also been shown to activate channels in the plasma membrane of some cells [150–152]. Like many signal transduction cascades, the PI pathway is capable of great signal amplification. One receptor may activate several GTP-binding proteins, many PIP_2 molecules may be hydrolyzed during the active lifetime of one PLC molecule and thousands of Ca^{2+} ions may enter the cytosol through a single opening of the $InsP_3R$ [153].

Light-activated PIP_2 hydrolysis and/or $InsP_3$ production has been reported in photoreceptors of cephalopods, *Drosophila* and *Limulus* [15,154–157]. The biochemistry of microvillar photoreceptors has been much informed by the use of cephalopod eyes. The microvillar outer segments of the photoreceptors in the retinae of these eyes form a continuous sheet of tissue which can be peeled away from the underlying inner segments and supporting cells in dark-adapted eyes, resulting in a membrane preparation that is highly enriched in microvilli. In squid outer segments, PIP_2 has been estimated as constituting 3% of the total phospholipid in the

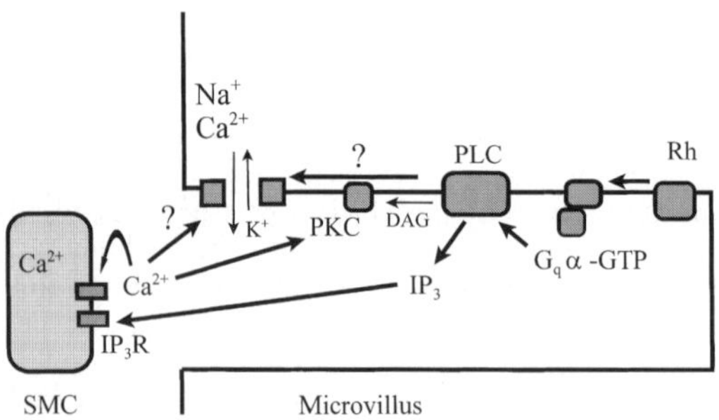

Fig. 11. Outline of a microvillus showing the elements of the PI cascade thought to be present. For abbreviations and explanation, see text.

microvillar membrane, 3–4 PIP_2 molecules per Rh [15]. Between 50 and 500 molecules of $InsP_3$ are produced per photoisomerization for flashes that photoisomerize 0.1–1% of the visual pigment, i.e. 1–10 Rh molecules per microvillus. Because of its tiny volume, light-induced elevations of $InsP_3$ may reach 100 μM–1 mM within a microvillus [15]. The enzymes responsible for light-activated PIP_2 hydrolysis have been extensively characterized. Alpha subunits of GTP-binding proteins of the G_q sub-family present in *Limulus*, squid, octopus and *Drosophila* photoreceptors have been sequenced or partially sequenced and in many cases immunolocalized to the microvilli [158–162]. In crayfish photoreceptors (Fig. 12), G_q-α has been identified immunologically [163,164], and appears to bind to the rhabdomeral membrane in the dark, but migrates to the cytoplasm in the light, possibly due to modification by fatty acid [163,164]. A soluble form of G_q-α has also been isolated from squid photoreceptors, but it cannot activate PLC [165]. Light-induced migration of G_q-α from the microvilli to the cytosol in crayfish photoreceptors might therefore reduce sensitivity to light during light adaptation. A ratio of G_q-α:Rh of about 1:12 has been estimated in the microvillar membrane of squid photoreceptors [166], comparable to that of the ratio of the G-protein transducin and Rh in the vertebrate rod outer segment. Interaction of photoisomerized rhodopsin, metarhodopsin, with G_q-α can be demonstrated by measuring light-activated GTP-ase activity resulting from the turnover of active G_q-α, by measuring light-activated binding of GTP-γS to G_q-α or by immunoprecipitation of G_q-α bound to metarhodopsin. These methods have been successfully applied to fly, cephalopod and crayfish photoreceptors [160,163,166–169]. In squid photoreceptors, each light-activated (meta)rhodopsin molecule has been shown to activate about 10 G_q-α [166].

β and γ GTP-binding protein subunits found in squid microvillar membrane have been sequenced [170,171], as has an eye-specific G-β subunit from *Drosophila* [172]. The G-β subunit of squid photoreceptors does not appear to be able to activate PLC [165], but the G-β subunit of *Drosophila* photoreceptors is necessary

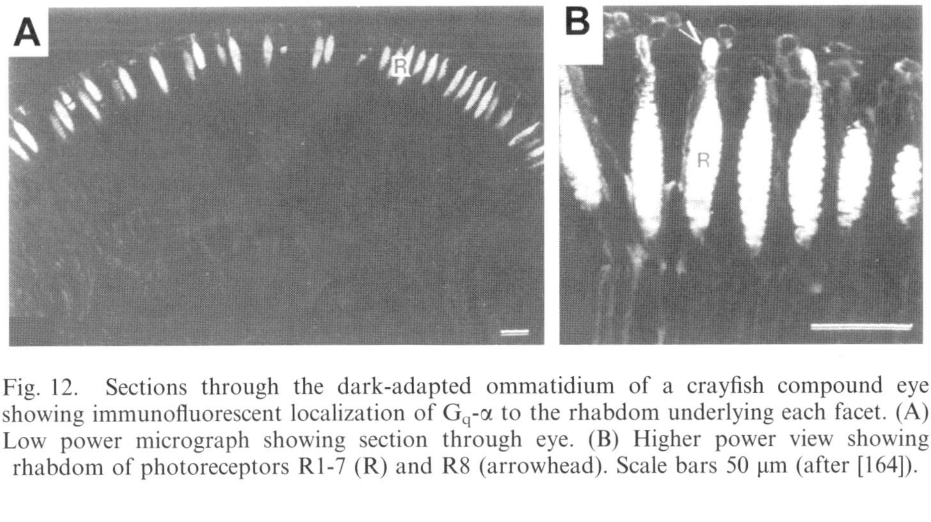

Fig. 12. Sections through the dark-adapted ommatidium of a crayfish compound eye showing immunofluorescent localization of G_q-α to the rhabdom underlying each facet. (A) Low power micrograph showing section through eye. (B) Higher power view showing rhabdom of photoreceptors R1-7 (R) and R8 (arrowhead). Scale bars 50 μm (after [164]).

for phototransduction, perhaps because it localizes G_q-α correctly [172]. In crayfish photoreceptors, a homolog of G-β undergoes translocation from rhabdomeral membrane to cytosol, similar to that of G_q-α [173].

PLC has been purified and biochemically assayed in squid, octopus and *Drosophila* photoreceptors. A PLC related to the mammalian β subfamily, which runs on SDS-PAGE gels with apparent MW 120,000, has been sequenced from squid photoreceptive membranes [174]. Other cephalopod preparations have yielded PLCs with similar apparent MWs [165,175,176]. *Drosophila* photoreceptor microvilli contain a homologous PLC of the β subfamily, NORPA, which is necessary for vision [177,178]. The full pathway leading to activation of a PLC has been functionally re-constituted from purified components derived from cephalopod photoreceptor membranes [160,176,179]. In cephalopod and *Drosophila* membranes, PLC activity is stimulated as Ca^{2+} ion concentration rises from 0.1 to 1 μM and, in cephalopods, is inhibited above 10 μM [175,176,179–181]. Since these concentrations are well within the physiological range found in other microvillar photoreceptors, it is possible that light-induced elevation of $[Ca^{2+}]_i$ feeds back to both positively and negatively regulate PLC activity, providing another possible biochemical site for the role of calcium in light adaptation.

5.4. Biochemical characterization of downstream targets of the PI pathway: Protein kinase C (PKC), the inositol trisphosphate receptor protein ($InsP_3R$) and calmodulin

Of the downstream targets of the phosphoinositide pathway, PKC, which is activated by both Ca^{2+} and DAG, has so far been thoroughly characterized only in the microvillar membrane of photoreceptors in the compound eyes of the flies *Drosophila* and *Calliphora*, where it may play an important role in light adaptation and response termination [182–185]. Antibodies to mammalian PKC isoforms do label *Limulus* lateral eye microvillar membrane, but the protein(s) to which the antibodies are binding has not been characterized on Western blots [186]. There is a similar lack of information concerning the $InsP_3R$. An immunological study in *Limulus* lateral eye photoreceptors indicates that a putative $InsP_3$ receptor with MW approx. 250,000 is present in the cytoplasm that underlies the microvillar membrane, the region that also contains the SMC, but is not present in the microvilli themselves (Fig. 13) [187]. In *Drosophila* photoreceptors, the $InsP_3R$ has been immunolocalized to perinuclear ER, but not the sub-microvillar region [188], consistent with the diminished role of Ca^{2+} release compared to Ca^{2+} influx in these cells.

Another prominent target for the action of Ca^{2+} is calmodulin, which is abundant in the microvilli, being bound to a number of proteins whose activity or presence modulates the visual cascade, including unconventional myosins in *Drosophila* (ninaC) [33] and *Limulus* [34], and the TRPL channel [189] and INAD complex in *Drosophila* [37]. In *Drosophila*, Ca^{2+}–calmodulin regulates the opening of TRPL channels and activates a kinase that phosphorylates arrestin (see Section 5.6). In *Limulus* ventral photoreceptors, Ca^{2+}–calmodulin activates a phophatase,

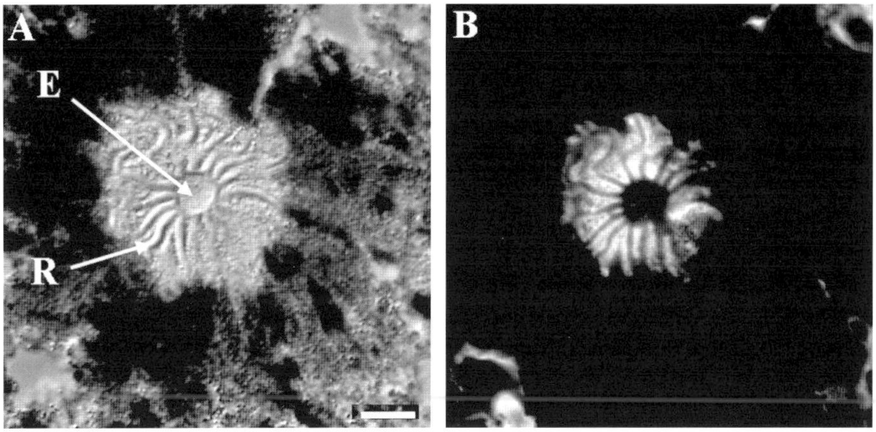

Fig. 13. Immunolocalization of putative $InsP_3$ receptor in cryosections of *Limulus* lateral eye photoreceptors. (A) Nomarski contrast micrograph of the cross section of an ommatidium. The star-like rhabdom (R) with the non-photoreceptive eccentric cell (E) at its center is clearly seen in the central area. (B) Cryosection of the lateral eye labeled with a primary rabbit anti-$InsP_3R$ serum and then with secondary FITC-conjugated goat anti-rabbit antibodies. The anti IP_3R antibody labels the cytoplasm in between the rhabdomeres but not within the rhabdomeres forming the star-shaped rhabdom nor that of the central non-photoreceptive eccentric cell (after [187]).

calcineurin [190]. Inhibition of calcineurin enhances arrestin phosphorylation and reduces bump amplitude and time-course.

5.5. Metarhodopsin phosphorylation and arrestin may terminate rhodopsin activity

As with vertebrate photoreceptors, Rh has been shown to become phosphorylated after illumination in cephalopod and fly photoreceptors [191–194]. A rhodopsin kinase has recently been isolated from octopus photoreceptors [195]. In re-constituted vesicles containing purified rhodopsin, the kinase only phosphorylated rhodopsin which had been illuminated. The octopus rhodopsin kinase is more homologous to vertebrate β-adrenergic receptor protein kinase than it is to vertebrate rhodopsin kinase, suggesting independent evolution. Consistent with this homology, the activity of octopus Rh kinase is enhanced, *in vitro*, by the presence of the α-subunits of G_q. The role played by rhodopsin phosphorylation in terminating the response to light is not yet clear. In preparations of squid microvillar membrane, ATP is not absolutely required for termination of light-activated GTP-ase activity [196].

Another event that follows illumination is the binding of a protein, arrestin, to metarhodopsin. Binding of arrestin terminates the ability of rhodopsin to activate G_q. Work on fly photoreceptors indicates that, unlike in vertebrate photoreceptors, binding of arrestin does not seem to depend upon phosphorylation of metarhodopsin [197]. However, little work has been performed on this termination reaction in other species. Invertebrate visual arrestins have been sequenced and localized only in *Limulus*, *Drosophila* and *Calliphora* [197–202]. Arrestin undergoes phosphorylation by a Ca/calmodulin-dependent protein kinase II (CaM-KII) [203–207]. Phosphorylation of arrestin is stimulated by light, as might be expected from the rapid light-induced elevation of $[Ca^{2+}]_i$. Arrestin phosphorylation is necessary for the detachment of *Drosophila* arrestin 2 from the microvillar membrane. Depletion of calmodulin or inhibition of CaM-KII prolongs the response of *Drosophila* and *Limulus* ventral photoreceptors, but the mechanism(s) are unknown [69,208].

5.6. *Evidence that the PI pathway participates in phototransduction in microvillar photoreceptors*

There is abundant evidence that microvillar photoreceptors possess and utilize the enzymes necessary to couple photoisomerized rhodopsin to the PI pathway. However, the question remains as to whether it is this pathway that actually couples Rh to the activation of the light-sensitive channels and how that coupling is achieved. The critical importance of the PI pathway for vision is demonstrated by the profound reduction in electrical sensitivity to light produced by a reduction of PLC expression in the microvilli of mutant *norpA* photoreceptors of *Drosophila* or *rpa* photoreceptors of *Calliphora* [109,154,210] and by genetically engineered reduction of Gq-α in *Drosophila* photoreceptors [211]. A wide body of physiological and pharmacological experiments performed upon other species, which will be reviewed below, complements these experiments in Diptera. Some of these experiments also hint that there may be additional pathways that activate microvillar photoreceptors and in no case is there yet an unequivocal identification of the messengers responsible for activating the light-sensitive channels.

5.7. *Involvement of GTP-binding proteins in generating the electrical response*

In addition to the Gq-α-deficient *Drosophila* result (see above) there is pharmacological evidence supporting a central role of GTP-binding proteins for vision in *Limulus* ventral photoreceptors. Introduction of the GTP-binding protein activator, GTP-γS, into the cytosol of *Limulus* ventral [212,213] and fly photoreceptors [167,214] results in the prolongation of electrical activity following bright light flashes. In *Limulus* ventral photoreceptors, the activity induced by GTP-γS takes the form of quantal events that are about eight times smaller in size than light (rhodopsin)-induced events [215,216], implying that only a handful of GTP-binding proteins are activated by a photoisomerized rhodopsin molecule. Introduction of the GTP-binding protein inhibitor GDP-βS into *Limulus* ventral photoreceptors results in a dramatic reduction in response sensitivity [217,218]. Light-induced bump size and the efficiency with which bumps are generated are progressively reduced following injection of GDP-βS [67,215]. Injection into *Limulus* ventral photoreceptors of antibodies raised against a highly conserved peptide in the sequence of G_q-α greatly reduces the C2 component of the response to a bright flash [219].

5.8. Involvement of PLC in generating the electrical response

In addition to the *norpA Drosophila* result (see above) there is pharmacological evidence supporting a central role of phosphoinositide hydrolysis for vision in *Limulus* and *Hermissenda* photoreceptors. Intracellular injection of the PLC inhibitor neomycin desensitizes the electrical response of *Limulus* ventral [220–223] and *Hermissenda* photoreceptors [224]. In *Limulus* ventral photoreceptors, neomycin injection blocked the transient response to sustained moderate illumination (Fig. 14B) and the C2 component of the response to bright flashes. The plateau phase of the response to sustained illumination [220] and the C1 and C3 components of the response to a bright flash were much less affected by neomycin [223]. Recent experiments by Johnson et al. [225] are also consistent with a requirement for inositol for generating the electrical response of *Limulus* ventral photoreceptors. Lithium is known to inhibit myo-inositol-1-phosphatase [226], preventing the recycling of inositol during prolonged activation of the PI pathway. Consistent with this blockade of inositol metabolism, replacement of half of the extracellular sodium surrounding *Limulus* ventral photoreceptors (normally 425 mM) with lithium

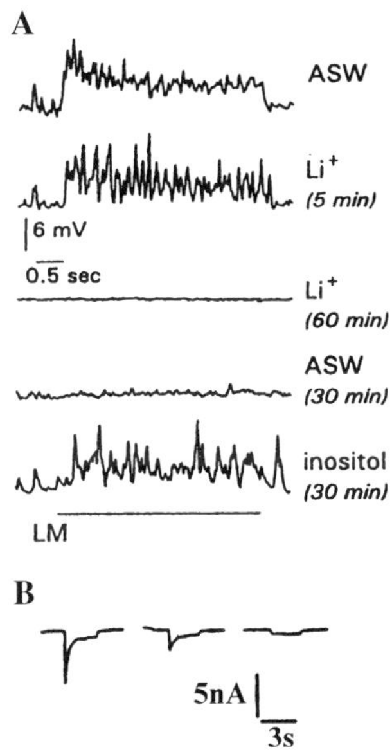

Fig. 14. (A) Effect of extracellular Li^+ on response of *Limulus* ventral photoreceptors to dim light. (traces from top) 5 min perfusion with 213 mM external Li^+ has little effect on quantum bumps but after 60 min quantum bumps are abolished. Subsequent perfusion with artificial seawater (ASW) containing no Li^+ only partially restores sensitivity. Full sensitivity recovers when 10 mM inositol is added to the ASW for 30 min light monitor (LM) (after [225]). (B) Neomycin desensitizes the response of *Limulus* ventral photoreceptors. Photocurrents recorded in response to a dim light flash recorded before neomycin injection (left-hand trace) after 1st (central trace) and 2nd injections (right-hand trace) of 5 mg/ml neomycin into the photoreceptor (after [220]).

resulted in a slowly-developing, but complete, suppression of the responses to dim flashes and of quantum bumps (Fig. 14A). The addition of 10 mM myo-inositol was required to restore sensitivity fully, after the lithium was removed.

6. Downstream targets of the PI pathway that might mediate the electrical response: The role of $InsP_3$-induced Ca^{2+} release in *Limulus* ventral photoreceptors

Light-induced release of Ca^{2+} from intracellular stores is a prominent feature of the response of some microvillar photoreceptors and is expected of the PI cascade. Although it may involve time delays, as $InsP_3$ and Ca^{2+} diffuse between the microvillar membrane and the Ca^{2+} stores beneath, it has potential for great amplification, since thousands of Ca^{2+} ions may pass through a single $InsP_3R$. There is substantial evidence from work on *Limulus* ventral photoreceptors that $InsP_3$-induced Ca^{2+} release contributes to visual excitation. We will review this evidence and then compare it with evidence for alternative pathways, mainly from *Drosophila* photoreceptors, where light-induced Ca^{2+} release is largely replaced by Ca^{2+} influx. In rationalizing the need for this apparent diversity, it is worth considering that the needs of different organisms might trade off high amplification, as exemplified by mechanisms driven by Ca^{2+} release, for high speed mechanisms that are driven directly by a product of PLC, such as DAG, which remains bound to the microvillar membrane.

6.1. $InsP_3$ releases Ca^{2+} ions from stores in Limulus *ventral photoreceptors*

The interest in $InsP_3$ as an excitatory messenger stems from the observation that intracellular injection of $InsP_3$ into *Limulus* mimics some aspects of the light response: injection of $InsP_3$ into the rhabdomeral lobe of *Limulus* ventral and lateral eye photoreceptors by brief pulses of pressure depolarizes the photoreceptors by activating an inward current (Fig. 15) and desensitizes (adapts) the cell to subsequent light flashes [155,227,228]. $InsP_3$-induced currents elicited by pulsed pressure injection can reach tens of nA in amplitude and have a similar reversal potential and selectivity for sodium ions to those of the light-induced current [155,227]. $InsP_3$ injections therefore appear to activate at least a sub-set of the light-activated channels.

Intracellular pulsed pressure injection of $InsP_3$ into *Limulus* ventral photoreceptors also results in a rapid, transient elevation of $[Ca^{2+}]_i$ (Fig. 15) [229,230]. Following the methodology established by [231], microphotolysis and confocal microscopy have been used to investigate the timing of Ca^{2+} release by $InsP_3$ in *Limulus* ventral photoreceptors. Photolysis of caged $InsP_3$ close to the microvillar plasma membrane results in Ca^{2+} release within 10–20 ms, 20–45 ms before the physiological release of Ca^{2+} by light would normally be detected under the same conditions (Fig. 16) [232]. Because the ability of $InsP_3$ to elevate $[Ca^{2+}]_i$ in *Limulus* ventral photoreceptors persists in the presence of nominally Ca^{2+}-free extracellular

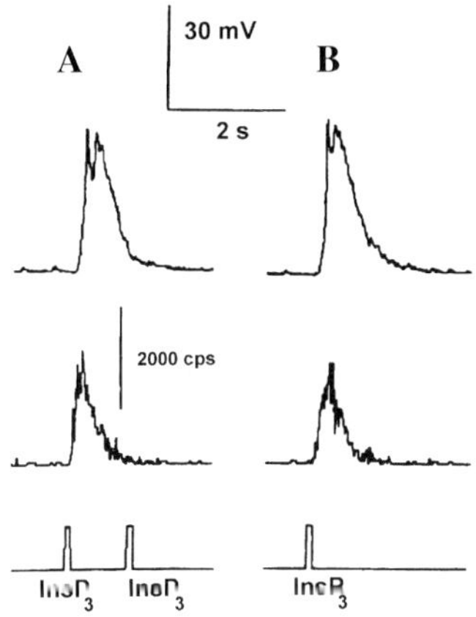

Fig. 15. $InsP_3$-induced depolarization and Ca^{2+} release in *Limulus* ventral photoreceptors. (A) Membrane potential (top trace) and aequorin luminescence (middle trace, indicating intracellular Ca^{2+} concentration) following two consecutive pulsed-pressure injections of 100 μM $InsP_3$ (bottom trace). Feedback desensitization greatly reduces the depolarization and Ca^{2+} elevation produced in response to the second injection. (B) Recovery from desensitization of $InsP_3$-induced depolarization (top trace) and Ca^{2+} elevation (middle trace), recorded 100 s after the traces in (A) (for details see [243]).

solutions, it is assumed that $InsP_3$ releases Ca^{2+} from intracellular stores. The SMC are the most likely candidate stores. The action of $InsP_3$ is stereospecific to the D-myo-inositol 1,4,5 trisphosphate isomer [233] and it seems likely that, as in mammalian cells, a specific $InsP_3$ receptor acting as an $InsP_3$-gated channel exists on the SMC membrane [149]. A putative $InsP_3$ receptor has recently been identified immunologically in *Limulus* ventral photoreceptors (Section 5.5) [187].

6.2. *Ca^{2+} ions released by $InsP_3$ rapidly activate a cation conductance in* Limulus *ventral photoreceptors resembling the light-activated conductance*

All of the effects of $InsP_3$ injected into *Limulus* ventral photoreceptors can be mimicked by Ca^{2+} injection and can be reversed by injection of Ca^{2+} chelators [230,234]. Therefore, in *Limulus* ventral photoreceptors, the known actions of $InsP_3$ appear to result from released Ca^{2+} and not from the direct gating of plasma membrane channels by $InsP_3$. Pulsed pressure injection of Ca^{2+} into the microvillar lobe activates a cation conductance in the plasma membrane with a similar reversal potential to that activated by light and $InsP_3$ [234], which depolarizes the photoreceptor (Fig. 10). The coupling between the elevation of $[Ca^{2+}]_i$ and the depolarization of the photoreceptor is rapid. Photolysis of caged Ca^{2+} (*o*-nitrophenyl EGTA) at the edge of the R-lobe activates an inward ionic current within 1.8 ± 0.7 ms (Fig. 16C) [232]. This current reversed at a membrane potential of 10 ± 6 mV, in the range typical of that of the light-activated current under physiological conditions [52,126].

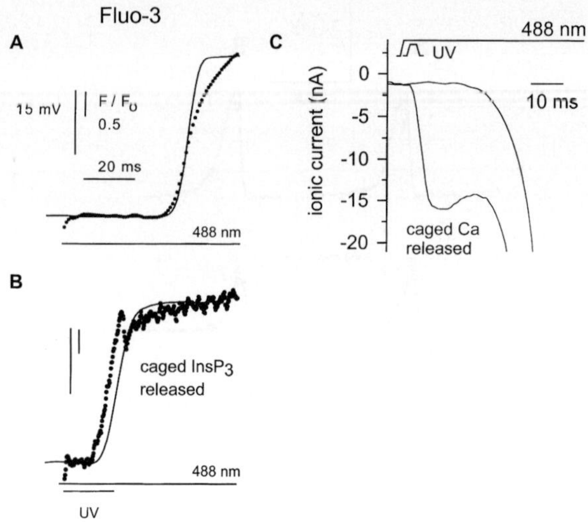

Fig. 16. Timing of Ca^{2+} release investigated using photolysis of caged $InsP_3$ and Ca^{2+} in *Limulus* ventral photoreceptors. Photoreceptors were loaded with the Ca^{2+}-indicator dye fluo-3, GDP-βS and caged $InsP_3$. 488 nm and UV (351/364 nm) laser beams were focused onto the edge of the R-lobe. (A) Membrane potential (solid line) and fluo-3 fluorescence (dots; expressed as the ratio of dye fluorescence change, ΔF to its initial level, F_0) recorded during illumination by the 488 nm laser, which stimulated Ca^{2+} release via rhodopsin photoisomerization. Ca^{2+} release and depolarization began about 40 ms after the onset of illumination. (B) Superimposition of a 20 ms duration UV flash, which released caged $InsP_3$, resulted in much faster Ca^{2+} release and accompanying depolarization of the membrane potential. (C) Similar experiment performed upon a cell loaded with caged Ca^{2+} (Ca^{2+}-dinitrophenyl EGTA) instead of $InsP_3$. Photocurrent in response to the 488 nm laser, initiated via photoisomerization of Rh, began after a latent period of approximately 30 ms. Superimposition of a UV flash to release caged Ca^{2+} activated an inward current within the 3 ms duration of the UV flash (for details see [232]).

6.3. *Pharmacological evidence that $InsP_3$-induced Ca^{2+} release contributes to excitation in* Limulus *ventral photoreceptors*

The response sensitivity of a dark-adapted *Limulus* ventral photoreceptor to a flash of light and the size of quantum bumps are greatly reduced by manipulations designed to interfere with the light-induced Ca^{2+} elevation, such as depletion of Ca^{2+} stores (Fig. 17A) [137,235,236], introduction of rapid Ca^{2+}-chelators like BAPTA (Fig. 17B) [220,223,237], or inhibition of the $InsP_3$ receptor by heparin [220–223]. Non-specific phosphatase inhibitors that prolong the excitatory response to injected $InsP_3$ also prolong the response to dim light flashes [238,239]. Although these pharmacological demonstrations support a pivotal involvement of $InsP_3$ and Ca^{2+} in visual excitation in *Limulus* ventral photoreceptors, a critical consideration concerns the timing of these two events. Certainly, the coupling between stimulation with $InsP_3$ and the resulting inward current is rapid. In experiments in which photoreceptors were filled with caged $InsP_3$ as well as fluorescent Ca^{2+} indicators,

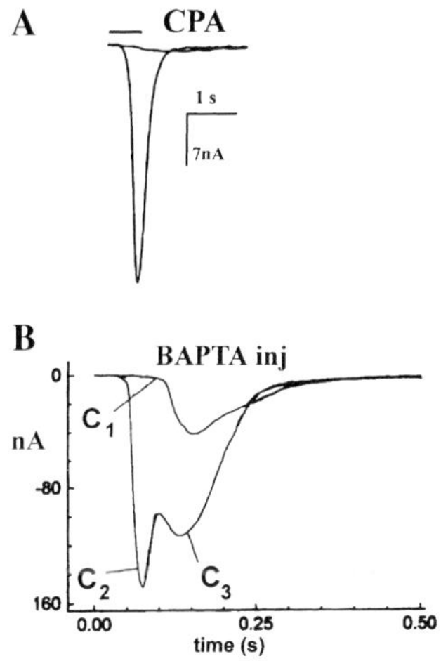

Fig. 17. Blockade of the light-induced elevation of Ca^{2+} desensitizes *Limulus* ventral photoreceptors. (A) Application of 100 μM cyclopiazonic acid (CPA), which depletes endoplasmic reticulum Ca^{2+} stores, greatly reduces the amplitude of the photocurrent generated by a dim flash of light (after [137]). (B) Intracellular injections of 2 mM BAPTA eliminated the C2 component of the photocurrent generated by a bright flash, reduced C3 and exposed the smallest component, C1 (after [223]).

photolysis of caged $InsP_3$ at spots within a few micrometers of the microvillar membrane resulted in inward ionic current flow and depolarization within 2.5 ± 3.3 ms of the first detection of $InsP_3$-induced Ca^{2+} release (Fig. 16). The question is whether the Ca^{2+} elevation resulting from stimulation by light is sufficiently fast to cause channel gating. Early measurements of Ca^{2+} changes pointed to an inconsistency in the proposed causal link between light-induced Ca^{2+} release and the visual response because of the failure to observe any Ca^{2+} release prior to the electrical response – delays of 10–25 ms being reported as typical [131,240,241].

There are several possible explanations for this failure. On the technical side, the slowness of some Ca^{2+} indicators and the use of optical measurements that average the Ca^{2+} concentration in the entire photoreceptor may have diminished the detectability of Ca^{2+} release initiated directly beneath the plasma membrane. It is also possible that, under the conditions used, the first component of the electrical response was, indeed, not mediated by Ca^{2+} release. This might be particularly true of measurements made using the absorbance indicator arsenazo III, in which cells were light-adapted by a continuous measuring light prior to the stimulation that elicited Ca^{2+} release. In this case, the C1 component, which is more resistant to light adaptation and may not be mediated by the PI pathway [223], might dominate the initial response of the cell [126].

The technical problems have recently been reduced by the use of confocal microscopy to both excite ventral photoreceptors at a spot as close as possible to their microvillar membrane and to measure Ca^{2+} release using fluorescent Ca^{2+}-indicator dyes at that spot. A total of 30 out of 76 recordings were obtained in which

However, they reported the activation of channels by cGMP in a minority of excised patches. These patches were carefully shielded from the extracellular solution as they were excised, possibly to avoid exposure to high Ca^{2+} ion concentrations. The channel events activated during application of cGMP had a similar conductance to light-activated channels (43 and 18 pS), similar reversal potential when bathed in media mimicking intracellular and extracellular ion concentrations, and a similar increase in open probability upon membrane depolarization. Since intracellular injections of cGMP or its analogs depolarize the photoreceptor [252,253], it was proposed that cGMP might be a terminal messenger in the visual cascade in *Limulus* ventral photoreceptors. Either an independent pathway of light-induced elevation of cGMP might exist [221], or Ca, released via the PI pathway, may act to increase cGMP levels either by stimulating guanylate cyclase or inhibiting cGMP phosphodiesterase [254]. Although the sequence of a putative cyclic nucleotide channel has been cloned from *Limulus* brain and ventral eye tissue [255], its localization to the photoreceptors has not so far been demonstrated.

There is some circumstantial evidence to support these proposals, but negative findings dominate. Application of cyclic-nucleotide phosphodiesterase (PDE) inhibitors, which might be expected to sustain elevated levels of cGMP, prolongs the transient response to a bright step of light [254], while intracellular injection of a PDE reduces the amplitude of the final C3 component observed after bright flashes as well as the small C1 component [221]. L-*cis* diltiazem, a blocker of some vertebrate cGMP-gated channels as well as those of molluskan ciliary (see below) and extra-ocular photoreceptors [256] reduces the small C1 component [257]. However, the response to moderate flashes of light and the C2 component of the response to bright flashes were not diminished by injection of PDE or L-*cis* diltiazem, which is not consistent with mediation by cGMP of the effects of the PI pathway. No light-induced elevation of cGMP has been detected in *Limulus* ventral photoreceptors [258], nor have early reports been confirmed of a light-induced increase in cGMP concentration in squid retinae [252,258]. Furthermore, in squid retinae, Ca^{2+} ions increase, rather than reduce, cGMP-phosphodiesterase activity [259]. A recent report on the effects of cyclic nucleotides dialyzed into *Drosophila* photoreceptors also failed to detect the activation of inward current by cGMP [260]. Thus, while it is clear that ciliary photoreceptors (see below), extra-ocular photoreceptive neurons [261] and other cells in invertebrates do express cGMP-gated channels [262,263] the evidence for their mediating the light response in rhabdomeric photoreceptors is not strong.

7. Coupling the PI pathway to channel activation in other microvillar photoreceptors

7.1. *The role of* $InsP_3$

In photoreceptors of the bee, where light-induced release of Ca^{2+} from the SMC has been demonstrated, the SMC also release Ca^{2+} in response to the application of $InsP_3$ to permeabilized photoreceptor preparations [264]. Thus it seems plausible that the PI pathway mediates light-induced Ca^{2+} release in bee photoreceptors. Release of Ca^{2+} may also be aided by caffeine-sensitive ryanodine receptors that mediate Ca^{2+}-induced Ca^{2+} release [250]. $InsP_3$ also depolarizes *Hermissenda* photoreceptors when pressure-injected into the rhabdomeral region [265] and activates bursts of inward current when dialyzed into *Lima* photoreceptors [133]. However, as with *Limulus* ventral photoreceptors, the question remains as to whether, or to what extent, release of Ca^{2+} or $InsP_3$ mediates the electrical response to light.

7.2. *The role of the DAG branch of the PI pathway*

The logical need for a pathway of excitation that is not dependent on Ca^{2+} release is more compelling in microvillar photoreceptors, such as those of *Drosophila* and *Balanus*, in which elevation of $[Ca^{2+}]_i$ occurs largely via influx from the extracellular space. In these cases, there is no doubt that Ca^{2+} plays an important role in facilitating and accelerating the light response [266,267], but there is no published evidence that elevation of $[Ca^{2+}]_i$ or $InsP_3$ can induce excitation. Photolytic release of Ca^{2+} from caged precursors does not directly excite *Drosophila* photoreceptors, although it does greatly accelerate the light response [266].

An alternative to the use of Ca^{2+} as a messenger would be provided if $InsP_3Rs$ could be localized to the microvillar membrane as well as the ER, creating $InsP_3$-gated channels in the plasma membrane. However, a null mutation of the $InsP_3$ receptor gene in *Drosophila* appears not to affect excitation by light [268]. Alternative possible mechanisms of excitation have therefore been proposed for *Drosophila*. One is that the activation of the light-sensitive channels in these species is physically or chemically coupled to the mechanism that releases Ca^{2+} from intracellular stores [97], eliminating the need for Ca^{2+} release, but not for the InsP3 receptor. Depletion of Ca^{2+} stores does not activate an inward current in *Limulus* ventral photoreceptors [137,236].

Little is known about another possible messenger, DAG, the other product of PLC-mediated hydrolysis of PIP_2. The major target of DAG in many cell types is PKC [148] and one line of experiment has therefore sought to implicate PKC in visual excitation. Injection of phorbol esters, DAG surrogates that activate PKC, into *Balanus* photoreceptors caused a transient depolarization having an I–V relation resembling that of the light-induced current [269]. The effect of chemically distinct DAG surrogates was subsequently examined in *Lima* photoreceptors [133]. Application of the phorbol ester PMA or the alkaloid (−)-indolactam elicited an inward current, several hundred pA in amplitude, accompanied by a pronounced increase in membrane conductance. The stereoisomers 4-PMA and (+)-indolactam were both inactive, arguing for the specificity of the effects. Elevation of $[Ca^{2+}]_i$ by intracellular dialysis accelerated this current, whereas chelerythrine antagonized it, suggesting mediation by PKC. The reversal potential of the current induced by PKC activators was in the vicinity of +10 mV and shifted in the negative direction upon removing Na^+, indicating a similar ionic dependency as the photocurrent, or a component thereof. Stimulation by PMA and (+)-indolactam was accompanied by a pronounced depression of light responsiveness; conversely, steady illumination reduced the size of the current elicited by these PKC activators. These results support a participation of the DAG branch of the PLC cascade in the activation of some light-dependent conductance, probably in synergy with $InsP_3$-mediated Ca^{2+} release. However, the lack of antagonistic effects of PKC inhibitors on the peak amplitude of the light-evoked current indicate that these mechanisms are probably not involved in the large early transient, but may concern a smaller, slower component of the response. Application of DAG surrogates to *Limulus* photoreceptors does not activate an inward current but, as in *Lima*, it does profoundly desensitize the photoreceptor [270], implying a role for PKC in regulating the sensitivity of *Limulus* photoreceptors, possibly connected with PKC-induced stimulation of rhabdomeral membrane turnover [186], but not in excitation.

The identification in *Drosophila* of TRP [271] and TRPL [189] as light-sensitive channels that mediate light-induced Ca^{2+} influx [97,272] should provide new structural information to help identify new messenger candidates. A homolog of these channels has been found to be a major protein within squid outer segments [273], lending credence to the idea that a solution to the activation of these channels will prove to be of general importance for microvillar photoreceptors. Genetic deletion of an eye-specific PKC (INAC) in *Drosophila* photoreceptors affects adaptation and de-activation of the light response, but it does not affect excitation [183,184]. Therefore, if DAG is to play a role in excitation in *Drosophila*, an alternative target to PKC must be found. Recently, the activation of TRP and TRPL by fatty acids, possible metabolites of DAG, has been reported in *Drosophila* photoreceptors and in a heterologous expression system [274]. It is also worth noting that, when expressed heterologously, TRP channels have been reported to be activated by the depletion of Ca^{2+} stores in the host cells [275]. Exogenously expressed TRPL channels have been reported as being activated by $InsP_3$ [276] and by Ca^{2+} influx via activation of calmodulin [277]. Thus teasing apart the mechanisms of TRP and TRPL gating may prove challenging.

Another solution to the problem of identifying messenger alternatives has been proposed for the *Hermissenda* photoreceptor. Hydrolysis of PIP_2 in the membrane, rather than the release of DAG or $InsP_3$ might constitute a signal that opens the light-sensitive channels [265]. Voltage-gated K^+ channels, for instance, have been shown to bind PIP_2 and to be modulated by its presence [278].

8. Shaping the light response: Voltage-dependent Na^+, K^+, Ca^{2+} channels

In addition to the light-sensitive channels, microvillar photoreceptors have evolved an array of voltage-gated conductances that shape the light response. A delayed rectifier potassium conductance provides sustained potassium current during depolarization while a rapidly inactivating potassium conductance increases the outward current during the first few tens of ms of the response [279–283]. In *Lima* photoreceptors, Ca^{2+}-dependent K^+ channels mediate a sizable portion of the depolarization-activated potassium current [125]. For a rapidly flying diurnal insect such as *Calliphora*, the delayed rectifier predominates [284,285]. The resulting large, sustained potassium currents and the concurrent reduction in membrane resistance antagonizes the effect of the light-activated conductance, reducing the sensitivity to low frequency components of the light response. This extends the dynamic range of the cell by avoiding saturation induced by steady background light, while maintaining sensitivity to rapid changes. The sustained potassium conductance also reduces the membrane time-constant, which increases the temporal resolution of the photoreceptor by matching the reduction in the timescale of phototransduction that occurs during light-adaptation [286]. For slowly moving, crepuscular or nocturnal animals, the inactivating potassium conductance predominates, resulting in a higher membrane resistance during sustained depolarization and allowing the summation of highly amplified, slow quantum bumps arising from weak sources [5]. In locust photoreceptors, diurnal modulation of potassium conductance occurs, by serotonin, so that an inactivating potassium conductance predominates by night and a sustained conductance by day [287].

In addition to voltage-dependent potassium channels, which reduce sensitivity to depolarization by light, some photoreceptors have evolved voltage-dependent sodium and calcium channels. Activation of these channels during the rising phase of the action potential creates a regenerative "spike", which enhances the leading edge of the light response and in some cells can lead to the generation of an action potential [123,280,288,289]. In the light-adapted state, voltage-dependent sodium channels in drone bee photoreceptors amplify small voltage changes arising from small brief changes in contrast [290–292]. Voltage-dependent Ca^{2+} channels may enhance light-adaptation, allowing Ca^{2+} to enter cells during the light-induced depolarization [124,293].

9. Maintaining the light response: Ion pumps and exchangers

The massive light-activated influx of sodium and efflux of potassium requires an extremely active Na^+/K^+ pump and a correspondingly high consumption of energy and oxygen [294]. In bee retinae, at least 60% of the ATP consumed in response to a light stimulus is the result of Na^+/K^+-ATP-ase activity [295]. Electrogenic activity of the Na^+/K^+ ATP-ase results in a transient hyperpolarization of the photoreceptor membrane following bright stimulation, as observed in *Limulus* ventral, bee and fly photoreceptors [296–299]. An inward rectifier potassium conductance in *Limulus* ventral photoreceptors, activated upon hyperpolarization, may act to limit this hyperpolarization [300]. Recent immunological surveys of Na^+/K^+ ATP-ase in insect species show that, in many insect species, the pump is excluded from the microvillar membrane [28,301,302] presumably so as to maximize the space available to transducing molecules. The moth appears to be the exception found so far [303].

Na^+/Ca^{2+} exchange plays a major role in maintaining low levels of $[Ca^{2+}]_i$ in darkness, during illumination and in rapidly restoring $[Ca^{2+}]_i$ after illumination [304]. Na^+/Ca^{2+} exchange activity has been observed in *Limulus* ventral, bee and fly

response and the depth of microelectrode penetration in intact retinas [326], without histological identification of the cells that had been impaled.

Ciliary photoreceptors appear to be relatively widespread among mollusks, and their hyperpolarizing photoresponse is thought to subserve an important biological function: by triggering a discharge of action potentials at the termination of illumination [316,318,327], these cells initiate defensive reflexes in response to dark objects and shadows appearing in the visual field of the animal ("shadow reflex").

10.3. Properties of the light-activated conductance of ciliary cells

In voltage-clamped, internally dialyzed *Lima* and *Pecten* ciliary photoreceptors the reversal potential of the light response is −80 mV, near the calculated value of E_K, and shifts in a way closely predicted by the Nernst equation when the concentration of extracellular K^+ is altered, indicating that light activates K^+-selective channels [325]. These properties are consistent with previous reports of intracellular measurements in distally located retinal cells *in situ* [328,329]. In *Pecten*, cell-attached recordings were also obtained from the ciliary appendages, presumed to be the site of phototransduction. In some patches outwardly directed single-channel currents could be activated by light but not by voltage (Fig. 19) [325]. The unitary conductance of these channels is ~27 pS, significantly smaller than that of the primary light-sensitive channels of depolarizing receptors [103,110], but similar to the conductance of light-suppressed single-channel currents in amphibian rods, measured in the absence of divalent cations [330].

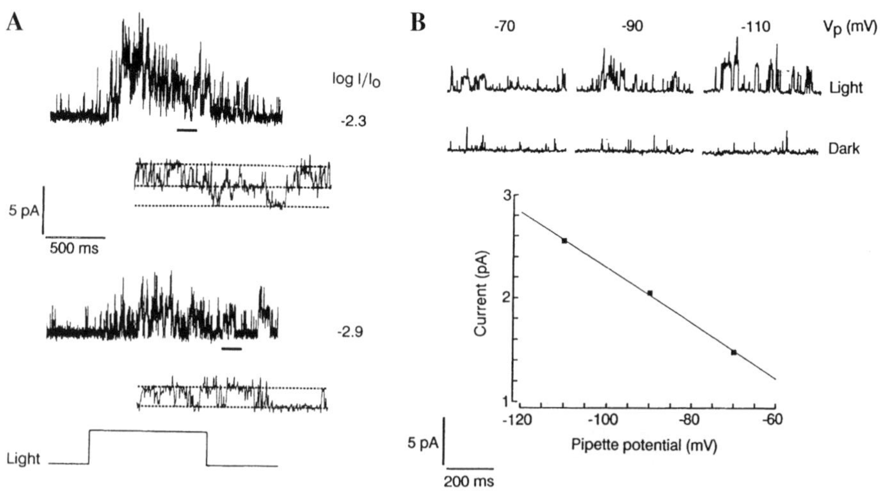

Fig. 19. (A) Single channel activity evoked by light in a ciliary photoreceptor of *Pecten* at two light intensities, indicated on the right. Insets show the underlined portion of the traces on an expanded timescale. (B) Relationship between membrane potential and single-channel currents evoked by light. The amplitude of the unitary current is plotted as a function of applied voltage in the bottom panel. The single-channel conductance is 27 pS (for details see [325]).

The high potassium selectivity of the light-dependent channels of these ciliary photoreceptors prompted an examination of the effects of conventional antagonists of voltage-gated K^+ channels [331]. Charybdotoxin, which blocks with high affinity large-conductance Ca^{2+}-activated potassium channels, had no effect on the photocurrent, but 50 mM TEA produced a reversible block that was weakly voltage-dependent (increasing 20% for a 20 mV hyperpolarization), suggestive of a site of interaction superficially located within the electric field of the membrane. The effects of superfusion with 4-aminopyridine (4-AP) were far more dramatic (Fig. 20). The light-evoked current was nearly abolished, with a half-maximal dose < 1 μM. The blockage had a rapid onset and was slowly reversible, with no significant use or voltage-dependency. The photoresponse kinetics and the light sensitivity of the cell were little affected by 4-AP, suggesting that the suppression of the photocurrent is due to blockage of the light-sensitive channels, rather than impairment of some of the activation steps. Light-dependent K^+-channels thus share significant similarities with voltage-activated K^+ channels. It was surprising, nevertheless, to find such a pronounced sensitivity to 4-AP, a trait that typically characterizes rapidly-inactivating K^+ channels of the *Shaker* class. An I_A-like depolarization-activated transient K^+ current sensitive to 4-AP has been recently described in distal photoreceptors of the scallop [332]. Light stimulation modulates this current, reducing its inactivation. This observation prompted the suggestion that these channels may be identified as the light-activated conductance, and that the mechanism of light-dependent gating may entail removal of inactivation by photostimulation.

10.4. The transduction cascade of ciliary photoreceptors

Two observations suggested that the biochemical pathway mediating transduction in hyperpolarizing photoreceptors may be more tractable than that of microvillar cells: (i) the onset of the photocurrent is remarkably fast, 5-fold shorter than the fastest responses obtained in microvillar photoreceptors from the same preparation; (ii) the sensitivity to light is considerably lower (≈ 2.5 log units) for ciliary photoreceptors than their microvillar counterparts [326]. This is consistent with the observation that dim illumination only produces smooth photocurrents graded with

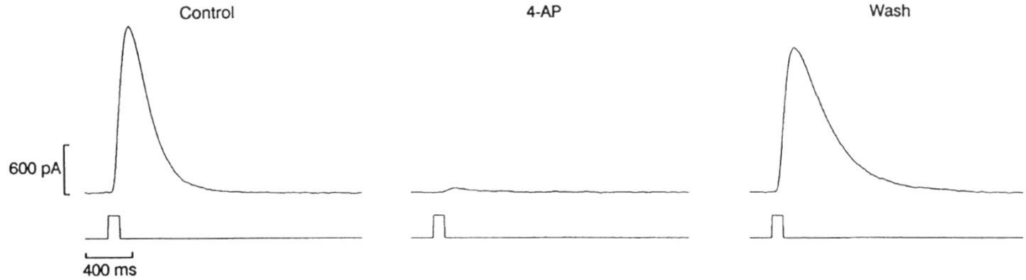

Fig. 20. Blockage of outward photocurrent in *Pecten* ciliary photoreceptors by 4-AP, applied extracellularly at a concentration of 100 μM.

light intensity, but lacking detectable quantum bumps [333,334]. One can calculate the relative transduction gain in microvillar vs ciliary photoreceptors taking into account (a) the magnitude of the macroscopic light-activated conductance, (b) the driving force under the recording conditions employed, (c) the unitary conductance of light-dependent channels, and (d) their fractional open times. The estimated number of channels activated per isomerized rhodopsin turns out to be at least 100-fold lower in ciliary cells. Short response latencies and low gain may be indicative of a simpler transduction cascade, perhaps involving fewer biochemical steps. In addition, the effector mechanism (namely, the light-activated conductance) of hyperpolarizing photoreceptors also seemed simpler than that of their depolarizing counterparts, because the photocurrent exhibits a unique reversal potential, with no hint of the biphasic behavior observed in microvillar photoreceptors, indicative of multiple conductance components [97,125,126]. This result is therefore consistent with the existence of a single population of channels underlying the light response.

Intracellular dialysis of putative messengers via the patch pipette was used to investigate the nature of the internal messenger. With this method, it was determined that small molecules in the electrode solution readily exchange with the cytosol and reach the internal compartment, where light-dependent channels are located [335]. A cursory examination indicated that, like in other photoreceptors, a GTP-binding protein is most likely involved in the early stages of transduction. Flashes delivered during intracellular perfusion with GTP-γS evoked a sustained photocurrent, so that successive stimuli gave rise to a staircase-like change in the holding current; conversely, GDP-βS induced an irreversible loss of responsiveness [336]. Recently, a GTP-binding protein highly homologous to mammalian G_o (83% aminoacid identity) was identified in scallop eyes, and immunolocalized to the distal retinal layer, which contains ciliary photoreceptors [337]. Nothing was known about the subsequent biochemical steps and the final internal messenger, but a simple and appealing transduction scheme suggested by Gorman and McReynolds [338] proposed a photo-induced elevation of cytosolic Ca, similarly to microvillar photoreceptors [98], coupled to the activation of Ca^{2+}-dependent K^+ channels. Surprisingly, Gomez and Nasi [335] found that internal perfusion with 1–100 μM Ca^{2+} did not appreciably change the membrane current in the dark nor did it affect light responsiveness. Furthermore, BAPTA (10 mM) failed to suppress the photocurrent and to slow down its kinetics, and the light response also survived prolonged (>1 h) superfusion in Ca^{2+}-free extracellular media, with repetitive photostimulation. Manipulations of the PI pathway were similarly ineffective: no currents were induced by internal application of $InsP_3$ (1–10 μM) (Fig. 21B), and the light response was seemingly unaffected by the antagonists heparin and decavanadate. These null results stand in sharp contrast with the dramatic effects that similar manipulations have on the light response of microvillar photoreceptors.

The consequences of applying guanine cyclic nucleotides, on the other hand, were quite dramatic. Dialysis of 8-Br-cGMP (10–250 μM) induced a pronounced dose-dependent reduction of the light-activated current (Fig. 21B). Such an outcome could be interpreted either as a direct antagonistic action on light transduction, or, conversely, as a sustained stimulatory effect on the cascade, resulting in a

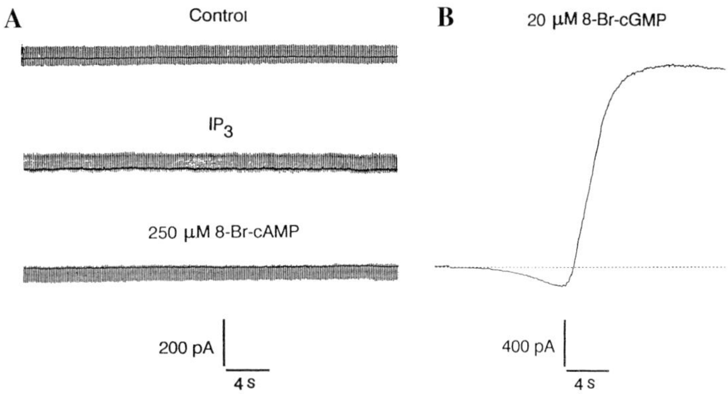

Fig. 21. (A) Intracellular dialysis of *Pecten* ciliary photoreceptors with $InsP_3$ or 8-Br-cAMP does not evoke an outward current at a holding potential of −30 mV. (B) Intracellular dialysis with 20 μM 8-Br-cGMP evokes a large outward current.

decreased residual responsiveness to light. The effects of cGMP analogs on the membrane current in the dark suggest the latter possibility. Shortly after accessing the cell interior and initiating the dialysis, an outward current consistently develops (up to 2 nA), graded with the concentration of 8-Br-cGMP and accompanied by a marked increase in membrane conductance. Similar effects were also obtained with cGMP, but at higher concentrations, whereas 8-Br-cAMP was ineffective (Fig. 21A), indicating that this response is nucleotide-specific. The membrane conductance activated by cGMP and the light-activated conductance share a number of key features: (a) the maximum conductance induced by cGMP is comparable to the saturating light-activated conductance; (b) the I–V relations for both currents have similar shape and reversal potential; (c) the cGMP-induced current, like the photocurrent, is carried by potassium ions.; (e) micromolar concentrations of 4-AP block both the photocurrent and the cGMP-activated current. In addition, activation of single channel currents by 8-Br-cGMP was observed in inside-out patches excised from the ciliary appendages previously screened for the presence of light-activated channels. Because cGMP was effective in the absence of nucleoside trisphophates, its action is likely to be directly on the channels, rather than mediated by some phosphorylation mechanism.

Taken together, these observations suggest that hyperpolarizing invertebrate photoreceptors utilize cGMP as an internal messenger for light transduction, whereas the PI pathway, which plays a pivotal role in the visual response of microvillar photoreceptors, is apparently not involved. The link between light stimulation and the increase in cGMP may involve a guanylate cyclase, rather than a PDE [336]. One unexpected complication to identify the intermediate light-sensitive steps is that some of the pharmacological inhibitors of cGMP phoshodiesterase also produce a direct blockage of the light-activated conductance.

The segregation of vertebrate and invertebrate photoreceptors on the basis of response polarity became untenable as hyperpolarizing receptor potentials were documented in several invertebrates. An alternative suggestion was that the polarity of the light response may be associated with the structure of the light-sensing organelles (cilia vs microvilli) [339], but exceptions have been found in the eyes of primitive chordates [340] and in the parietal eye of lizards [341]. The above results indicate instead that these different structural arrangements correlate with the type of messenger system that couples rhodopsin to ion channels, a proposition also suggested by Finn et al. [342]. Considering that cyclic nucleotides also function as internal messengers for chemoreception in the ciliary neurons of the olfactory epithelium, an extension of this proposition is that the utilization of cyclic nucleotides-based signalling is a common feature shared by a variety of sensory cells of ciliary origin.

10.5. Similarities in the light-dependent conductance of ciliary photoreceptors and rods

The commonality of the internal messenger between invertebrate hyperpolarizing photoreceptors and vertebrate photoreceptors raises the issue of whether their respective light-dependent ion channels may also be related, prompting a search for additional functional similarities. L-*cis*-diltiazem, a stereoisomer of a Ca^{2+}-channel antagonist, is one of the few effective blockers of the light-dependent conductance of vertebrate photoreceptors. Its efficacy has been extensively documented on the photocurrent of intact cells [343,344] and on cGMP-induced currents in patches excised from the outer segment [343–346]. L-*cis*-diltiazem applied by local extracellular perfusion to voltage-clamped ciliary *Pecten* photoreceptors rapidly induced a reversible, dose-dependent suppression of the photocurrent, with a $K_{1/2}$ = 400 μM (Fig. 22) [347]. This figure compares favorably with that obtained in amphibian rods ($K_{1/2}$ = 150 μM) [344]. Direct intracellular dialysis was effective at lower doses (1/5 the concentration). The blockage was weakly voltage-dependent, increasing with membrane depolarization (15% per 20 mV) regardless of the side of application of the drug, as previously reported in rods and cones [345,346]. On the assumption that l-*cis*-diltiazem is preferentially active in the positively charged form, one interpretation of the effects of voltage is that the drug interacts with a site accessible only from the intracellular compartment. Other effective antagonists of the light-sensi-

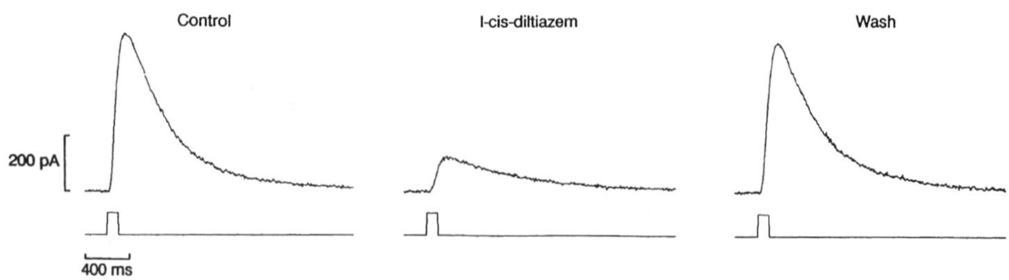

Fig. 22. Inhibition of outward photocurrent in a *Pecten* ciliary photoreceptor by local superfusion with 1 mM 1-*cis*-diltiazem, a blocker of cGMP-gated channels.

tive, cGMP-dependent channels of rods are 3′,4′-dichlorobenzamil (DCPA), a derivative of amiloride, potent at micromolar concentrations [348], and amiloride itself [349]. Extracellular administration of DCPA suppressed the photocurrent of *Pecten* hyperpolarizing receptors ($K_{1/2} = 5$ μM), without affecting its kinetics or operating range, whereas amiloride itself was also effective at ~10× higher concentrations [350].

A conspicuous property of the photocurrent of ciliary photoreceptors of *Pecten* and *Lima* is that the I–V relation markedly rectifies in the outward direction, in a manner similar to rods. After removal of divalent cations this rectification is abolished, and conduction becomes linear [325]. This implies that the mechanism of rectification is also a voltage-dependent blockage of the pores by extracellular Ca^{2+} and Mg^{2+}, again in striking analogy to vertebrate photoreceptors. Because in *Pecten* light-activated single-channel currents can be recorded under physiological ionic conditions, this blockage must be significantly slower. This feature was exploited to resolve the kinetics of the interactions of divalent ions with the channels [350]. Voltage pulses applied during illumination induce instantaneous jumps in the photocurrent, followed by slower relaxations (time constant $\tau = 10$ to 20 ms), which disappear in the absence of divalent cations, reflecting the re-equilibration of the blocking site occupancy after a voltage perturbation. Because the photocurrent is unaffected by voltage steps terminated before the onset of light stimulation, even if the temporal lag is shorter than τ, it appears that channels must open in order for divalent cations to access the blocking site. Consistently with this suggestion, the apparent kinetics of the decline of the photocurrent elicited by brief flashes become faster at more negative holding voltages, but only in the presence of extracellular Ca^{2+} and Mg^{2+}. Again, this effect suggests that hyperpolarization enhances blockage, but the occupancy of the binding site does not equilibrate until the channels are opened by the light stimulus.

Finally, the unitary conductance of the light-dependent ion channels in ciliary photoreceptors is 27 pS [325], a value that is considerably lower than that of microvillar cells (48 pS) [110] but is similar to the single-channel conductance in rods [330]. Taken together, these results suggest the existence of a relationship between the light-dependent, cGMP-gated channels of vertebrate photoreceptors and those found in hyperpolarizing invertebrate photoreceptors.

10.6. cGMP-gated, K^+-selective channels: A missing link?

In recent years, cyclic nucleotide-activated channels have been cloned from vertebrate photoreceptors [351] and olfactory neurons [352]. Unexpectedly, their primary structure lacks any resemblance to other ligand-gated channels, but displays a remarkable homology to voltage-gated K^+ channels, especially of the *Shaker* and *eag* types [353–355]. This finding was surprising not only because of the fundamentally different gating mechanisms involved (electrical vs chemical), but also because of the profound differences in ion conduction properties: cyclic nucleotide-gated channels from vertebrate sensory cells are cationic non-selective, whereas depolarization activated potassium channels are usually exquisitely ion-selective.

Such discrepancy is not necessarily damaging to the suggestion of a kinship between the two channel families, because recent mutagenesis experiments in *Shaker* demonstrated that a switch from potassium selectivity to cationic non-selectivity can occur with only minor structural changes, involving substitution of two amino acid residues located in the pore region [356]. The light-sensitive channels of invertebrate ciliary photoreceptors appear to bridge the gap between these two classes, as they retain both the ion selectivity and susceptibility to blockers of K^+ channels, while also displaying a gating mechanism and a pharmacological profile similar to those of other cyclic nucleotide-activated channels. This lends functional support to the conjecture of a common evolutionary origin of these two superfamilies of channel proteins.

10.7. Light adaptation in ciliary cells: Ca^{2+} independence

The observation that in ciliary *Pecten* cells cGMP mediates transduction without a participation of the $InsP_3$/Ca pathway not only underscores an essential divergence in excitation mechanisms across photoreceptor lines, but also raises questions about mechanisms of response modulation. In all other known visual receptors, light adaptation is chiefly controlled by $[Ca^{2+}]_i$ (reviews [357,358]). In ciliary photoreceptors, the absence of IP_3-induced Ca^{2+} release and the K^+-selectivity of the light-activated conductance implies that the light response may not be coupled to changes in intracellular Ca^{2+}, possibly implying a deficiency of normal light adaptation processes. In fact, McReynolds [333] had suggested that these cells may be constitutively in a state akin to light adaptation, as indicated by their low light sensitivity, absence of resolvable quantum bumps and fast photoresponse kinetics. Furthermore, McReynolds and Gorman [326] reported that ciliary photoreceptors are capable of responding without attenuation to closely spaced pairs of flashes, whereas microvillar cells required minutes to regain responsiveness. Although this phenomenon may indicate that ciliary cells lack the capability to desensitize, one cannot rule out the possibility that light-adaptation does occur, but recovers rapidly.

To clarify this issue, Gomez and Nasi [334] systematically examined the effects of background and conditioning lights on the photocurrent, and demonstrated the presence of all the typical manifestations of sensory adaptation. (a) The response amplitude to a test flash is decreased in a graded way by background or conditioning lights. This attenuation of the response develops with a time constant of 200–800 ms, inversely related to background intensity. (b) Adapting stimuli shift the stimulus-response curve and reduce the size of the saturating photocurrent. (c) The fall kinetics of the photoresponse are accelerated by light adaptation, and the roll-off of the modulation transfer function is displaced to higher frequencies. This light-induced desensitization exhibits a rapid recovery, on the order of a few seconds.

Based on the notion that Ca^{2+} mediates light adaptation in other cells, the consequences of manipulating this ion were examined. Removal of external Ca^{2+} reversibly increased the photocurrent amplitude, without affecting light sensitivity, photoresponse kinetics, or susceptibility to background adaptation. The effect, therefore, concerns ion permeation, rather than the regulation of the visual response. Intracellular dialysis with BAPTA did not reduce the peak-to-plateau decay of the photocurrent elicited by prolonged light steps, nor the background-induced compression of the response amplitude range and the acceleration of its kinetics. Conversely, high levels of buffered $[Ca^{2+}]_i$ (10 μM) only marginally shifted the sensitivity curve and spared all manifestations of light adaptation. These results indicate that hyperpolarizing invertebrate photoreceptors adapt to light, but the underlying mechanisms must utilize a pathway – as yet unidentified – that operates independently of changes in cytosolic Ca. Such a regulatory scheme may not be unique to molluskan hyperpolarizing photoreceptors. In spite of the powerful case for Ca^{2+} modulation of visual function in vertebrate rods, evidence has recently emerged that some manifestations of light adaptation occur independently of Ca^{2+} changes. For instance, adapting lights can induce an acceleration of the rising phase of the photoresponse, even under conditions in which $[Ca^{2+}]_i$ is manipulated over a wide range of concentrations [359]. Conversely, the acceleration of flash response recovery induced by background illumination does not obtain if an equivalent change of $[Ca^{2+}]_i$ is imposed in the dark [360]. Furthermore, the state of phosphorylation of light dependent channels, which determines their responsiveness to cyclic nucleotides, is controlled by endogenous phosphatases in a manner that is insensitive to Ca^{2+} over a concentration range spanning four orders of magnitude [361]. Thus, alternative Ca^{2+}-independent modulatory mechanisms of visual transduction may be of wide generality.

Abbreviations

4-AP, 4-aminopyridine
$[Ca^{2+}]_i$, intracellular free Ca^{2+} concentration
$[Ca^{2+}]_o$, extracellular Ca^{2+} concentration
$[K^+]_o$, extracellular potassium ion concentration
DAG, diacylglycerol
BAPTA, (1,2-*bis*(*o*-Aminophenoxy)ethane-*N*,*N*,*N′*,*N′*-tetraacetic acid
DCPA, 3′,4′-dichlorobenzamil
E_x, equilibrium (Nernst) potential for ion x
$InsP_3$, D-myo-inositol 1,4,5 trisphosphate
$InsP_3R$, D-myo-inositol 1,4,5 trisphosphate receptor protein
PDE, cyclic-nucleotide phosphodiesterase
PI, phosphoinositide
PLC, phospholipase C
PKC, protein kinase C
SMC, submicrovillar cisternae of smooth endoplasmic reticulum
TEA, tetraethylammonium ion
V_{rev}, photocurrent reversal potential
MW, molecular weight
Rh, rhodopsin
G_q, G-protein

Acknowledgements

Supported by NIH grants RO1 EY-07559 and EY-07743. We thank Monika Deshpande for her critical reading of a draft of this manuscript.

References

1. Howard, J., Dubs, A. and Payne, R. (1984) J. Comp. Physiol. A **154**, 707–718.
2. Borsellino, A., Fuortes, M.G.F. and Smith, T.G. (1965) Cold Spring Harbor Symp. Quant. Biol. **30**, 429–443.
3. Barlow, H.B. (1965) Cold Spring Harb. Symp. Quant. Biol. **30**, 539–546.
4. Lillywhite, P.G. (1977) J. Comp. Physiol. **122**, 189–200.
5. Laughlin, S.B. and Weckström, M. (1993) J. Comp. Physiol. A **172**, 593–609.
6. Autrum, H. (1950) Z. Vergl. Physiol. **32**, 176–227.
7. De Souza, J.M. and Ventura, D.F. (1989) J. Comp. Physiol. A **165**, 237–245.
8. Laughlin, S.B. (1996) Vision Res. **36**, 1529–1541.
9. Brown, J.E. and Coles, J.A. (1979) J. Physiol. (Lond.) **296**, 373–392.
10. Baylor, D.A., Lamb, T.D. and Yau, K.W. (1979) J. Physiol. (Lond.) **288**, 589–611.
11. Schnapf, J.L. and McBurney, R.N. (1980) Nature **287**, 239–241.
12. Fein, A. and Charlton, J.S. (1975) J. Gen. Physiol. **66**, 823–836.
13. Fein, A. and Charlton, J.S. (1975) Nature **258**, 250–252.
14. Bacigalupo, J. and Johnson, E.C. (1992) Vis. Neurosci. **8**, 41–47.
15. Szuts, E.Z. (1993) Vis. Neurosci. **10**, 921–929.
16. Eguchi, E. and Waterman, T.H. (1976) Cell Tissue Res. **169**, 419–434.
17. Boschek, B.C. and Hamdorf, K. (1976) Z. Naturforsch. **31c**, 763.
18. Harris, W.A., Ready, D.F., Lipson, E.D., Hudspeth, A.J. and Stark, W.S. (1977) Nature **266**, 648–650.
19. Schinz, R.H., Lo, M.V., Larrivee, D.C. and Pak, W.L. (1982) J. Cell Biol. **93**, 961–967.
20. Lisman, J.E. and Bering, H. (1977) J. Gen. Physiol. **70**, 621–633.
21. Hamdorf, K. (1979) in: Handbook of Sensory Physiology, Vol. VII/6A, ed H. Autrum. pp. 145–224, Springer, Berlin.
22. Saibil, H.R. (1982) J. Mol. Biol. **158**, 435–456.
23. Blest, A.D., Stowe, S. and Eddey, W. (1982) Cell Tissue Res. **223**, 553–573.
24. De Couet, H.G., Stowe, S. and Blest, A.D. (1984) J. Cell. Biol. **98**, 834–846.
25. Calman, B.G. and Chamberlain, S.C. (1992) Vis. Neurosci. **9**, 365–375.
26. Arikawa, K., Hicks, J.L. and Williams, D.S. (1990) J. Cell Biol. **110**, 1993–1998.
27. Hafner, G.S., Tokarski, T.R. and Kipp, J. (1992) J. Neurocytol. **21**, 94–104.
28. Sturmer, K., Baumann, O. and Walz, B. (1995) J. Cell. Sci. **108**, 2273–2283.
29. Walz, B. (1979) Eur. J. Cell Biol. **20**, 83–91.
30. Baumann, O. (1998) Cell Motil. Cytoskeleton **41**, 74–86.
31. Baumann, O. (1998) Cell Tissue Res. **291**, 351–361.
32. De Velasco, B., Martinez, J.M., Ochoa, G.H., Miller, A.M., Clark, Y.M., Matsumoto, B. and Robles, L.J. (1999) Exp. Eye Res. **68**, 725–737.
33. Porter, J.A., Yu, M., Doberstein, S.K., Pollard, T.D. and Montell, C. (1993) Science **262**, 1038–1042.
34. Battelle, B.A., Andrews, A.W., Calman, B.G., Sellers, J.R., Greenberg, R.M. and Smith, W.C. (1998) J. Neurosci. **18**, 4548–4559.
35. Shieh, B.H. and Niemeyer, B. (1995) Neuron **14**, 201–210.
36. Huber, A., Sander, P., Gobert, A., Bahner, M., Hermann, R. and Paulsen, R. (1996) EMBO J. **15**, 7036–7045.
37. Xu, X.Z., Wes, P.D., Chen, H., Li, H.S., Yu, M., Morgan, S., Liu, Y. and Montell, C. (1998) J. Biol. Chem. **273**, 31297–31307.

38. Wes, P.D., Xu, X.Z., Li, H.S., Chien, F., Doberstein, S.K. and Montell, C. (1999) Nature Neurosci. **2**, 447–453.
39. Anderson, R.E., Benolken, R.M., Kelleher, P.A., Maude, M.B. and Wiegand, R.D. (1978) Biochim. Biophys. Acta **510**, 316–326.
40. Akino, T. and Tsuda, M. (1979) Biochim. Biophys. Acta **556**, 61–71.
41. Paulsen, R., Zinkler, D. and Delmelle, M. (1983) Exp. Eye Res. **36**, 47–56.
42. Corda, D. and Shinitzky, M. (1985) Exp. Eye Res. **40**, 575–583.
43. Goldsmith, T.H. and Wehner, R. (1977) J. Gen. Physiol. **70**, 453–490.
44. Ryba, N.J., Hoon, M.A., Findlay, J.B., Saibil, H.R., Wilkinson, J.R., Heimburg, T. and Marsh, D. (1993) Biochemistry **32**, 3298–3305.
45. Walz, B. (1982) J. Ultrastruct. Res. **81**, 240–248.
46. Matsumoto-Suzuki, E., Hirosawa, K. and Hotta, Y. (1989) J. Neurocytol. **18**, 87–93.
47. Walrond, J.P. and Szuts, E.Z. (1992) J. Neurosci. **12**, 1490–1501.
48. Skalska-Rakowska, J.M. and Baumgartner, B. (1985) Experientia **41**, 43–45.
49. Feng, J.J., Carson, J.H., Morgan, F., Walz, B. and Fein, A. (1994) J. Comp. Neurol. **341**, 172–183.
50. Walz, B. (1982) J. Cell Biol. **93**, 839–848.
51. Payne, R., Walz, B., Levy, S. and Fein, A. (1988) Philos. Trans. R. Soc. London, B **320**, 359–379.
52. Millecchia, R. and Mauro, A. (1969) J. Gen. Physiol. **54**, 331–351.
53. Yeandle, S. (1958) Amer. J. Ophth. **46**, 82.
54. Fuortes, M.G.F. and Yeandle, S. (1964) J. Gen. Physiol. **47**, 443–462.
55. Walther, J.B. (1965) in: The Functional Organisation of the Compound Eye, ed C.G. Bernhard. pp. 329–336, Pergamon, Oxford.
56. Scholes, J. (1964) Nature **202**, 572–573.
57. Laughlin, S.B., Blest, A.D. and Stowe, S. (1980) J. Comp. Physiol. A **141**, 53–65.
58. Takeda, T. (1982) Vision Res. **22**, 303–309.
59. Doujak, F.E. (1984) J. Comp. Physiol. A **154**, 587–605.
60. Reuss, H., Mojet, M.H., Chyb, S. and Hardie, R.C. (1997) Neuron **19**, 1249–1259.
61. Henderson, S.R., Reuss, H. and Hardie, R.C. (2000) J. Physiol. (Lond.) **524**, 179–194.
62. Spiegler, J.B. and Yeandle, S. (1974) J. Gen. Physiol. **64**, 494–502.
63. Reuss, H. and Stieve, H. (1993) J. Comp. Physiol. **173**, 377–383.
64. Schnakenberg, J. (1989) Biol. Cybern. **60**, 421–430.
65. Howard, J. (1983) Biophys. Struct. Mech. **9**, 341–348.
66. Stieve, H. and Bruns, M. (1983) Biophys. Struct. Mech. **9**, 329–339.
67. Kirkwood, A. and Lisman, J.E. (1994) J. Gen. Physiol. **103**, 679–690.
68. Goldring, M.A. and Lisman, J.E. (1994) J. Gen. Physiol. **103**, 691–727.
69. Scott, K., Sun, Y., Beckingham, K. and Zuker, C.S. (1997) Cell **91**, 375–383.
70. Srebro, R. and Behbehani, M. (1972) J. Physiol. (Lond.) **224**, 349–361.
71. Yeandle, S. and Spiegler, J.B. (1973) J. Gen. Physiol. **61**, 552–571.
72. Wu, C.F. and Pak, W.L. (1975) J. Gen. Physiol. **66**, 149–168.
73. Laughlin, S.B. and Lillywhite, P.G. (1982) J. Physiol. (Lond.) **332**, 25–45.
74. Barlow, R.B., Birge, R.R., Kaplan, E. and Tallent, J.R. (1993) Nature **366**, 61–66.
75. Barlow, R.B. Jr, Kaplan, E., Renninger, G.H. and Saito, T. (1987) J. Gen. Physiol. **89**, 353–378.
76. Laughlin, S.B. (1975) Z. Naturforsch. **30c**, 306–308.
77. Herzog, E.D., Powers, M.K. and Barlow, R.B. Jr (1996) Vis. Neurosci. **13**, 31–41.
78. Lisman, J.E. and Brown, J.E. (1975) J. Gen. Physiol. **66**, 473–488.
79. Stieve, H., Bruns, M. and Gaube. H. (1983) Z. Naturforsch. **38c**, 1043–1054.
80. Wong, F., Knight, B.W. and Dodge, F.A. (1980) J. Gen. Physiol. **76**, 517–537.
81. Wong, F., Knight, B.W. and Dodge, F.A. (1982) J. Gen. Physiol. **79**, 1089–1113.
82. Wu, C.F., Pak, W.L. (1978) J. Gen. Physiol. **71**, 249–268.
83. Stieve, H., Reuss, H., Hennig, H.T. and Klomfass, J. (1991) Z. Naturforsch. **46(c)**, 461–486.
84. Fuortes, M.G.F. and Hodgkin, A.L. (1964) J. Physiol. (Lond.) **172**, 239–263.
85. Grzywacz, N.M. and Hillman, P. (1988) Biophys. J. **53**, 337–348.
86. Weckström, M., Kouvalainen, E. and Jarvilehto, M. (1988) Acta Physiol. Scand. **57**, 733–743.

87. Pece, A.E., French, A.S., Korenberg, M.J. and Kuster, J.E. (1990) Biophys. J. **132**, 103–113.
88. Brown, J.E. and Mote, M.I. (1974) J. Gen. Physiol. **63**, 337–350.
89. Brown, H.M., Hagiwara, S., Koike, H. and Meech, R.W. (1971) Fed. Proc. **30**, 69–78.
90. Detwiler, P.B. (1976) J. Physiol. **256**, 691–708.
91. Pinto, L. and Brown, J.E. (1977) J. Comp. Physiol. **122**, 241–250.
92. Chinn, K.S. and Gillary, H.L. (1985) Comp. Biochem. Physiol. **80**, 233–245.
93. Gomez, M.D. and Nasi, E. (1996) J. Gen. Physiol. **107**, 715–730.
94. Brown, H.M. (1976) J. Gen. Physiol. **68**, 281–296.
95. Holt, C. and Brown, J. (1972) Biochim. Biophys. Acta **274**, 140–157.
96. Hardie, R.C. (1991) Proc. R. Soc. London, B **245**, 203–210.
97. Hardie, R.C. and Minke, B. (1992) Neuron. **8**, 643–651.
98. Brown, J.E. and Blinks, J.R. (1974) J. Gen. Physiol. **64**, 643–665.
99. Peretz, A., Suss-Toby, E., Rom-Glas, A., Arnon, A., Payne, R. and Minke, B. (1994) Neuron **12**, 1257–1267.
100. Ranganathan, R., Bacskay, B.J., Tsien, R.Y. and Zuker, C.S. (1994) Neuron **13**, 837–848.
101. Wong, F. (1978) Nature **275**, 76–79.
102. Stern, J., Chinn, K., Bacigalupo, J. and Lisman, J. (1982) J. Gen. Physiol. **80**, 825–837.
103. Bacigalupo, J. and Lisman, J.E. (1983) Nature **304**, 268–270.
104. Bacigalupo, J., Chinn, K. and Lisman, J. (1986) **87**, 73–89.
105. Hamill, O.P., Marty, A., Neher, E., Sakmann, B. and Sigworth, F.J. (1981) Pflugers Archiv. **391**, 85–100.
106. Ruknudin, A., Song, M.J. and Sachs, F. (1991) J. Cell Biol. **112**, 125–143.
107. Millton, R.L. and Caldwell, J.H. (1990) J. Neurosci. **10**, 885–893.
108. Nagy, K. and Stieve, H. (1990) Eur. Biophys. J. **18**, 221–224.
109. Nagy, K. (1990) Eur. Biophys. J. **19**, 47–54.
110. Nasi, E. and Gomez, M. (1992) J. Gen. Physiol. **99**, 747–769.
111. Colquhoun, D. and Hawkes, A.G. (1981) Proc. R. Soc. London, B **211**, 205–235.
112. Bodoia, R.D. and Detwiler, P.B. (1984) J. Physiol. (Lond.) **367**, 183–216.
113. Dirnberger, G., Keiper, W., Schnakenberg, J. and Stieve, H. (1985) J. Membr. Biol. **83**, 39–43.
114. Postma, M., Oberwinkler, J. and Stavenga, D.G. (1999) Biophys. J. **77**, 1811–1823.
115. Lisman, J.E. and Brown, J.E. (1971) J. Gen. Physiol. **58**, 544–561.
116. Leonard, R.J. and Lisman, J.E. (1981) Science **212**, 1273–1275.
117. Chinn, K. and Lisman, J. (1984) J. Gen. Physiol. **84**, 447–462.
118. Dennis, M.J. (1967) J. Neurophys. **30**, 1439–1465.
119. Alkon, D.L. and Fuortes, M.G.F. (1972) J. Gen. Physiol. **60**, 631–649.
120. Wulff, V.J. and Mueller, W.J. (1973) Vision Res. **13**, 661–671.
121. Maaz, G., Nagy, K., Stieve, H. and Klomfa, J. (1981) J. Comp. Physiol. A **141**, 303–310.
122. Lisman, J.E. and Strong, J.A. (1979) J. Gen. Physiol. **73**, 219–243.
123. Nasi, E. (1991) J. Gen. Physiol. **97**, 17–34.
124. Nasi, E. (1991) J. Gen. Physiol. **97**, 55–72.
125. Nasi, E. (1991) J. Gen. Physiol. **97**, 35–54.
126. Deckert, A., Nagy, K., Helrich, C.S. and Stieve, H. (1992) J. Physiol. **453**, 69–96.
127. Johnson, E.C., Bacigalupo, J., Vergara, C. and Lisman, J.E. (1991) J. Gen. Physiol. **97**, 1187–1205.
128. Hardie, R.C. (1996) J. Neurosci. **16**, 2924–2933.
129. Brown, H.M., Rydquist, B. and Moser, H. (1988) Cell Calcium **9**, 105–119.
130. O'Day, P.M., Gray-Keller, M.P. and Lonergan, M. (1991) J. Gen. Physiol. **97**, 369–391.
131. Ukhanov, K.Y., Flores, T.M., Hsiao, H.-S., Mohapatra, P., Pitts, C.H. and Payne, R. (1995) J. Gen. Physiol. **105**, 95–116.
132. Rydqvist, B., Brown, H.M. (1986) Acta Physiol. Scand. **127**, 499–506.
133. Gomez, M. and Nasi, E. (1998) J. Neurosci. **18**, 5253–5263.
134. Ziegler, A. and Walz, B. (1989) J. Comp. Physiol. A **165**, 697–709.
135. Walz, B., Zimmermann, B. and Seidl, S. (1994) J. Comp. Physiol. A **174**, 421–431.

136. Baumann, O., Walz, B., Somlyo, A.V. and Somlyo, A.P. (1991) Proc. Natl. Acad. Sci. USA **88**, 741–744.
137. Ukhanov, K. and Payne, R. (1995) Cell Calcium **18**, 301–313.
138. Oberwinkler, J. and Stavenga, D.G. (1998) J. Gen. Physiol. **112**, 113–124.
139. Oberwinkler, J. and Stavenga, D.G. (2000) J. Neurosci. **20**, 1701–1709.
140. Brown, J.E., Brown, P.K. and Pinto, L.H. (1977) J. Physiol. (Lond.) **267**, 299–320.
141. Levy, S. and Fein, A. (1985) J. Gen. Physiol. **85**, 805–841.
142. Lisman, J.E. and Brown, J.E. (1972) J. Gen. Physiol. **59**, 701–719.
143. Raggenbass, M. (1983) J. Physiol. (Lond.) **344**, 525–548.
144. Brown, J.E. and Lisman, J.E. (1975) Nature **258**, 252–254.
145. Fein, A. and Charlton, J.S. (1977) J. Gen. Physiol. **70**, 591–600.
146. Martinez, J.M. II and Srebro, R. (1976) J. Physiol. (Lond.) **261**, 535–562.
147. Berridge, M.J. (1993) Nature **365**, 388–389.
148. Nishizuka, Y. (1988) Nature **334**, 661–665.
149. Furuichi, T., Yoshikawa, S., Miyawaki, A., Wada, K., Maeda, N. and Mikoshiba, K. (1989) Nature **342**, 32–38.
150. Khan, A.A., Steiner, J.P., Klein, M.G., Schneider, M.F. and Snyder, S.H. (1992) Science **257**, 815–818.
151. McDonald, T.V., Premack, B.A. and Gardner, P. (1993) J. Biol. Chem. **268**, 3889–3896.
152. Hatt, H. and Ache, B.W. (1994) Proc. Natl. Acad. Sci. USA **91**, 6264–6268.
153. Bezprozvanny, I. and Ehrlich, B.E. (1994) J. Gen. Physiol **104**, 821–856.
154. Yoshioka, T., Inoue, H. and Hotta, Y. (1983) Biochem. Biophys. Res. Commun. **111**, 567–573.
155. Brown, J.E., Rubin, L.J., Ghalayini, A.J., Tarver, A.L., Irvine, R.F., Berridge, M.J. and Anderson, R.E. (1984) Nature **311**, 160–162.
156. Baer, K.M. and Saibil, H.R. (1988) J. Biol. Chem. **263**, 17–20.
157. Szuts, E.Z., Wood, S.F., Reid, M.A. and Fein, A. (1986) Biochem. J. **240**, 929–932.
158. Lee, Y.J., Dobbs, M.B., Verardi, M.L. and Hyde, D.L. (1990) Neuron **5**, 889–898.
159. Ryba, N.J., Findlay, J.B. and Reid, J.D. (1993) Biochem. J. **292**, 333–341.
160. Kikkawa, S., Tominaga, K., Nakagawa, M., Iwasa, T. and Tsuda M. (1996) Biochemistry **35**, 15857–15864.
161. Munger, S.D., Schremser-Berlin, J.-L., Brink, C.M. and Battelle B.-A. (1997) Invert. Neurosci. **2**, 175–182.
162. Dorlochter, M., Klemeit, M. and Stieve, H. (1977) Vis. Neurosci. **14**, 287–292.
163. Terakita, A., Hariyama, T., Tsukahara, Y., Katsukura, Y. and Tashiro, H. (1993) FEBS Lett. **330**, 197–200.
164. Terakita, A., Takahama, H., Tamotsu, S., Suzuki, T., Hariyama, T. and Tsukahara, Y. (1996) Vis. Neurosci. **13**, 539–547.
165. Suzuki, T., Terakita, A., Narita, K., Nagai, K., Tsukahara, Y. and Kito, Y. (1995) FEBS Lett. **372**, 333–337.
166. Nobes, C., Baverstock, J. and Saibil, H. (1992) Biochem. J. **287**, 545–548.
167. Devary, O., Heichal, O., Blumenfeld, A., Cassel, D., Suss, E., Barash, S., Rubinstein, C.T., Minke, B. and Selinger, Z. (1987) Proc. Natl. Acad. Sci. USA **84**, 6939–6943.
168. Yarfitz, S.L., Running Deer, J.L., Froelick, G., Colley, N.J. and Hurley, J.B. (1994) J. Biol. Chem. **269**, 30340–30344.
169. Lee, Y.J., Shah, S., Suzuki, E., Zars, T., O'Day, P.M. and Hyde, D.R. (1994) Neuron **13**, 1143–1157.
170. Ryba, N.J., Pottinger, J.D., Keen, J.N. and Findlay, J.B. (1991) Biochem. J. **273**, 225–228.
171 Lott, J.S., Ryba, N.J., Pottinger, J.D., Keen, J.N., Carne, A. and Findlay, J.B. (1992) FEBS Lett. **312**, 241–244.
172. Dolph, P.J., Man-Son-Hing, H., Yarfitz, S., Colley, N.J., Running Deer, J., Spencer, M., Hurley, J.B. and Zuker, C.S. (1994) Nature **370**, 59–61.
173. Terakita, A., Takahama, H., Hariyama, T., Suzuki, T. and Tsukahara, Y. (1998) J. Comp. Physiol. A **183**, 411–417.

174. Carne, A., McGregor, R.A., Bhatia, J., Sivaprasadarao, A., Keen, J.N., Davies, A. and Findlay, J.B. (1995) FEBS Lett. **372**, 243–248.
175. Mitchell, J., Gutierrez, J. and Northup, J.K. (1995) J. Biol. Chem. **270**, 854–859.
176. Bhatia, J., Davies, A., Gaudoin, J.B. and Saibil, H.R. (1996) J. Photochem. Photobiol. B **35**, 19–23.
177. Bloomquist, B.T., Shortridge, R.D., Schneuwly, S., Perdew, M., Montell, C., Steller, H., Rubin, G. and Pak, W.L. (1988) Cell **54**, 723–733.
178. Schneuwly, S., Burg, M.G., Lending, C., Perdew, M.H. and Pak, W.L. (1991) J. Biol. Chem. **266**, 24314–24319.
179. Suzuki, T., Narita, K., Yoshihara, K., Nagai, K. and Kito, Y. (1995) Vision Res. **35**, 1011–1017.
180. Running Deer, J.L., Hurley, J.B. and Yarfitz, S.L. (1995) J. Biol. Chem. **270**, 12623–12628.
181. Rack, M., Xhonneux-Cremers, B.A., Schraermeyer, U. and Stieve, H. (1994) Exp. Eye Res. **58**, 659–664.
182. Smith, D.P., Ranganathan, R., Hardy, R.W., Marx, J., Tsuchida, T. and Zuker, C.S. (1991) Science **254**, 1478–1484.
183. Ranganathan, R., Harris, G.L., Stevens, C.F. and Zuker, C.S. (1991) Nature **354**, 230–232.
184. Hardie, R.C., Peretz, A., Suss-Toby, E., Rom-Glas, A., Bishop, S.A., Selinger, Z. and Minke, B. (1993) Nature **363**, 634–637.
185. Huber, A., Sander, P., Bahner, M. and Paulsen, R. (1998) FEBS Lett. **425**, 317–322.
186. Jinks, R.N., White, R.H. and Chamberlain, S.C. (1996) J. Photochem. Photobiol. B **35**, 45–52.
187. Ukhanov, K., Ukhanova, M., Taylor, C. and Payne, R. (1998) Neuroscience **86**, 23–28.
188. Raghu, P. and Hasan, G. (1995) Dev. Biol. **171**, 564–577.
189. Phillips, A.M., Bull, A. and Kelly, L.E. (1992) Neuron **8**, 631–642.
190. Kass, L., Ellis, D.Z., Pelletier, J., Tableman, N.E. and Edwards, S.C. (1998) Vis. Neurosci. **15**, 1039–1049.
191. Paulsen, R. and Hoppe, I. (1978) FEBS Lett. **96**, 55–58.
192. Vandenberg, C.A. and Montal, M. (1984) Biochemistry **23**, 2347–2352.
193. Bentrop, J. and Paulsen, R. (1986) Eur. J. Biochem. **161**, 61–67.
194. Byk, T., Bar-Yaacov, M., Doza, Y.N., Minke, B. and Selinger, Z. (1993) Proc. Natl. Acad. Sci. USA **90**, 1907–1911.
195. Kikkawa, S., Yoshida, N., Nakagawa, M., Iwasa, T. and Tsuda, M. (1998) J. Biol. Chem. **273**, 7441–7447.
196. Kahana, A., Robinson, P.R., Lewis, L.J., Szuts, E.Z. and Lisman, J.E. (1992) Vis. Neurosci. **9**, 595–602.
197. Plangger, A., Malicki, D., Whitney, M. and Paulsen, R. (1994) J. Biol. Chem. **269**, 26969–26975.
198. Smith, D.P., Shieh, B.H. and Zuker, C.S. (1990) Proc. Natl. Acad. Sci. USA **87**, 1003–1007.
199. LeVine, H. III, Smith, D.P., Whitney, M., Malicki, D.M., Dolph, P.J., Smith, G.F., Burkhart, W. and Zuker, C.S. (1990) Mech. Dev. **33**, 19–25.
200. Yamada, T., Takeuchi, Y., Komori, N., Kobayashi, H., Sakai, Y., Hotta, Y. and Matsumoto, H. (1990) Science **248**, 483–486.
201. Hyde, D.R., Mecklenburg, K.L., Pollock, J.A., Vihtelic, T.S. and Benzer, S. (1990) Proc. Natl. Acad. Sci. USA **87**, 1008–1012.
202. Smith, W.C., Greenberg, R.M., Calman, B.G., Hendrix, M.M., Hutchinson, L., Donoso, L.A. and Battelle, B.A. (1995) J. Neurochem. **64**, 1–13.
203. Matsumoto, H., Kurien, B.T., Takagi, Y., Kahn, E.S., Kinumi, T., Komori, N., Yamada, T., Hayashi, F., Isono, K., Pak, W.L., et al. (1994) Neuron **12**, 997–1010.
204. Kahn, E.S. and Matsumoto, H. (1998) J. Neurochem. **68**, 169–175.
205. Edwards, S.C., Wishart, A.C., Wiebe, E.M. and Battelle, B.A. (1989) Vis. Neurosci. **3**, 95–105.
206. Wiebe, E.M., Wishart, A.C., Edwards, S.C. and Battelle, B.A. (1989) Vis. Neurosci. **3**, 107–118.
207. Calman, B.G., Andrews, A.W., Rissler, H.M., Edwards, S.C. and Battelle, B.A. (1996) J. Photochem. Photobiol. B **35**, 33–44.
208. Kass, L. and Bray, W.O. (1996) J. Photochem. Photobiol. B **35**, 105–113.
209. Pak, W.L., Ostroy, S.E., Deland, M.C. and Wu, C.-F. (1976) Science **194**, 956–959.

210. McKay, R.R., Miller, K., Weckström, M., Torkkeli, P., Järvilehto, M. and Shortridge, R.D. (1994) J. Neurogenet. **9**, 177–187.
211. Scott, K., Becker, A., Sun, Y., Hardy, R. and Zuker, C. (1995) Neuron **15**, 919–927.
212. Fein, A. and Corson, D.W. (1981) Science **212**, 555–557.
213. Bolsover, S.R. and Brown, J.E. (1982) J. Physiol. (Lond.) **332**, 325–342.
214. Suss, E., Barash, S., Stavenga, D.G., Stieve, H., Selinger, Z. and Minke, B. (1989) J. Gen. Physiol. **94**, 465–491.
215. Kirkwood, A., Weiner, D. and Lisman J.E. (1989) Proc. Natl. Acad. Sci. USA **86**, 3872–3876.
216. Reuss, H., Rack, M. and Stieve, H. (1992) Z. Naturforsch. **47c**, 932–936.
217. Fein, A. (1986) Science **232**, 1543–1545.
218. Stieve, H., Niemeyer, B., Aktories, K. and Hamm, H.E. (1992) Z. Naturforsch. **47c**, 915–921.
219. Contzen, K. and Nagy, K. (1995) NeuroReport **6**, 1905–1908.
220. Frank, T.M. and Fein, A. (1991) J. Gen. Physiol. **97**, 697–723.
221. Nagy, K. (1993) Neurosci. Lett. **152**, 1–4.
222. Faddis, M.N. and Brown, J.E. (1993) J. Gen. Physiol. **101**, 909–931.
223. Contzen, K., Richter, K.-H. and Nagy, K. (1995) J. Comp. Physiol. A **177**, 601–610.
224. Sakakibara, M., Alkon, D.L., Kouchi, T., Inoue, H. and Yoshioka, T. (1994) Biochem. Biophys. Res. Commun. **202**, 299–306.
225. Johnson, E.C., Gray-Keller, M.P. and O'Day, P.M. (1998) Vis. Neurosci. **15**, 105–112.
226. Berridge, M.J., Downes, C.P. and Hanley M.R. (1982) Biochem. J. **206**, 587–595.
227. Fein, A., Payne, R., Corson, D.W., Berridge, M.J., Irvine, R.F. (1984) Nature **311**, 157–160.
228. Payne, R. and Fein, A. (1987) Biol. Bull. **173**, 447–448.
229. Brown, J.E. and Rubin, L.J. (1984) Biochem. Biophys. Res. Commun. **125**, 1137–1142.
230. Payne, R., Corson, D.W., Fein, A. and Berridge, M.J. (1986) J. Gen. Physiol. **88**, 127–142.
231. Faddis, M.N. and Brown, J.E. (1992) J. Gen. Physiol. **100**, 547–570.
232. Ukhanov, K. and Payne, R. (1997) J. Neurosci. **17**, 1701–1709.
233. Payne, R. and Potter, B.V.L. (1991) J. Gen. Physiol. **97**, 1165–1186.
234. Payne, R., Corson, W.D. and Fein, A. (1986) J. Gen. Physiol. **88**, 107–126.
235. Bolsover, S.R. and Brown, J.E. (1985) J. Physiol. **364**, 381–393.
236. Dorlochter, M., Yuan, W.J. and Stieve, H. (1999) Z. Naturforsch. **54c**, 446–455.
237. Shin, J., Richard, E.A. and Lisman, J.E. (1993) Neuron **11**, 845–855.
238. Vallet, A.M. and Fein, A. (1997) Brain Res. **768**, 91–101.
239. Corson, D.W. and Fein, A. (1983) J. Gen. Physiol. **82**, 639–657.
240. Stieve, H. and Benner, S. (1992) Vision Res. **32**, 403–416.
241. Payne, R. and Flores, T.M. (1992) J. Comp. Physiol. A **170**, 311–316.
242. Payne, R. (1999) Biophys. J. **76**, A243.
243. Payne, R., Flores, T.M. and Fein, A. (1990) Neuron **4**, 547–555.
244. Levitan, I., Hillman, P. and Payne, R. (1993) Biophys. J. **64**, 1354–1360.
245. Levy, S. and Payne, R. (1993) J. Gen. Physiol. **101**, 67–84.
246. Bacigalupo, J., Johnson, E.C., Vergara, C. and Lisman, J.E. (1991) Proc. Natl. Acad. Sci. USA **88**, 7938–7942.
247. Nagy, K. and Contzen, K. (1997) Vis. Neurosci. **14**, 995–998.
248. Mason, M.J., Garcia-Rodriguez, C. and Grinstein, S. (1991) J. Biol. Chem. **266**, 20856–20862.
249. Demareux, N., Lew, D.P. and Krause, K.-H. (1992) J. Biol. Chem. **267**, 2318–2324.
250. Walz, B., Baumann, O., Zimmerman, B. and Ciriacy-Wantrup, E.V. (1995) J. Gen. Physiol. **105**, 537–567.
251. Rubin, L.J., Womble, M., Brown, J.E. and Finger, T.E. (1989) Vis. Neurosci. **2**, 89–96.
252. Johnson, E.C., Robinson, P.R. and Lisman, J.E. (1986) Nature **324**, 468–470.
253. Feng, J.J., Frank, T.M. and Fein, A. (1991) Brain Res. **552**, 291–294.
254. Johnson, E.C. and O'Day, P.M. (1995) J. Neurosci. **15**, 6586–6591.
255. Chen, F.H., Ukhanova, M., Thomas, D., Afshar, G., Tanda, S., Battelle, B.-A. and Payne, R. (1999) J. Neurochem. **72**, 461–471.
256. Gotow, T., Nishi, T. and Murakami, M. (1997) Brain Res. **745**, 303–308.

257. Nagy, K. (1994) NeuroReport **5**, 847–849.
258. Brown, J.E., Faddis, M. and Combs, A. (1992) Exp. Eye Res. **54**, 403–410.
259. Brown, J.E. and Kelman, E.S. (1996) Biochem. Biophys. Res. Commun. **224**, 684–689.
260. Chyb, S., Hevers, W., Forte, M., Wolfgang, W.J., Selinger, Z. and Hardie, R.C. (1999) J. Neurosci. **19**, 8799–8807.
261. Gotow, T., Nishi, T. and Kijima, H. (1994) Brain Res. **662**, 268–272.
262. Baumann, A., Frings, S., Godde, M., Seifert, R. and Kaupp, U.B. (1994) EMBO J. **13**, 5040–5050.
263. Coburn, C.M. and Bargmann, C.I. (1996) Neuron **17**, 695–706.
264. Baumann, O. and Walz, B. (1989) J. Comp. Physiol. A **165**, 627–636.
265. Sakakibara, M., Inoue, H. and Yoshioka, T. (1998) J. Biol. Chem. **273**, 20795–20801.
266. Hardie, R.C. (1995) J. Neurosci. **15**, 889–902.
267. Werner, U., Suss-Toby, E., Rom, A. and Minke, B. (1992) J. Comp. Physiol. A **170**, 427–434.
268. Acharya, J.K., Jalink, K., Hardy, R.W., Hartenstein, V. and Zuker, C.S. (1997) Neuron **18**, 881–887.
269. Brown, H.M., Burnham, J. and Smolley, J. (1991) Biophys. J. **59**, 530a.
270. Dabdoub, A. and Payne, R. (1999) J. Neurosci. **19**, 10262–10269.
271. Montell, C. and Rubin, G.M. (1989) Neuron **2**, 1313–1323.
272. Niemeyer, B.A., Suzuki, E., Scott, K., Jalink, K. and Zuker, C.S. (1996) Cell **85**, 651–659.
273. Monk, P.D., Carne, A., Liu, S.-H., Ford, J.W., Keen, J.N. and Findlay, J.B.C. (1996) J. Neurochem. **67**, 2227–2235.
274. Chyb, S., Raghu, P. and Hardie, R.C. (1999) Nature **397**, 255–259.
275. Vaca, L., Sinkins, W.G., Hu, Y., Kunze, D.L. and Schilling, W.P. (1994) Am. J. Physiol. **267**, C1501–C1505.
276. Dong, Y., Kunze, D.L., Vaca, L. and Schilling, W.P. (1995) Am. J. Physiol. **269**, C1332–C1339.
277. Estacion, M., Sinkins, W.G. and Schilling, W.P. (1999) Biochem. J. **341**, 41–49.
278. Huang, C.L., Feng, S. and Hilgemann, D.W. (1998) Nature **391**, 803–806.
279. Pepose, J.S. and Lisman, J.E. (1978) J. Gen. Physiol. **71**, 101–120.
280. Lisman, J.E., Fain, G.L. and O'Day, P.M. (1982) J. Gen. Physiol. **79**, 187–209.
281. Hardie, R.C., Voss, D., Pongs, O. and Laughlin, S.B. (1991) Neuron **6**, 477–486.
282. Hardie, R.C. (1991) J. Neurosci. **11**, 3079–3095.
283. Acosta-Urquidi, J. and Crow, T. (1995) J. Neurosci. **15**, 319–332.
284. Weckström, M., Hardie, R.C. and Laughlin, S.B. (1991) J. Physiol. (Lond.) **440**, 635–657.
285. Juusola, M. and Weckström, M. (1993) Neurosci. Lett. **154**, 84–88.
286. Weckström, M. and Laughlin, S.B. (1995) Trends Neurosci. **18**, 17–21.
287. Cuttle, M.F., Hevers, W. and Laughlin, S.B. (1995) J. Comp. Physiol. A **176**, 307–316.
288. Autrum, H. and Von Zwehl, V. (1964) Z. Vergl. Physiol. **48**, 357.
289. Baumann, F. (1968) J. Gen. Physiol. **52**, 855–875.
290. Coles, J.A. and Schneider-Picard, G. (1989) J. Comp. Physiol. A **165**, 109–118.
291. Vallet, A.M., Coles, J.A., Eilbeck, J.C. and Scott, A.C. (1992) J. Physiol (Lond.) **456**, 303–324.
292. Vallet, A.M. and Coles, J.A. (1993) J. Comp. Physiol. A **173**, 163–168.
293. O'Day, P.M., Lisman, J.E. and Goldring, M. (1982) J. Gen. Physiol. **79**, 211–232.
294. Widmer, H., Poitry, S. and Tsacopoulos, M. (1990) J. Gen. Physiol. **96**, 83–108.
295. Tsacopoulos, M., Orkand, R.K., Coles, J.A., Levy, S. and Poitry, S. (1983) Nature **301**, 604–606.
296. Baumann, F. and Hadjilazaro, B. (1971) Vision Res. **11**, 1198.
297. Brown, J.E. and Lisman, J.E. (1972) J. Gen. Physiol. **59**, 720–733.
298. Jansonius, N.M. (1990) J. Comp. Physiol. A **167**, 461–467.
299. Gerster, U., Stavenga, D.G. and Backhaus, W. (1997) J. Comp. Physiol. A **180**, 113–122.
300. Phillips, C.L., Baciagalupo, J. and O'Day, P.M. (1992) Vis. Neurosci. **8**, 19–25.
301. Baumann, O. and Takeyasu, K. (1993) J. Cell. Sci. **105**, 287–301.
302. Baumann, O., Lautenschlager, B. and Takeyasu, K. (1994) Cell Tissue Res. **275**, 225–234.
303. Baumann, O. and Lautenschlager, B. (1995) Cell Tissue Res. **281**, 119–126.
304. O'Day, P.M. (1991) Ann. N. Y. Acad. Sci. **639**, 285–299.
305. Minke, B., Armon, E. (1984) Vision Res. **24**, 109–115.

306. Minke, B. and Tsacopoulos, M. (1986) Vision Res. **26**, 679–690.
307. O'Day P.M. and Gray-Keller, M P. (1989) J. Gen. Physiol. **93**, 473–492.
308. Gerster, U. (1997) Vision Res. **37**, 2477–2485.
309. Bauer, P.J., Schauf, H., Schwarzer. A. and Brown, J.E. (1999) Am. J. Physiol. **276**, C558–C565.
310. Tokuyasu, K. and Yamada, E. (1959) J. Biophys. Biochem. Cytol. **6**, 225–230.
311. Tomita, T. (1965) Cold Spring Harb. Symp. Quant. Biol. **30**, 559–566.
312. Yau, K.-W. and Baylor, D.A. (1989) Ann. Rev. Neurosci. **12**, 289–327.
313. Dakin, W.J. (1910) Quart. J. Micro. Sci. **55**, 49–112.
314. Miller, W.H. (1958) J. Biophys. Biochem. Cytol. **4,** 227–228.
315. Barber, V.C., Evans, E.M. and Land, M.F. (1967) Z. Zellforsch. Mikrosk. Anat. **76**, 295–312.
316. Hartline, H.K. (1938) J. Cell. Comp. Physiol. **2**, 465–478.
317. Barber, V.C. and Land, M.F. (1967) Experientia **23**, 677–681.
318. Mpitsos, G.J. (1973) J. Neurophys. **36**, 371–383.
319. Gillary, H.L. and Gillary, E.W. (1979) J. Morphol. **159**, 89–116.
320. Kennedy, D. (1960) J. Gen. Physiol. **44**, 277–299.
321. Wiederhold, M.L., MacNichol, E.F. and Bell, A.L. (1973) J. Gen. Physiol. **61**, 24–55.
322. Gorman, A.L.F. and McReynolds, J.S. (1969) Science **165**, 309–310.
323. Wilkens, L.A. (1984) Nature **309**, 446–448.
324. Toyoda, J.-I. and Shapley, R.M. (1967) Biol. Bull. **133**, 490.
325. Gomez, M. and Nasi, E. (1994) J. Gen. Physiol. **103**, 939–956.
326. McReynolds, J.S. and Gorman, A.L.F. (1970) J. Gen. Physiol. **56**, 392–406.
327. Cornwall, M.C. and Gorman, A.L.F. (1979) J. Physiol. (Lond.) **291**, 207–232.
328. Cornwall, M.C. and Gorman A.L.F. (1983) Cell. Mol. Neurobiol. **3**, 311–328.
329. Cornwall, M.C. and Gorman, A.L.F. (1983) J. Physiol. **340**, 287–305.
330. Matthews, G. and Watanabe, S.-I. (1987) J. Physiol. **389**, 691–715.
331. Gomez, M. and Nasi, E. (1994b) J. Gen. Physiol. **104**, 487–505.
332. Shimatani, Y. and Katagiri, Y. (1995) J. Neurosci. **15**, 6489–6497.
333. McReynolds, J.S. (1976) in: Neural Principles in Vision, eds F. Zettler and R. Weiler. pp. 394–409, Springer, Berlin.
334. Gomez, M. and Nasi, E. (1997b) J. Gen. Physiol. **109**, 371–384.
335. Gomez, M. and Nasi, E. (1995) Neuron **15**, 607–618.
336. Gomez, M. and Nasi, E. (2000) J. Neurosci. **20**, 5254–5263.
337. Kojima, D., Terakita, A., Ishikawa, T., Tsukahara, Y., Maeda, A. and Shichida, Y. (1997) J. Biol. Chem. **272**, 22979–22982.
338. Gorman, A.L.F. and McReynolds, J.S. (1974) Science **185**, 620–621.
339. Eakin, R.M. (1965) Cold Spring Harb. Symp. Quant. Biol. **30**, 363–370.
340. Gorman, A.L.F., McReynolds, J.S. and Barnes, S.N. (1971) Science **172**, 1052–1054.
341. Solessio, E. and Engbreston, G.A. (1993) Nature **364**, 442–445.
342. Finn, J.T. Solessio, E.C. and Yau, K.W. (1997) Nature **385**, 815–819.
343. Stern, J.H., Kaupp, U.B. and MacLeish, P.R. (1986) Proc. Natl. Acad. Sci. USA **83,** 1163–1167.
344. Rispoli, G. and Menini, A. (1988) Eur. Biophys. J. **16**, 65–71.
345. Haynes, L.W. (1992) J. Gen. Physiol. **100**, 783–801.
346. McLatchie, L.M. and Matthews, H.R. (1992) Proc. R. Soc. London, B **247**, 113–119.
347. Gomez, M. and Nasi, E. (1997a) J. Physiol. **500**, 367–378.
348. Nicol, G.D., Schnetkamp, P.P.M., Saimi, Y., Cragoe, E.J. and Bownds, M.D. (1987) J. Gen. Physiol. **90**, 651–669.
349. McLatchie, L.M. and Matthews, H.R. (1994) Proc. R. Soc. London **255**, 231–236.
350. Nasi, E. and Gomez, M. (1999) J. Gen. Physiol. **114**, 653–672.
351. Kaupp, B.U., Niidome, T., Tanabe, T., Terada, S., Bönigk, W., Stühmer, W., Cook, N.J., Kangawa, K., Matsuo, H., Hirose, T., Miyata, T. and Numa, S. (1989) Nature **342,** 762–766.
352. Dhallan, R.S., Yau, K.-W., Schrader, K. and Reed, R.R. (1990) Nature **347**, 184–189.
353. Kaupp, B.U. (1991) Trends Neurosci. **14**, 150–157.
354. Guy, H.R., Durell, S.R., Warmke, J., Drysdale, R. and Ganetzky, B. (1991) Science **254**, 730.

355. Yau, K.-W. (1994) Proc. Natl. Acad. Sci. USA **91**, 3481–3483.
356. Heginbotham, L., Abramson, T. and MacKinnon, R. (1992) Nature **258**, 1152–1155.
357. Fain, G.L. and Matthews, H.R. (1990) Trends Neurosci. **13**, 378–384.
358. Fein, A. and Szuts, E.Z. (1982) Photoreceptors, their Role in Vision. Cambridge University Press, Cambridge.
359. Nicol, G.D. and Bownds, M.D. (1989) J. Gen. Physiol. **94**, 233–259.
360. Gray-Keller, M.P. and Detwiler P.B. (1996) Neuron **17**, 323–331.
361. Gordon, S.E., Brautigan, D.L. and Zimmerman, A.L. (1992) Neuron **9**, 739–748.
362. Baumann, O. (1997) Int. Rev. Cytol. **176**, 307–348.
363. Hanna, W.J., Johnson, E.C., Chaves, D. and Renninger, G.H. (1993) Vis. Neurosci. **10**, 609–620.

CHAPTER 9

Genetic Dissection of *Drosophila* Phototransduction

B. MINKE

Department of Physiology,
The Hebrew University,
Jerusalem

R.C. HARDIE

Department of Anatomy,
University of Cambridge

Handbook of Biological Physics
Volume 3, edited by D.G. Stavenga, W.J. DeGrip and E.N. Pugh Jr

Contents

1. Introduction

Vertebrates and invertebrates share the same visual world, and in many respects their photoreceptors display similar functional attributes, representing cells exquisitely adapted for the rapid and sensitive transduction of environmental light signals. Both, for example have achieved the ultimate in sensitivity in being able to respond to single photons of light and have evolved powerful adaptational mechanisms to regulate gain in the face of literally astronomical variations in background intensity. However, whilst vertebrate photoreceptors hyperpolarize in response to light by closing ion channels, most invertebrate photoreceptors respond to illumination with a more conventional depolarization, mediated via an increase in conductance. Other major differences, e.g. in structure (vertebrates have ciliary photoreceptors, invertebrates microvillar) and development (vertebrate photoreceptors are neural in origin, invertebrates epithelial), suggest that any common ancestral photoreceptor must be very remote and that the functional similarities are largely a result of convergent evolution within different molecular and structural constraints.

This dichotomy is also apparent in terms of the mechanism of phototransduction: the basic mechanism of excitation in vertebrate rods became generally accepted over 10 years ago with the demonstration of a G-protein-coupled enzyme cascade linking rhodopsin photoisomerization to hydrolysis of cGMP and experiments showing that this cyclic nucleotide was responsible for gating the light-sensitive channels. At about the same time it became clear that most invertebrate photoreceptors utilize a distinct class of G-protein-coupled cascade involving phospholipase C (PLC) rather than phosphodiesterase as the effector enzyme. This so-called phosphoinositide (PI) signaling pathway is probably the most widespread of the G-protein-coupled signaling cascades; it plays a crucial role in regulating cellular Ca^{2+} and is characterized by the production of the second messengers inositol 1,4,5 trisphosphate ($InsP_3$) and diacylgycerol (DAG). However, whilst the essential role of PLC is undisputed, how its activation is linked to the opening of ion channels remains unknown and controversially discussed – arguably representing the major unanswered question in sensory transduction.

A complete understanding of invertebrate phototransduction may require cloning and sequencing of all the gene products that have a role in this process and physiological studies which determine the effect of their loss or misfunction. At the present time *Drosophila* phototransduction is the only available system in which to apply such an ambitious program, because only in this system it is possible to apply a multidisciplinary approach allowing the application of sophisticated genetic, molecular, biochemical and electrophysiological techniques to the same cells. This approach, which is the subject of this chapter, has yielded unique information on the molecular components and the machinery of phototransduction, allowing, for example, both *in vivo* analysis of specific mutations as well as the discovery of many novel proteins of importance not only to phototransduction but to cell signaling generally.

2. Morphology of the compound eye

The *Drosophila* eyes include the compound eyes and ocelli. The ocelli are simple eyes located on the vertex of the head and have a unique photopigment. Each of the two

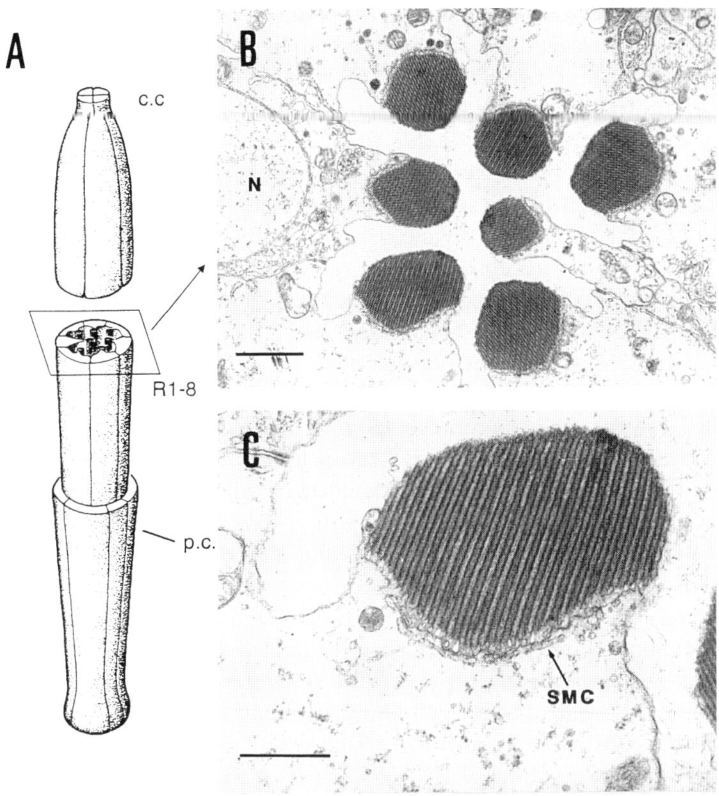

Fig. 1. Microvillar photoreceptors in *Drosophila*. (A) An ommatidium, the cluster of eight photoreceptor cells and surrounding pigment cells (PC), which lies behind every facet of the *Drosophila* compound eye. The visual pigment and most of the phototransduction machinery is localized in the light-sensitive rhabdomeres (R), each composed of $\sim 10^5$ microvilli forming a pencil-like waveguide. (B) Electron micrographs (EM) of a cross-section through the ommatidium (only seven of the eight cells are sectioned at this level). The seven rhabdomeres can be clearly seen projecting into a central extracellular cavity. (C) Detail of one rhabdomere showing the electron-dense microvilli. Each microvillus contains ~ 1000 rhodopsin molecules and is connected by a narrow neck to the cell body. Notice also the extensive system of sac-like cisternae near the base of the microvilli (submicrovillar cisternae, SMC). These are presumed to represent Ca^{2+} stores. The EM sections are from the *trpl* mutant [58], which has an ommatidial structure similar to that of wild-type fly (EM photos courtesy of Hagit Cohen).

compound eyes is made up of many repeat units called ommatidia. There are about 700 ommatidia in the *Drosophila* eye, ~3000 in *Musca*, and ~5000 in *Calliphora* [1]. An ommatidium consists of 20 cells, eight of which are photoreceptor cells (see Fig. 1). Each ommatidium contains a dioptric apparatus composed of transparent chitinous cuticle and an extracellular fluid-filled cavity, the pseudocone. The floor of the cavity is formed by four Semper cells and the walls by primary pigment cells, which together circle the pseudocone. Below this rigid structure of the optical apparatus lie the photoreceptor cells. Each photoreceptor cell has a specialized organelle consisting of a stack of microvilli, known as the rhabdomere, where a large fraction of the phototransduction machinery is localized. The rhabdomere provides the photoreceptor cell with a high directional sensitivity as well as a high light sensitivity, because it acts as an optical waveguide. This characteristic is widely exploited for optical methods [2]. The two short central rhabdomeres in each ommatidium are located one above the other, forming one continuous waveguide ([3], for review see Ref. [4]).

The eight photoreceptors can be divided into three classes according to their spectral sensitivity, position of the rhabdomere within the ommatidium, and synaptic connections in the optic lobes (reviewed by Hardie [1]). The R1–R6 cells represent the major class of photoreceptors in the retina (see Fig. 1). These cells have peripherally located rhabdomeres, and express a single opsin called Rh1 which forms a blue-absorbing rhodopsin (for reviews see Refs. [5,6]). This rhodopsin represents over 90% of the visual pigment present in the compound eyes. R1–R6 cells send axons that synapse in the first optic lobe (lamina). The other two classes of photoreceptors, R7 and R8, are each represented by a single cell per ommatidium and have centrally located rhabdomeres. The R7 cell is located distally in the retina and expresses one of two opsins, Rh3 or Rh4, both of which form UV-absorbing rhodopsins (Section 5.1; for reviews see Refs. [4,6]). Along the dorsal margin of the eye all R7 (R7 marg) cells and also the underlying R8 cells contain Rh3 and have an enhanced sensitivity to polarized light (for reviews see Refs. [1,4]). In a specific region of the compound eye of *Musca* males, both R7 and R8 express the Rh1 photopigment [7]. The majority of R7 cells are, therefore, ultraviolet-sensitive and send axons to the second optic lobe (medulla). The R8 photoreceptor cell is located proximally in the retina, just beneath the R7 cell and also synapses in the medulla. The opsins expressed in R8 cells have recently been isolated. A blue absorbing photopigment (Rh5) is expressed in a subset of R8 cells in which the overlaying R7 cells express Rh3, a pattern which is disrupted in the sevenless (*sev*) mutant flies [8]. Rh6, the photopigment of the other set of R8 cells absorbs in the green and associates with R7 cells expressing Rh4 [9,291]. These intriguing data have very interesting implications on the developmental mechanism of the ommatidia that still need to be explored.

3. Characteristics of the electrophysiological responses to light

3.1. The electroretinogram (ERG)

The ERG measures the extracellular current flow in the eye following or during illumination. It is composed of several components (Fig. 2, for review see Ref. [10]).

The main component is a corneal negative potential arising from summation of the extracellularly recorded receptor potential and a slow response of the pigment cells. The latter is a depolarization initiated by accumulation of K^+ in the extracellular space during the receptor potential in the photoreceptors (for review see Ref. [11]). This response of the pigment cells is substantial only during bright light. Additional components in the ERG are the "on" and "off" transients (Fig. 2C) appearing shortly after the onset and offset of the receptor potential [10]. The polarity of the "on" transient is reversed relative to the receptor potential because of the sign inverting synapse between the R1–R6 axons and the large monopolar neurons in the lamina. This synapse uses histamine as a neurotransmitter which directly gates chloride channels on the postsynaptic target cells [12]. The ERG of the fly is very large relative to similar signals in other eyes (it can reach 30 mV in the fly compared to several μV in vertebrate eyes). It is, therefore, a convenient signal for many studies, especially in screening for visual mutants.

3.2. The receptor potential and membrane properties

The receptor potential represents the photoreceptor's physiological response to light and is used as a monitor for the response of a single cell in the intact fly (for review see Ref. [11]). As the final output of the cell, the receptor potential is not a direct or accurate reflection of the activity of the phototransduction cascade. Although arising from the light-induced opening of ionic channels in the plasma membrane (see Sections 3.5 and 7), the receptor potential is further shaped by passive membrane properties and a variety of voltage-gated channels and electrogenic ion transporters. In the intact animal, the input resistance of photoreceptors R1–R6 in *Drosophila* has been reported as ca. 100 MΩ at resting potential (ca. −60 mV), although this may be underestimated due to a shunt resistance of electrode penetration. Under whole-cell recording conditions resistances of up to 1 GΩ are routinely measured. The cell capacitance, estimated at ca. 60 pF in whole-cell recordings of dissociated ommatidia from newly eclosed flies and 100 pF in intracellular recordings from intact adult flies, is dominated by the microvillar membrane and limits the membrane time constant. Resistance and, consequently, membrane time constant drop markedly as the cell depolarizes in response to light, not only because of the light-sensitive conductance, but also because *Drosophila* photoreceptors express at least three classes of voltage-gated potassium channels. In R1–R6 these include a rapidly activating and inactivating A current (I_A) encoded by the *Shaker* gene [13], and two "delayed rectifier" currents of unknown molecular composition: I_{KS} which inactivates with a time constant of ca. 1 s and I_{KF} with an inactivation time constant of ca. 50–100 ms [14,15]. The voltage operating ranges of these channels, which broadly speaking cover the physiological range (ca. −70 to + 10 mV), can be shifted by up to 30 mV by bath application of serotonin, apparently via a G-protein-coupled receptor as it can be mimicked by internal application of GTPγS [15]. In addition, a non-inactivating voltage gated inward current is found, probably carried by Ca^{2+} channels, with a threshold for activation at ca. −65 mV [16,17]. Photoreceptors R7 and R8 have a slightly different set of voltage-gated channels: only two of the three classes of

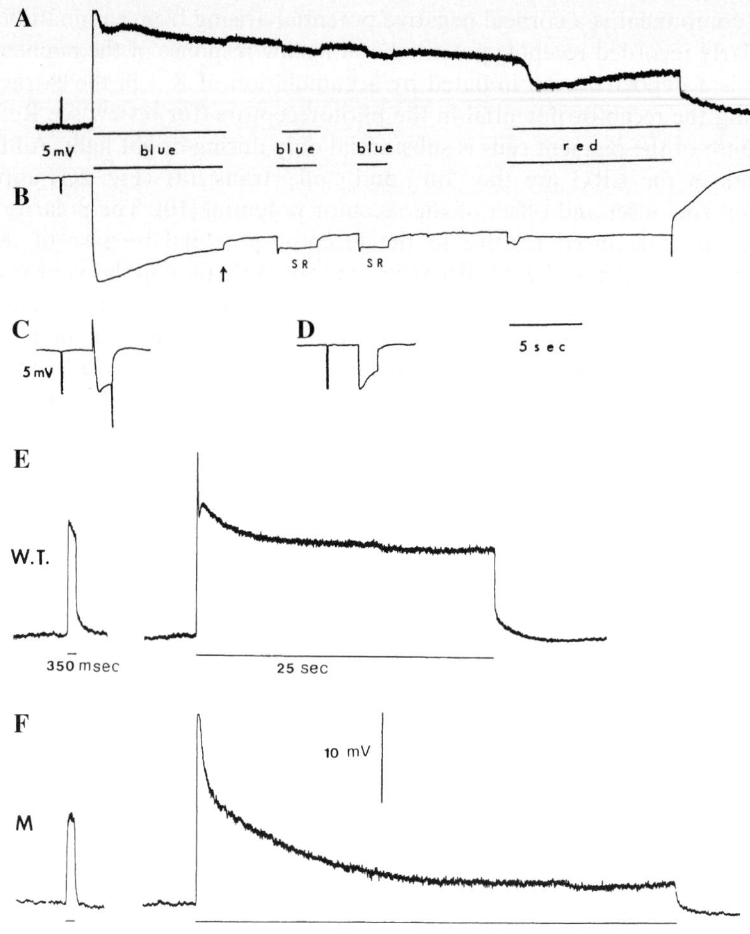

potassium are found (I_A and I_{KF}), and the voltage-gated inward current is larger and also inactivating [17]. Other conductances reported in R1–R6 cells include an inward current activated at voltages below −70 mV probably carried by chloride ions [18] and a cationic current activated by micromolar concentrations of Ca^{2+} which has been revealed after blocking Na/Ca exchange [19].

Two electrogenic ion transporters can also make a substantial contribution to the receptor potential. In intracellular recordings from the larger flies *Calliphora* and *Musca* their activity has been revealed as afterpotentials following bright flashes of light. An electrogenic Na/Ca exchanger generates a net inward current leading to an immediate, but short-lived depolarizing afterpotential [20–22], whilst a Na/K ATP-ase generates a more slowly developing and long-lasting afterhyperpolarization [23]. In *Drosophila* an electrogenic Na/Ca exchange current of up to 100 pA has also been directly measured under voltage-clamp conditions in response to release of caged Ca^{2+} [24]. As described below (see Section 3.3) there is a large amplification in the

Opposite: Fig. 2. The intracellularly recorded receptor potential and electroretinogram (ERG) of *Drosophila* eye in normal and mutant fly under various illuminations. (A) Intracellularly recorded responses of a red adapted R1–R6 cell to three saturating blue stimuli followed by a red stimulus. The intense blue light induced a prolonged depolarizing afterpotential (PDA) which persisted in the dark and inactivated the cell, which did not respond to further stimuli. The subsequent red light suppressed the PDA and re-established sensitivity to light. Illuminances at the level of the preparation were 4.0×10^{15} and 2.5×10^{16} photons/cm^2 s for the blue and red stimuli, respectively. (B) ERG recording obtained under the same stimulus paradigm as in (Fig. 2A). In contrast to the above case, there was a reduction in the response size at the cessation of the blue light (arrow), and each subsequent blue light elicited an additional superimposed response (SR) which arises from R7 and R8 cells. These cells do not have synaptic input in the lamina and therefore do not show the on- and off-transients (trace C). (C) and (D): Comparison of the ERG (C) obtained from a red adapted fly, which represents the summed response of all R1–R8 cells, and the SR (D) obtained from R7,8 only. Note that the on and off transients appear in the response to the red-adapted eye only. Blue stimuli (480 nm) were used. The stimulus duration was 1.3 s, and illuminances were 1.6×10^{10} photons/cm^2 s for the ERG of C and 7.1×10^{13} photons/cm^2 s for the SR. These illuminances were chosen to yield responses of comparable amplitude. The illumination paradigm in A and B have been widely exploited to screen and characterize visual mutants [37] (from [143]). (E,F): Intracellularly recorded receptor potentials from a white-eyed normal *Drosophila* fly (WT), showing the typical receptor potential of *Drosophila* (E), and from white-eyed trp^{CM} mutant (M) showing the typical phenotype of this mutant (F, see Section 7). The monochromatic 540 nm green light intensity was 3.6×10^{14} photons/cm^2 s for the upper right trace, attenuated 0.5 log unit in the upper left trace and by 2.0 and 2.5 log units in the lower, right and left traces, respectively (adapted from [56]).

phototransduction cascade enabling the detection of the voltage response to the absorption of a single photon by a single rhodopsin molecule [25]. The resulting quantum bumps reach an amplitude of ca. 1–2 mV in the intact *Drosophila* and *Lucilia* retina [26–28] (see Section 3.3). Accordingly, in *Drosophila* mutants with a 10^6-fold reduction in photopigment level, the remaining 10^2 molecules are still capable of giving close to a saturating response [27,29].

3.3. Single photon responses (quantum bumps)

Dim light stimulation induces discrete voltage (or current) fluctuations in most invertebrate species, which are called quantum bumps (Fig. 3, [25]). Each bump is assumed to be evoked by the absorption of a single photon. The bumps vary significantly in latency, time course and amplitude even when the stimulus conditions are identical (for review see Ref. [30]). Bumps are due to synchronized activation of many light-sensitive channels (for review see Ref. [31]). The number of channels which are activated to produce a bump vary greatly in different species: few tens in *Drosophila* and up to several thousands in *Limulus* ventral photoreceptors [31]. Bump generation is a stochastic process described by Poisson statistics where each effectively absorbed photon elicits only one bump. However in at least two *Drosophila* mutants (*cam* and *arr*, see Sections 5.3 and 10.3.2), absorption of a single photon elicits a train of bumps with a rather fixed interval in *cam* mutants, which are not independent of each other. This oscillatory phenomenon in bump generation is observed when one rhodopsin molecule triggers a train of bumps (*cam* mutant, [32]). This interesting phenomenon is

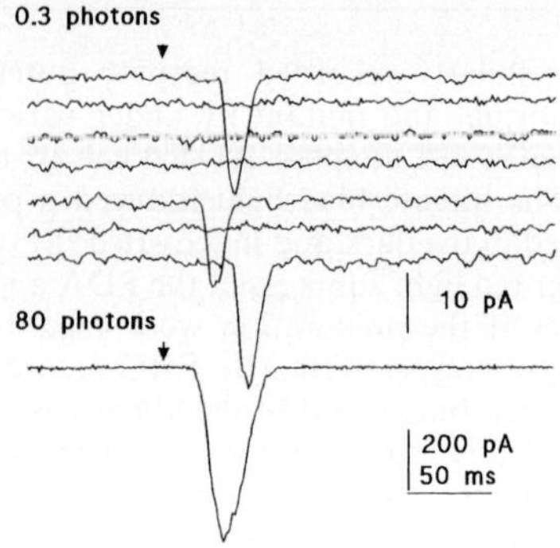

Fig. 3. Quantum bumps measured during patch clamp whole-cell recording. Every effectively absorbed photon evokes a discrete event known as a quantum bump, which represents the concerted opening of many light-sensitive channels. In *Drosophila* the average amplitude of these events under physiological conditions is ca. 10 pA, corresponding to ca. 15 light-sensitive channels simultaneously open at the peak of the bump. *Top*: with dim flashes (0.4 ms containing on average 0.3 photons), many presentations result in failure. The bumps that are generated occur with a variable latency and waveform. The response to a brighter flash (containing ca. 80 photons, below) is therefore broader although at these intensities it still represents the linear summation of the underlying bumps (Hardie RC, unpubl.).

probably related to the mechanism of bump generation which might be explained by the existence of mechanisms similar to those found in the generation of action potentials, namely, positive and negative feedback and refractory period.

A detailed study [290] has indicated that, as in *Limulus* and locust photoreceptors, the latency of bumps in *Drosophila* is not correlated with the bump waveform, thus strongly suggesting that the triggering mechanism of the bump arises from molecular processes different from those that determine the bump waveform. These findings are partly explained by models in which the amplification process is preceded by a series of non-amplifying latency producing steps (see Ref. [31] for a review). To produce realistic bumps by such a model means that no step in the transduction cascade could have a life time greater than the duration of a bump generating mechanism which includes the latency, bump duration and bump refractory period. Accordingly, absorption of a single photon produces a single bump and not a train of bumps in wild-type flies. The single photon-single bump relationship requires that each step in the cascade must have not only an efficient "turn on" mechanism, but also an equally effective "turn-off" mechanism. The functional advantage of such a transduction mechanism is obvious; it produces a very sensitive photon counter with a fast transient response, very well suited for both the sensitivity and the temporal resolution required by the visual system. Unitary Ca^{2+} signaling events with kinetics and waveform similar to the bumps have been recorded in non-photoreceptor cells. Examples are quantum emission domains (QED) in presynaptic terminals, the sparks in cardiac cells, puffs and blips in *Xenopus* oocytes (for reviews see Refs. [33,34]. All these events depend upon brief opening of either single or a small localized group of Ca^{2+} channels located on the calcium stores.

3.4. The prolonged depolarizing afterpotential (PDA)

The PDA, like the receptor potential, arises from light-induced opening of ionic channels in the plasma membrane. However, in contrast to the receptor potential, which quickly declines to baseline after the cessation of the light stimulus, the PDA is a depolarization that continues long after light offset (Fig. 2A) (see Refs. [5,35,36] for reviews). The PDA has been a major tool to screen for visual mutants of *Drosophila* [37].

In *Drosophila*, as in many other invertebrate photoreceptors, the chromophore remains bound to opsin following photoisomerization, forming a dark-stable photoproduct, acid metarhodopsin which absorbs at longer wavelengths and can be converted back to rhodopsin by absorption of another photon. The PDA is observed only when a considerable amount of pigment ($>20\%$) is converted from rhodopsin (R) to metarhodopsin (M) using blue light; the larger the amount of R to M conversion, the longer the PDA. The duration of the PDA depends in a supra-linear manner on the amount of light. The PDA can be depressed at any time by pigment conversion from M to R using orange or red light. The degree of PDA depression depends on the amount of M to R conversion [38]. After the depression of maximal PDA, an additional PDA can be induced immediately by R to M conversion. The situation is more complex when the PDA-depressing light is given following the decline of the PDA (for review see Ref. [36]). When R to M conversion is maximal, a maximal PDA is induced. Additional strong light stimuli, which do not change the distribution of the pigment between R and M, do not affect the duration of the PDA, but only induce light-coincident receptor potentials, which are superimposed on the PDA [5]. In the fly R1–R6 cells, the wide separation of absorption spectrum maxima between R and M allows a relatively large conversion from R to M which is essential for induction of the PDA ([39], Fig. 2A,B). The duration of a maximal PDA varies greatly among different species. Even in different species of fly, such as *Calliphora* and *Drosophila*, the duration of maximal PDA can be very different: a few minutes in *Calliphora* [5] as compared to a few hours in *Drosophila* [36,39]. The duration of a maximal PDA in a specific photoreceptor is tightly linked to the total amount of its visual pigment. Elimination of carotenoids, the precursors of the rhodopsin chromophore, from the diet markedly reduced the concentration of rhodopsin in the photoreceptor cells and completely eliminated the PDA response [5,40]. Similarly, reduction of the photopigment by mutations which affect the opsin level, also eliminated the PDA (see Sections 4 and 5; [41]). The duration of the maximal PDA is reduced in correlation with the amount of pigment present until the PDA is completely abolished.

3.5. The light-induced current (LIC)

The development of the isolated ommatidium preparation [42,43], in which the pigment cells are stripped off the photoreceptor cells, made it possible to apply the powerful patch clamp technology to *Drosophila* photoreceptors. Although the photoreceptors are still physically joined within the ommatidium, whole-cell voltage clamp recordings from single photoreceptors show no evidence of electrical or dye coupling. The photoreceptor axons are broken off at the basement membrane and calculations suggest that currents of up to ca. 5 nA can be effectively voltage clamped in the remaining cell bodies. However, the maximum light-induced currents exceed 20 nA at resting potential and cannot be accurately measured. The use of the whole-cell voltage clamp procedure made it possible, for the first time, to record the light-induced current (LIC) as a function of membrane potential in the small photoreceptor cells of *Drosophila* and to calculate the permeability to the main ions (Figs. 8, 9, Table 2). This method also allows recordings of the LIC under controlled composition of the intracellular and extracellular environment and to apply Ca^{2+} indicators to measure light-induced changes in cellular Ca^{2+}. The use of the LIC, with its very large signal to noise ratio, in *Drosophila* has had a great impact on our understanding of invertebrate phototransduction, because it allows the application of two of the most powerful techniques in neurobiology, molecular genetics and patch clamp recordings, to *Drosophila* photoreceptors.

4. Genetic screening for phototransduction mutants of *Drosophila*

Extensive genetic studies of the fruit fly *Drosophila*, initiated at the turn of the century by Morgan [44], and since then greatly extended and advanced by many other laboratories have established *Drosophila* as an extremely useful experimental system for genetic analysis of complex biological phenomena. The relatively small size of the *Drosophila* genome, ease of growth, and rapid generation time make this system ideally suited for screening large numbers of mutagenized individuals for defects in any phenotypically observable or measurable trait. The creation of balancer chromosomes, containing dominant markers and multiple inversions, which eliminate recombination with wild-type chromosomes, allows any mutation, once recognized, to be rapidly isolated and maintained (for review see Ref. [6]). Importantly, germ line transformation using P-element transposition [45], combined with the availability of tissue specific promoters [46], allows the introduction of cloned genes back into the organism, thus providing a way to study *in vitro* modified gene products in their natural cellular environment [47,48].

4.1. A screen based on visual responses

In principle, it is possible to systematically screen each of the four chromosomes (chromosomes II and III contain about 80% of the total genes) of *Drosophila* for mutations that produce any identified phenotype. Systematic genetic screens that identify genes encoding products involved in visual behavior or photoreceptor cell function have been used by several laboratories over many years. These screens have been designed to isolate mutants with defects in a number of visual functions such as optomotor response, phototaxis or electrophysiological response to light. A screen based on the optomotor response measuring the ability of the fly to adjust its orientation and keep moving an object fixed in his visual field has been developed by Heisenberg and Goetz [49]. As expected, most (but not all) of these behavioral mutants affect the function of the optic lobes and visual processing centers rather than genes involved in phototransduction [49]. Phototaxis screens were initiated by Benzer and co-workers using countercurrent distribution devices [50]. This procedure separates phototactic from non-phototactic flies by allowing the flies to run towards a light source. These phototaxis screens generated a collection of mutants that eventually were shown to define genes necessary for development of the visual system, photoreceptor cell integrity, or phototransduction (for review see Ref. [51]).

To isolate mutants that are specifically defective in the visual transduction pathway several groups have initiated the use of electroretinogram (ERG) recording either for genetic screen or characterization of existing mutants [52,53]. Despite the need to apply the assay to individual flies, which is extremely labor intensive, the simplicity of the ERG measurement and particularly its direct relationship to the output signal of the photoreceptor cells indeed yielded mutants that were specifically defective in the phototransduction process [54,55]. However, even the ERG screen did not yield the large number of mutants that are expected from a multistep and complex process such as the phototransduction cascade. Because of the very high gain of the phototransduction process and the presence of a ceiling to the depolarization signal, this ceiling is reached when only a fraction of a percent of the rhodopsin molecules are excited ([27], for reviews see Refs. [36,56]). The presence of the phototransduction components in great abundance ensures that this ceiling will be achieved even in mutations that result in significant reduction in concentration or subtle malfunction of the phototransduction components, thus necessitating more sophisticated assays.

4.2. A screen based on the PDA

In order to find mutants that specifically affect the phototransduction cascade and to identify those with subtle phenotypes which may evade the screening strategies described above, Pak and co-workers began to use the PDA screen (see Fig. 2A,B). The PDA, which is discussed in Sections 3.4 and 5.3, is elicited by a large net conversion of rhodopsin to metarhodopsin. This is achieved in the fly by genetically removing the red screening pigment and by application of colored light (blue) which is preferentially absorbed by the R state of the Rh1 opsin. Large net conversion of rhodopsin to metarhodopsin brings about disruption of the termination process at the photopigment level which results in sustained excitation long after the light is turned off. Throughout this period the photoreceptors are partially desensitized (inactivated) and are less sensitive to subsequent test lights ([39], Fig. 2A). Since the PDA tests the maximal capacity of the photoreceptor cell to maintain excitation for an extended period and is strictly dependent on the presence of high concentrations of rhodopsin, it detects even minor defects in rhodopsin biogenesis and easily scores deficient replenishment of phototransduction components which results in temporary inactivation of the phototransduction process. Indeed, the PDA screen yielded a plethora of new and most interesting visual mutants (for reviews see Refs. [37,56]). By functional criteria one group of PDA mutants lost all the features of the PDA as they were no longer able to maintain the prolonged depolarizing after potential and its associated features. They were termed *nina* mutants, which stands for neither inactivation nor afterpotential. Among the *nina* PDA mutants, 10 different allelic groups were isolated (belonging to different mutual complementation groups) marked A–K. All the *nina* mutants were found to have reduced levels of Rh1 rhodopsin; they are described in Section 5. The second group of PDA mutants lost the ability to produce the voltage response associated with the PDA but were still inactivated by strong blue light and the inactivation was removed by orange light. These mutants consisting of six allelic groups are termed *ina* A–F which stands for inactivation but no afterpotential. The *ina* mutants have a normal rhodopsin level; they are discussed in Sections 9 and 10.

4.3. A screen based on the loss of a specific antigen on Western blots

A novel and powerful screening strategy has been recently developed by Zuker and colleagues [57]. This screen is based on the loss of specific antigen on Western blots of chemically mutagenized homozygous second chromosome flies. This method, which is highly labor intensive, is very useful to screen for mutants of known proteins without prior assumptions as to the nature of the expected phenotype. The isolation of the arrestin (*arr*) and *trpl* (*trpl*) mutants (see Sections 5.3 and 7) are striking examples for the power of this method [57,58].

4.4. A screen based on preparation of antibodies against eye-specific proteins

This method which was introduced by Benzer and colleagues [59] utilizes monoclonal antibodies which are prepared blindly against all eye proteins. Then, specific proteins are identified on the basis of specific staining in the retina. The gene encoding the photoreceptor-specific antigen is cloned and sequenced later on. This method identified the cellular localization of proteins with known mutation (e.g. *trp*, [60]). It was also used to identify novel proteins (e.g. calphotin, [61]) and to make mutants, which were isolated on the basis of the cloned gene (e.g. the calphotin, *cap*, mutants, [62]).

5. The photopigment cycle

5.1. Six different opsins are expressed in the Drosophila *eyes*

The *ninaE* gene (which is defective in the *ninaE* mutants, [63]) encodes the opsin expressed in the R1–R6 photoreceptor cells, Rh1 [63,64]. The peak absorption spectrum (λ_{max}) of the rhodopsin (R) state is ~480 nm and of the metarhodopsin (M) state is ~570 nm [5]. The availability of an opsin gene and a corresponding null mutant in *Drosophila* makes it possible to carry out *in vitro* mutagenesis studies in a whole organism (for review see Ref. [6]).

Rh1 is a 373-amino-acid polypeptide that is homologous to bovine and human rhodopsin, in both primary sequence (22% identity at the amino acid level), and in structure, which includes the seven transmembrane domains typical for all G-protein-coupled receptors (for review see Ref. [65]). Among the conserved residues between vertebrate rhodopsin and Rh1 is the lysine in the seventh transmembrane segment where the chromophore retinal is attached via a Schiff base [66]. Unlike vertebrates and many invertebrate species, *Drosophila* (and all dipteran flies) use 3-hydroxy-retinal [67] (for reviews see Refs. [5,35,36]). The fly has a secondary mechanism for sensitivity in UV light, namely a second photostable chromophore, which absorbs in the UV range and transfers the absorbed energy into the photopigment molecule which then activates the normal cascade [68]. The sensitizing second chromophore is 3-hydroxy retinol [67]. Screening *Drosophila* genomic and cDNA libraries for Rh1-related sequences has yielded five additional opsin genes (see Table 1). Rh2, which is 67% identical to Rh1 [69], is expressed specifically in ocellar photoreceptor cells with a peak absorption (λ_{max}) of ~420 nm [48,70]. Rh3 and Rh4 encode the visual pigments of the two classes of R7, with λ_{max} of ~345 nm and ~375 nm (see Table 1 for further details). These opsin genes encode highly related opsins that share 69% amino acid identity between themselves and 35% amino acid identity to the Rh1 and Rh2 opsins [71,72]. Rh3 and Rh4 are expressed in non-overlapping sets of R7 cells. The last two opsins, Rh5 and Rh6 are expressed in R8 cells. Rh5, which has a λ_{max} at ca. 437 nm is 31% identical to Rh1, 30% identical to Rh2, 41% identical to Rh3 and 44% identical to Rh4. In this respect Rh5 does not belong to the two groups of Rh1, Rh2 with 67% identity between themselves and Rh3, Rh4 with 69% identity between themselves (see Table 1). The opsins Rh1, Rh3, and Rh4 are also expressed in the larval photoreceptor organ [70]. The recently isolated Rh6 pigment is expressed in the second subset of R8 cells; it has an opsin absorbing maximally at ca. 508 nm, which is equally homologous to Rh1 and Rh2 (51% amino acid identity) but shows only 32% and 33% amino acid identity with Rh3 and Rh4, respectively [9]. Uniquely amongst the *Drosophila* rhodopsins, Rh6 metarhodopsin has a λ_{max} shifted to shorter wavelengths, and appears to be unstable [291]; see also Chapter 10.

Both *Calliphora* and *Musca* have been shown to have different classes of R7 photoreceptors. The major two types are referred to as R7 yellow (R7y) and R7 pale (R7p), based on their appearance under blue illumination [73] or fluorescence [74]; the R8 cells (8y and 8p) are named according to the overlying R7. R7p has a UV

Table 1
Characteristics of the various opsins in *Drosophila* eyes

Opsin	Cellular location	λ_{max} R state (nm)	λ_{max} M state (nm)	% Identity with Rh1	Rh2	Rh3	Rh4	Refs.
Rh1	R1–R6	480	570	–	–	–	–	[62,63]
Rh2	Ocelli	420	520	67	–	–	–	[48,69,70]
Rh3	R7	345	460	35	35	–	–	[71,72]
Rh4	R7	375	460	35	35	69	–	[71,72]
Rh5	R8	437	494	31	30	41	44	[8,291]
Rh6	R8	508	468	51	51	32	33	[9,291]

photopigment similar to that of *Drosophila* R7, whereas the R7y cell contains a blue sensitive rhodopsin that utilizes a sensitizing pigment similar to that found in R1–R6 cells. The Rh3-based rhodopsin has a similar spectral sensitivity and expression pattern to the pigment found in the R7p and R7marg (see Section 2; for a review and summary see Refs. [4,8]); whilst Rh5 and Rh6 appear to correspond to *Musca* and *Calliphora* 8y and 8p, respectively [291].

Since Rh2, Rh3, Rh4, Rh5 and Rh6 are expressed in minor classes of photoreceptor cells, it is difficult to analyze their properties *in vivo*, although the spectral properties of their presumptive orthologues have been investigated by a combination of intracellular recordings and microspectrophotometry in larger flies (for review see Ref. [1]). Studies of transgenic flies ectopically expressing one of the minor opsins in the major class of photoreceptor cells (R1–R6) has allowed a more thorough characterization of these molecules ([47,48]; for review see Ref. [6]). Ectopic expression is achieved by generating transgenic animals that express a chimeric gene construct using the Rh1 upstream regulatory sequences to direct the expression of the structural gene for one of the minor opsins. By using *ninaE* mutant host flies, which contain no Rh1, one effectively replaces one of the minor opsins for Rh1 in the major class of photoreceptors. Interestingly, several minor opsins that have been studied are capable of activating the phototransduction cascade of the R1–R6 cells [8,48,291]. The spectral properties of the transgenic R1–R6 cells are thus changed according to which opsin they are expressing [48], and the flies' electrophysiology and behavior is tuned to the new absorption spectra of the ectopically expressed photopigments (for review see Ref. [6]).

The ability to express a UV-absorbing rhodopsin (Rh4) in place of the major rhodopsin (Rh1) in R1–R6 cells was exploited to image intracellular Ca^{2+} concentration changes, so that long wavelength light was used to measure Ca^{2+} concentration changes while minimally exciting the UV absorbing photopigment [75]. This is an elegant solution for an inherent problem of measuring the fluorescence of Ca^{2+} indicators in photoreceptor cells, in which the excitation light of the indicator also excites the photoreceptor cell.

5.2. Mutations affecting rhodopsin biogenesis

From the 10 different complementation groups that composed the *nina* mutants only one (*ninaE*) is directly defective in opsin. The remaining *nina* genes affect the biogenesis of the photopigment.

5.2.1. Mutations affecting the chromophore

Two of the *nina* genes, *ninaB* and *ninaD* (which are still uncloned), appear to play a role in the metabolism of retinal, because these mutant strains can be rescued with dietary supplements of retinal [29,37]. *NinaB* mutant flies can be rescued only with retinal, whereas *ninaD* flies can be rescued with a number of different retinoids. These results suggest that *ninaD* acts prior to *ninaB*. It is not known whether these genes encode photoreceptor-specific proteins or ubiquitous proteins required for vitamin A metabolism in general (for review see Ref. [6]).

5.2.2. The NINAA protein participates in rhodopsin folding and transport

The power of the unbiased genetic approach for functional analysis of genes that are involved in rhodopsin biogenesis became apparent upon cloning and characterization of the NINAA protein. The *ninaA* gene has been found to encode an eye-specific 237-amino-acid polypeptide with 43% amino acid identity to the human cyclophylin protein, the intracellular receptor that binds the immunosuppressant drug cyclosporin A [76,77]. The *Drosophila* cyclophylin *ninaA* gene product is larger than the human and *Neurospora* cyclophylins having an N-terminal signal sequence and a C-terminal hydrophobic domain that can serve as membrane anchor [76,77]. Cyclophylins of various sources have been found to have peptidyl prolyl *cis–trans* isomerase activity and have therefore been implicated in protein folding. Recent findings, however, are inconsistent with inhibition of the cyclophylin prolyl *cis–trans* isomerase activity by cyclosporin A as the basis for its immunosuppressant effect. This is apparently not the case for the physiological role of the *Drosophila ninaA* cyclophylin gene product. Studies by Zuker and colleagues [78] have shown that the NINAA cyclophylin is required in the secretory pathway which transports the newly synthesized rhodopsin from the endoplasmic reticulum to its final target in the microvilli of the rhabdomeres. In the *ninaA* mutant alleles, unprocessed opsin molecules are present in the endoplasmic reticulum and colocalization of the *ninaA* gene product and opsin are found in secretory vesicles that apparently do not reach their final destination. Interestingly, *ninaA* mutants have a dramatic increase in the amount of endoplasmic reticulum (ER) membranes in their cytoplasm [78]. This build-up of the ER is dependent on the continued accumulation of improperly processed Rh1 rhodopsin. Extensive studies of the *ninaA* locus by saturation mutagenesis and *in vitro* test of prolyl *cis–trans* isomerase activity are consistent with participation of the NINAA cyclophylin in protein folding and transport of rhodopsin [79]. It seems likely that opsin is a substrate for the NINAA protein; isomerization about one out of several Xaa-Pro peptide bonds may be crucial for the proper membrane intercalation, folding, stability, or transport of opsins (for review see Ref. [56].

Analysis of transgenic flies ectopically expressing minor opsins in the R1–R6 photoreceptor cells of wild-type or *ninaA* mutants shows that NINAA acts only on Rh1 and Rh2 opsins [80]. The more distantly related Rh3, Rh4, Rh5 and Rh6 opsins probably use other cyclophilin-like molecules.

5.3. The mechanism underlying termination of metarhodopsin activity

5.3.1. Light-dependent phosphorylation of the photopigment

Rhodopsin belongs to the superfamily of G-protein-coupled receptors (see Section 6). Two features which are unique to the visual system are the presence of extremely high concentrations of rhodopsin molecules in the photoreceptor cells and the isomerization reaction of the retinal chromophore which activates the photopigment. In the fly, this reaction is virtually irreversible unless another photon hits the chromophore and regenerates the metarhodopsin back to rhodopsin (for review see Ref. [36]). These features emphasize the need for a rapid and efficient termination reaction that will inactivate the metarhodopsin immediately after activation and prevent excitation in the dark (for review see Ref. [56]). A common mechanism of inactivation of G-protein-coupled receptors is via phosphorylation of Ser/Thr sites on the cytoplasmic tail and on the third intracellular loop. Fly rhodopsin is known to undergo light-dependent phosphorylation [81,82] and *in vitro* studies have shown that phosphorylation partially inactivates the metarhodopsin by decreasing its ability to activate the G-protein (see Fig. 6, [83]). However, *in vivo* studies in a transgenic *Drosophila* mutant (*ninaE* P[Rh1Δ*356*]), in which the last 18 amino acids residues that include the serine/threonine phosphorylation sites at the C terminal domain of Rh1 were eliminated, revealed no effect on the LIC [84]. This unexpected result suggests that phosphorylation of the C terminal part of Rh1 opsin has no obvious physiological role. In contrast, in vertebrate cells the formation of rhodopsin–arrestin complex is largely determined by the phosphorylation state of rhodopsin [85,86] and truncation of the C terminal led to severe defects in response termination [87]. *In vitro* studies in *Calliphora* have demonstrated that the 49 kDal arrestin (see below) can bind to unphosphorylated metarhodopsin with comparable affinity to that of the phosphorylated state [88].

The fly rhodopsin kinase (RoK, Fig. 4) was partially purified and characterized [82]. In several aspects it resembles the bovine rhodopsin kinase and the β-adrenergic receptor kinase (for review see Ref. [89]). Rhodopsin kinase phosphorylates the

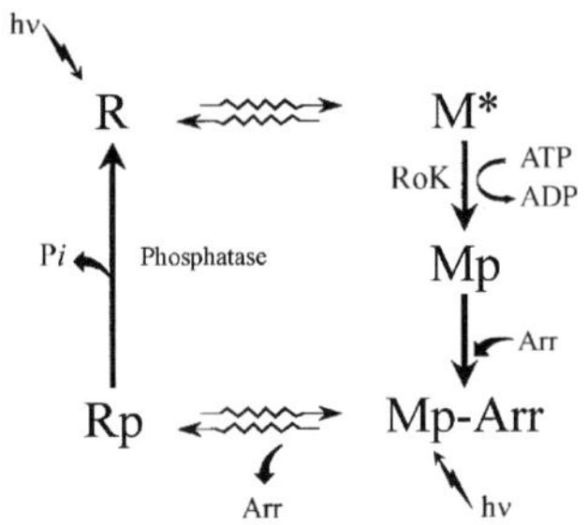

Fig. 4. Regulatory arrestin cycle secures the fidelity of phototransduction. The fly photoreceptor accommodates both a highly active rhodopsin kinase and an equally effective rhodopsin phosphatase (RDGC) which together with a 49 kDa arrestin (Arr) secure the maintenance of the photoreceptor cell and the inactivation of phototransduction in the dark after light off. This is accomplished by concerted operation of the following reactions. Photoconversion of rhodopsin (R) to a physiologically active metarhodopsin (M*) results in rapid phosphorylation of M by rhodopsin kinase (RoK) to give Mp and binding of 49Arr (Mp-Arr). It is still not entirely clear if the phosphorylation precedes 49Arr binding. Phosphorylation of M decreases its ability to activate the G-protein. Binding of 49Arr quenches the ability of M to activate G-protein and protect it from phosphatase activity. Absorption of a photon by the Mp-Arr complex regenerates it to Rp with concomitant release of 49Arr. The Rp then becomes a substrate for RDGC rhodopsin phosphatase (Phosphatase) that reintroduces it to the excitable rhodopsin pool without reinitiating of phototransduction in the dark. Binding and release of 49Arr to Mp and Rp, respectively, confines the rhodopsin phosphatase activity only to Rp, thereby securing the fidelity of phototransduction (adapted from [56]).

receptor only after photoconversion of rhodopsin to metarhodopsin. The phosphorylation is optimal in the presence of EGTA, as under these conditions the Ca^{2+} dependent rhodopsin phosphatase is inhibited (see Fig. 6 and see below). As other kinases of G-protein-coupled receptors, the fly rhodopsin kinase binds to the membranes when the receptor is in the active state (metarhodopsin), and translocates to the cytosol after regeneration of metarhodopsin to the inactive state (rhodopsin) [82]. The mechanism of translocation of the bovine rhodopsin kinase and of the β-adrenergic receptor kinase has been found to depend on isoprenylation of the rhodopsin kinase [90] or in the case of the β-adrenergic receptor kinase, on binding to the free βγ subunits of the activated G-protein [91]. The recognition of metarhodopsin by rhodopsin kinase seems to include several sites, as peptides derived from the sequence of the *Drosophila* rhodopsin carboxyl tail were not recognized by the rhodopsin kinase. In contrast, a peptide used as substrate for the bovine rhodopsin kinase [92], without any resemblance to the sequence of the fly rhodopsin was phosphorylated by the fly rhodopsin kinase, although the K_m was high (at mM levels) and the turnover was slow. The fly rhodopsin kinase is inhibited by heparin, in a similar manner to the β-adrenergic receptor kinase, and is not inhibited specifically by sangivamycin, a specific inhibitor of the bovine rhodopsin kinase ([82], for review see Ref. [56]). Recently, a new rhodopsin kinase with a molecular mass of 80 kDa was purified from octopus eyes. Unlike vertebrate rhodopsin kinase the octopus rhodopsin kinase was directly activated by the βγ subunits of the photoreceptor G-protein. The deduced amino acid sequence of the cloned octopus rhodopsin kinase was highly homologous to the β-adrenergic receptor kinase over the entire molecule, while the homology to the mammalian rhodopsin kinase was significantly lower [93]. The octopus rhodopsin kinase is expressed solely in the retina like the fly rhodopsin kinase activity [93].

5.3.2. Binding of arrestin quenches the activity of metarhodopsin and protects it from phosphatase activity

To complete the immediate inactivation of the bovine metarhodopsin upon phosphorylation, an arrestin molecule must bind to the phosphorylated receptor. *Drosophila*, has two proteins homologous to the vertebrate arrestin [94–97]: the 49 kDa arrestin (49Arr, also referred to as Arr2 [95] or phosrestin 1 [94,99]) and the 39 kDa arrestin (39Arr, also referred to as Arr1 [95] or phosrestin 2 [98,99]); both undergo light-dependent phosphorylation. The phosphorylation of 49Arr is already evident at dim light, while phosphorylation of 39Arr requires much stronger light intensities. Both arrestins are phosphorylated by a Ca^{2+}–calmodulin-dependent kinase (CaM kinase [83,100]). The 39Arr was originally isolated by subtractive-hybridization techniques designed to identify genes preferentially expressed in the eye and was shown to encode a 364-amino-acid protein that displays over 40% amino acid identity with human and bovine arrestins [96,97]. The 49Arr protein was purified on the basis that it is one of the most abundant light-dependent phosphoproteins in the eye. Using N-terminal sequence analysis, the gene encoding 49Arr was cloned [94,95]. Like 39Arr, 49Arr also shows extensive amino acid similarity throughout its sequence with human and bovine arrestin (>40%

identity), but it contains the extended C-terminal region that is missing in 39Arr. The functional significance of variation in this C-terminal domain is not understood. Quantitative Northern blot analysis [95] and densitometric scanning of two-dimensional (2D) protein gels [99] have shown that the 49Arr protein is five- to sevenfold more abundant than 39Arr. Interestingly, rhodopsin appears to be approximately fivefold more abundant than 49Arr (for review see Ref. [101]).

The fact that two arrestin genes expressed in *Drosophila* photoreceptors have been isolated raises the issue of whether both arrestins are functionally redundant or whether each arrestin has a distinct function in regulating phototransduction. *In vivo* phosphorylation studies have shown that although both arrestins become phosphorylated upon photoreceptor stimulation, the time course of 49Arr phosphorylation is much faster than that of 39Arr [99,100] (for review see Ref. [101]).

One of the most important studies which clarified the regulatory role of arrestin was carried out by Selinger and colleagues using *in vitro* assays of arrestin binding to metarhodopsin in *Drosophila* and *Musca* eyes [83]. Upon conversion of rhodopsin to metarhodopsin, the fly 49Arr translocates from the cytosol to the membrane, and is released to the cytosol when metarhodopsin is regenerated back to rhodopsin (Fig. 4). 39Arr, on the other hand, always remains membrane bound [83]. These studies indicated that the functional role of metarhodopsin phosphorylation and the binding of 49Arr is to terminate the activity of metarhodopsin [83]. Binding of 49Arr also protects the phosphorylated metarhodopsin against phosphatase activity. Only upon regeneration of phosphorylated metarhodopsin to rhodopsin, arrestin is released and the phosphorylated rhodopsin is exposed to phosphatase activity (Fig. 4). These combined actions are crucial for preventing reinitiation of phototransduction in the dark as the reversible binding of 49Arr directs the protein phosphatase only toward the inactive rhodopsin state [83] (for reviews see Refs. [56,101,102]).

5.3.3. The PDA arises by the presence of metarhodopsin in excess over 49Arr

The use of an illumination protocol that results in a PDA for isolation of *Drosophila* visual mutants has been very successful but the underlying mechanism for the PDA response was obscure for many years. The first clue to the step in the phototransduction cascade which is disrupted upon initiation of the PDA response came as a result of analysis of light dependent GTPase activity. It was found that the *in vitro* assay of light dependent GTPase activity closely resembles the PDA phenomenology, indicating that the reactions that give rise to the PDA must take place at the level of the interaction of metarhodopsin with the G-protein [103] (Fig. 5). With the recognition of the role of 49Arr in termination of metarhodopsin activity and its light-dependent reversible binding to the rhodopsin containing membranes, it was possible to show that after a short pulse of blue light, membrane depleted of 49Arr shows undiminished GTPase activity, while membranes containing 49Arr undergo a time-dependent progressive decline of their GTPase activity (Fig. 6). Furthermore, re-addition of purified 49Arr to the membranes that had been previously depleted of 49Arr restored the time dependent decline of their light dependent GTPase activity ([83], for review see Ref. [56], Fig. 6). The most

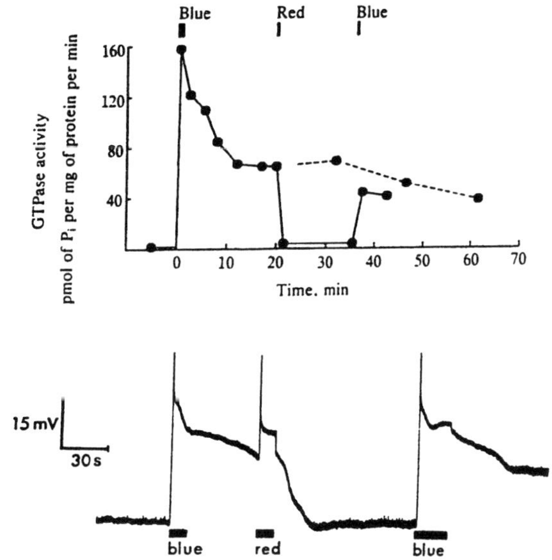

Fig. 5. Effect of colored lights on the PDA and GTPase activity in *Musca* photoreceptor and in eye membrane preparations, respectively. (A) GTPase activity of *Musca* eye membranes (upper curve). *Musca* eye membranes suspended in a solution of 40 mM MOPS buffer (pH 6.7), 6 mM $MgCl_2$, and 1 mM dithiothreitol were incubated at 20°C and illuminated with saturating blue and red light as indicated. At each point, an aliquot was removed for 1 min assay of GTPase activity at 4°C as described previously [103]. (B) Intracellular recordings from *Musca* photoreceptor cells (lower curve). Intracellular recordings were performed on living *Musca* flies. Blue and red illuminations were obtained by a Xenon light beam (XBO, 150 W) passed through either a blue filter or a red-orange 610 nm edge filter as indicated (from [115]).

straightforward explanation of these results is that cellular 49Arr is present at concentrations which are insufficient to titrate and inactivate all the metarhodopsin generated by a large net conversion of rhodopsin to metarhodopsin, leaving the excess of metarhodopsin persistently active. This explanation easily accounts for the elimination of the PDA response by reducing the cellular level of rhodopsin and the need to convert large net amounts of rhodopsin to metarhodopsin to elicit the PDA response (for review see Ref. [56]).

With regard to arrestin function *in vivo*, significant insight has been gained from the isolation and characterization of *Drosophila* mutants defective in arrestin expression [57]. Because of the difficulty in predicting a reliable and easily scorable phenotype that defines the loss of arrestin function, Zuker and colleagues [57] used a genetic screen based on the loss of arrestin antigen on immunoblots to isolate several mutant alleles of the two arrestins, *arr2* and *arr1* (see Section 4.3). Electrophysiological analysis of strong *arr2* and *arr1* alleles revealed a set of phenotypes consistent with the stoichiometric requirement of arrestin binding for metarhodopsin inactivation *in vivo*. In whole-cell voltage-clamp experiments, a significant reduction in arrestin levels leads to abnormally slow termination of light-activated currents with

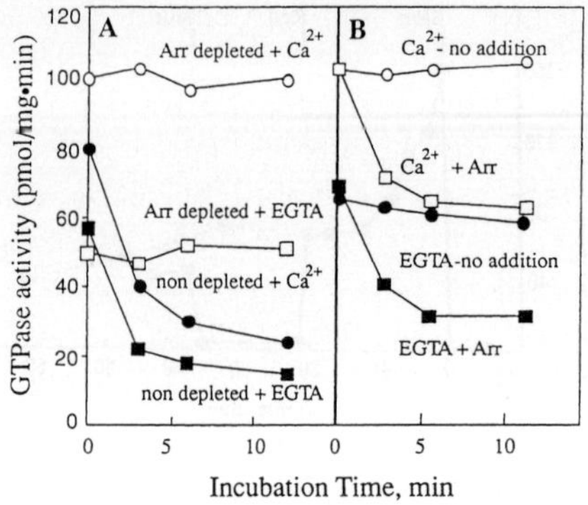

Fig. 6. Effect of 49Arr on metarhodopsin-stimulated GTPase. At zero time, *Musca* eye membranes in homogenization buffer were illuminated for 10 s with blue light, ATP was added to 0.5 mM and the systems were incubated at 25°C in the dark. Where indicated, 0.1 mM Ca^{2+} or 1 mM EGTA was added together with ATP. At the times indicated, aliquots were taken for 1-min assay of GTPase activity at 4°C in the dark [103]. (A) Open and filled symbols depict GTPase activities of membranes derived from eyes illuminated with orange (arrestin-depleted) and blue (non-depleted) light, respectively. Data are from a single experiment with single experiment with similar results obtained in four other experiments. (B) Membranes derived from orange-illuminated *Musca* eyes were incubated with ATP together with 0.1 mM Ca^{2+} or 1 mM EGTA as in A. A partially purified arrestin was prepared by two rounds of release and binding to orange- and blue-illuminated membranes, respectively. The arrestin was estimated to be 90% pure by SDS/PAGE. Each GTPase assay mixture contained 10 μg of eye membrane proteins (equivalent to two eyes) and, when indicated, the amount of arrestin (Arr) added at zero time was derived from an equivalent number of eyes. Open symbols are GTPase activities of membranes incubated in the presence of 0.1 mM Ca^{2+} and filled symbols are GTPase activities of membranes incubated with 1 mM EGTA. Data are from a single experiment. Similar results were obtained in two other experiments with different preparations of arrestin.

each photon apparently eliciting a continuous train of quantum bumps [32,57] (for review see Ref. [101]).

In vivo evidence for defects in metarhodopsin inactivation in arrestin mutants came with the analysis of the PDA. If arrestin is involved in metarhodopsin inactivation, as shown *in vitro* by measurements of GTPase activity, then defects in arrestin function may be expected to show defects in the PDA process. Interestingly, *arr2* mutant photoreceptors (defective in 49Arr) undergo a PDA with 10 times less photoconversion of rhodopsin to metarhodopsin [57]. This finding fits with the *in vitro* studies and suggests that 49Arr interacts stoichiometrically to inactivate metarhodopsin and further suggests that saturation of 49Arr function by excess metarhodopsin may represent the basis for the PDA ([57], for review see Ref. [101]). Several findings are consistent with this model. First, the ratio of arrestin to rhodopsin in wild-type photoreceptors roughly predicts the requirement for photoconversion of at least 20% R to M for PDA induction [36]. Second, the *Drosophila ninaA* mutant with reduced rhodopsin levels is unable to generate a PDA (see Sections 4.2 and 5.2.), but a corresponding decrease in 49Arr levels restores the PDA phenotype [57]. Thus, arrestin is required for termination of the light response and its lower concentration relative to rhodopsin enables production of the PDA (for reviews see Ref. [56,101]).

Until recently the role of arrestin phosphorlyation was unclear. However, a recent study utilizing an *arr2* mutant allele (S366A), in which the CaMkinase phosphorylation site (serine 366) is replaced by alanine, indicated that arrestin phosphorylation is required in order for arrestin to dissociate from rhodopsin, and in the absence of phosphorylation remains permanently bound. Although the S366A mutant had a similar phenotype to a null *arr2* allele, this appeared to be accounted for by all the arrestin becoming bound to rhodopsin and hence unavailable for further action [292].

6. Coupling of photoexcited rhodopsin to inositol phospholipid hydrolysis

6.1. Light-activated G-protein

It has been well established that photoexcited rhodopsin activates a heterotrimeric G-protein in invertebrate photoreceptors. In fly photoreceptors application of pharmacological agents that are known to activate G-proteins, such as hydrolysis-resistant GTP analogues and fluoride mimic light-dependent activation of the photoreceptor cells [104]. Noise analysis of the chemically induced noise has suggested that there is only little gain at the G-protein stage [104]. The light-dependent G-protein of fly (*Musca*) eye was first demonstrated in membrane preparations using the α-^{32}P-labeled azidoanilido analog of GTP [105]. Polyacrylamide gel electrophoresis and autoradiography revealed a labeled 41-kDa protein band in the blue-illuminated membranes that was not seen in controls illuminated with red light. Binding assays, which show strict dependence of GTPγS binding to eye membrane preparation on production of metarhodopsin with blue light, also reveal involvement of G-protein in fly phototransduction [105]. Strong evidence for the participation of G-protein in fly phototransduction came from a study which demonstrated that induction and suppression of the PDA by blue and red lights, respectively, can be mimicked *in vitro*, by measuring GTPase activity during blue and red illumination of *Musca* eyes [103] (Fig. 5). The blue/red dichotomy of G-protein activation was also demonstrated in *Calliphora* eyes [81].

6.2. Mutations affecting genes which encode visual system-specific G-proteins

Two genes encoding visual system-specific G-protein subunits have been isolated in *Drosophila* [106,107]. These genes, *dgq* and *gbe*, encode a $G_q\alpha$ (called DG_q or $DG_q\alpha$ in *Drosophila* [106], and $G_q\beta$ subunit, respectively. The eye-specific $G_q\beta$ ($G_q\beta e$) homologue shares only 50% amino acid identity with other Gβ-proteins and may therefore represent a unique member of this highly conserved gene family. Analysis of recently generated mutants defective in $G_q\beta e$ (*Gβe¹* and *Gβe²*) showed a large reduction in light-stimulated [^{35}S]GTPγS binding in the *Gβe¹* mutant [108,109] and that the $G_q\beta e$ protein is essential for coupling $G_q\alpha$ with metarhodopsin [108]. The *Gβe* mutants show slow deactivation of the light response, suggesting a role for the

$G_q\beta e$ subunit in terminating some active state of the signaling cascade [108]. It cannot be excluded, however, that the absence of $G_q\beta e$ may reduce the $G_q\alpha$ level and that this is the cause of slow response termination (see below).

The eye-specific DG_q is most homologous to a vertebrate G-protein that activates phospholipase C (PLC); this homology is consistent with its role as activator of PLC in *Drosophila* photoexcitation (see below). The *dgq* gene produces two alternatively spliced transcripts that should generate two slightly different $G\alpha$ products [106,110]. However, only one of these forms (DG_{q1}) seems to function in phototransduction *in vivo* [111]. An affinity purified anti $DG_q\alpha$ immunoglobulin blocked the light-dependent GTPase activity of *Drosophila* head membranes. The dominantly-active DG_{q1}*-203* mutant (also carrying a wild-type copy of $Dg_q\alpha$) exhibited constitutively active GTPase, reduced sensitivity to light and an abnormally slow and small ERG [110]. These data indicate that $DG_q\alpha$ mediates phototransduction.

Additional mutants which are defective in $DG_q\alpha$ were generated by the loss of immunoreactivity on Western blots [111] (see Section 4.3). Reduction of $DG_q\alpha$ to ~1% by the mutation $G\alpha q^1$ had virtually no effect on the level of $G\beta$, PLC or rhodopsin, but reduced sensitivity to light by more than a 1000-fold. Flies having only one copy of mutant $G\alpha q^1$ (the other chromosome having a lethal deficiency eliminating DG_q entirely) were almost totally insensitive to light thus strongly suggesting that there is no parallel pathway mediated by another G-protein [111], as has been suggested for the *Limulus* ventral photoreceptor [31]. Manipulation of the $DG_q\alpha$ level by expressing different amounts of DG_{q1} or by using transgenic flies expressing DG_{q1} under the control of the inducible heat-shock promoter showed that the DG_{q1} level was strongly correlated with the sensitivity to light and quantum bump production. Strikingly, the average amplitude of the quantum bump was not affected by a highly reduced DG_{q1} level. A similar effect was found by reduction in functional G protein in *Limulus* eye by GDPβS inhibition [112], thus strongly suggesting that there is little [104,113] or no [111] gain in the G-protein stage of the phototransduction cascade.

6.3. *Light-activated phospholipase C*

Evidence for a light-dependent PLC in fly photoreceptors came from biochemical experiments in membrane preparations of *Musca* and *Drosophila* eyes [105]. An eye membrane preparation was developed in which light-dependent phosphoinositide hydrolysis could be studied under defined conditions allowing the effects of activators and inhibitors to be analyzed. *Musca* eyes or *Drosophila* heads containing intact cells were preincubated with [^{3}H]inositol to label the phosphoinositides. The membrane prepared from these cells was used to analyze the hydrolysis of phosphoinositides. The *Musca* eye membrane preparation responded to illumination by an increase in the accumulation of $InsP_3$ and inositol bisphosphate ($InsP_2$), the respective products of polyphospoinositide hydrolysis by a phospholipase-C-type enzyme. Addition of 2,3-diphosphoglycerate (DPG), a known inhibitor of $InsP_3$ phosphatase [114], substantially decreased the accumulation of $InsP_2$ and concurrently increased the accumulation of $InsP_3$, which thus becomes the major product of the light-induced phosphoinositide hydrolysis (Fig. 7). It is thus apparent that fly membranes are endowed with the enzymatic system necessary to produce $InsP_3$ and to eliminate it after it has been produced. Both the robust light-dependent preferential production of $InsP_3$ and the efficient turn-off mechanism to stop its action are consistent with an internal messenger role of $InsP_3$ in fly phototransduction (for review see Ref. [115]).

6.4. Production of inositol phosphates is controlled by a G-protein

There are a number of indications that light-dependent PLC activity is controlled by a G-protein (fly, [105]; squid, [116,117]). *In vitro* measurements on the squid retina using exogenous phosphatidylinositol 4,5, bisphosphate (PIP_2) demonstrated that light-activated PLC is GTP dependent [116]. In the fly, addition of 0.1 mM $GDP_{\beta}S$ completely inhibited blue light-induced production of inositol phosphates. In addition, application of GTPγS followed by illumination with a pulse of blue light persistently enhanced the production of inositol phosphates. Also, increased accumulation of $InsP_2$ and $InsP_3$ was observed in *Musca* eye membranes incubated with fluoride ions (F^-) in the dark, and this effect was $GDP_{\beta}S$ sensitive [105]. This effect resembles the ability of F^- to bypass the receptor in other systems and directly activate enzymes such as adenylate cyclase and cGMP phosphodiesterase through its actions on the stimulatory G-protein and transducin, respectively [118,119].

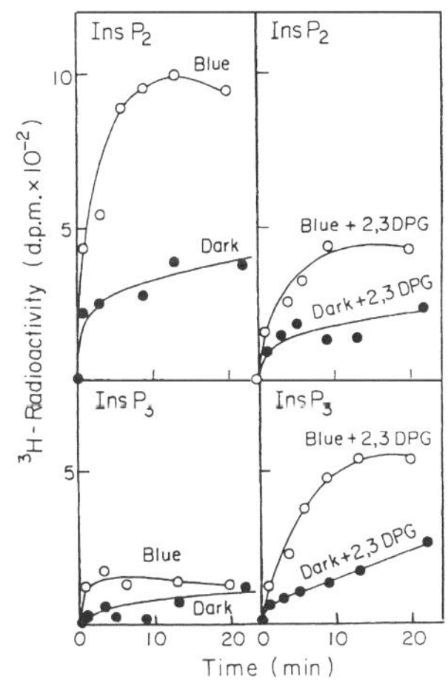

Fig. 7. Light-induced production of $InsP_3$ and $InsP_2$ in *Musca* eye membranes and effect of 2,3 diphosphoglycerate (DPG). Equivalents of 100 *Musca* eyes each cut into two halves were incubated in the dark for 4 h at 30°C with [^{3}H] inositol. Preparation of eye membranes and measurement of light-dependent phosphoinositide hydrolysis were carried out as described previously [105]. The medium was adjusted with Ca-EGTA to give 50 nM free Ca^{2+}. The reaction was initiated by the addition of membranes (500 μg protein/ml). The right column shows the effect of the $InsP_3$ phosphatase inhibitor DPG (10 mM) added together with $MgCl_2$ (10 mM) (from [105]).

6.5. *A mutation in the* norpA *gene, which encodes for PLC, blocks phototransduction*

One of the key pieces of evidence that light-activated PLC is a necessary step in visual excitation in the fly was provided by the isolation and analysis of a PLC gene of *Drosophila* – the no receptor potential A (*norpA*) (for reviews see Refs. [120,121]). The *norpA* gene encodes a β-class PLC [122] that is predominately expressed in the rhabdomeres of the compound eye [123]. It was one of the first PLC cDNA sequences to be elucidated.

The *norpA* mutant [54,124] has long been a strong candidate for a transduction-defective mutation because of the drastically reduced receptor potential. The reduction in receptor potential amplitude in the *norpA* mutant is variable in different alleles: in the strongest alleles the light response is totally abolished while the photopigment is normal (for reviews see Refs. [121,125]). A correlation between the *norpA* mutant and PLC was first suggested by Inoue et al. [126] who reported that PLC activity, which is abundant in normal *Drosophila* eyes, is drastically reduced in *norpA* mutants. Moreover, in several *norpA* alleles tested, the degree of reduction in PLC activity was found to be correlated with behavioral defects as well as with a reduced ERG amplitude [126]. However, to make the above findings more definitive, a demonstration of deficiency in light-activated PLC was necessary. The correlation between light-activated PLC and the *norpA* mutation was provided by electrophysiological [127] and biochemical studies of the *norpA* allele $norpA^{H52}$ [115] which was found to be a reversibly temperature-sensitive mutant. At permissive temperature (19°C) the ERG of this allele is normal. However, raising the temperature above 35°C abolishes the ERG instantaneously in a reversible manner [127]. Light-dependent phosphoinositide hydrolysis in either $norpA^{H52}$ or wild-type derived head membranes reveals that in wild-type membranes, there was practically no difference in phosphoinositide hydrolysis at 21°C or 37°C. On the other hand, incubation of $norpA^{H52}$ membranes at the restrictive temperature of 37°C considerably reduced the rate and extent of $InsP_2$ accumulation, as compared to membranes incubated at 21°C. Control experiments indicated that the temperature dependence of PLC activity in $norpA^{H52}$ membranes was solely due to the effect on PLC [115].

The most conclusive evidence that the *norpA* gene encodes light-activated PLC came from the cloning and sequencing of the *norpA* locus of *Drosophila* [122]. The *norpA* protein is composed of 1095 amino acid residues and has extensive sequence similarity to a PLC amino acid sequence from bovine brain [128]. Recent studies using transgenic *Drosophila* carrying the *norpA* gene on null *norpA* background have shown that the transformed flies are rescued from all the physiological, biochemical and morphological defects which are associated with the *norpA* mutations [129,130]. The *norpA* mutant thus provides essential evidence in favor of the inositol–lipid signaling system as the pathway of excitation in invertebrates, by showing that no excitation takes place in the absence of functional PLC. By contrast, the events downstream of PLC activation, including the question of whether $InsP_3$ is an essential messenger of excitation, remain unresolved and controversially discussed. This issue is covered separately (see Section 8).

6.6. *Mammalian homologues of NORPA*

PLCs have been isolated from a variety of mammalian tissues (for review see Ref. [131]), and complementary DNAs encoding some of these proteins, as well as complementary DNAs encoding two *Drosophila* PLCs have been cloned and sequenced. Based on deduced amino acid sequences and immunological data, PLCs have been grouped into three classes: β, γ, and δ [131] that differ in primary structure, molecular weight, and mode of regulation. Activation of β class PLCs is mediated through a G-protein-dependent mechanism, whereas γ PLCs are activated by tyrosine residue phosphorylation [132]. All known eukaryotic PLCs have two regions of extensive amino acid sequence homology, which are approximately 120 and 150 amino acids in length, respectively [131]. The sequence conservation, along with *in vitro* biochemical data [133], suggests that it contains the catalytic site of these enzymes.

Conserved regions of *norpA* cDNA were used to isolate bovine cDNAs that would encode four alternative forms of phospholipase C of the β class that are highly homologous to the NORPA protein and expressed preferentially in the retina. Two of the variants are highly unusual in that they lack much of the N-terminal region present in all other known phospholipases C. The sequence conservation between these proteins and the NORPA protein is higher than that between any other known phospholipases C. G-protein-activated sequence motifs found in proteins of this superfamily are found conserved in all four variants of the bovine retinal protein as well as the NORPA protein, but not in other phospholipases C. The results suggest that these proteins together with the NORPA protein constitute a distinctive subfamily of phospholipases C that are closely related in structure, function, and tissue distribution [134].

Localization of the bovine retinal PLC transcripts by *in situ* hybridization revealed that the mRNAs were localized in the outer nuclear layer, inner nuclear layer and ganglion cell layers [134]. Immunolocalization further showed that the NORPA homologue PLCs are localized to the cones but not to rod photoreceptor cells [134].

7. The light-sensitive channels

Immuno-gold labeling and immunofluorescence studies indicate that the putative light-sensitive channels are confined to the microvillar membrane which presents considerable difficulties for access with patch electrodes. Correspondingly there are no reports of *in situ* channel recordings in patches in *Drosophila* although such recordings have been reported in *Limulus* and mollusk photoreceptors (Chapter 8). The light-sensitive conductance in *Drosophila* has therefore been studied under whole-cell voltage clamp conditions using the dissociated ommatidial preparation ([42,43]; see Section 3.5). It has, however, proven possible to record single channel activity in patches of at least one class of channel when heterologously expressed [135–137].

The light-sensitive channels in invertebrate photoreceptors have generally been considered to be non-selective cation channels, and indeed the channels in *Droso-*

phila readily permeate a variety of monovalent ions including Na, K, Cs, Li and even large organic cations such as TRIS and TEA [42,43]. However, the reversal potential of the LIC (see Fig. 8) shows a marked dependence on extracellular Ca^{2+}. Assuming Goldman Hodgkin Katz (GHK) constant field theory, the quantitative dependence implies a relative Ca^{2+} permeability (P_{Ca}:P_{Na}) of ca. 50:1 [42,138,139]. Consistent with these electrophysiological determinations, the ability of the light-sensitive channels to permeate Ca^{2+} has been directly demonstrated by measurements of very large light-induced Ca^{2+} influx signals with Ca^{2+} indicator dyes [75,140,141].

7.1. trp *and* trpl *are putative light-sensitive channel genes*

The transient receptor potential (*trp*) mutant of *Drosophila* [142,143] and the no steady state (*nss*) mutant of the sheep blowfly *Lucilia* [144–146] are so-called because the response decays to baseline during prolonged light stimuli of medium intensity (Fig. 2F). Due to the ability of La^{3+}, a blocker of Ca^{2+} channels and transporters, to mimic the *trp* and *nss* phenotype in wild-type (WT) flies [147,148], it was originally suggested by Minke and Selinger [149] that the *trp* gene product (TRP) is a Ca^{2+} transporter and that the decay might be due to depletion of calcium in intracellular stores [149]. This explanation for the decay phenotype is still debated (see Section 10.3); however, the identification of *trp* as a Ca^{2+} transporter or channel is now generally accepted. Thus when investigated under whole-cell voltage clamp conditions, the fundamental defect in *trp* was revealed as a change in the ionic selectivity of the light-sensitive conductance. Specifically, the relative Ca^{2+} permeability of the light-sensitive conductance in *trp* was reduced by a factor of ca. 10 [138,139], along with a significant change in the relative permeability to different

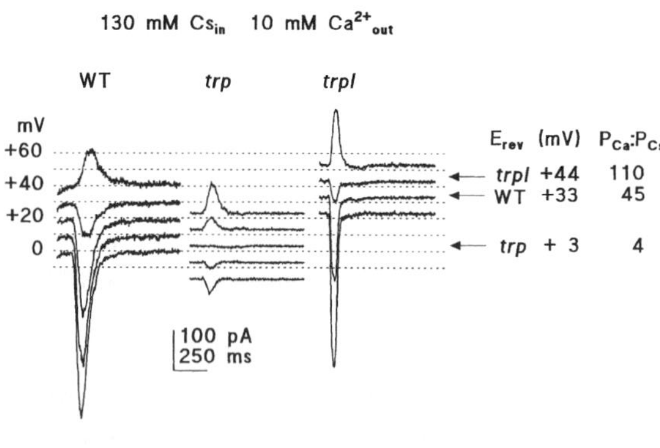

Fig. 8. Reversal potentials determined under bi-ionic conditions with 10 mM external Ca^{2+} and 130 mM internal Cs^+ in WT, *trp*^{P301} and *trpl* photoreceptors. Responses to 50 ms flashes were recorded as the holding potential was stepped in 10 mV intervals as indicated (WT from −2 to +38 mV; *trp* from −17 to +23 mV, and *trpl* from +23 to +53 mV). In addition to 130 mM Cs-gluconate the pipette solution contained 10 mM EGTA, 1 mM $CaCl_2$ to buffer Ca^{2+} influx. Average reversal potentials from several cells and calculated permeability ratios indicated on right (adapted from [139]).

monovalent ions (Table 2). The reduced Ca^{2+} permeability in the *trp* mutant has also been corroborated by the demonstration of a reduced light-induced Ca^{2+} influx signal using Ca^{2+} indicator dyes [140,141] or Ca^{2+} selective microelectrodes [150]. The effect of the *trp* mutation on ionic selectivity can be precisely mimicked in WT photoreceptors by bath application of La^{3+} in micromolar quantities [138,147].

The effects of the *trp* mutation on the pore properties of the light-sensitive channels allow at least two interpretations. Either the pore of the channel is altered in the mutant, or there are two classes of channel in WT flies: one with the permeation properties of the channels remaining in the *trp* mutant, and a second class with high Ca^{2+} permeability which is absent in *trp* and blocked by La^{3+}. Clearly the results also raise the possibility that the channel may be encoded, at least in part, by the *trp* gene [138]. This suggestion is supported by molecular analysis (see Section 7.3). First cloned by Montell and Rubin [151], *trp* was originally described as encoding a novel membrane protein; however, subsequent reanalysis of the sequence [152] revealed significant homologies to voltage-gated Ca^{2+} channels (dihydropyridine receptors, DHPr). Unlike the DHPr, which contains four repeated transmembrane domains, each with six putative membrane spanning helices, the *trp* sequence encodes only one such domain. This suggests that channels formed by the TRP protein are likely to represent multimeric assemblies (see further in Section 7.3.2).

Table 2
Biophysical properties of TRP and TRPL channels

A			
Permeability ratios	P_{TRP} (in *trpl*)	P_{TRPL} (in *trp*)	P_{WT}
$P_{Ca}:P_{Cs}$	110	4.3	45.1
$P_{Ba}:P_{Cs}$	55	4.2	49
$P_{Mg}:P_{Cs}$	13.0	1.4	5.7
$P_{Mn}:P_{Cs}$	11.8	1.5	6.8
$P_{Na}:P_{Cs}$	1.27	0.84	1.16
$P_{Li}:P_{Cs}$	0.89	0.80	0.89

B	γ (pS) (divalent free)	γ (pS)	τ (ms)	IC_{50} Mg (mM) 0 Ca	1.5 Ca	La^{3+} block (10 μM)
TRP (in *trpl*)	35	8	2, 0.2	(0.28)	(1.3)	total
TRPL (in *trp*)	70	35	0.4		4	none
WT			2, 0.2	0.28	1.3	90%

A: Permeability ratios ($P_x:P_{Cs}$) derived from bi-ionic reversal potential data. Values expressed with respect to internal Cs^+ (130 mM) and determined in respective mutants, i.e. TRPL channels measured in the *trp* mutant, TRP-dependent channels in *trpl* (data from [139] B: Estimates of single channel characteristics and ion block. Single channel conductances (γ, pS) derived from noise analysis in both divalent free and physiological Ringer's. Time constant (τ) refer to Lorentzian fits to power spectra. Mg^{2+} block has been determined only in WT (dominated by TRP channels) and *trp* mutant. (Data from Refs. [16,58,137,289] and Hardie and Reuss, unpublished).

A second putative channel gene, *trp*-like (*trpl*) was isolated using a screen for calmodulin binding proteins, and was found to show ca. 40% overall identity with *trp* ([152] for review see Refs. [153–155]). This obvious homology suggested that *trpl* might encode either a second class of light-sensitive channel, responsible for the residual current in the *trp* mutant or a second subunit of a heteromultimeric channel [24,152]. To test this, Niemeyer et al. [58] isolated a null mutant of the *trpl* gene using a screen based on Western blots (see Section 4.3). Under physiological conditions (i.e. dark-adapted, and at resting potential) the *trpl* mutation has no detectable influence on the light response (e.g. macroscopic response kinetics and bump amplitude are indistinguishable); however, the response to light in *trpl* is completely abolished by La^{3+} or in the double mutant combination (*trp*;*trpl*), indicating that the response in the *trpl* mutant is carried entirely by *trp*-dependent channels [32,58,139]. In the meantime various other subtle phenotypic manifestations of the *trpl* mutation have also been revealed by careful determinations of ionic selectivity ([139], Table 2) and noise analysis (see Section 7.2). These results currently represent the only reliable data on the biophysical properties of the TRP-dependent channels in isolation *in vivo*.

By analogy e.g. with voltage-gated K channels and the cyclic nucleotide gated channels, both *trp* and *trpl* gene products represent subunits of putative tetrameric channels. *A priori* it could be hypothesized that they assemble as homomultimers or as heteromultimers. Since null *trp* and *trpl* mutants both respond to light, each can clearly function without the other; however, a recent study based on heterologous co-expression studies and co-immunoprecipitation led to the suggestion that in WT flies the major conductance was a TRP–TRPL heteromultimer [156]. Detailed *in situ* measurements of biophysical properties including ionic selectivity and single channel conductance questioned this conclusion since it was found that the WT conductance could be quantitatively accounted for by the sum of the conductances determined in the *trp* and *trpl* mutants respectively [139]. There are however, indirect indications that there may be additional channel subunit(s) to be found [139,156] and further studies will be required to unequivocally determine the subunit composition of the light-sensitive channels.

7.2. *Biophysical properties*

What is known of the biophysical properties of the *trp*- and *trpl*-dependent channels, derives for the most part from measurements of the *in situ* conductance in the respective mutants.

7.2.1. *Ionic selectivity*

Permeability ratios to monovalent and divalent ions in WT and *trp* were originally described by Hardie and Minke [138] based upon GHK solutions to reversal potential data obtained under various ionic conditions. More recently permeability ratios have been determined more directly under bi-ionic condition where Cs^+ was the only internal cation and the bath contained only one of a variety of monovalent or divalent cations [139]. Table 2 summarizes these data for TRP- and TRPL-dependent channels (isolated in the *trpl* and *trp* mutants respectively) as well as for WT. The WT values are generally well explained by the approximately equal summation of TRP and TRPL conductances. Whilst TRPL represents a non-selective cation channel, TRP-dependent channels are highly selective for divalent cations (e.g. $P_{Ca}:P_{Cs} > 100:1$). On the (unproved) assumption of independent mobility of ions, these permeability data imply that, e.g. in WT ca. 45% of the LIC at resting potential is mediated by Ca^{2+}, 25% by Mg^{2+} and 30% by Na^{+} (in physiological Ringer's containing 1.5 mM Ca^{2+}, 4 mM Mg^{2+}, 120 mM Na^{+} and 5 mM K^{+}).

7.2.2. Single channel properties

Single channel recordings have not yet been made from native TRP or TRPL channels because of their inaccessible location in the microvilli. However, noise analysis has been used to estimate single channel properties from the so-called "run down current" (RDC). The RDC represents the spontaneous opening of light-sensitive channels which develops after several minutes of whole-cell recording in the absence of ATP or GTP in the electrode [157]. More recently similar results have been obtained from non-stationary noise analysis of the light-induced current recorded under Ca^{2+}-free conditions to slow down the kinetics of transduction thereby enabling extraction of the high frequency channel noise [137,139].

In the presence of physiological concentrations of external Mg^{2+} (4 mM), TRPL channels have an apparent conductance of 35 pS and TRP channels ca. 8 pS. Both channels however, are sensitive to block by divalent ions including Mg^{2+}, and under divalent free conditions values increase to ca. 70 pS (TRPL) and 35 pS (TRP) respectively ([137,139,157] and unpublished). The recently revised estimate of 8 pS for TRP [290] allows an estimate of the minimum number of channels recruited during a quantum bump. Thus the peak amplitude (10 pA) of an average quantum bump recorded at −70 mV represents the simultaneous opening of ca. 15 TRP channels. To estimate the total number involved also requires knowledge of the open probability for which there is no reliable information.

Power spectra have also been derived via noise analysis providing information on channel kinetics. Power spectra of TRPL channels can be fitted by two Lorentzians (time constant ca. 2 and 0.2 ms) . This would be consistent with a channel with two open states with mean open times of 2 ms and 0.2 ms, a channel with a rapid flickering behavior or a combination of the two. Noise analysis of heterologously expressed TRPL channels were indistinguishable both in terms of estimated single channel conductance and power spectra, and in this case the quantitative validity could be directly confirmed by recordings of single channels [137]. TRP channel power spectra (in *trpl*) are characterized by even higher frequencies and can be fitted with a single Lorentzian with a time constant of ca. 0.4 ms (Reuss and Hardie, unpublished). The extremely short open times of TRP and to lesser extent TRPL channels is probably of some functional significance as it should ensure that channel open time will not limit the temporal response of the photoreceptors.

7.2.3. Magnesium block and current–voltage relation

In common with a variety of channels (e.g. Ca^{2+} channels, NMDA receptors, CNG channels) both TRP and TRPL channels are blocked by Mg^{2+} ions at physiological concentrations [16]. For TRPL channels (in *trp* mutants or when heterologously expressed) the block is moderate (maximally ca. three fold with an IC_{50} of ca. 4 mM), and has little if any dependence on either voltage or extracellular Ca^{2+}. In WT flies, where the current is dominated by the TRP channels, the block is much more severe and shows a complex voltage dependence, being relatively relieved at both hyperpolarized and depolarized potentials thereby generating a characteristic S-shaped inward and outward rectification (Fig. 9). In common with voltage-gated

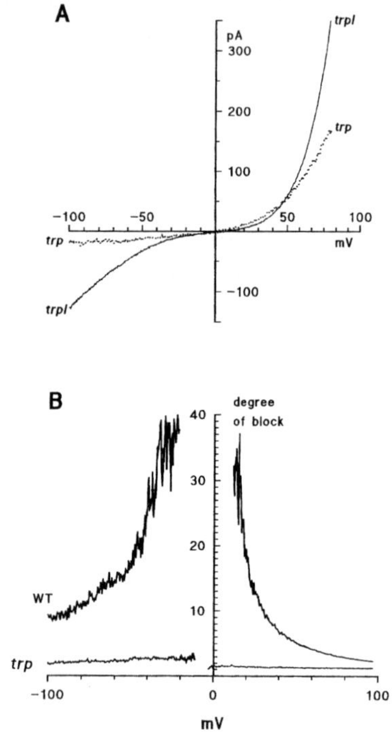

Fig. 9. Current–voltage relationships and divalent ion block. (A) Current–voltage (*i–V*) relationships of TRP and TRPL determined from voltage ramps in trp^{P301} and *trpl* photoreceptors in physiological Ringer's containing 0 Ca and 4 mM Mg^{2+}. The *i–V* curves were derived using ramps from −100 to +80 mV applied during steady state light responses, subtracting a template recorded immediately before in the dark. The current in *trp* shows a simple outward rectification; however, in *trpl* flies there is a conspicuous S-shaped inward and outward rectification which is due to a voltage dependent Mg^{2+} block of the TRP channels. (B) Voltage dependence of Mg^{2+} block. In WT flies the block by Mg^{2+} is relieved as the cell is hyperpolarized or depolarized. In *trp* there is a modest (ca. two and threefold) block which shows little voltage dependence. The voltage dependences were derived by recording voltage ramps (as in (A)) in the presence and absence of Mg^{2+} and dividing one by the other. Data were obtained in the absence of external Ca^{2+} (adapted from Refs. [16,139]).

Ca^{2+} channels, the block is also stronger in the absence of extracellular Ca^{2+}. Under these conditions there is a maximal ca. 20-fold block at resting potential with an IC_{50} of ca. 280 μM Mg^{2+}. In the presence of normal (1.5 mM) extracellular Ca^{2+} the maximum block (in 20 mM Mg^{2+}) is ca. 10-fold (as estimated from single channel conductances) with an IC_{50} of 1.3 mM. Voltage dependence of block has only been accurately determined in the absence of Ca^{2+}. The block increases ca. five fold as the cell is depolarized over the range of voltages experienced during light adaptation (ca. −70 to −20 mV) and thus has the potential to make an important contribution to the gain reduction associated with light adaptation. The block is reflected in a reduction in single channel conductance which may be an important factor in improving signal to noise ratio. A similar argument has been proposed as the functional rationale for the extremely low (ca. 100 fS) conductance of the vertebrate cGMP-gated channels [158], which is also achieved via a voltage dependent block by divalent ions (for review see Refs. [159,160]). Ca^{2+} ions themselves probably also contribute to the divalent ion block; however, this is difficult to quantify since Ca^{2+} influx via the channels also profoundly modulates channel open probability of both TRP and TRPL channels via feedback at various stages of the transduction cascade (see Sections 8 and 10).

The S-shaped rectification seen in the current–voltage (*i–V*) relationship of the WT conductance is due in part to the voltage dependent Mg^{2+} block. A similar *i–V* function is found in *trpl* mutants indicating that this is a property of the *trp*-dependent channels. Under divalent free conditions the *i–V* relationship shows a simple exponential outward rectification. The cause for this has not been determined but may reflect a voltage dependence of channel open probability. The *i–V* relationship of TRPL channels (in the *trp* mutant or in heterologously expressed channels) shows a simple but steep outward rectification both in the presence and absence of divalent ions (Fig 9). This is presumably due to an increase in channel open probability as single channel currents are approximately ohmic [136], whilst open times estimated by noise analysis, relaxation analysis or, for the case of heterologously expressed channels, by direct measurements in patches, increase at depolarized potentials [18,135,137].

7.3. Molecular analysis of TRP and TRPL

The *trp* gene was cloned by Montell and Rubin [151] by chromosome walking and found to encode a novel 1275-amino-acid membrane protein. Reintroduction of *trp* genomic cDNA into a functionally null mutant (trp^{CM}) by germ-line transformation rescued the mutant phenotype [161] indicating that the gene was indeed responsible for the *trp* mutant phenotype. The significance of the *trp* sequence was however, only first appreciated following cloning of a *trp* homologue, *trp-like* (*trpl*) by Kelly and co-workers [152]. Overall *trpl* shares 40% identity with *trp*, with much greater similarity (ca. 70%) in the putative transmembrane regions. Significantly, both *trp* and *trpl* share about 40% identity with vertebrate voltage-gated Ca^{2+} channels in the limited region of the last four transmembrane helices, S3–S6. However, the charged residues in the putative S4 helix believed to be responsible for voltage gating are replaced by non-charged residues. The N- and C-termini of TRP and TRPL both contain a number of recognizable motifs, summarized in Fig. 10.

TRPL was originally identified as a calmodulin (CaM) binding protein and analysis of the sequence indicated two regions capable of forming amphiphilic α-helices which might represent CaM binding sites [152]. In the meantime the two CaM binding sites (CBS1 and CBS2) have been localized more precisely by CaM overlays on fusion protein fragments [162,293]. CBS1 corresponds to one of the originally identified amphiphilic helices and is contained within residues 710–728; but the putative site for CBS2 was incorrect, and it is now believed to lie within residues 853–895, where there is no predicted amphiphilic helix (see Section 10.3.2.1). A C-terminus fragment (residues 628–977) of the TRP sequence also has been reported to bind CaM in a Ca^{2+}-dependent manner [163]. Sequence analysis suggests that the most likely CaM binding site in this region would be residues 705–725, i.e. the equivalent location to CBS1 in TRPL, even though there is no amino acid identity in this region between the two proteins.

The C-termini of TRP and TRPL show very little homology (17%). The TRPL sequence has no further recognizable motifs. The TRP sequence contains a long proline-rich region, within which there is a multiple (ninefold) tandem repeat of the sequence DKDKKPG/AD. Such proline-rich sequences with tandem repeats occur widely, and whilst their function is often unclear they are predicted to form extended structures and are often believed to be involved in binding interactions (e.g forming interlocking structures or binding to other proteins including cytoskeletal elements, such as actin, for review see Ref. [164]). Previous authors have pointed to similarities to the bacterial protein (TonB) in which a proline-rich tail bridges the gap between outer and inner membranes (e.g. [165]). However, it should be noted that the short proline repeats $(XP)_n$ in TonB are of a qualitatively different sort from the longer tandem repeats in TRP. The TRP sequence also includes a PEST sequence which is commonly found in CaM binding proteins and is believed to target the Ca^{2+}-dependent protease calpain for rapid degradation [166]. Finally the C-terminal

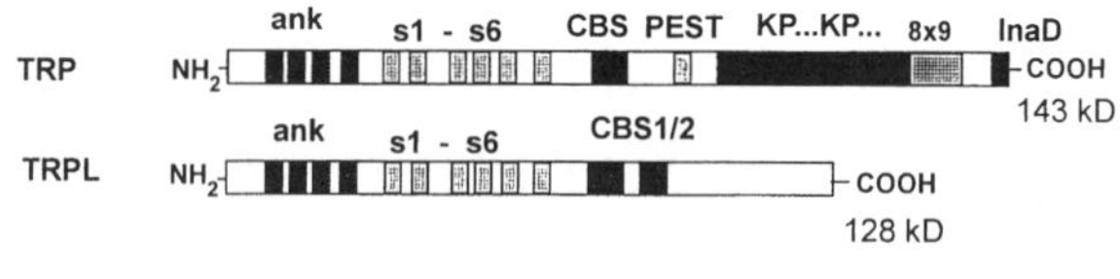

Fig. 10. Structural features of TRP and TRPL. Both sequences have the overall structure of a channel subunit with six putative transmembrane helices (S1–S6) with homologies to L-type Ca^{2+} channels. The native channels are likely to be hetero- or homo-tetramers. Both sequences have multiple ankyrin repeats on the N terminal. TRP contains one and TRPL two CaM binding sites (CBS). TRP has an extended C terminal with a PEST sequence (usually found on proteins targeted for rapid degradation), and an extended (ca. 250 residues) proline rich region within which the dipeptide KP is repeated 27 times. Towards the end of this region the hydrophilic sequence (DKDKKPG/AD) is repeated nine times (8 × 9). The final 19 amino-acids of the C-terminal bind to the *Drosophila* InaD protein (adapted from [190] and [154]).

includes a sequence which is required for binding to one of the PDZ domains of the scaffolding molecule INAD (Section 9).

By making TRP–TRPL chimeras in which the C-termini were swapped, Sinkins et al. [167] provided evidence that the C-terminus was critical for gating in heterologous expression systems. Thus a chimera with a TRP N-terminal and pore but a TRPL C-terminal had the characteristic constitutive activity of TRPL but the permeation properties of TRP. Conversely a protein with TRPL pore and TRP C-terminus could be activated by thapsigargin (see Section 8.2).

Outside the transmembrane regions, the most conserved regions of the TRP and TRPL proteins are the N-terminals which show overall ca. 40% identity. The only recognizable motifs in this region are 4 ankyrin (ANK) repeats. ANK repeats, which are usually 33 residues long, are found not only in ankyrin but in a wide variety of proteins and are believed to mediate protein–protein interactions both with cytoskeleton and other proteins, possibly including the $InsP_3$ and ryanodine receptors [168,169]. Their role in TRP and TRPL is unknown.

7.4. Heterologous expression of trp *and* trpl

A number of groups have now successfully expressed *trp* and *trpl* cDNA in a variety of expression systems, including two insect cell lines (*Spodoptera* Sf9 cells and *Drosophila* S2 cells), mammalian cell lines (HEK293T and CHO cells) and *Xenopus* oocytes [135,137,156,165,170–173]. In general the results confirm that expression of both proteins results in the appearance of novel conductances thereby providing support for their identification as ion channels, although the relationship to the native conductances is not always clear.

The evidence is most compelling for the case of TRPL which has been reported to lead to the appearance of a non-selective cation conductance after expression in Sf9 cells [165,172], S2 cells [137], HEK293 cells [156], CHO cells [173] and *Xenopus* oocytes [171,174]. In several cases the channels were shown to be activated by co-expressed receptors which activate endogenous G-protein-coupled PI pathways. A common feature of these studies is a constitutive spontaneous activity which increases with time during whole-cell recording, although Xu et al. [156] and Gillo et al. [171] reported that this was prevented when co-expressed with TRP. When examined under identical ionic conditions in *Drosophila* S2 cells, a variety of properties including single channel conductance and open times (estimated by noise analysis), ionic selectivity for monovalent and divalent ions, block by Mg^{2+} and current–voltage relationship were found to be indistinguishable from those of the *trpl*-dependent light-sensitive conductance determined in the *trp* mutant [137]. Although not investigated in the same detail, the properties of TRPL channels expressed in other expression systems appear similar. The *trpl* gene product alone thus appears sufficient to account for the *in situ* conductance, at least in terms of pore properties. This does not, however, rule out the possibility that other proteins/subunits may be involved, for example in channel gating or regulation, and that some factor is required to prevent constitutive activity which does not occur under normal conditions *in vivo*.

Heterologous expression of *trp* was first reported by Vaca et al. [165] who found that expression of *trp* cDNA in Sf9 cells led to the appearance of a novel conductance that could be activated by thapsigargin (see also Section 8.2.3). The permeability properties of this conductance were distinct from the endogenous capacitative Ca^{2+} entry pathway in the same cells; however, when compared to the properties of *trp*-dependent channels *in situ* (i.e. the light-sensitive conductance in the *trpl* mutant) there were major quantitative discrepancies e.g. in relative permeability to Mg^{2+} and Ba^{2+}, and in rectification characteristics [139]. Functional *trp* expression was also reported by Xu et al. [156], who found a novel conductance, which could be activated by thapsigargin in HEK293T cells; again the properties of this current showed several quantitative discrepancies from the *in situ* current (e.g. less permeable to Ca^{2+}, and less sensitive to block by either La^{3+} or Mg^{2+}). Two other groups have also reported apparent functional expression of *trp*: Petersen et al. [170] reported that expression of *trp* cDNA led to an enhanced capacitative Ca^{2+} entry signal in *Xenopus* oocytes; however, the properties were monitored by the Ca^{2+}-dependent Cl^- current and were not distinguished from the endogenous pathway. Gillo et al. [171] reported that co-expression of TRP and TRPL led to a *de novo* conductance that could be activated by complete depletion of the Ca^{2+} stores by bathing the cells in Ca-free medium in the presence of thapsigargin, or by injection of hydrolysis resistant $InsP_3$; a similar conductance could also be generated by expression of TRPL alone (but not TRP, even when expressed at higher levels), although now constitutively active and not activated by thapsigargin. The conductance did not however resemble the TRPL channel either *in situ* (i.e. in *trp* mutants) or in other expression systems, and, as recognized by the authors, the possibility should be considered that it represents activation of some endogenous current by an unknown mechanism. The inability to reconstruct the *in situ trp*-dependent conductance by *trp* expression leads us to suggest that TRP may require another as yet unidentified component for its normal function – and/or that TRP may interact with endogenous channels in the various expression systems [139,171].

7.5. TRP-related proteins are found throughout the animal kingdom

Not surprisingly, TRP homologues are also found in other invertebrate photoreceptors, including the closely related blowfly *Calliphora* [170,175] and the squid *Loligo* [176]. The *Calliphora* sequence shows 77% sequence identity with *Drosophila* TRP, clearly indicating it is the orthologue of TRP rather than TRPL. The greatest difference is in the C-terminal in which the proline-rich sequence in *Calliphora*, although still recognizably present, is somewhat shorter [175]. The squid sequence (sTRP) is more divergent with only 46% identity to TRP and TRPL; it has no proline-rich domain suggesting it may be more closely related to TRPL (see Figs. 10 and 11). In common with both TRP and TRPL, sTRP and *Calliphora* TRP (cal-TRP) have CaM binding domains on the C-terminus. In common with all TRP homologues reported, including those in vertebrates (see below) multiple ANK repeats are found in the N-termini of both squid and blowfly sequences.

In virtually all cells which use phospholipase C signaling, activation of PLC, whether by G-protein-coupled receptors or tyrosine kinase receptors, results in the influx of Ca^{2+}. However, the mechanism of activation of this influx and the molecular identity of the channels is obscure ([177]; for reviews see Refs. [33,178,294]). Although different mechanisms may exist in different cells, the most widely accepted model for activation suggests that influx is triggered by depletion of Ca^{2+} stores, and it is consequently referred to as capacitative Ca^{2+} entry (CCE), or store-operated Ca^{2+} entry (SOC). How the signal (reduction in intraluminal [Ca]) might be relayed to the plasma membrane is unclear, but experimentally such Ca^{2+} influx can be triggered by agents which deplete the stores, such as Ca^{2+} ionophores and thapsigargin, which inhibits the smooth endoplasmic reticulum Ca ATPase (SERCA pumps). The mounting evidence that the TRP protein was responsible for a PLC-regulated Ca^{2+} influx pathway in *Drosophila* photoreceptors led to the proposal that TRP-related proteins might also be responsible for such influx pathways in other cells as well [138,149,179].

The first report that TRP-related proteins might also be found outside invertebrate photoreceptors came from Petersen et al. [170] who used degenerate PCR to identify partial sequences of TRP homologues from *Xenopus* oocyte and murine brain cDNA libraries. Shortly thereafter the full sequence of a human homologue (TRPC1, see Fig. 11) was reported, identified by homology searches of EST databases [180]. Further homologues now sequenced include six non-allelic murine homologues (mTRP, Fig. 11) [181,182], whilst the recently completed *C. elegans* genome contains a total of 11 identifiable homologues [295], and thus sets the minimum number of isoforms present in any one organism. In addition at least four human sequences have now been reported (TRPC1–4, alternatively referred to as hTrp1–4), at least one of which (TRPC1/hTrp1) exists in alternative splice variants, as well as bovine (bCCE = bTrp4, bTRP4 in Fig. 11) and rat sequences (R-Trp) (for reviews see Refs. [155,182] and Fig. 11).

The functional roles of vertebrate TRP homologues remains debated; whilst showing very widespread tissue distribution, their properties have thus far been mainly inferred from heterologous expression studies. Several studies have reported that expression of certain homologues (TRPC1A and btrp4) in cell lines result in an increase in thapsigargin-sensitive Ca^{2+} influx, and Zhu et al. [183] reported that an anti-sense RNA cocktail covering partial sequences of all six murine homologues, blocked endogenous CCE in a murine cell line. However, from the limited electrophysiology performed thus far, none of the TRP homologues lead to the appearance of conductances with the properties of I_{crac}, which is the most thoroughly studied CCE conductance [184]. Perhaps the closest is bCCE [185] which shows a similar inward rectification, but it lacks the extremely high Ca^{2+} selectivity of I_{crac}. In addition, at least two homologues (TRPC3 and the closely related mTrp6, mTRPC6 in Fig. 11) were reported to generate non-selective cation channels which, reminiscent of TRPL, appear to be regulated by G-protein-coupled signaling pathways and/or cytosolic Ca^{2+} but not by store-depletion [186,187]. It should also be noted that TRPC3 (= hTrp3) was originally proposed as a CCE channel on the basis of an enhanced CCE signal monitored by Ca^{2+} fluorimetry [183]; however, subse-

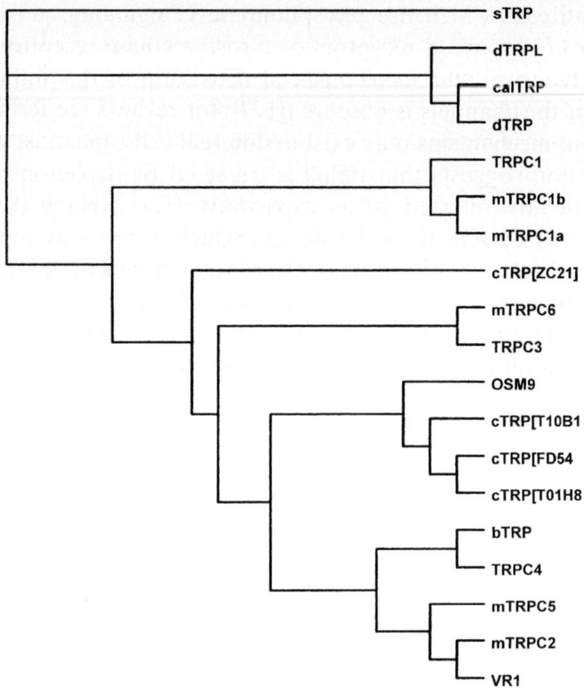

Fig. 11. Pairwise similarity phylogenetic tree of the TRP family. The tree was prepared on the basis of the sequences referred to in the Human TRP Web Review using the program Clustal W(1.6). The drawing was prepared by TreeView (Version 1.5, 1998 R.D.M. Page). The Tree was prepared on the basis of the protein amino acid sequences of the proteins as described and submitted to the databases and not on the basis of the DNA sequence. The protein sequences vary in length and in completeness which may result in certain deviation of the tree. sTRP – from squid, dTRP, dTRPL from *Drosophila* fruit fly, TRPC – from human, bTRP – from bovine, mTRPC from mouse, cTRP – from *Caenorhabditis elegans* (first four letters of the clone name are in block parenthesis), VR1 Capsaicin receptor from rat, and OSM9 is the olfactory channel from *C. elegans*. (This figure and The Human TRP Web Review are the work of Dr. Ofer Markman. http://bioinformatics.weizmann.ac.il/cards/webreview/TRP.html. The Page, R.D.M. (1996) TreeView: is an application to display phylogenetic trees on personal computers. Computer Applications in the Biological Sciences, 12, 357–358.).

quent studies using patch clamp techniques suggested that it was in fact not regulated by store depletion but by the associated rise in Ca^{2+} [186]. This highlights a common problem: namely that conclusive demonstration of CCE cannot be made by thapsigargin-activated signals using fluorimetric methods alone, without adequate controls to show that the Ca^{2+} entry is not triggered or regulated by the associated Ca^{2+} rise. Most recently it has been suggested that TRPC3 may be activated by direct protein–protein interaction with the $InsP_3$ receptor and/or DAG [296,297].

Recently the identification of somewhat more distantly related TRP homologues has indicated that the TRP family may be even more diverse than originally thought.

One of these is the capsaicin receptor (vanilloid receptor 1, or VR1), which was isolated by expression cloning. Expression of VR1 in oocytes led to the appearance of a cation-selective ion conductance activated both by the vanilloid capsaicin and also directly by heat [188]. The authors suggest that this may represent a membrane ion channel responsible for detection of noxious thermal stimuli as well as the active ingredient of chilies.

Finally, Colbert et al. [189] reported that OSM-9, a *C. elegans* gene required for normal olfactory, mechanosensory and osmosensory responses encodes a protein with ankyrin repeats and multiple predicted transmembrane domains with weak, but unmistakable homology to the TRP family (see OSM9 in Fig. 11). Using this new sequence as the basis for further database searches, the authors reported yet further new TRP homologous, including one partial sequence expressed in vertebrate retina (Fig. 11).

A detailed survey of the structures of the various members of the TRP family is beyond the scope of this review (but see Refs. [154,155,182,190]). Hydropathy analysis suggests they have 6–8 transmembrane helices, and multiple ANK repeats are found in the N-terminal. Beyond the first 50 residues however, the C-terminals of different subclasses show very little homology and a range of different motifs suggest a diverse regulatory repertoire. Given Sinkins et al.'s [167] evidence that the C-terminal may be crucial for gating (see Sections 7.3 and 8.2.3), it need not be surprising that different members may be gated by different mechanisms. As noted by several authors, the potential for heteromultimerization may further increase the diversity of function in this novel ion channel family, which seems certain to reveal yet further members in the coming years.

8. Mechanism of activation

8.1. In situ *experiments*

The essential role of PLC in phototransduction has been widely accepted (see Section 6), but despite extensive biochemical, electrophysiological and molecular genetic approaches, how activation of PLC is linked to the opening of light-sensitive channels in *Drosophila* remains controversial. It has often been assumed on the basis of evidence from other species, that $InsP_3$ is an essential messenger of excitation. For example, in *Limulus* ventral photoreceptors there is convincing evidence that light induces release of Ca^{2+} from intracellular stores via $InsP_3$ receptors in the SMC and that Ca^{2+} can activate at least one class of light-sensitive channel directly or indirectly (for review see Ref. [31] and Chapter 8). Briefly summarized, this body of evidence includes direct measurements of light-induced Ca^{2+} release coincident with, or a few ms prior to the light-induced current [191], the ability of both $InsP_3$ and Ca^{2+} to excite the light-sensitive channels [192–194] and the inhibition of the light response by Ca^{2+} chelators [195]. However, similar approaches in *Drosophila* have led to largely negative results, although importantly $InsP_3$ has been reported to mimic excitation in larger flies ([105,146], see Section 8.1.4).

8.1.1. Does Ca^{2+} mimic excitation?

Photolytically released caged Ca^{2+} (DM-nitrophen) raises Ca^{2+} intracellularly into the high micromolar range, but failed to activate the light-sensitive channels [24]. The released Ca^{2+} did however induce a substantial electrogenic Na/Ca exchange current and was capable of profoundly modulating the light-sensitive current after it had previously been activated by light. In WT flies the modulation was facilitatory when the Ca^{2+} was released during the rising phase of the light response but inhibitory when released during the falling phase. Intriguingly exactly the opposite behavior was found in the *trp* mutant. Both facilitation and inhibition occurred with a sub-millisecond latency suggesting it occurred at the level of the channels themselves and indicating that the released Ca^{2+} had direct access to the channels. Alternative methods used to raise cytosolic Ca^{2+} have included perfusion of up to 0.5 mM Ca^{2+} from the whole-cell patch pipette [19], and also application of the Ca^{2+} ionophore, ionomycin ([196]; see also Fig. 12). None of these procedures appear to activate any light-sensitive channels (but see Section 8.1.6), although again

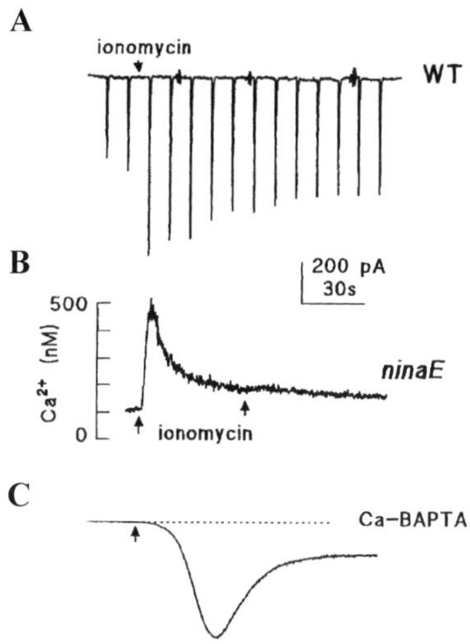

Fig. 12. Effects of the Ca^{2+} ionophore ionomycin. (A) Responses to repeated flashes of light delivered to a wild-type (WT) photoreceptor clamped at resting potential and bathed in Ca^{2+} free Ringer's solution (2 mM EGTA). A brief pulse of ionomycin (14 μM, 2 s) applied from a puffer pipette led to a transient facilitation of the responses which returned to near control levels after 1–2 min. Notice that no currents were activated by ionomycin under these conditions (but see C below). (B) Measurement of intracellular Ca^{2+} using INDO-1 in the rhodopsinless mutant *ninaE* under identical conditions. Ionomycin induced a rapid, transient rise in Ca^{2+} due to release from an intracellular compartment which also decayed over a period of 1–2 min (i.e. the facilitation in A could probably be accounted for by transient rise in cytosolic Ca^{2+}). (C) When the photoreceptor was loaded with Ca-BAPTA buffering Ca^{2+} at ca. 10 nM Ca^{2+} (4 mM BAPTA: 0.25 mM Ca^{2+}) and bathed in Ca^{2+} free Ringer's application of a ionomycin (14 μM) induced an inward current with properties indistinguishable from those of the light-sensitive conductance (adapted from [196]).

they could activate an electrogenic Na/Ca exchange current and modulate the response to light itself.

8.1.2. Does light induce Ca^{2+} release?

Measurements with Ca^{2+} indicator dyes show that there is a massive light-induced Ca^{2+} influx associated with the LIC in the presence of normal extracellular Ca^{2+}. However, in Ca^{2+}-free conditions most measurements have shown no, or virtually negligible Ca^{2+} rise suggesting there is very little if any light-induced Ca^{2+} release from internal stores [75,140,141]. Using quantitative measurements made with INDO-1, Hardie [141] suggested that the maximum light-induced rise in Ca^{2+} that could be attributed to release was of the order of 20 nM. However, Arnon et al. [197] have reported that the putative Ca^{2+} release signal is larger if photoreceptors are first perfused with the calmodulin antagonist M5 and suggest that under normal conditions Ca^{2+} release is tightly regulated by Ca/CaM-dependent negative feedback.

A potential difficulty in measuring light or $InsP_3$-induced release of Ca^{2+} is the tiny dimensions of the sub-microvillar cisternae in *Drosophila*, which might easily lose their Ca^{2+} content in Ca^{2+}-free medium. To overcome this difficulty Cook and Minke [298] have used short incubation times in Ca^{2+}-free medium, sufficient to eliminate Ca^{2+} influx. Under these conditions a significant light-induced release of Ca^{2+} was observed. Interestingly, this release was correlated to responsivity to light in the *trp* mutant.

8.1.3. Effects of Ca^{2+} chelators

The effects of intracellular Ca^{2+} chelators in *Drosophila* have been studied by inclusion in the patch pipette used to record the whole-cell currents [19]. The highest concentrations of EGTA tested (18 mM) had only weak effects upon voltage-clamped light responses in normal Ringer's solution (1.5 mM Ca^{2+}), and no apparent effect in Ca^{2+}-free solutions. BAPTA, which differs from EGTA primarily in the speed with which it chelates Ca^{2+} (microsecond cf. millisecond time scales), was considerably more potent, with the highest concentration tested (14 mM) reducing sensitivity by ca. 5000-fold in normal Ringer's solution. However, in Ca^{2+} free solution, the effects of BAPTA were much more modest with only a ca. 2 log unit reduction in sensitivity with 14 mM BAPTA, and with concentrations below 3 mM no effect could be detected at all.

The effects of BAPTA add independent support for the role of Ca^{2+} influx in regulating sensitivity and response kinetics (see Section 10); however, in view of the high concentrations required for significant attenuation of responses in Ca^{2+}-free Ringer's solution, the role of Ca^{2+} release in excitation remains questionable. It should also be noted that the specificity of BAPTA has been questioned, and in particular it has been shown to be a competitive antagonist of the $InsP_3$ receptor [198].

The effects of attempting to deplete putative light-sensitive Ca^{2+} stores in *Drosophila* have also been investigated using Ca-ATPase inhibitors (cyclopiazonic acid and thapsigargin) and the Ca^{2+} ionophore ionomycin [75,196]. Measurements using Ca^{2+} indicator dyes show that all these agents release Ca^{2+} from internal stores in a saturating manner; however, when applied in Ca^{2+} free Ringer's solution they actually lead to a facilitation of the response to light, presumably due to the raised cytosolic Ca^{2+} levels [196].

8.1.4. $InsP_3$

Attempts to introduce and release caged $InsP_3$ into photoreceptors via the whole-cell recording pipette have not resulted in excitation of the light-sensitive channels [299]. However, similar negative results have also been obtained with GTPγS, which is expected to have an effect according to every current model of *Drosophila* phototransduction. It is thus possible that these negative results may be attributed to the highly compartmentalized location of the phototransduction machinery being inaccessible to internal dialysis.

An alternative method of application has been successfully attempted in the larger flies, *Lucilia* and *Musca* [105,146]. These studies exploited the finding that lucifer yellow applied extracellularly, i.e. in the intact retina, is taken up by the photoreceptors following intense illumination, possibly via receptor-mediated pinocytosis [199]. Minke and Stephenson [104] subsequently demonstrated that extracellular application of GTPγS and guanosine 5′-[β,γ-imido]triphosphate in *Musca* induced a noisy depolarization in the dark following strong illumination in the presence of the nucleotides. Similarly, introduction of extracellularly applied $InsP_3$ into *Musca* or *Lucilia* photoreceptors by intense white light resulted in a large increase in baseline noise in the dark, whilst responses to test flashes were enhanced and deactivated more slowly than controls. 2,3-diphosphoglycerate (DPG), an inhibitor of $InsP_3$ hydrolysis, introduced by the same method, greatly enhanced the effect of exogenously applied $InsP_3$ when applied together and also enhanced and prolonged the physiological response to light when applied alone, suggesting an endogenous production of $InsP_3$ during illumination. Noise analysis suggested that the $InsP_3$-induced noisy depolarization is similar to the noisy depolarization induced by dim light [105,146].

Although extensive biochemical and molecular genetic evidence supports the essential role of PLC in fly phototransduction (see Section 6), these experiments remain the only direct evidence for an excitatory role of $InsP_3$ in flies, although the possible involvement of $InsP_3$ receptors has also been suggested by the ability of heparin to block the light response [200].

8.1.5. $InsP_3$ receptor mutants

There is believed to be only one $InsP_3$ receptor gene in the *Drosophila* genome and not surprisingly mutations in this gene are lethal [201,202]. However, it is possible to generate mutant photoreceptors in mosaic patches by inducing mitotic recombination in heterozygotes. Intracellular recordings from photoreceptors in such mosaic patches revealed no differences from wild-type leading the authors to conclude that the $InsP_3$ receptor played no role in phototransduction [202]. This is potentially a very important argument against the role of $InsP_3$ receptors in *Drosophila* phototransduction. However, since photoreceptors were assayed using intracellular recordings it was not possible to determine whether both TRP and TRPL channel activation were unaffected (since, e.g., a *trpl* mutant's response would be indistinguishable under these conditions), whilst evidence confirming the absence of $InsP_3R$ mRNA and protein in the mosaic patches was not presented. In view of the potential significance of this result it was recently repeated using an alternative, and demonstrably null allele of the $InsP_3R$ and whole-cell recording techniques which, inter alia, allowed resolution and distinction of both TRP and TRPL channel activity. The results confirmed that phototransduction was quantitatively unaffected in photoreceptors from mosaic eyes showing no trace of $InsP_3R$ mRNA or protein [300].

8.1.6. Capacitative Ca^{2+} entry

The high Ca^{2+} selectivity of the light-sensitive channels led to the suggestion that excitation in *Drosophila* might be analogous to the process of PI-regulated Ca^{2+} influx described in a wide variety of mammalian cell types [149,153,179]. The most generally accepted mechanism underlying this influx pathway in vertebrates is the so-called capacitative Ca^{2+} entry (CCE; also referred to as store-operated Ca^{2+} entry) in which depletion of store Ca^{2+} (rather than the associated rise in cytosolic Ca^{2+}) is supposed to trigger influx (for reviews see Refs. [33,177,203]; see Section 7.5). Experiments designed to test whether a CCE like mechanism might operate in *Drosophila* phototransduction have been equivocal. Both ionomycin and thapsigargin release Ca^{2+} from intracellular compartment(s), as witnessed by their ability to rapidly raise cytosolic Ca^{2+} (measured using INDO-1) in the absence of external Ca^{2+} (Fig. 12; [196]; see also Ref. [75]). Under normal conditions however, store depletion by these agents is not associated with the activation of any currents, nor do these agents block the response to light during several minutes of exposure. Instead responses recorded in Ca^{2+}-free media are enhanced as might be expected from the raised cytosolic Ca^{2+}, which facilitates the response to light under these conditions.

It is still however, possible that the photoreceptors contain specialized stores which are unusually resistant to depletion by these agents; consequently, the experiments were also performed under conditions designed to further promote depletion by loading the cells with Ca-BAPTA, buffering Ca^{2+} at concentrations at a level of ca. 10 nM, thereby providing a Ca^{2+} sink. Under these conditions, ionomycin (but not thapsigargin) did indeed activate an inward current indistinguishable from the light-activated conductance. However, ionomycin was no longer effective when cells were loaded with BAPTA but no Ca^{2+}. Apparently, if this protocol is activating the light-sensitive channels by depletion of stores, there is in an addition a requirement for some permissive level of cytosolic Ca^{2+}. However, an alternative interpretation is that this paradigm facilitates the access of Ca^{2+} (albeit only 10 nM) to some protected zone or compartment in the cell which is normally completely free of Ca^{2+}, and that under these conditions at least, access of Ca^{2+} to this zone can cause excitation [196].

In summary, most protocols which would be expected to deplete internal stores do not lead to activation of the light-sensitive conductance; although conditions have been found which do, it is questionable whether they are acting via store depletion and even if this is the case, it is by no means certain that this represents the physiological route of excitation. The only other evidence supporting CCE as the mechanism of excitation is the ability of thapsigargin to activate heterologously expressed TRP (see Section 8.2); however, as discussed above (see Section 7.2) the properties of heterologously expressed TRP are distinct from TRP dependent channels *in situ* raising doubts as to whether the same channels are being investigated. Ironically, whilst TRP may have led to the discovery of vertebrate SOCs, it currently seems questionable whether TRP or TRPL are activated by a store-depletion *in situ*.

8.1.7. Diacylglycerol and other bioactive lipids

As well as generating $InsP_3$, PLC also produces diacylglycerol (DAG). This arm of the PI pathway has been largely ignored as a possible route of excitation. The most familiar action of DAG is to activate PKC synergistically with Ca^{2+}; however, mutations in an eye-specific PKC (*inaC*) lead to defects in response termination (see Section 10) with no apparent effects on activation. A second role of DAG in *Drosophila* photoreceptors is as the starting point for regeneration of PIP_2 (see Fig. 13) via conversion to phosphatidic acid (via DAG kinase). Mutations in DAG kinase (*rdgA*) actually result in a most severe form of light-independent degeneration, although the reason for this is unknown [204]. Other mutations in the same pathway (*cds* and *rdgB*) also lead to less severe forms of light-dependent degeneration (see Section 11). DAG is also a potential precursor for formation of arachidonic acid and other polyunsaturated fatty acids (PUFAs) via DAG lipase. Although the presence of this enzyme has not been demonstrated in *Drosophila* photoreceptors, PUFAs, including arachidonic and linolenic acid, have recently been shown to excite both TRP and TRPL channels when applied to dissociated photoreceptors, and to directly activate heterologously expressed TRPL channels in inside-out patches with an effective EC_{50} of ca. 10 μM. Furthermore a variety of inhibitors of lipoxygenases (a class of enzyme which metabolises PUFAs), potently activate quantum bump like activity within seconds of application to dissociated photoreceptors [301]. These results raise the possibility that a poly-unsaturated fatty acid may be a messenger of excitation in *Drosophila* phototransduction. Curiously arachidonic acid is reported to be absent in *Drosophila* [302]; however, other PUFAs including linolenic acid, are present and might reward further investigation as putative excitatory messengers.

8.1.8. Cyclic GMP

In *Limulus* ventral photoreceptors, injections of cGMP can mimic excitation both in intact cells and excised patches ([205]; see Chapter 8). The possible involvement of cGMP in *Drosophila* was first suggested by Baumann et al. [206], who cloned a cGMP-gated channel that was expressed in *Drosophila* eyes, although it was not more precisely localized. When expressed, this channel was found to have several properties in common with the *Drosophila* light-sensitive conductance, including high Ca^{2+} permeability and voltage-dependent Mg^{2+} block. However, both these properties of the LIC are dependent upon the presence of the *trp* gene product suggesting that the similarity may be coincidental [16]. Subsequently Bacigalupo et al. [207] reported that 8-bromo cGMP could activate an inward current in *Drosophila* photoreceptors and facilitate the response to light. Hardie and Mojet [16]

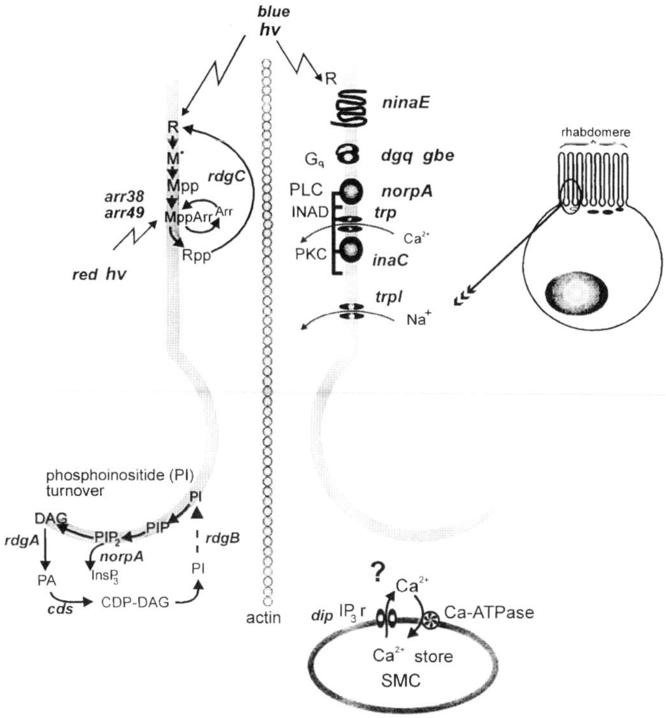

Fig. 13. Phototransduction cascade in *Drosophila*. The major components of the phototransduction machinery are shown superimposed on a schematic diagram of a microvillus, each of which has a central actin filament. Cloned genes (for all of which mutants are available) are shown in italics, alongside their corresponding proteins. Upon absorption of light rhodopsin (R, *ninaE* gene) is converted to the active metarhodopsin state (M), which is subsequently inactivated by phosphorylation and binding to arrrestin (top left, see (Fig. 4) for further details). M activates a heterotrimeric G-protein (G_q) for which both an α (*dgq*) and β (*gbe*) subunit have been identified. This leads to activation of phospholipase C (PLC, *norpA* gene) and subsequent opening of two classes of light-sensitive channels encoded at least in part by *trp* and *trpl* genes, by an as yet unknown mechanism. Deactivation of channel activity is regulated by, amongst others, protein kinase C (*inaC* gene). PLC, PKC and the TRP ion channel form a supramolecular complex with the scaffolding protein INAD (*inaD* gene). Along with R, G_q and TRPL all these components have been immunolocalized to the microvillar membrane. PI turnover: PLC catalyses the hydrolysis of PIP_2 into the soluble second messenger $InsP_3$ and the membrane bound diacyl glycerol (DAG). DAG is recycled to PIP_2 by the PI cycle shown on the bottom left. DAG is converted to phosphatidic acid (PA) via DAG kinase (*rdgA* gene) and to CDP-DAG via CD synthetase (*cds* gene). After conversion to phosphatidyl inositol (PI), PI is presumed to be transported back to the microvillar membrane by the PI transfer protein RDGB. Both RDGA and RDGB have been immunolocalized to the SMC. PI is converted to PIP_2 via sequential phosphorylation (PI kinase and PIP kinase). At the base of the microvilli a system of submicrovillar cisternae have been proposed, by analogy to other insects, to represent SER Ca^{2+} stores endowed with $InsP_3$ receptors (IP_3R) and SER Ca-ATPases; however, there is no direct evidence for this in *Drosophila* photoreceptors.

and Chyb et al. [303] were however, unable to find any action of 8-bromo cGMP in *Drosophila* photoreceptors. The reason for this discrepancy is not clear, however inspection of Bacigalupo et al.'s records indicate that the cells were probably in a "run down" condition. A possible role for cGMP unrelated to phototransduction in *Drosophila* photoreceptors has been suggested by the finding that an NO-sensitive guanylate cyclase is active during a narrow temporal time window of pupal development and is required for the stabilization of photoreceptor axon growth cones during synaptic assembly [208]. In addition, recent work in adult locusts has revealed that second order neurons (large monopolar cells, LMCs) are nitrergic and that NO donors lead to an increase in cGMP immunoreactivity in the photoreceptor axon terminals and, to a lesser extent, cell bodies [209]. According to these results it can be suggested that NO is a retrograde messenger from the LMC's that modulates synaptic transmission at the photoreceptor synapse.

8.2. Activation in heterologous expression systems

An alternative approach taken by a number of groups has been to investigate the mechanism of activation of TRP and TRPL channels in heterologous expression systems. All of these studies should however be prefaced with a caveat. Firstly the biophysical properties of heterologously expressed TRP channels do not closely resemble the *in situ trp*-dependent channels raising doubts as to whether the same channels are being investigated under the two conditions. Secondly, although the biophysical properties of native and expressed TRPL channels appear indistinguishable [137], in most cases heterologously expressed TRPL channels appear to be constitutively activated, so that any reported "activation" may in fact represent modulation of activity, just as, for example, Ca^{2+} can clearly modulate both TRP and TRPL channels in situ although it appears not to be the primary messenger of excitation [24]. Encouragingly however, in common with the *in situ* mechanism of activation, expressed TRPL channels can be reliably activated by stimulation of co-expressed heptahelical receptors which couple to endogenous [137,172,210] or overexpressed [135] G_q proteins.

8.2.1. InsP$_3$ and G-protein

Dong et al. [210] reported that TRPL channels expressed in Sf9 cells using the baculovirus expression system could be activated by inclusion of $InsP_3$ in the patch pipette. Although heparin blocked activation by $InsP_3$, it could only partially counteract activation by a co-expressed bradykinin receptor suggesting an additional alternative route of excitation. The authors concluded that the effect of $InsP_3$ was not mediated via a rise in Ca^{2+}, since it was unaffected by inclusion of 1 mM EGTA in the pipette. They therefore suggested that $InsP_3$ might directly gate TRPL – possibly via a conformationally coupled interaction with the $InsP_3$ receptor. It should be noted however, that EGTA is notoriously ineffective in controlling local Ca^{2+} transients, and several authors have now reported that $InsP_3$ has no effect on TRPL channels beyond indirect effects caused by the released Ca^{2+} [135,173,299,303].

Obukhov et al. [135] reported that activated $G_{11\alpha}$ subunits could activate TRPL channels expressed in Sf9 cells either when overexpressed or, in a few cases, when applied to inside-out patches. They concluded that TRPL channels were directly regulated by the G-protein. The authors did not however, control for the possibility of downstream membrane delimited effects of the G-protein, in particular via PLC. By contrast, Hardie and Raghu [299] found that receptor mediated activation of TRPL expressed in S2 cells was blocked by the PLC inhibitors U-73122 and 4-bromo-phenacyl bromide (see also Hu and Schilling [211]). It should also be noted that direct activation by G-protein would seem unable to account for *in situ* activation since both TRP and TRPL activation are blocked in null mutants of PLC (see Section 6).

8.2.2. Ca–calmodulin

The TRPL channel is a calmodulin binding protein [152] with two CaM binding sites (CBS1 and CBS2) ([32,162]; see Section 7.3). CBS1 is a conventional calmodulin binding site which binds to CaM in the presence of raised Ca^{2+} (100–500 nM). The second site CBS2 was originally reported to bind CaM in the absence of Ca^{2+} [162]; however, more recent experiments concluded that CBS2 also binds Ca^{2+} in a Ca^{2+}-dependent manner, but with a K_d of ca. 3 μM [293]. In light of these findings both dissociation of CaM from CBS2 and Ca^{2+}-dependent CaM binding to CBS1 have been suggested as a mechanism of TRPL activation. In support of a role for calmodulin, Lan et al. [174] reported that TRPL activity (monitored by fluo-3 measurements of Ca^{2+} influx in *Xenopus* oocytes) was increased by low concentrations of CaM and inhibited at high concentrations. Zimmer et al. [173] reported that TRPL activity in CHO cells could be enhanced by infusion of cells with Ca^{2+} with an EC_{50} of 500 nM, and that this could be reduced by the CaM antagonist calmidazolium. These results contrast however with experiments performed *in situ*. Scott et al. [32] generated transgenic flies in which TRP channels were absent and the native TRPL channels were replaced with constructs with CBS1 or CBS2 deleted. Activation in these flies was essentially unaltered, however, there were major defects in deactivation (see Section 10). Furthermore, under most conditions, TRPL channels appear to be negatively regulated by Ca^{2+}; e.g. when extracellular Ca^{2+} is removed the response to light in the *trp* mutant is increased ca. threefold [139]. Similarly caged Ca^{2+} released during the rising phase of the light response inhibited the light response in *trp*, although it enhanced the response during the decaying phase of the response [24].

Recent experiments using caged compounds in S2 cells expressing TRPL channels co-expressed with a muscarinic receptor [299] may provide a partial resolution to this confusion. Releasing caged Ca^{2+} loaded via the patch pipette normally resulted in a pronounced facilitation of the spontaneous TRPL activity. Particularly after several minutes of recording, Ca^{2+} could also inhibit the channel activity (even in cells in which the same concentration of Ca^{2+} had previously facilitated the response). Releasing caged $InsP_3$ released Ca^{2+} from internal stores had at most a small effect on TRPL activity and none at all when BAPTA was included in the pipette solution. By contrast, excitation of TRPL via the co-expressed muscarinic receptor resulted in a rapid excitation of TRPL channels which was not blocked by prior depletion of stores (by $InsP_3$ or thapsigargin), or by inclusion of BAPTA in the electrode. We suggest therefore that heterologously expressed TRPL channels can be activated by a process requiring G protein and PLC, but not $InsP_3$, and further that the channels, once activated can be modulated by Ca^{2+}/CaM. This modulation may be either positive or negative depending, in an as yet undefined manner, on the state of the channels and/or cells. Apparently in expression systems the channels are usually facilitated by Ca^{2+} but inhibited under most conditions *in situ*.

In summary, heterologous expression studies are unanimous in reporting that TRPL channels can be activated by stimulation of receptors coupled to endogenous PI pathways. As *in situ* there is no consensus on the final mechanism of activation; the balance of evidence suggests that activation does not require $InsP_3$.

8.2.3. Capacitative Ca^{2+} entry

As already mentioned, Vaca et al. [165] reported activation of heterologously expressed TRP channels by thapsigargin. This has been widely taken as an indication that *Drosophila* TRP is a *bona fide* store-operated channel or SOC. Apart from the *caveat* already raised that the biophysical properties of in situ and heterologously expressed TRP are different, it should also be noted that the stringent controls required to demonstrate unequivocal activation by store-depletion were lacking. Nevertheless, the same authors found a clear distinction between TRP and TRPL in this respect as TRPL channels expressed using the same baculovirus Sf9 cell system were insensitive to thapsigargin, and in a significant study using TRP–TRPL chimeras they demonstrated that the thapsigargin sensitivity resided in the TRP C terminal [167]. More recently TRPL has also been reported to be activated by thapsigargin [156,304], but the possibility that the regulation was due, e.g. to the associated increase in Ca^{2+} was again not rigorously excluded.

9. Transduction complexes and *InaD*

In vertebrate photoreceptors, excitation is believed to involve random and sequential diffusional encounters of elements of the transduction cascade (e.g. [212]). One of the most recent and exciting developments in *Drosophila* phototransduction has been the realization that some of the key elements of the cascade are incorporated into supramolecular signaling complexes via a newly discovered "scaffolding protein", INAD.

9.1. PDZ domains

InaD^{P215} (inactivation but no afterpotential D) was a dominant mutation isolated by Pak and colleagues on the basis of an abnormal PDA [37]. The *InaD* gene was subsequently isolated from a collection of eye-specific clones identified by subtractive hybridization and found to encode a 674-amino-acid protein with limited homology with a class of cytoskeleton associated proteins which includes PSD-95 (implicated in clustering of NMDA receptors and K channels), *Drosophila discs large* and the epithelial tight junction protein ZO-1 [213] (for review see Ref. [214]). These proteins all include a variable number of so-called PDZ domains, now recognized as protein modules which anchor a variety of receptors ion channels and other signaling molecules by specific binding to target sequences typically, though not always, in the final three residues of the carboxy-terminal. Shieh and Zhu [215] subsequently found that the *InaD*^{P215} mutation represents a point mutation (M442K) in one such PDZ domain (now identified as the third domain or PDZ3) and, by overlay assays and co-immunoprecipitation, that this mutation specifically disrupted binding to the TRP protein via an interaction within the last 19 amino acid residues of TRP C-terminal. Immuno-gold labeling further indicated that this interaction was necessary for the normally strict localization of the TRP protein to the microvillar membrane; thus in *InaD*^{P215} flies TRP protein was found to be delocalized, occurring in both the plasma as well as the microvillar membrane [163,216].

The association of INAD with two further elements of the transduction machinery, namely PLC (*norpA*) and PKC (*inaC*), was first described by Huber et al. [175,217] using co-immunoprecipitation studies of the homologous proteins in the blowfly *Calliphora*; in addition they found that INAD could be an *in vitro* target for phosphorylation by PKC [217]. The interaction of INAD with NORPA and INAC was subsequently confirmed in *Drosophila* [163,216]. Tsunoda et al. [216] also recognized that the *InaD* sequence contained five consensus PDZ domains rather than the two originally suggested [215], and using GST fusion protein fragments they identified specific interactions between PKC and PDZ4, TRP and PDZ3, and PLC with PDZ5. By screening for antigenicity they also isolated additional *InaD* mutants including a null allele (*inaD*^{1}) and *inaD*^{2} which has a point mutation in the fifth PDZ domain (PDZ5) that specifically disrupted interaction with NORPA. Shieh et al. [218] disrupted the INAD–NORPA interaction by the alternative strategy of identifying and mutating the C terminal sequence of the NORPA protein required for binding to INAD. In all three mutants the NORPA protein is no longer localized in the microvilli and sensitivity to light was substantially reduced as witnessed by severe ERG defects (see Section 9.2). Most recently van Huizen et al. [219] reported that NORPA binds to INAD via two distinct sites. According to their results the C-terminal of NORPA actually binds to PDZ1, whilst PDZ5 binds to an internal region overlapping with the putative site for G-protein interaction. Transgenic flies expressing truncated versions of NORPA lacking both PDZ binding regions completely lacked a response to light although basal PLC activity was similar to WT controls.

Although Chevesich et al. [163] reported that rhodopsin also co-immunoprecipitated with INAD this was refuted by Tsunoda et al. [216]. Given its extremely high concentration in the microvillar membrane it would clearly be impossible for more than a small fraction of rhodopsin to be bound, whilst INAD and its more established partners (TRP, NORPA and PKC) are all reported to occur in approximately equal molar ratios (all ca. 0.1 per mol rhodopsin) [175]. G_q also fails to immuno-precipitate with the INAD complex and is presumably free to diffuse between active rhodopsin (M*, Fig. 4) and form transient interactions with PLC. Whether or not the second light-sensitive channel, TRPL, associates with INAD complex is also debated: Xu et al. [305] reported binding of TRPL to INAD; however, Tsunoda et al. [216] concluded that TRPL was not a member of the complex, since unlike PKC, NORPA and TRP it remains strictly localized to the microvilli in the *inaD*^{1} null mutant [216].

In summary, the INAD protein has five PDZ domains; PDZ1 and PDZ5 interact with PLC (*norpA*), PDZ3 with TRP and PDZ4 with PKC (*inaC*). PKC may also have the potential to regulate INAD by phosphorylation. This leaves one site (PDZ2) for which a partner has yet to be found. It will be of interest to find this partner as it may provide vital clues to the mechanism of excitation. There is also now evidence that the INAD complexes may multimerise thereby forming "rafts" or clusters of transduction complexes [305]. Recently Shieh and co-workers [306] have challenged the suggestion that INAC binds to PDZ4 and report that in fact it binds to PDZ2 via the last three residues in the carboxyl-terminal tail of INAC.

9.2. Electrophysiological phenotypes and the function of INAD

As in all the *ina* mutants, the originally described ERG phenotype of *InaD*^{P215} was the lack of a PDA following blue illumination, although inactivation due to pigment conversion was normal. This phenotype is reminiscent of the *trp* phenotype and it has subsequently been found that in older *InaD*^{P215} flies TRP protein levels decline and that the flies develop a *trp*-like ERG phenotype [216]. In young *InaD*^{P215} flies, TRP levels are near normal, and when investigated in whole-cell recording Shieh and Niemeyer [213] found an apparent defect in response deactivation, reminiscent of the *inaC* phenotype. Like the *inaC* phenotype, this defect was masked in Ca^{2+} free Ringer's, indicative of a defect in Ca^{2+} dependent response termination. However, Tsunoda et al. [216] reported that quantum bumps in *InaD*^{P215} were normal and that the prolonged macroscopic response was in fact due to a finite probability of bumps occurring with a longer latency. They therefore interpreted the *InaD*^{P215} phenotype as a defect in excitation. However, closer examination of quantum bumps in *InaD*^{P215} reveal that they do not in fact terminate normally, and are characterized by a noisy tail [290], similar to those recorded in *inaC* [220]. In any case, the initial phase of both macroscopic response and bump waveform appear essentially wild-type in character suggesting that the association of TRP with INAD is not required for excitation (or that the *in situ* effect of the *InaD*^{P215} mutation on TRP binding is not as severe as indicated by co-immunoprecipitation and overlay assays). By contrast, in the *inaD*^{2} or *norpA*^{C1094S} mutant, which both disrupt association of INAD with NORPA and lead to random localization of NORPA throughout the cytoplasm, sensitivity to light (assayed by ERG) is substantially reduced and response latency and kinetics slowed. In the null allele (*inaD*^{1}), in the double mutant *InaD*^{P215}; *norpA*^{C1094S} or *norpA* mutants lacking the PDZ binding domains, sensitivity to light is decreased even further [216,218,219].

Several authors have commented that the close association of elements of the transduction cascade may be essential for the rapid response kinetics which characterize *Drosophila* phototransduction by minimizing or eliminating diffusional delays [216–218]. The specific (albeit debated) defect in response termination in *InaD*^{P215} also leads to speculation that one function of the INAD complex is to target Ca^{2+} influx via the TRP-dependent channels directly to PKC [215], which appears to be required for Ca^{2+} dependent inactivation of the light response ([179,221]; see Section 10.3.1), and possibly to PLC which may also be subject to feedback regulation by Ca–calmodulin [222,223]. With respect to activation the situation is less clear: any *norpA* or *inaD* mutation which prevents association of PLC and INAD leads to severe defects in excitation, but they also lead to a large reduction of PLC in the microvillar membranes which would be expected to be essential for normal phototransduction on any model. However, the specific dissociation of TRP from the INAD complex via the $InaD^{P215}$ mutation leads to very minor, if any defect in excitation except in older flies where TRP levels begin to decline (note that, although also found outside the rhabdomere, substantial levels of TRP protein are still found within the microvilli of young $InaD^{P215}$, although presumably no longer associated with the INAD complex). Nevertheless several authors [216,218] have speculated that the INAD complex may provide a mechanism for direct gating of the light-sensitive channels by protein–protein interactions.

10. Ca^{2+}-dependent feedback and adaptation

One of the more impressive features of the performance of *Drosophila* photoreceptors is that, despite producing quantum bumps that are both larger and faster than their counterparts in vertebrate rods, the photoreceptors are still capable of light-adapting over the full environmental range of light intensities. Fig. 14, which shows responses to steps of increasing intensity, gives some indication of this performance with the rapid and pronounced transition from initial peak to adapted plateau response occurring within ca. 200 ms at high intensities. Noise analysis, performed mainly in larger flies ([28,145,146]; but see Ref. [224] for *Drosophila*) and other invertebrates [225] indicates that during light adaptation the single photon responses (quantum bumps) become progressively smaller, shorter in duration and are generated with shorter latency. In some species of flies, light-adapted impulse responses have latencies of 2–3 ms and time-to-peak of ca. 5–10 ms, representing the fastest G-protein-coupled signaling system known [226]. We are only slowly beginning to understand some of the molecular mechanisms responsible for this performance, but, as in vertebrate photoreceptors, Ca^{2+} has long been recognized as an essential messenger of adaptation, and at least some of its actions can be understood as an acceleration of the mechanisms of response termination.

10.1. Ca^{2+} influx and homeostasis

Although in some invertebrates the rise in Ca^{2+} derives at least partly from release from intracellular stores (see Chapter 8), in *Drosophila*, most [197,298] if not all the light-induced rise in Ca^{2+} is mediated via Ca^{2+} influx through the light-sensitive channels [75,140,141]. In whole-cell voltage-clamped photoreceptors loaded with Ca^{2+} indicator dyes, light induces a rapid rise in global cytosolic Ca^{2+} levels which can reach 10–50 μM during the bright illumination generated by the light used to measure dye fluorescence. Recent *in situ* measurements in the larger fly *Calliphora* have estimated that cytosolic levels approaching 50 μM are also reached and

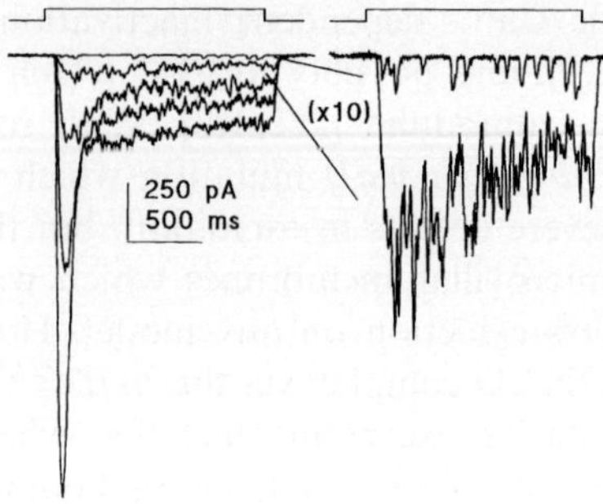

Fig. 14. Whole-cell recordings of light induced currents in *Drosophila* photoreceptors in response to 1.5 s light flashes of increasing intensities (10, 500, 3000, 9000, 30000 effective absorbed photons s^{-1}) delivered to a cell clamped at a holding potential of −60 mV in physiological Ringer's solution. With dim light, responses to single absorbed photons (quantum bumps) can clearly be resolved. At higher intensities, the responses rapidly adapt to a steady-state plateau. The two smallest responses on the left are shown again at 10× higher amplification on the right. Saturating responses reach values greater than 20 nA (not shown).

maintained during light adaptation in intact retinae, whilst in the rhabdomere Ca^{2+} may reach mM levels [307,308]. For small currents up to ca. 500 pA, Ca^{2+} rises approximately linearly as a function of the total charge carried by the light-sensitive current at a rate of ca. 3 nM/pC [141]. In WT flies, the major source of Ca^{2+} influx is via the TRP-dependent channels and correspondingly Ca^{2+} influx is reduced in the *trp* mutant [140,141,150]. Nevertheless the TRPL channels are still significantly permeable to Ca^{2+} and can raise Ca^{2+} into the micromolar range before the response decays. Assuming a fractional Ca^{2+} current estimated from GHK analysis, quantitative data suggest that ca. 99.5–99.9% of the Ca entering via the light-sensitive channels is buffered or extruded. It seems probable that Na/Ca exchange is the major mechanism of Ca^{2+} extrusion, and large (up to ca. 100 pA) exchange currents have been directly measured in response to Ca^{2+} loads imposed via release of caged Ca^{2+} [24]. Interestingly, a quantitative model of ionic homeostasis suggests that the Na/Ca exchanger may even represent the major source of Na influx into the cell during the light response [22]. Little is known of Ca^{2+} buffering mechanisms within the cell; however, both calmodulin, which is present in mM concentrations in the rhabdomeres [227], and a novel Ca^{2+} binding protein (calphotin) localized to a large, though poorly defined intracellular compartment [228] have been suggested as molecular buffers. In addition there are thapsigargin and ionomycin sensitive Ca^{2+} stores [75,196] as well as abundant mitochondria, which in other cells are believed to play a major role in Ca^{2+} homeostasis.

10.2. Ca^{2+} mediates positive and negative feedback

Hardie [42] and Ranganathan et al. [43] demonstrated the profound importance of Ca^{2+} influx in regulating response kinetics both by lowering extracellular Ca^{2+} and also by depolarizing the cell to reduce the driving force for Ca^{2+} entry (Fig. 15). In the presence of normal extracellular Ca^{2+}, response kinetics show a strong voltage dependence with both activation and deactivation being greatly delayed (ca. five-fold) at positive holding potentials. In the absence of external Ca^{2+}, responses are ca. 5× slower and no longer show any marked voltage dependence. That these effects are mediated intracellularly is further supported by the ability of caged Ca^{2+} to facilitate the light response when released during the rising phase of the response and to inhibit during the falling phase [24]. Conversely, buffering cytosolic Ca^{2+} with high concentrations of Ca^{2+} chelators such as BAPTA also largely mimics the effect of lowering extracellular Ca^{2+} [75,196]. As well as slowing the activation and deactivation kinetics of flash responses, lowering external Ca^{2+} or buffering cytosolic Ca^{2+} prevents or slows the peak to plateau transition that is a direct manifestation of light adaptation (Fig. 15). With the exception of the Ca/CaM dependent inactivation of TRPL channels (see Section 10.3.2) most indications of positive and negative feedback by Ca^{2+} appear to be absent in *trp* mutants [18,138]. This implies either that most feedback is mediated via modulation, directly or indirectly, of TRP channel activity, or that only Ca^{2+} influx via the TRP channels is capable of activating feedback mechanisms.

10.3. Molecular targets of Ca^{2+}-dependent feedback

10.3.1. Protein kinase C (PKC)

The $inaC^{P209}$ mutant was originally isolated by the PDA screen (see Section 4.2, [37]) and later shown to represent a null mutant in an eye-specific PKC [229] now also shown to be a member of the INAD complex (see Section 9). When investigated under whole-cell voltage clamp, *inaC* photoreceptors were discovered to have a

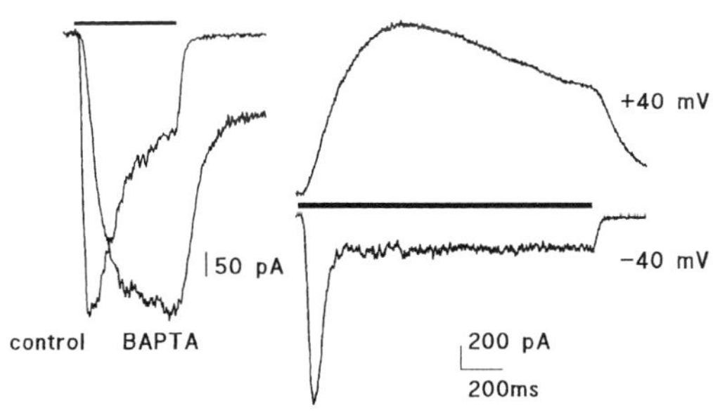

Fig. 15. Ca^{2+} influx regulates kinetics and mediates adaptation. The rapid transition from peak to plateau (e.g. control, left; −40 mV right) during a prolonged light step (indicated by the black bar) is a direct manifestation of light adaptation. This transition is abolished or greatly slowed in the absence of external Ca^{2+} (not shown). In addition, it can be blocked both by internal perfusion with Ca^{2+} chelators such as BAPTA (4 mM free BAPTA, Ca buffered at ca. 250 nM; left) or by depolarizing the cell (+40 mV, right) thereby reducing the driving force for Ca^{2+} influx. Notice also that the response rise time is greatly slowed under both these conditions indicating that Ca^{2+} influx also mediates positive feedback. All responses recorded in physiological Ringer's solution containing 1.5 mM Ca^{2+} (Hardie R.C., unpublished data).

specific defect in inactivation, with responses to flashes of light decaying over a period of ca. 1–2 s. This phenotype was masked in the absence of external Ca^{2+} (i.e. WT and *inaC* response now have similar kinetics) indicating a defect in Ca^{2+}-dependent inactivation [43,229]. Hardie et al. [220] further showed that, in the intact retina under physiological conditions, classical manifestations of light adaptation such as peak–plateau transition and parallel shift of V–log I curves were absent and localized the defect to a failure of individual quantum bumps to terminate normally, suggesting an action of eye-PKC at a late stage of transduction (probably after PLC). The original *ina* phenotype in *inaC* reflects a decay of the macroscopic response to baseline during prolonged bright illumination, reminiscent of the *trp* phenotype. On the (disputed) assumption that this is caused by depletion of intracellular Ca^{2+} stores (see Section 10.3.2), this led us to suggest that PKC might be required for rapid termination of Ca^{2+} release [220]. Further progress in understanding the precise nature by which PKC regulates response termination will require identification of its substrate(s). Thus far the only evidence in this respect comes from *in vitro* phosphorylation studies. Huber et al. [217,309] reported that the *Calliphora* homologues of both the INAD and TRP proteins are phosphorylated in a Ca^{2+} and phorbol ester dependent manner, but what the functional consequence of this might be is unknown. Secondly, Warr and Kelly [162] found that a peptide fragment representing the first calmodulin binding site of the TRPL ion channel served as an *in vitro* substrate for both PKA and PKC dependent phosphorylation and that PKA dependent phosphorylation inhibited the ability of the peptide to bind CaM. Although PKC dependent phosphorylation did not affect CaM binding directly, pre-phosphorylation by PKC inhibited phosphorylation by PKA and might therefore be expected to promote CaM binding and hence inactivation of the channel (see below).

10.3.2. Calmodulin

Calmodulin has been implicated at a variety of stages within the phototransduction cascade including, as already discussed, both the TRP and TRPL ion channels and arrestin which is phosphorylated in a CaM dependent manner via CaM kinase II [100]. Recent functional insight into the role of calmodulin in regulating these, and perhaps other targets has come from the generation of mutants with reduced CaM levels, transgenic flies expressing proteins with deleted CaM binding sites and from the effects of CaM antagonists.

10.3.2.1. TRPL CaM binding sites

Responses to light mediated by TRPL channels in the *trp* mutant are subject to Ca^{2+}-dependent inactivation resulting, e.g. in larger and longer-lasting responses when recorded in Ca^{2+}-free solutions [19,32,139]. This Ca^{2+}-dependent inactivation appears to be mediated at least in part via the CaM binding sites on TRPL (see Section 7.3) since flies expressing TRPL ion channels in which one or other of the two CaM binding sites (CBS1 or CBS2) had been deleted had abnormally prolonged responses with a reduced Ca^{2+} dependency. Similar results were found by combining the *trp* mutation with a *cam* hypomorph mutation in which CaM levels were reduced to ca. 10% [32].

The role of Ca/CaM-dependent inactivation of TRPL channels in light adaptation is unclear since in WT flies the response appears to be dominated by TRP-dependent channels. Scott et al. [32] concluded that the Ca^{2+}-dependent inactivation of TRPL underlies the *trp* decay phenotype since they failed to detect this decay either in Ca^{2+}-free solutions or in the transgenic flies with TRPL CBS deletions. This would appear to be a premature conclusion however, since the light response in *trp* flies does indeed decay in Ca^{2+}-free solutions, and although it does so rather more slowly than in the presence of Ca^{2+}, the inactivation under these conditions is in fact much more profound with responses recovering only slowly if at all unless Ca^{2+} is returned to the bath. This result shows that suppression of excitation does not require any Ca^{2+} influx in *trp* flies [298]. Furthermore, whilst releasing caged Ca^{2+} during the rising phase of the response inhibits the response in *trp*, during the decay phase, Ca^{2+} actually facilitates the response [24]. Ca^{2+}-dependent inactivation and the decay of the response in *trp* thus appear to be separable, and possibly unrelated phenomena.

The role of the CaM binding site in the TRP channel has not yet been investigated with comparable techniques; however, the TRP channels are also subject to Ca^{2+} dependent inactivation that can be revealed by hyperpolarizing voltage steps or by release of caged Ca^{2+}. The time course of this inactivation is extremely fast (relaxation time constants of 1–2 ms) [24,230].

10.3.2.2. CaM mutants

Analysis of the light response in the *cam* hypomorph mutants (now on a WT background) has revealed further *in vivo* evidence for the role of calmodulin [32]. In the absence of external Ca^{2+}, WT and *cam* responses were indistinguishable; however, in the presence of external Ca^{2+} *cam* photoreceptors again showed profound defects in response deactivation although the rising phase of the response appeared normal. Investigation of quantum bumps in *cam* flies revealed no obvious abnormalities in quantum bump shape; however, in marked contrast to WT flies, where a single photon absorption generates only one bump, in *cam* photorecceptors a single photon elicited a train of 10 or more bumps over a period of 1 or 2 s. This defect was further localized to site upstream from the G_q protein by showing that the rare bumps in mutants with low levels of G-protein ($G\alpha q^1$) did not develop into trains of bumps in the *cam*; $G\alpha q^1$ double mutant combination. The most obvious explanation of this phenotype seems to be a defect in arrestin phosphorylation which is known to be mediated by CaM kinase. In support of this explanation phosphorylation of arrestin was found to be blocked in *cam* mutants and *arr2* mutants show a similar multiple bump phenotype to *cam* [32]. However, as discussed in Section 5.3.3, CaM-dependent phosphorylation appears to be required for arrestin to dissociate from metarhodopsin [292]. Hence the *cam* phenotype may reflect the unavailability of arrestin due to it being permanently bound to rhodopsin, rather than any immediate requirement of Ca/CaM for effective rhodopsin inactivation.

10.3.2.3. NINAC

An alternative approach to manipulating CaM levels in photoreceptors makes use of *ninaC* mutants. The *ninaC* gene encodes two photoreceptor-specific proteins of 174 and 132 kDa [231], which are localized to the rhabdomeres and to the cell body, respectively [232]. These two polypeptides (p174, p132) are translated from separate transcripts generated by alternative RNA processing. The NINAC proteins have multiple domains. Near the N-terminus is a stretch of 266 amino acid residues that shows a high degree of sequence homology with serine/threonine protein kinases [231]. Adjacent to this domain is a 725-amino-acid long region homologous to the head portion of portion of myosin heavy chain, typical of myosin I-like proteins (reviewed by [233]). The third domain, located near the C-terminus, is not shared between the two NINAC polypeptides and does not show homology to other known proteins. Like other myosin I-like proteins, NINAC binds calmodulin. The p174 protein is responsible for CaM localization in the rhabdomere and p132 binds calmodulin in the cell body. Accordingly, NINAC constitutes the major CaM binding protein in the photoreceptor cell [227]. Two physiological phenotypes are associated with mutations in *ninaC*. Null mutations result in an abnormally large steady-state component of the LIC [200,234] and slow termination of the light response at light off [200,216,234,235]. The abnormally large steady state of the LIC seems to arise from the reduced CaM level in the cell because it can be mimicked in a transgenic mutant (P[$ninaC^{\Delta B}$]), in which the CaM binding sites were removed. Also application of CaM and CaM inhibitor suppressed and enhanced this phenotype respectively [200].

ninaC mutants also show abnormalities in screening pigment migration (pupil mechanism); mutants lacking the rhabdomere specific p174 isoform showed reduced sensitivity to light, but null mutants, or mutants lacking the cytoplasmic isoform p132 showed increased sensitivity, although as in all alleles tested the amplitude of the response was distinctly smaller than that in WT flies [234]. The diverse phenotypes of *ninaC* flies may suggest multiple regulatory roles of NINAC, and by implication, calmodulin, in phototransduction and in regulation of pigment migration.

10.3.2.4. Ryanodine receptor

In muscle, neurons and many types of non-excitable cells [236] including insect photoreceptors [237–239], the endoplasmic reticulum (ER) contains both $InsP_3$ receptors and another Ca^{2+} release channel, the ryanodine receptor (RyR [236,240]). The RyR responds to an increase in cellular Ca^{2+} by Ca^{2+}-induced Ca^{2+}-release. *In situ* microphotometric measurements of Ca^{2+} fluxes across the ER membrane in permeabilized slices of drone retina show that caffeine induces ryanodine-sensitive (and heparin-insensitive) Ca^{2+} release from ER. Electrophysiological recordings have demonstrated that caffeine mimics all aspects of Ca^{2+}-mediated facilitation and adaptation in drone photoreceptors [239]. All known RyR have CaM binding sites [236]. In rabbit skeletal muscle the RyR binds CaM with high affinity. At micromolar to millimolar Ca^{2+} concentrations, CaM inhibited Ca^{2+} release via the RyR channels suggesting an important role for CaM in regulating ER Ca^{2+} release at both resting and elevated Ca^{2+} concentrations [241]. The *Drosophila* RyR shows

45–47% homology with the amino acid sequence of the mammalian ryanodine receptors, including the preserved CaM binding sites [237,238]. Prolonged exposure of *Drosophila* photoreceptors to ryanodine and caffeine abolishes the LIC, which could be rescued by inclusion of CaM in the recording pipette. Thus, ryanodine-sensitive stores may be a component of *Drosophila* phototransduction and CaM negatively regulates these stores [197,200].

10.3.2.5. CaM antagonist peptides

The functional role of CaM in phototransduction has been recently investigated by intracellular application of Ca^{2+}-CaM binding peptides (CCBP) derived from the pseudosubstrate domain of several CaM-regulated kinases [200,222]. In *Drosophila* photoreceptors CCBP blocks the LIC in transgenic *ninaC* mutant with reduced CaM, or in wild-type cells bathed in Ca^{2+}-free medium [197]. In *Limulus* photoreceptors CCBP strongly inhibited the light response [222]. Interestingly, although some of the inhibition occurred at late stages of the cascade as revealed by inhibition of the response to injection of $InsP_3$ and Ca^{2+}, a large fraction of inhibition appeared to be at the PLC level. This inhibition seems to arise from the existence of a CaM-like structure in some forms of PLC sequences. Consistent with this structure, *in vitro* essays of PLCδ and γ isozymes were strongly inhibited by two forms of CCBPs.

11. Photoreceptor degeneration

Photoreceptor degeneration is a phenomenon which appears in strong alleles of almost every phototransduction-defective mutant of *Drosophila*. This fact is not surprising given the abundance of the phototransduction molecules in the microvilli of the photoreceptor cell and the crucial role of these molecules in the assembly of the phototransduction machinery, which constitutes a large fraction of the microvilli proteins. In addition, this highly active cascade of molecular events can quickly lead to toxic conditions for the cell upon misfunction. Because of the great interest in neuronal cell death in general and in human diseases that are associated with neuronal cell death and in retinal degeneration, there is a great interest in these mutants, and a search for mammalian homologues of *Drosophila* genes has been recently pursued with much effort. Despite extensive investigation rather little is known about the molecular mechanisms underlying neural or retinal degeneration in either vertebrates or invertebrates. Substantial evidence suggests that the final pathway in several vertebrate animal models of retinal dystrophy is via apoptosis [242] and recent evidence suggests this may be the case in *Drosophila* as well ([243], see Section 11.2).

In this section we describe mutants which exhibit photoreceptor degeneration due to misfunction of eye-specific proteins. These mutants can be divided into two main categories: those which degenerate with no obvious relation to activation of the transduction cascade by light and those in which the degeneration depends on light-activation of the transduction cascade.

11.1. Mutations causing degeneration in the dark

These mutations include gene products that are important for the normal development and assembly of the photoreceptor cells and thus strongly affect the structure of the cell (mainly the rhabdomere) upon absence or misfunction. In addition, they include mutations that cause constitutive activity of the light sensitive channels in the dark. In this section we discuss only those mutations which are largely photoreceptor-specific and, therefore, are not lethal. Recently, new mutants which eliminate calmodulin [32] or the $InsP_3$ receptor [202] have been isolated. As expected, these mutations, which are not eye-specific, have been shown to be essential for the development and the survival of the fly and therefore will not be discussed in this section.

11.1.1. Mutations affecting rhodopsin

Null alleles of *ninaE* in which rhodopsin does not form [63,244] or does not reach the surface membrane (*ninaA* [245]) lead to abnormal structure and degeneration. In $ninaE^{ora}$ or $ninaE^{I117}$ the microvilli are abnormally short at the pupal stages upon eclosion and then disappear a few days later [60,246,247].

There is a striking similarity between vertebrates and invertebrates with respect to the effects of dominant mutations in rhodopsin. There is a great interest in these mutations, because more than 25% of all retinitis pigmentosa diseases in humans appear to be caused by dominant mutations in rhodopsin [245]. In both species, a variety of single amino acid substitutions perturb opsin stability, folding, or transportation and produce an autosomal dominant slow, age-dependent form of photoreceptor degeneration [244,245]. These dominant rhodopsin mutations also suppress the rapid light-dependent degeneration seen in *rdgC* and *norpA* flies [244] thus supporting the hypothesis that degeneration in the latter result from a defective photopigment cycle [83] (see Section 11.2.1).

11.1.2. Chaoptin

The *chaoptic* gene of *Drosophila* encodes a glycoprotein required to generate the normal tightly-packed array of microvilli in the rhabdomeres. The *chaoptic* gene product, chaoptin, is a 127,000-kd protein which is associated with photoreceptor membranes by covalent binding to the extracellular surface of cells via a glycosyl-phosphatidylinositol linkage [248].

Mutations in the *chaoptic* gene lead to highly disorganized or missing rhabdomeres [249]. Morphological analysis of developing rhabdomeres showed that the microvilli form in mutant photoreceptor cells, but fail to produce the normal rhabdomere structure suggesting that chaoptin is a cell-adhesion molecule. In consistence with this suggestion, chaoptin causes cells to aggregate in a homophilic manner, when expressed in tissue culture of *Drosophila* cell line [248] (for review see Ref. [6]).

11.1.3. Calphotin

Calphotin is a *Drosophila* photoreceptor cell-specific protein. It is a 85-kDa protein of unusual amino acid composition, having a C-terminal segment that binds Ca^{2+}.

The location of calphotin within a distinct cytoplasmic region below the SMC suggests that it might function as a Ca^{2+} sequestering "sponge" to regulate the amount of Ca^{2+} in a strategic location [61]. Calphotin is expressed very early in eye development. Two classes of calphotin (*cap*) mutants have been isolated and both of them have strong effects on the morphology of the photoreceptor cell. Class I have highly disorganized rhabdomeres which appear similar to those of the *chaoptic* mutant [62]. Class II, which belongs to the same complementation group, have rough eyes and cause photoreceptor degeneration. Analysis of the *cap* mutants suggest that calphotin plays important roles in both rhabdomere development and maintenance of the photoreceptor cell [62].

11.1.4. Retinal degeneration A (rdgA) *mutant*

One of the most interesting light-independent retinal degeneration mutants is the retinal degeneration A (*rdgA*) mutant [124]. This mutant already has a largely abnormal structure at the pupal stage. Upon eclosion it has: (i) short and irregularly arranged microvilli; (ii) abundant rough endoplasmic reticulum; (iii) absence of SMC; (iv) reduced lysosome level. A biochemical search of the defect in *rdgA* mutants demonstrated that these mutants are deficient in diacylglycerol (DAG) kinase activity [250–252]. The highly reduced DAG kinase activity results from absence of a photoreceptor specific isoform of diacylglycerol kinase, designated DGK2 [204]. DGK2 , like porcine DGK, has two cysteine-rich zinc-finger motifs. A unique feature for DGK2 is the four ankyrin-like repeats at the C-terminal region. DGK2 has been immunolocalized to the sub-microvillar cisternae [253]. DAG-kinase phosphorylates diacylglycerol to form phosphatidic acid (PA, reviewed by [254]). Phosphorylation of DAG by DGK is thought to be a cellular mechanism for PKC inactivation [255,256]. Two possible mechanisms underlying degeneration in *rdgA* flies have been suggested: (i) Chronic activation of protein kinase C (PKC) due to accumulation of DAG is a possible mechanism. However, the double mutant *rdgA;inaC* which lacks the eye PKC [229] does not rescue the degeneration of *rdgA* flies [253], and no indication of DAG accumulation was found by thin layer chromatography in *rdgA* mutants [252]. Also the double mutant *rdgA;norpA* in which light-production of DAG is prevented still degenerates, consistent with the dark degeneration of *rdgA* [204]. (ii) Alternatively, or in synergism with constitutive action of PKC, it has been suggested that degeneration might arise from the lack of phosphatidic acid (PA), a component of cellular membranes. PA is the first step in phosphoinositide recycling, which finally leads to the production of PIP_2, the substrate of PLC . Defects in this cycle are known to induce light-dependent degeneration (see below), but in synergism with other mechanisms they can also lead to degeneration in the dark. In this regard, autoradiographic studies of the incorporation of radiolabeled amino acids and sugars into photoreceptors suggest that *rdgA* mutant cells are defective in the turnover of rhabdomeric membrane [204,257,258], (for review see Refs. [6,101]). Recent studies in yeast have shown that DAG is essential for protein transport from the Golgi complex [259]. Lack of DAG can, thus, lead to defect in vesicle trafficking and reduction in some crucial component for microvilli structure or the phototransduction itself.

11.1.5. A new trp *mutant*

A new mutant, Trp^{P365}, was recently described. This mutant does not display the transient receptor potential phenotype and is characterized by a substantial level of the TRP protein and rapid, semi-dominant, degeneration of photoreceptors in the dark already at pupae stage. This phenotype is markedly different from other *trp* mutants which have a transient receptor potential, absence or near absence of TRP protein and do not degenerate until old age. In spite of its unusual phenotypes, Trp^{P365} is a *trp* allele because a Trp^{P365} transgene induces the mutant phenotype in a wild-type background, and a wild-type *trp* transgene in a Trp^{P365} background suppresses the mutant phenotype. Moreover, amino acid alterations that could cause the Trp^{P365} phenotype are found in the transmembrane segment region of the mutant channel protein. Whole-cell recordings clarified the mechanism underlying the retinal degeneration by showing that the TRP channels of Trp^{P365} are constitutively active. Heteroallelic combination of Trp^{P365} and other *trp* alleles that slow the degeneration process indicate that the constitutive activity of TRP channels precedes the degeneration [310].

11.2. Mutations causing light-dependent retinal degeneration

Two further retinal degeneration mutants, *rdgB* and *rdgC* have been found in which degeneration only occurs following illumination [124,247,260,261]. In addition, a number of mutants originally isolated on the basis of a phototransduction phenotype have also been found to undergo light-dependent retinal degeneration as a secondary defect of the mutation, including the *norpA*, *ninaC*, *inaD*, *arr*, *eye-CDS* mutants. To account for the conditional phenotype of retinal degeneration, it has been suggested that in wild-type fly there is a counterpart molecule of the retinal degeneration gene product, which counteracts one of the steps in the phototransduction cascade to turn its activity off [260,262]. When this gene is mutated, it leads to unbalanced activity of the transduction step with which the retinal degeneration gene product normally interacts and this, in turn, gives rise to light dependent retinal degeneration.

11.2.1. Dephosphorylation of rhodopsin, which is defective in rdgC *and* norpA *mutants is required for maintenance of the photoreceptor cell*

11.2.1.1. The retinal degeneration C mutant (*rdgC*)

The *rdgC* mutant undergoes light-dependent retinal degeneration which requires several days of illumination to be completed [261]. Originally, no electrophysiological phenotype was associated with *rdgC* flies. However, a recent study has shown that *rdgC* flies have a slow response termination in intact flies but not in whole cell recordings of isolated ommatidia [84]. The *rdgC* mutant has been genetically characterized as acting upstream of phospholipase C [261]. Molecular cloning and sequencing revealed homology to serine/threonine protein phosphatases with an appended domain containing at least two EF hand motifs which are putative Ca^{2+} binding domains [263]. Recent functional analysis of *rdgC* and wild-type flies, revealed that phosphorylated rhodopsin in wild-type eye membrane preparations is rapidly dephosphorylated and that the dephosphorylation of rhodopsin was highly dependent on Ca^{2+}, but not on calmodulin [83]. In contrast in *rdgC* membrane Ca^{2+}-dependent dephosphorylation of rhodopsin was completely eliminated leading to hyperphosphorylation of metarhodopsin [83]. This result indicates that phosphorylated rhodopsin is a major substrate for the Ca^{2+}-dependent RDGC protein phosphatase [56,83,84]. This conclusion was supported by the generation of a transgenic *Drosophila* mutant (*ninaE* P[*Rh1Δ356*]), in which the last 18 amino acids residues that include the serine/threonine phosphorylating sites at the C terminal domain of Rh1, was eliminated. Interestingly, the double mutant *rdgC*; *ninaE* P[*Rh1Δ356*] neither reveals a hyperphosphorylated metarhodopsin nor degenerates [84].

11.2.1.2. The *norpA* mutant

The dependence of the RDGC phosphatase on Ca^{2+}, which is one of the end products of the inositol lipid phototransduction cascade, is interesting as it bears on the unexplained observation that the *norpA*, which codes for phospholipase C [122], undergoes age and light dependent retinal degeneration resulting in clear degeneration 6 days after eclosion in light/dark raised flies at 24°C [264,265]. Retinal degeneration of the *norpA* mutant which was studied extensively showed that strong *norpA* alleles which have no electrical response to light demonstrated more prominent retinal degeneration than weak alleles that have some light dependent electrical activity. Surprisingly, in *norpA* alleles, which do not respond electrically to light, retinal degeneration was still light dependent [265]. It was suggested [83] that the mechanism of retinal degeneration in the strong *norpA* alleles is similar to the mechanism of retinal degeneration in the *rdgC* mutant. In both mutants, retinal degeneration is initiated by light-dependent phosphorylation of the photopigment which is not adequately followed by dephosphorylation. In the *rdgC* mutant, this is due to deficiency in rhodopsin phosphatase, while in the *norpA* mutant deficient dephosphorylation of rhodopsin is due to the inability to generate the Ca^{2+} signal which is required for activation of rhodopsin phosphatase [83] (for review see Ref. [56]). This mechanism can also account for light-dependent retinal degeneration in other transduction mutants which block the increase of cellular Ca^{2+}. This putative mechanism of light-dependent retinal degeneration was corroborated by phosphorylation experiments in the intact fly *in vivo* [83,266]. In these experiments rhodopsin was found to be hyperphosphorylated both in the *rdgC* and in the *norpA* mutant, while it was minimally phosphorylated in the wild-type fly [266]. A strong support for the hypothesis that hyperphosphorylation of the photopigment is at the root of degeneration in *norpA* and *rdgC* flies has been provided by the dominant rhodopsin mutants mentioned above. When Rh1 level is reduced below 15–20% by the dominant rhodopsin mutation, retinal degeneration is suppressed in *norpA* and *rdgC* flies [244]. Degeneration in a dominant rhodopsin mutant ($ninaE^{RH27}$) and in *rdgC* has recently been rescued by expression of the anti-apoptotic gene p35 (a baculoviral cell survival factor) under the control of a photoreceptor-specific promoter, suggesting that the degeneration proceeds via an apoptotic pathway [243].

Since a reduction in the photopigment level is sufficient to rescue *rdgC* and dominant *ninaE* flies, it is essential to examine if the expression of p35 is not associated with a reduction in Rh1 level. An alternative possible mechanism of degeneration in *norpA* and similar mutants lacking light response has emerged with the findings of Alloway and Dolph [292] that constitutive association of 49Arr to metarhodopsin causes degeneration. In conditions when activation of the photopigment cycle is not accompanied by a rise in cellular Ca^{2+}, 49Arr will fail to phosphorylate and thus remain bound to metarhodopsin.

Further studies will have to test whether hyperphosphorylation of rhodopsin is a tag which marks it as a molecule which is destined for degradation, thereby increasing the rate of rhodopsin degradation beyond the capacity to replenish it by rhodopsin biogenesis (for review see Ref. [56]).

11.2.1.3. Mammals and *C. elegans* homologues of RDGC

By a combination of large-scale sequencing of human retina cDNA clones and searches of expressed sequence tag and genomic DNA databases, three mammal [267,268] and one *C. elegans* (PPEF) homologue of *Drosophila* RDGC have been identified [267]. One of them (PPEF-1) was found to be localized to the inner segment of vertebrate rods [267]. Together with the recently described localization of PPEF-1 transcript to primary somatosensory neurons and inner ear cells in developing mouse [269], these data suggest that *rdgC* is a multi-gene family of Ca^{2+}-dependent serine/threonine phosphatases with a specific and conserved role in diverse sensory neurons [267].

11.2.2. Light-dependent retinal degeneration in rdgB *flies*

The *Drosophila* retinal degeneration B mutant (*rdgB*) was one of the first retinal degeneration mutants to be isolated [124] and most of the studies on retinal degeneration in *Drosophila* have been done on this mutant. Nevertheless, the mechanism underlying photoreceptor degeneration in this mutant remains obscure. In addition, some *rdgB* alleles demonstrate olfaction defects [270], indicating that RDGB is not confined to the visual system. The RDGB protein seems to play a critical role in the fly photoreceptor cell. The *rdgB* mutant phenotype is characterized by retinal degeneration whose onset, while discernible in dark-reared flies, is greatly accelerated by raising the flies in light [260,271] (see Fig. 16). Typically, *rdgB* mutant flies begin to exhibit the morphological hallmarks of photoreceptor cell degeneration several days after eclosion [260,271]. In the photoreceptor cell, RDGB localizes to both the axon and the submicrovillar cisternae (SMC) [272].

11.2.2.1. RDGB operates at a late stage of the phototransduction cascade

Genetic analyses suggest that RDGB functions downstream of both rhodopsin and phospholipase C in the visual transduction cascade. This is because both the *ninaE* and *norpA* mutations suppress the *rdgB*-dependent, light-enhanced retinal degeneration [260,271], suggesting that malfunctioning of a step in the PI cascade downstream of PLC activation causes retinal degeneration in *rdgB* flies. Consistent with this view, constitutive activation of the *Drosophila* G-protein (DGq), either by application of non-hydrolyzable GTP analogs and fluoride [273], or by the expression of a constitutively activated G_α subunit (*DGq1* [110]) effects a rapid degeneration of *rdgB* retinas in the absence of light. RDGB apparently functions downstream of diacyl glycerol (DAG)-activated protein kinase C (PKC), because (i) application of phorbol ester to *rdgB* mutant retinas, which presumably activates PKC, stimulates retinal degeneration in the absence of light [274] (see Fig. 16), and (ii) the *rdgB* retinal degeneration is weakly suppressed by the eye-specific PKC mutant, *inaC* [229]. Thus, the available evidence identifies an action of RDGB downstream of protein kinase C in the visual transduction cascade.

11.2.2.2. The electrophysiological phenotype of *rdgB* flies

Several electrophysiological phenotypes are characteristic of the *rdgB* mutant. These mutant flies exhibit an abnormal light response, as recorded by the rapid deterioration of the electroretinogram (ERG), shortly after the fly's initial exposure to light. This ERG defect is manifested prior to any obvious physical signs of retinal degeneration [260,271], which suggests that the defect in the light response may precede the course of retinal degeneration. The phototransduction defects are also evidenced by abnormal termination of the light response and profound loss of the electroretinogram (ERG) amplitude shortly after initial light exposure [260,275]. The ERG changes occur before any morphological evidence of retinal degeneration, which becomes evident several days after eclosion.

At young ages, before any obvious signs of degeneration, light stimuli evoke Ca^{2+} spikes from the axon terminals of R1–R6 photoreceptors of the *rdgB* mutant fly not normally seen in wild-type flies [262]. The Ca^{2+} spikes can be blocked by *d-cis* diltiazem, verapamil hydrochloride and cadmium, and application of these Ca^{2+} channel blockers to the eyes of *rdgB* flies over a 7-day period also largely inhibited light-dependent degeneration. Retarding the light-induced photoreceptor degeneration in the mutant by Ca^{2+}-channel blockers thus suggests that a toxic increase in intracellular Ca^{2+} by means of voltage-gated Ca^{2+} channels, at the synaptic terminal, may lead to degeneration in the *rdgB* mutant [276]. The localization of RDGB to the axon and axon terminals (in addition to its presence in or near the SMC) is consistent with this view and the finding that the degeneration in the *rdgB* mutant begins at the axon terminals of the photoreceptors [271] also supports the above hypothesis.

The calcium content of wild-type and mutant photoreceptor of *Drosophila* was measured using rapid freezing of the eyes and energy-dispersive X-ray analysis (e.d.x.) of cryosections and semi-thin section of cryosubstituted material. The results indicate that there is a large increase in cellular calcium in specific compartments during light-induced photoreceptor degeneration in several *Drosophila* mutants. The fact that this accumulation of calcium is observed even when induced by very different mutations, such as *rdgB*, *rdgC*, *norpA* suggests that the calcium accumulation is a secondary rather than a primary effect in the degeneration process [276].

11.2.2.3. The RDGB protein

RDGB is a novel 116 kDa membrane polypeptide with six potential transmembrane domains [277]. Additionally, the amino-terminal 281 RDGB residues share 42%

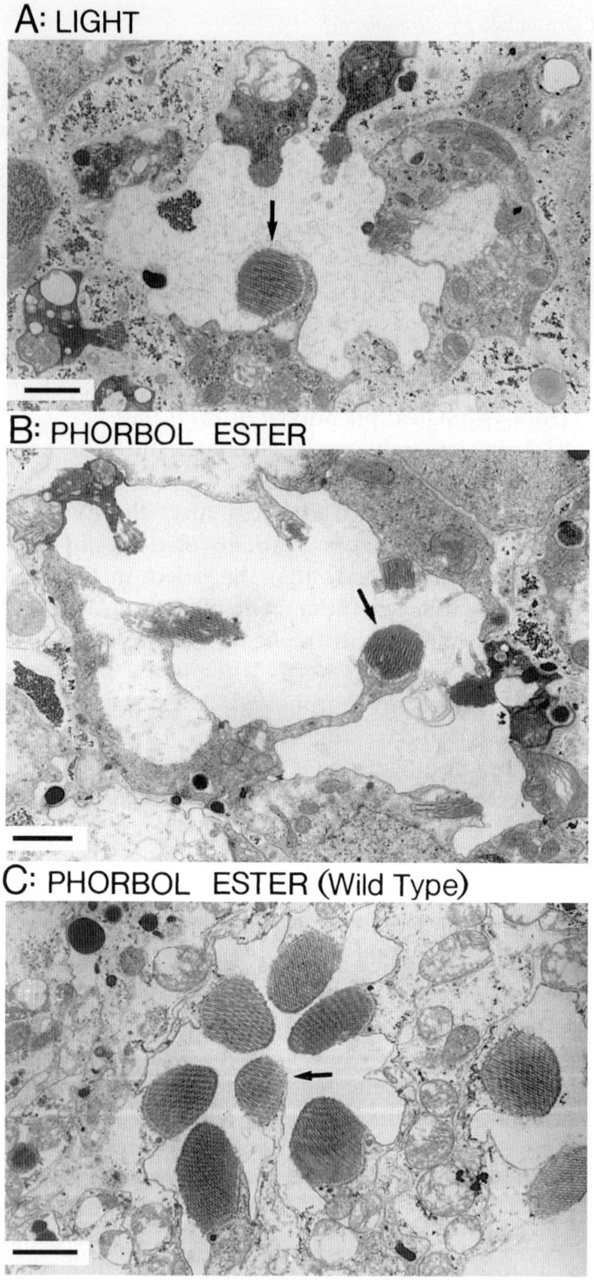

Fig. 16. The ultrastructure of a highly degenerated ommatidium induced by light (A) compared to a highly degenerated ommatidium of the *rdgB* mutant exposed to phorbol ester and raised in the dark (B). Phorbol ester was applied for 8 days. The intact appearance of wild-type ommatidium treated with phorbol ester is included as a control (C). R1–R6 cells are almost completely degenerated in (A) and (B) but the central cell remained intact in mutants exposed to light (arrows). (A) or phorbol ester (B) ML, multilamellar bodies. No sign of degeneration was observed in any of the ommatidia of 11 wild-type (white-eyed) flies treated with phorbol ester in the same manner as the mutant (C), which were used in the present study (bars = 1.0 μm; from [274]).

amino acid identity with the rat brain phosphatidylinositol transfer protein isoform (PITPα) [272]. Unlike all previously characterized PITPs, which are 32 to 35 kDa soluble proteins [278,279], RDGB is a large integral membrane protein. Thus, the *Drosophila* RDGB protein defines a new class of functionally equivalent transmembrane PITPs. While PITPs are operationally defined by their ability to catalyze the transfer of either phosphatidylinositol (PI) or phosphatidylcholine (PC) monomers between membrane bilayers *in vitro* [278,279], how the phospholipid transfer activity pertains to *in vivo* function is less clear. The PITP domain of RDGB, when expressed as a soluble protein in *E. coli*, is able to effect intermembrane transfer of PI *in vitro* [272]. Recently, Hyde and colleagues [275] have demonstrated that the complete repertoire of RDGB functions essential for normal phototransduction resides in the N terminus, which includes the PITP domain. Consistent with this result, expression of RDGB's PITP domain as a soluble protein (RDGB-PITP) in *rdgB*^{2} mutant flies is sufficient to completely restore the wild-type electrophysiological light response and prevent the degeneration. However, introduction of the T59E mutation (in which thr 59 was substituted for glu), which does not affect RDGB-PITP's PI and PC transfer *in vitro*, into the soluble (RDGB-PITP-T59E) or full-length (RDGB-T59E) proteins eliminated rescue of retinal degeneration in *rdgB*^{2} flies, while the light response was partially maintained. Substitution of the rat brain PITPα, a classical PI transfer protein, for RDGB's PITP domain (PITPα or PITPα–RDGB chimeric protein) neither restored the light response nor maintained retinal integrity when expressed in *rdgB*^{2} flies. Therefore, although all essential RDGB functions appear to reside in the N-terminus which includes the PITP domain, other PITPs possessing PI/PC transfer activity *in vitro* cannot supplant RDGB function *in vivo*. Expression of either RDGB-T59E or PITPα-RDGB in *rdgB*$^+$ flies produced a dominant retinal degeneration phenotype. While RDGB-T59E functioned in a dominant manner to significantly reduce steady-state levels of rhodopsin, PITPα–RDGB was defective in the ability to recover from prolonged light stimulation and caused photoreceptor degeneration through an unknown mechanism. Taken together, the data indicate an underlying complexity to the mechanism of RDGB function and its role in the photoreceptor cell that is not easily reconciled with a simple role in potentiating signal transduction via PITP function [275]. Therefore, in spite of postulated *in vivo* activities for PITPs, the other functions of RDGB in the photoreceptor cell remain unknown.

11.2.2.4. Mammalian homologues of RDGB

A screening for conserved mammalian genes that are differentially expressed in the retina and retinal pigment epithelium (RPE) have been recently conducted in an attempt to find novel genes involved in retinal development and function as well as to provide new candidate genes for the study of inherited retinal diseases. In this process, mammalian homologues of *rdgB* were isolated [280] from mouse (*mrdgB*) bovine and humans (see also Ref. [281]). Based on strong sequence conservation and similarity of the expression pattern at both the RNA and protein levels, it was suggested that the mammalian homologues are in fact the orthologues of *rdgB* because they contain similar organization with an amino-terminal PITP domain, multiple hydrophobic domains and long carboxy terminous [280,281,311,312]. Recently, three human rdgB homologues (Nir1, Nir2, and Nir3) were identified by their interaction with the protein tyrosine kinase PYK2 [313]. The Nir2 and Nir3 proteins correspond to the two mouse rdgB homologues, however, the Nir1 protein lacks the corresponding amino-terminal PITP domain. This interesting observation suggests that RDGB has multiple functions and some of them are not related to PI transfer. Interestingly, the human *rdgB* gene maps at or near the site of four retinal diseases. Most importantly, and perhaps surprisingly given the significant differences between mammalian and invertebrate phototransduction, expression of murine *mrdgB* in *rdgB* mutant flies fully rescues the mutant phenotypes. These results suggest the existence of novel aspects of vertebrate photoreceptor signal transduction that were not appreciated previously. The importance of PITP function for the maintenance and possibly function of vertebrate neurons has been revealed by the recent finding that the mouse *vibrator* mutation represents a hypomorphic mutation in the *pitpn* gene, which encodes PITPα [282].

11.2.3. The eye-CDS *mutant*

CDP-diacylglycerol synthase (CDS) is an enzyme required for the generation of CDP–DAG from PA, which is an essential step in the PI cycle for regenerating PIP_2 (Fig. 13). A photoreceptor-specific CDS protein and a mutant (*eye-CDS*) have been recently isolated [283]. The ERG response to prolonged intense light of the mutant has an *ina* phenotype, namely the response to light declines slowly towards baseline, while overexpression of CDS increased the ERG amplitude above normal. Calcium deprivation eliminated the LIC in the eye-CDS mutant [283] like in WT cells [284]. In Ca^{2+}-deprived mutant cells, illumination followed by application of Ca^{2+} did not rescue the LIC in contrast to WT cells. However, application of PI, PIP, PIP_2 mixture or PIP_2 alone rescued the LIC of the mutant [283]. These findings suggest that the PIP_2 level is a limiting factor for excitation in *Drosophila*. Interestingly, the *eye-CDS* mutants undergo light-dependent degeneration which requires about 10 days to be completed. This light-dependent degeneration is protected by the *norpA* mutation, suggesting that the events responsible for degeneration occur downstream of PLC activation [283].

11.2.4. The ninaC *mutant*

The rhabdomeres of *ninaC* mutant flies are greatly reduced in size in flies raised in the light, primarily because the electron dense core present in normal rhabdomeres is greatly reduced or missing [285].

The null *ninaC* photoreceptors undergo light and age-dependent retinal degeneration [232,286]. This degeneration seems to depend entirely on illumination as no degeneration appears in dark raised mutants. This phenotype of *ninaC* is associated with p174 but not with p132 [232]. Transgenic flies in which the myosin domain of p174 was deleted caused mislocalization of the protein, degeneration and altered ERG suggesting that the myosin domain is required for several functions. It is likely that the myosin domain functions as a motor which is required to move the functionally-important segments of this molecule and this capability is important for maintenance of the photoreceptor cell. The role and substrate of the kinase domain is unknown.

11.2.5. The arrestin mutant (arr)
Photoreceptors of the *arr2* mutant (which affects the 49Arr) with severe reduction in arrestin levels show light-dependent degeneration 5 days after exposure to 12 h light–12 h dark cycle [57]. The *norpA* mutation protects *arr2* mutants from the light-dependent degeneration suggesting that the prolonged response of the *arr* mutant, due to inability to inactivate metarhodopsin is the cause of degeneration. These data argue that sustained activation of the transduction cascade may irreversibly damage photoreceptors, possibly through calcium-dependent cytotoxicity. In support of this model, the *trp* mutation, which reduces much of the light-dependent rise in intracellular calcium [140,141], also protects photoreceptor degeneration in *arr2* mutants (PJ Dolph and CS Zuker, unpublished data in [101]).

12. Conclusion and outlook

This review of our current understanding of phototransduction in *Drosophila* has highlighted the importance of genetic analysis combined with sophisticated physiological and biochemical techniques for dissection of function. The characterization of mutants with defects in phototransduction and the molecular cloning of the responsible genes has identified the key roles of many of the elements of the PI cascade in phototransduction. It has also been instrumental in the identification of novel molecules, which have defined new families of signaling molecules such as TRP, NINAA, INAD, RDGC, RDGB and NINAC, which are likely to play important roles in a broader context. Despite the resulting advances in our understanding of phototransduction there are still many unanswered questions. The major outstanding problem concerns the mechanism of activation of the light-sensitive channels; a question now all the more urgent since it may also provide insight into the equally mysterious process of PI-regulated Ca^{2+} influx which plays important roles in a wide variety of cells throughout the animal kingdom. There are still also many problems to be addressed concerning the mechanisms of response inactivation and adaptation. As in vertebrates Ca^{2+} appears to play a major role acting at multiple downstream targets, but many molecular details remain to be clarified. It will also be of great interest to understand the processes underlying the generation of quantum bumps, and how invertebrate photoreceptors have managed to achieve a sensitivity to single photons that exceeds that of vertebrates, whilst being capable of adapting to respond over the entire range of environmental light intensities.

Finally, it is clear that mutations in many of the genes associated with the transduction cascade lead to retinal degeneration in both vertebrates and *Drosophila*, but the underlying pathological mechanisms are only poorly understood. Retinal degeneration in *Drosophila* has many proven parallels with degenerative processes in vertebrates, in several cases with the involvement of homologous genes.

The ease, for example, of epistatic genetic analysis in *Drosophila* provides unique tools for exploring the underlying mechanisms, whilst there is much active interest in using degeneration genes identified in *Drosophila* as candidate genes for degenerative disorders in humans (e.g. [268,281,287,288]).

Note added in proof

A number of relevant papers have appeared since preparation of this review. Some of the most important recent studies are summarised here in order of the relevant section number

1. A variety of alleles of both *norpA* and *InaD*, all of which reduce the amount of PLC in the microvillar membrane, also resulted in reduced GTP-ase activity and slowed response termination due to quantum bumps being generated with long latencies. The results indicated that PLC-INAD interactions are not required to produce normal bumps, and that PLC is a GTP-ase activating protein (GAP) essential for rapid inactivation of G-protein *in vivo*. (Section 6.5; Ref. [314])
2. Single channel recordings have now been reported from isolated *Drosophila* rhabdomeral membranes (Section 7; Ref. [315])
3. A third putative light sensitive channel gene, *trpγ* has been identified and proposed to form a heteromultimer with TRPL. (Section 7.1; Ref. [316])
4. A lethal hypomorphic mutation in the only ryanodine receptor gene in the *Drosophila* genome was reported to have no effect on phototransduction in eye mosaics (Section 10.3.2.4; Ref. [317])
5. A recent study has found that the light-sensitive TRP channels are constitutively active in the diacylglycerol kinase mutant *rdgA*, suggesting the Ca^{2+} influx may be responsible for degeneration. Accordingly, degeneration was rescued in the *rdgA;trp* double mutant. Responses to light in the rescued flies show a defect in response inactivation – consistent with a role for DAG in excitation. (Section 11.1.4; Ref. [318])

Abbreviations

CaM, calmodulin
cam, calmodulin mutant
nina (A–F), neither inactivation nor afterpotential mutant
ina (A–K), inactivation but not afterpotential mutant
eye cds, eye specific CDP-diacylglycerol syntase mutant
arr1, 39 kDal arrestin (39Arr) mutant
arr2, 49 kDdal arrestin (49Arr) mutant
norpA, no receptor potential A mutant
cap, chaoptin mutant
cal, calphotin mutant
trp, transient receptor potential mutant
trpl, transient receptor potential like mutant
rdg (A–C), retinal degeneration mutant
PDA, prolonged depolarizing afterpotential
SOC, store-operated current
$InsP_3$, D myo inositol-1,4,5 trisphosphate
RyR, ryanodine receptor
CCE, capacitative Ca^{2+} entry
CICR, Ca^{2+}-induced Ca^{2+}-release
RDC, rundown current

EGTA, ethylene glycol-bis(β-aminoethyl ether)*N*,*N*,*N*′,*N*′-tetraacetic acid BAPTA, 1,2-bis(2-aminophenoxy)ethan-*N*,*N*,*N*′,*N*′-tetraacetic acid
$InsP_3R$, inositol 1,4,5-trisphosphate receptor
DPG, 2,3, diphosphoglycerate
TRPC(1–4) human TRP channel-related proteins 1–4
sTRP, squid TRP protein
dTRP, *Drosophila* TRP protein
dTRPL, *Drosophila* TRPL protein
bTRP, bovine TRP protein
cTRP, *Caenorhabditis elegans* TRP protein
mTRPC1(a,b), C3, C5, C6, mouse TRP proteins
VR1, capsaicin receptor
OSM9, olfactory channel from *C. elegans*
R, rhodopsin
M, metarhodopsin
Rh(1–6), *Drosophila* opsins 1–6
PLC, phospholipase C
DAG, diacylglycerol
GTPγS, guanosine 5′-[γ-thio]triphosphate
GDPβS, guanosine 5′-[β-thio]diphosphate
LIC, light-induced current
ER, endoplasmic reticulum
RoK, rhodopsin kinase
PIP_2, phosphatidylinositol 4′:5′-bisphosphate
PITP, phosphatidylinositol transfer protein
cGMP, guanosine 3′:5′-cyclic monophosphate

Acknowledgements

Work reported from the authors' laboratories was supported by grants from the Minerva Foundation, the National Institutes of Health (EY 03529), the Israel Science Foundation (ISF) and the German Israel Foundation (GIF) for BM and the Wellcome Trust and the BBSRC for RCH.

References

1. Hardie, R.C. (1985) in: Progress in Sensory Physiology, Vol. 5, ed D. Ottoson. pp. 1–79, Springer, Berlin.
2. Franceschini, N. and Kirschfeld, K. (1971) Kybernetik. **9**, 159–182.
3. Kirschfeld, K. and Snyder, A.W. (1976) Vision Res. **16**, 775–778.
4. Hardie, R.C. (1986) Trends Neurosci. **99**, 419–423.
5. Hamdorf, K. (1979) in: Handbook of Sensory Physiology, Vol VII/6A, ed H. Autrum. pp. 145–224, Springer, Berlin.
6. Smith, D.P., Stamnes, M.A. and Zuker, C.S. (1991) Annu. Rev. Cell Biol. **7**, 161–190.
7. Franceschini, N., Hardie, R.C., Ribi, W. and Kirschfeld, K. (1981) Nature **291**, 241–244.
8. Chou, W.H., Hall, K.J., Wilson, D.B., Wideman, C.L., Townson, S.M., Chadwell, L.V. and Britt, S.G. (1996) Neuron **17**, 1101–1115.

9. Huber, A., Schulz, S., Bentrop, J., Groell, C., Wolfrum, U. and Paulsen, R. (1997) FEBS Lett. **406**, 6–10.
10. Pak, W.L. (1975) in: Handbook of Genetics, Vol. 3, ed R.C. King. pp. 703–733, Plenum, New York.
11. Minke, B. and Selinger, Z. (1992) in: Neuromethods, eds A. Boulton, G. Baker, and C. Taylor. pp. 517–563, The Humana Press, Clifton, NJ.
12. Hardie, R.C. (1989) Nature **339**, 704–706.
13. Hardie, R.C., Voss, D., Pongs, O. and Laughlin, S.B. (1991) Neuron **6**, 477–486.
14. Hardie, R.C. (1991) J. Neurosci. **11**, 3079–3095.
15. Hevers, W. and Hardie, R.C. (1995) Neuron **14**, 845–856.
16. Hardie, R.C. and Mojet, M.H. (1995) J. Neurosci. **74**, 2590–2599
17. Anderson, J. and Hardie, R.C. (1996) J. Comp. Physiol. A **178**, 513–522.
18. Hardie, R.C. and Minke, B. (1994) J. Gen. Physiol. **103**, 409–427.
19. Hardie, R.C. (1995) J. Comp. Physiol. A **177**, 707–721.
20. Minke, B. and Armon, E. (1984) Vision Res. **24**, 109–115.
21. Hochstrate, P. (1991) Z. Naturforsch. **46c**, 451–460.
22. Gerster, U. (1997) Vision Res. **37**, 2477–2485.
23. Jansonius, N.M. (1990) J. Comp. Physiol. A **167**, 461–467.
24. Hardie, R.C. (1995) J. Neurosci. **15**, 889–902
25. Yeandle, S. (1958) Am. J. Ophthal. **46**, 82–87.
26. Wu, C.F. and Pak, W.L. (1978) J. Gen. Physiol. **71**, 249–268.
27. Johnson, E.C. and Pak, W.L. (1986) J. Gen. Physiol. **88**, 651–673.
28. Barash, S. and Minke, B. (1994) Comments Theoret. Biol. **3**, 229–263.
29. Stephenson, R.S., O'Tousa, J., Scavarda, N.J., Randall, L.L. and Pak, W.L. (1983) Symp. Soc. Exp. Biol. **36**, 477–501.
30. Stieve, H. (1986) in: The Molecular Mechanism of Photoreception, ed H. Stieve. pp. 199–230, Springer, Berlin.
31. Dorloechter, M. and Stieve, H. (1997) Prog. Neurobiol. **53**, 451–515.
32. Scott, K., Sun, Y., Beckingham, K. and Zuker, C.S. (1997) Cell **91**, 375–383.
33. Berridge, M.J. (1995) Biochem. J. **312**, 1–11.
34. Parker, I., Choi, J. and Yao, Y. (1996) Cell Calcium **20**, 105–121.
35. Hillman, P., Hochstein, S. and Minke, B. (1983) Physiol. Rev. **63**, 668–772.
36. Minke, B. (1986) in: The Molecular Mechanism of Photoreception, ed H. Stieve. pp. 241–265, Springer, Berlin.
37. Pak, W.L. (1979) in: Neurogenetics, Genetic Approaches to the Nervous System, ed X. Breakfield. pp. 67–99, Elsevier, Amsterdam.
38. Hochstein, S., Minke, B. and Hillman, P. (1973) J. Gen. Physiol. **62**, 105–128.
39. Minke, B., Wu, C. and Pak, W.L. (1975) J. Comp. Physiol. **98**, 345–355.
40. Stark, W.S. and Zitzmann, W.G. (1976) J. Comp. Physiol. **105**, 15–27.
41. Larrivee, D.C., Conrad, S.K., Stephenson, R.S. and Pak, W.L. (1981) J. Gen. Physiol. **78**, 521–545.
42. Hardie, R.C. (1991) Proc. R. Soc. London B **245**, 203–210.
43. Ranganathan, R., Harris, G.L., Stevens, C.F. and Zuker, C.S. (1991) Nature **354**, 230–232.
44. Morgan, T.H. (1910) Science **32**, 120–122.
45. Spradling, A.C. and Rubin, G.M. (1982) Science **218**, 341–347.
46. Mismer, D. and Rubin, G.M. (1987) Genetics **116**, 565–578.
47. Zuker, C.S., Mismer, D., Hardy, R. and Rubin, G.M. (1988) Cell **53**, 475–482.
48. Feiler, R., Harris, W.A., Kirschfeld, K., Wehrhahn, C. and Zuker, C.S. (1988) Nature **333**, 737–741.
49. Heisenberg, M. and Goetz, K.G. (1975) J. Comp. Physiol. **98**, 217–241.
50. Benzer, S. (1967) Proc. Natl. Acad. Sci., USA **58**, 1112–1119.
51. Hall, J.C. (1982) Q. Rev. Biophys. **15**, 223–479.
52. Pak, W.L., Grossfield, J. and Arnold, K.S. (1970) Nature **227**, 518–520.
53. Koenig, J. and Merriam, J. (1977) Drosoph. Inf. Ser. **52**, 50–51.
54. Pak, W.L., Grossfield, J. and White, N.V. (1969) Nature **222**, 351–354.
55. Pak, W.L. and Pinto, L.H. (1976) Annu. Rev. Biophys. Bioeng. **5**, 397–448.

56. Selinger, Z., Doza, Y.N. and Minke, B. (1993) Biochim. Biophys. Acta **1179**, 283–299.
57. Dolph, P.J., Ranganathan, R., Colley, N.J., Hardy, R.W., Socolich, M. and Zuker, C.S. (1993) Science **260**, 1910–1916.
58. Niemeyer, B.A., Suzuki, E., Scott, K., Jalink, K. and Zuker, C.S. (1996) Cell **85**, 651–659.
59. Zipursky, S.L., Venkatesh, T.R. and Benzer, S. (1985) Proc. Natl. Acad. Sci., USA **82**, 1855–1859.
60. Pollock, J.A., Assaf, A., Peretz, A., Nichols, C.D., Mojet, M.H., Hardie, R.C. and Minke, B. (1995) J. Neurosci. **15**, 3747–3760.
61. Martin, J.H., Benzer, S., Rudnicka, M. and Miller, C.A. (1993) Proc. Natl. Acad. Sci., USA **90**, 1531–1535.
62. Yang, Y. and Ballinger, D. (1994) Genetics **138**, 413–421.
63. O'Tousa, J.E., Baehr, W., Martin, R.L., Hirsh, J., Pak, W.L. and Applebury, M.L. (1985) Cell **40**, 839–850.
64. Zuker, C.S., Cowman, A.F. and Rubin, G.M. (1985) Cell **40**, 851–858.
65. Stryer, L. (1986) Annu. Rev. Neurosci. **9**, 87–119.
66. Applebury, M.L. and Hargrave, P.A. (1986) Vis. Res. **26**, 1881–1895.
67. Vogt, K. and Kirschfeld, K. (1984) Naturwissenschaften **77**, 211–213.
68. Kirschfeld, K., Franceschini, N. and Minke, B. (1977) Nature **269**, 386–390.
69. Cowman, A.F., Zuker, C.S. and Rubin, G.M. (1986) Cell **44**, 705–710.
70. Pollock, J.A. and Benzer, S. (1988) Nature **333**, 779–782.
71. Zuker, C.S., Montell, C., Jones, K., Laverty, T. and Rubin, G.M. (1987) J. Neurosci. **7**, 1550–1557.
72. Montell, C., Jones, K., Zuker, C. and Rubin, G. (1987) J. Neurosci. **7**, 1558–1566.
73. Kirschfeld, K., Feiler, R. and Franceschini, N. (1978) J. Comp. Physiol. **125**, 275–284.
74. Franceschini, N., Kirschfeld, K. and Minke, B. (1981) Science **213**, 1264–1267.
75. Ranganathan, R., Bacskai, B.J., Tsien, R.Y. and Zuker, C.S. (1994) Neuron **13**, 837–848.
76. Shieh, B.H., Stamnes, M.A., Seavello, S., Harris, G.L. and Zuker, C.S. (1989) Nature **338**, 67–70.
77. Schneuwly, S., Shortridge, R.D., Larrivee, D.C., Ono, T., Ozaki, M. and Pak, W.L. (1989) Proc. Natl. Acad. Sci., USA **86**, 5390–5394.
78. Colley, N.J., Baker, E.K., Stamnes, M.A. and Zuker, C.S. (1991) Cell **67**, 255–263.
79. Ondek, B., Hardy, R.W., Baker, E.K., Stamnes, M.A., Shieh, B.H. and Zuker, C.S. (1992) J. Biol. Chem. **267**, 16460–16466.
80. Stamnes, M.A., Shieh, B.H., Chuman, L., Harris, G.L. and Zuker, C.S. (1991) Cell **65**, 219–227.
81. Bentrop, J. and Paulsen, R. (1986) Eur. J. Biochem. **161**, 61–67.
82. Doza, Y.N., Minke, B., Chorev, M. and Selinger, Z. (1992) Eur. J. Biochem. **209**, 1035–1040.
83. Byk, T., Bar Yaacov, M., Doza, Y.N., Minke, B. and Selinger, Z. (1993) Proc. Natl. Acad. Sci., USA **90**, 1907–1911.
84. Vinos, J., Jalink, K., Hardy, R.W., Britt, S.G. and Zuker, C.S. (1997) Science **277**, 687–690.
85. Wilden, U., Hall, S.W. and Kuhn, H. (1986) Proc. Natl. Acad. Sci., USA **83**, 1174–1178.
86. Palczewski, K., Rispoli, G. and Detwiler, P.B. (1992) Neuron **8**, 117–126.
87. Chen, J., Makino, C.L., Peachey, N.S., Baylor, D.A. and Simon, M.I. (1995) Science **267**, 374–377.
88. Plangger, A., Malicki, D., Whitney, M. and Paulsen, R. (1994) J. Biol. Chem. **269**, 26969–26975.
89. Palczewski, K. and Benovic, J.L. (1991) Trends Biochem. Sci. **16**, 387–391.
90. Inglese, J., Koch, W.J., Caron, M.G. and Lefkowitz, R.J. (1992) Nature **359**, 147–150.
91. Pitcher, J.A., Inglese, J., Higgins, J.B., Arriza, J.L., Casey, P.J., Kim, C., Benovic, J.L., Kwatra, M.M., Caron, M.G. and Lefkowitz, R.J. (1992) Science **257**, 1264–1267.
92. Onorato, J.J., Palczewski, K., Regan, J.W., Caron, M.G., Lefkowitz, R.J. and Benovic, J.L. (1991) Biochemistry **30**, 5118–5125.
93. Kikkawa, S., Yoshida, N., Nakagawa, M., Iwasa, T. and Tsuda, M. (1998) J. Biol. Chem. **273**, 7441–7447.
94. Yamada, T., Takeuchi, Y., Komori, N., Kobayashi, H., Sakai, Y., Hotta, Y. and Matsumoto, H. (1990) Science **248**, 483–486.
95. LeVine, H. ed., Smith, D.P., Whitney, M., Malicki, D.M., Dolph, P.J., Smith, G.F., Burkhart, W. and Zuker, C.S. (1990) Mech. Dev. **33**, 19–25.

96. Hyde, D.R., Mecklenburg, K.L., Pollock, J.A., Vihtelic, T.S. and Benzer, S. (1990) Proc. Natl. Acad. Sci., USA **87**, 1008–1012.
97. Smith, D.P., Shieh, B.H. and Zuker, C.S. (1990) Proc. Natl. Acad. Sci., USA **87**, 1003–1007.
98. Matsumoto, H. and Pak, W.L. (1984) Science **223**, 184–186.
99. Matsumoto, H. and Yamada, T. (1991) Biochem. Biophys. Res. Commun. **177**, 1306–1312.
100. Matsumoto, H., Kurien, B.T., Takagi, Y., Kahn, E.S., Kinumi, T., Komori, N., Yamada, T., Hayashi, F., Isono, K., Pak, W.L., et al. (1994) Neuron **12**, 997–1010.
101. Ranganathan, R., Malicki, D.M. and Zuker, C.S. (1995) Annu. Rev. Neurosci. **18**, 283–317.
102. Hurley, J.B. (1994) Curr. Opin. Neurobiol. **4**, 481–487.
103. Blumenfeld, A., Erusalimsky, J., Heichal, O., Selinger, Z. and Minke, B. (1985) Proc. Natl. Acad. Sci., USA **82**, 7116–7120.
104. Minke, B. and Stephenson, R.S. (1985) J. Comp. Physiol. A **156**, 339–356.
105. Devary, O., Heichal, O., Blumenfeld, A., Cassel, D., Suss, E., Barash, S., Rubinstein, C.T., Minke, B. and Selinger, Z. (1987) Proc. Natl. Acad. Sci., USA **84**, 6939–6943.
106. Lee, Y.J., Dobbs, M.B., Verardi, M.L. and Hyde, D.R. (1990) Neuron **5**, 889–898.
107. Yarfitz, S., Niemi, G.A., McConnell, J.L., Fitch, C.L. and Hurley, J.B. (1991) Neuron **7**, 429–438.
108. Dolph, P.J., Man Son Hing, H., Yarfitz, S., Colley, N.J., Deer, J.R., Spencer, M., Hurley, J.B. and Zuker, C.S. (1994) Nature **370**, 59–61.
109. Yarfitz, S.L., Running Deer, J.L., Froelick, G., Colley, N.J. and Hurley, J.B. (1994) J. Biol. Chem. **269**, 30340–30344.
110. Lee, Y.J., Shah, S., Suzuki, E., Zars, T., O'Day, P.M. and Hyde, D.R. (1994) Neuron **13**, 1143–1157.
111. Scott, K., Becker, A., Sun, Y., Hardy, R. and Zuker, C. (1995) Neuron **15**, 919–927.
112. Fein, A. (1986) Science **232**, 1543–1545.
113. Kirkwood, A., Weiner, D. and Lisman, J.E. (1989) Proc. Natl. Acad. Sci., USA **86**, 3872–3876.
114. Downes, C.P., Mussat, M.C. and Michell, R.H. (1982) Biochem. J. **203**, 169–177.
115. Selinger, Z. and Minke, B. (1988) Cold Spring Harb. Symp. Quant. Biol. **53 Pt 1**, 333–341.
116. Baer, K.M. and Saibil, H.R. (1988) J. Biol. Chem. **263**, 17–20.
117. Wood, S.F., Szuts, E.Z. and Fein, A. (1989) J. Biol. Chem. **264**, 12970–12976.
118. Northup, J.K., Sternweis, P.C., Smigel, M.D., Schleifer, L.S., Ross, E.M. and Gilman, A.G. (1980) Proc. Natl. Acad. Sci., USA **77**, 6516–6520.
119. Bigay, J., Deterre, P., Pfister, C. and Chabre, M. (1985) FEBS Lett. **191**, 181–185.
120. Pak, W.L. (1991) in: Molecular Biology of the Retina: Basic and Clinical Relevant Studies, eds J. Piatigorsky, T. Shinohara and P.S. Zelenka. pp. 1–32, Wiley-Liss, New York.
121. Pak, W.L. (1995) Invest. Ophthalmol. Vis. Sci. **36**, 2340–2357.
122. Bloomquist, B.T., Shortridge, R.D., Schneuwly, S., Perdew, M., Montell, C., Steller, H., Rubin, G. and Pak, W.L. (1988) Cell **54**, 723–733.
123. Schneuwly, S., Burg, M.G., Lending, C., Perdew, M.H. and Pak, W.L. (1991) J. Biol. Chem. **266**, 24314–24319.
124. Hotta, Y. and Benzer, S. (1970) Proc. Natl. Acad. Sci., USA **67**, 1156–1163.
125. Minke, B. and Selinger, Z. (1992) in: Sensory Transduction, eds D. Corey and S.D. Roper. pp. 202–217, The Rockefeller University Press, New York.
126. Inoue, H., Yoshioka, T. and Hotta, Y. (1985) Biochem. Biophys. Res. Commun. **132**, 513–519.
127. Deland, M.C. and Pak, W.L. (1973) Nature New Biol. **244**, 184–186.
128. Stahl, M.L., Ferenz, C.R., Kelleher, K.L., Kriz, R.W. and Knopf, J.L. (1988) Nature **332**, 269–272.
129. McKay, R.R., Chen, D.M., Miller, K., Kim, S., Stark, W.S. and Shortridge, R.D. (1995) J. Biol. Chem. **270**, 13271–13276.
130. Pearn, M.T., Randall, L.L., Shortridge, R.D., Burg, M.G. and Pak, W.L. (1996) J. Biol. Chem. **271**, 4937–4945.
131. Rhee, S.G., Suh, P.G., Ryu, S.H. and Lee, S.Y. (1989) Science **244**, 546–550.
132. Morris, A.J., Waldo, G.L., Boyer, J.L., Hepler, J.R., Downes, C.P. and Harden, T.K. (1990) Soc. Gen. Physiol. Ser. **45**, 61–75.

133. Bristol, A., Hall, S.M., Kriz, R.W., Stahl, M.L., Fan, Y.S., Byers, M.G., Eddy, R.L., Shows, T.B. and Knopf, J.L. (1988) Cold Spring Harb. Symp. Quant. Biol. **53 Pt 2**, 915–920.
134. Ferreira, P.A., Shortridge, R.D. and Pak, W.L. (1993) Proc. Natl. Acad. Sci., USA **90**, 6042–6046.
135. Obukhov, A.G., Harteneck, C., Zobel, A., Harhammer, R., Kalkbrenner, F., Leopoldt, D., Luckhoff, A., Nurnberg, B. and Schultz, G. (1996) EMBO J. **15**, 5833–5838
136. Kunze, D.L., Sinkins, W.G., Vaca, L. and Schilling, W.P. (1997) Am. J. Physiol. **272**, C27–C34
137. Hardie, R.C., Reuss, H., Lansdell, S.J. and Millar, N.S. (1997) Cell Calcium **21**, 431–440.
138. Hardie, R.C. and Minke, B. (1992) Neuron **8**, 643–651.
139. Reuss, H., Mojet, M.H., Chyb, S. and Hardie, R.C. (1997) Neuron **19**, 1249–1259.
140. Peretz, A., Suss-Toby, E., Rom-Glas, A., Arnon, A., Payne, R. and Minke, B. (1994) Neuron **12**, 1257–1267.
141. Hardie, R.C. (1996) J. Neurosci. **16**, 2924–2933
142. Cosens, D.J. and Manning, A. (1969) Nature **224**, 285–287.
143. Minke, B., Wu, C. and Pak, W.L. (1975) Nature **258**, 84–87.
144. Howard, J. (1984) J. Exp. Biol. **113**, 471–475.
145. Barash, S., Suss, E., Stavenga, D.G., Rubinstein, C.T., Selinger, Z. and Minke, B. (1988) J. Gen. Physiol. **92**, 307–330.
146. Suss, E., Barash, S., Stavenga, D.G., Stieve, H., Selinger, Z. and Minke, B. (1989) J. Gen. Physiol. **94**, 465–491.
147. Hochstrate, P. (1989) J. Comp. Physiol. A **166**, 179–187.
148. Suss Toby, E., Selinger, Z. and Minke, B (1991) J. Gen. Physiol. **98**, 849–868.
149. Minke, B. and Selinger, Z. (1991) in: Progress in Retinal Research, eds N.A. Osborne and G.J. Chader. pp. 99–124, Pergamon, Oxford.
150. Peretz, A., Sandler, C., Kirschfeld, K., Hardie, R.C. and Minke, B. (1994) J. Gen. Physiol. **104**, 1057–1077.
151. Montell, C. and Rubin, G.M. (1989) Neuron **2**, 1313–1323.
152. Phillips, A.M., Bull, A. and Kelly, L.E. (1992) Neuron **8**, 631–642.
153. Hardie, R.C. and Minke, B. (1995) Cell Calcium **18**, 256–274.
154. Minke, B. and Selinger, Z. (1996) Curr. Opin. Neurobiol. **6**, 459–466
155. Montell, C. (1997) Mol. Pharmacol. **52**, 755–763
156. Xu, X.Z., Li, H.S., Guggino, W.B. and Montell, C. (1997) Cell **89**, 1155–1164.
157. Hardie, R.C. and Minke, B. (1994) J. Gen. Physiol. **103**, 389–407.
158. Fesenko, E.E., Kolesnikov, S.S. and Lyubarsky, A.L. (1985) Nature **313**, 310–313.
159. Yau, K.W. and Baylor, D.A. (1989) Annu. Rev. Neurosci. **12**, 289–327.
160. Yau, K.-W. (1994) Invest. Ophthalmol. Vis. Sci. **35**, 9–32.
161. Montell, C., Jones, K., Hafen, E. and Rubin, G. (1985) Science **230**, 1040–1043.
162. Warr, C.G. and Kelly, L.E. (1996) Biochem. J. **314**, 497–503
163. Chevesich, J., Kreuz, A.J. and Montell, C. (1997) Neuron **18**, 95–105.
164. Williamson, M.P. (1994) Biochem. J. **297**, 249–260.
165. Vaca, L., Sinkins, W.G., Hu, Y., Kunze, D.L. and Schilling, W.P. (1994) Am. J. Physiol. **267**, C1501–C1505.
166. Rogers, S., Wells, R. and Rechsteiner, M. (1986) Science **234**, 364–368.
167. Sinkins, W.G., Vaca, L., Hu, Y., Kunze, D.L. and Schilling, W.P. (1996) J. Biol. Chem. **271**, 2955–2960.
168. Bourguignon, L.Y., Jin, H., Iida, N., Brandt, N.R. and Zhang, S.H. (1993) J. Biol. Chem. **268**, 7290–7297.
169. Bourguignon, L.Y., Chu, A., Jin, H. and Brandt, N.R. (1995) J. Biol. Chem. **270**, 17917–17922.
170. Petersen, C.C.H., Berridge, M.J., Borgese, M.F. and Bennett, D.L. (1995) Biochem. J. **311**, 41–44.
171. Gillo, B., Chorna, I., Cohen, H., Cook, B., Manistersky, I., Chorev, M., Arnon, A., Pollock, J.A., Selinger, Z. and Minke, B. (1996) Proc. Natl. Acad. Sci., USA **93**, 14146–14151.
172. Harteneck, C., Obukhov, A.G., Zobel, A., Kalkbrenner, F. and Schultz, G. (1995) FEBS Lett. **358**, 297–300.

173. Zimmer, S., Trost, C., Cavalie, A., Philipp, S. and Flockerzi, V. (1997) Naunyn-Schmiedebergs Arch. Pharmacol. **355**, 238.
174. Lan, L., Bawden, M.J., Auld, A.M. and Barritt, G.J. (1996) Biochem. J. **316**, 793–803
175. Huber, A., Sander, P., Gobert, A., Bahner, M., Hermann, R. and Paulsen, R. (1996) EMBO J. **15**, 7036–7045.
176. Monk, P.D., Carne, A., Liu, S.H., Ford, J.W., Keen, J.N. and Findlay, J.B.C. (1996) J. Neurochem. **67**, 2227–2235
177. Putney, J.W. Jr (1990) Cell Calcium **11**, 611–624
178. Clapham, D.E. (1996) Neuron **16**, 1069–1072.
179. Hardie, R.C. and Minke, B. (1993) Trends Neurosci. **16**, 371–376.
180. Wes, P.D., Yu, M. and Montell, C. (1996) EMBO J. **15**, 5839–5848.
181. Zitt, C., Zobel, A., Obukhov, A.G., Harteneck, C., Kalkbrenner, F., Luckhoff, A. and Schultz, G. (1996) Neuron **16**, 1189–1196.
182. Birnbaumer, L., Zhu, X., Jiang, M.S., Boulay, G., Peyton, M., Vannier, B., Brown, D., Platano, D., Sadeghi, H., Stefani, E. and Birnbaumer, M. (1996) Proc. Natl. Acad. Sci., USA **93**, 15195–15202.
183. Zhu, X., Jiang, M.S., Peyton, M., Boulay, G., Hurst, R., Stefani, E. and Birnbaumer, L. (1996) Cell **85**, 661–671.
184. Hoth, M. and Penner, R. (1992) Nature **355**, 353–356.
185. Philipp, S., Cavalie, A., Freichel, M., Wissenbach, U., Zimmer, S., Trost, C., Marquart, A., Murakami, M. and Flockerzi, V. (1996) EMBO J. **15**, 6166–6171
186. Zitt, C., Obukhov, A.G., Strubing, C., Zobel, A., Kalkbrenner, F., Luckhoff, A. and Schultz, G. (1997) J. Biol. Chem. **138**, 1333–1341
187. Boulay, G., Zhu, X., Peyton, M., Jiang, M.S., Hurst, R., Stefani, E. and Birnbaumer, L. (1997) J. Biol. Chem. **272**, 29672–29680
188. Caterina, M.J., Schumacher, M.A., Tominaga, M., Rosen, T.A., Levine, J.D. and Julius, D. (1997) Nature **389**, 816–824.
189. Colbert, H.A., Smith, T.L. and Bargmann, C.I. (1997) J. Neurosci. **17**, 8259–8269
190. Hardie, R.C. (1996) Curr. Biol. **6**, 1371–1373.
191. Ukhanov, K. and Payne, R. (1997) J. Neurosci. **17**, 1701–1709
192. Fein, A., Payne, R., Corson, D.W., Berridge, M.J. and Irvine, R.F. (1984) Nature **311**, 157–160.
193. Brown, J.E., Rubin, L.J., Ghalayini, A.J., Tarver, A.P., Irvine, R.F., Berridge, M.J. and Anderson, R.E. (1984) Nature **311**, 160–163.
194. Payne, R., Corson, D.W., Fein, A. and Berridge, M.J. (1986) J. Gen. Physiol. **88**, 127–142.
195. Shin, J., Richard, E.A. and Lisman, J.E. (1993) Neuron **11**, 845–855.
196. Hardie, R.C. (1996) Cell Calcium **20**, 315–327.
197. Arnon, A., Cook, B., Montell, C., Selinger, Z. and Minke, B. (1997) Science **275**, 1119–1121.
198. Richardson, A. and Taylor, C.W. (1998) J. Biol. Chem. **268**, 11528–11533
199. Wilcox, M. and Franceschini, N. (1984) Science **225**, 851–854.
200. Arnon, A., Cook, B., Gillo, B., Montell, C., Selinger, Z. and Minke, B. (1997) Proc. Natl. Acad. Sci., USA **94**, 5894–5899.
201. Venkatesh, K. and Hasan, G. (1997) Curr. Biol. **7**, 500–509
202. Acharya, J.K., Jalink, K., Hardy, R.W., Hartenstein, V. and Zuker, C.S. (1997) Neuron **18**, 881–887.
203. Irvine, R.F. (1992) FASEB J. **6**, 3085–3091.
204. Masai, I., Okazaki, A., Hosoya, T. and Hotta, Y. (1993) Proc. Natl. Acad. Sci., USA **90**, 11157–11161.
205. Nagy, K. (1991) Q. Rev. Biophys. **24**, 165–226.
206. Baumann, A., Frings, S., Godde, M., Seifert, R. and Kaupp, U.B. (1994) EMBO J. **13**, 5040–5050.
207. Bacigalupo, J., Bautista, D.M., Brink, D.L., Hetzer, J.F. and O'Day, P.M. (1995) J. Neurosci. **15**, 7196–7200.
208. Gibbs, S.M. and Truman, J.W. (1998) Neuron **20**, 83–93
209. Bicker, G. and Schmachtenberg, O. (1997) Eur. J. Neurosci. **9**, 189–193
210. Dong, Y., Kunze, D.L., Vaca, L. and Schilling, W.P. (1995) Am. J. Physiol. **269**, C1332–C1339.
211. Hu, Y. and Schilling, W.P. (1995) Biochem. J. **305**, 605–611.

212. Lamb, T.D. (1996) Proc. Natl. Acad. Sci., USA **93**, 566–570.
213. Shieh, B.H. and Niemeyer, B. (1995) Neuron **14**, 201–210.
214. Ranganathan, R. and Ross, E.M. (1997) Curr. Biol. **7**, 770–773.
215. Shieh, B.H. and Zhu, M.Y. (1996) Neuron **16**, 991–998.
216. Tsunoda, S., Sierralta, J., Sun, Y., Bodner, R., Suzuki, E., Becker, A., Socolich, M. and Zuker, C.S. (1997) Nature **388**, 243–249.
217. Huber, A., Sander, P. and Paulsen, R. (1996) J. Biol. Chem. **271**, 11710–11717.
218. Shieh, B.H., Zhu, M.Y., Lee, J.K., Kelly, I.M. and Bahiraei, F. (1997) Proc. Natl. Acad. Sci., USA **94**, 12682–12687
219. van Huizen, R., Miller, K., Chen, D.-M., Li, Y., Lai, Z.-C., Raab, R.W., Stark, W.S., Shortridge, R.D. and Li, M. (1998) EMBO J. **17**, 2285–2297.
220. Hardie, R.C., Peretz, A., Suss Toby, E., Rom Glas, A., Bishop, S.A., Selinger, Z. and Minke, B. (1993) Nature **363**, 634–637.
221. Ranganathan, R., Harris, W.A. and Zuker, C.S. (1991) Trends Neurosci. **14**, 486–493.
222. Richard, E.A. and Lisman, J.E. (1997) Invest. Ophthalmol. Vis. Sci. **38**, 2775.
223. Richard, E.A., Ghosh, S., Lowenstein, J.M. and Lisman, J.E. (1997) Proc. Natl. Acad. Sci., USA **94**, 14095–14099.
224. Wu, C.F. and Pak, W.L. (1975) J. Gen. Physiol. **66**, 149–168.
225. Dodge, F.A. Jr, Knight, B.W. and Toyoda, J. (1968) Science **160**, 88–90.
226. Laughlin, S.B. and Weckström, M. (1993) J. Comp. Physiol. A **172**, 593–609.
227. Porter, J.A., Yu, M., Doberstein, S.K., Pollard, T.D. and Montell, C. (1993) Science **262**, 1038–1042.
228. Ballinger, D.G., Xue, N. and Harshman, K.D. (1993) Proc. Natl. Acad. Sci., USA **90**, 1536–1540.
229. Smith, D.P., Ranganathan, R., Hardy, R.W., Marx, J., Tsuchida, T. and Zuker, C.S. (1991) Science **254**, 1478–1484.
230. Hardie, R.C. and Minke, B. (1994) J. Gen. Physiol. **103**, 409–427.
231. Montell, C. and Rubin, G.M. (1988) Cell **52**, 722–757.
232. Porter, J.A., Hicks, J.L., Williams, D.S. and Montell, C. (1992) J. Cell. Biol. **116**, 683–693.
233. Korn, E.D. and Hammer, J.A. eds (1990) Curr. Opin. Cell Biol. **2**, 57–61.
234. Hofstee, C.A. and Stavenga, D.G. (1996) Vis. Neurosci. **13**, 257–263.
235. Porter, J.A., Minke, B. and Montell, C. (1995) EMBO J. **14**, 4450–4459.
236. Furuichi, T., Kohda, K., Miyawaki, A. and Mikoshiba, K. (1994) Curr. Opin. Neurobiol. **4**, 294–303.
237. Hasan, G. and Rosbash, M. (1992) Development **116**, 967–975.
238. Takeshima, H., Nishi, M., Iwabe, N., Miyata, T., Hosoya, T., Masai, I. and Hotta, Y. (1994) FEBS Lett. **337**, 81–87.
239. Walz, B., Baumann, O., Zimmermann, B. and Ciriacy Wantrup, E.V. (1995) J. Gen. Physiol. **105**, 537–567.
240. Ehrlich, B. (1995) Curr. Opin. Neurobiol. **5**, 304–309.
241. Tripathy, A., Xu, L., Mann, G. and Meissner, G. (1995) Biophys. J. **69**, 106–119.
242. Gregory, C.Y. and Bird, A.C. (1995) Br. J. Ophthalmol. **79**, 186–190.
243. Davidson, F.F. and Steller, H. (1998) Nature **391**, 587–591.
244. Kurada, P. and O'Tousa, J.E. (1995) Neuron **14**, 571–579.
245. Colley, N.J., Cassill, J.A., Baker, E.K. and Zuker, C.S. (1995) Proc. Natl. Acad. Sci., USA **92**, 3070–3074.
246. O'Tousa, J.E., Leonard, D.S. and Pak, W.L. (1989) J. Neurogenet. **6**, 41–52.
247. Harris, W.A., Stark, W.S. and Walker, J.A. (1976) J. Physiol. **256**, 415–439.
248. Krantz, D.E. and Zipursky, S.L. (1990) EMBO J. **9**, 1969–1977.
249. Van Vactor, D. Jr, Krantz, D.E., Reinke, R. and Zipursky, S.L. (1988) Cell **52**, 281–290.
250. Yoshioka, T., Inoue, H. and Hotta, Y. (1983) Biochem. Biophys. Res. Commun. **111**, 567–573.
251. Yoshioka, T., Inoue, H. and Hotta, Y. (1984) Biochem. Biophys. Res. Commun. **119**, 389–395.
252. Inoue, H., Yoshioka, T. and Hotta, Y. (1989) J. Biol. Chem. **264**, 5996–6000.
253. Masai, I., Suzuki, E., Yoon, C.S., Kohyama, A. and Hotta, Y. (1997) J. Neurobiol. **32**, 695–706.
254. Bishop, W.R. and Bell, R.M. (1988) Annu. Rev. Cell Biol. **4**, 579–610.

255. Berridge, M.J. and Irvine, R.F. (1984) Nature **312**, 315–321.
256. Nishizuka, Y. (1984) Nature **308**, 693–698.
257. Matsumoto, E., Hirosawa, K., Takagawa, K. and Hotta, Y. (1988) Cell Tissue Res. **252**, 293–300.
258. Suzuki, E., Hirosawa, K. and Hotta, Y. (1990) J. Electron Microsc. Tokyo **39**, 50–53.
259. Kearns, B.G., McGee, T.P., Mayinger, P., Gedvilaite, A., Phillips, S.E., Kagiwada, S. and Bankaitis, V.A. (1997) Nature **387**, 101–105.
260. Harris, W.A. and Stark, W.S. (1977) J. Gen. Physiol. **69**, 261–291.
261. Steele, F. and O'Tousa, J.E. (1990) Neuron **4**, 883–890.
262. Rubinstein, C.T., Bar Nachum, S., Selinger, Z. and Minke, B. (1989) Vis. Neurosci. **2**, 529–539.
263. Steele, F.R., Washburn, T., Rieger, R. and O'Tousa, J.E. (1992) Cell **69**, 669–676.
264. Ostroy, S.E. (1978) J. Gen. Physiol. **72**, 717–732.
265. Meyertholen, E.P., Stein, P.J., Williams, M.A. and Ostroy, S.E. (1987) J. Comp. Physiol. A **161**, 793–798.
266. Minke, B. and Selinger, Z. (1995) in: Degenerative Diseases of the Retina, eds R.E. Anderson, M.M. LaVail, and J.G. Hollyfield. pp. 227–234, Plenum, New York.
267. Sherman, P.M., Sun, H., Macke, J.P., Williams, J., Smallwood, P.M. and Nathans, J. (1997) Proc. Natl. Acad. Sci., USA **94**, 11639–11644.
268. Huang, X. and Honkanen, R.E. (1998) J. Biol. Chem. **273**, 1462–1468.
269. Montini, E., Rugarli, E.I., Van de Vosse, E., Andolfi, G., Mariani, M., Puca, A.A., Consalez, G.G., den Dunnen, J.T., Ballabio, A. and Franco, B. (1997) Hum. Mol. Genet. **6**, 1137–1145.
270. Woodard, C., Alcorta, E. and Carlson, J. (1992) J. Neurogenet. **8**, 17–31.
271. Stark, W.S. and Carlson, S.D. (1982) Cell Tissue Res. **225**, 11–22.
272. Vihtelic, T.S., Goebl, M., Milligan, S., O'Tousa, J.E. and Hyde, D.R. (1993) J. Cell Biol. **122**, 1013–1022.
273. Rubinstein, C., Bar-Nachum, S., Selinger, Z. and Minke, B. (1989) Vis. Neurosci. **2**, 541–551.
274. Minke, B., Rubinstein, C., Sahly, I., Bar-Nachum, S., Timberg, R. and Selinger, Z. (1990) Proc. Natl. Acad. Sci., USA **87**, 113–117.
275. Milligan, S.C., Alb, J.G., Elagina, R.B., Bankaitis, V.A. and Hyde, D.R. (1997) J. Cell Biol. **139**, 351–363.
276. Sahly, I., Bar Nachum, S., Suss Toby, E., Rom, A., Peretz, A., Kleiman, J., Byk, T., Selinger, Z. and Minke, B. (1992) Proc. Natl. Acad. Sci., USA **89**, 435–439.
277. Vihtelic, T.S., Hyde, D.R. and O'Tousa, J.E. (1991) Genetics **127**, 761–768.
278. Cleves, A.E., McGee, T.P., Whitters, E.A., Champion, K.M., Aitken, J.R., Dowhan, W., Goebl, M. and Bankaitis, V.A. (1991) Cell **64**, 789–800.
279. Wirtz, K.W. (1991) Klin. Wochenschr. **69**, 105–111.
280. Chang, J.T., Milligan, S., Li, Y., Chew, C.E., Wiggs, J., Copeland, N.G., Jenkins, N.A., Campochiaro, P.A., Hyde, D.R. and Zack, D.J. (1997) J. Neurosci. **17**, 5881–5890.
281. Guo, J. and Yu, F.X. (1997) Dev. Genet. **20**, 235–245.
282. Hamilton, B.A., Smith, D.J., Mueller, K.L., Kerrebrock, A.W., Bronson, R.T., van Berkel, V., Daly, M.J., Kruglyak, L., Reeve, M.P., Nemhauser, J.L., Hawkins, T.L., Rubin, E.M. and Lander, E.S. (1997) Neuron **18**, 711–722.
283. Wu, L., Niemeyer, B., Colley, N., Socolich, M. and Zuker, C.S. (1995) Nature **373**, 216–222.
284. Hardie, R.C., Peretz, A., Pollock, J.A. and Minke, B. (1993) Proc. R. Soc. London B. Biol. Sci. **252**, 223–229.
285. Matsumoto, H., Isono, K., Pye, Q. and Pak, W.L. (1987) Proc. Natl. Acad. Sci., USA **84**, 985–989.
286. Porter, J.A. and Montell, C. (1993) J. Cell Biol. **122**, 601–612.
287. Chang, J.T., Milligan, S., Li, Y., Chew, C.E., Wiggs, J., Copeland, N.G., Jenkins, N.A., Campochiaro, P.A., Hyde, D.R. and Zack, D.J. (1997) J. Neurosci. **17**, 5881–5890.
288. Banfi, S., Borsani, G., Rossi, E., Bernard, L., Guffanti, A., Rubboli, F., Marchitiello, A., Giglio, S., Coluccia, E., Zollo, M., Zuffardi, O. and Ballabio, A. (1996) Nat. Genet. **13**, 167–174.
289. Hardie, R.C. and Minke, B. (1994) J. Gen. Physiol. **103**, 389–407.
290. Henderson, S., Reuss, H. and Hardie, R.C. (2000) J. Physiol. **524**, 179–194.

291. Salcedo, E., Huber, A., Heinrich, S., Chadwell, L.V., Chou, W.-H., Paulsen, R. and Britt, S.G. (1999) J. Neurosci. **19**, 10716–10726.
292. Alloway, P.G. and Dolph, P.J. (1999) Proc. Natl Acad. Sci., USA **96**, 6072–6077.
293. Trost, C., Marquart, A., Zimmer, S., Philipp, S., Cavalie, A. and Flockerzi, V. (1999) FEBS Lett. **451**, 257–263.
294. Putney, J.W. and McKay, R.R. (1999) Bioessays **21**, 38–46.
295. Bargmann, C.I. (1998) Science **282**, 2028–2033.
296. Kiselyov, K., Xu, X., Mozhayeva, G., Kuo, T., Pessah, I., Mignery, G., Zhu, X., Birnbaumer, L. and Muallem, S. (1998) Nature **396**, 478–482.
297. Kiselyov, K. and Muallem, S. (1999) Trends Neurosci. **22**, 334–337.
298. Cook, B. and Minke, B. (1999) Cell Calcium **25**, 161–171.
299. Hardie, R.C. and Raghu, P. (1998) Cell Calcium **24**, 153–163.
300. Raghu, P., Colley, N.J., Webel, R., James, T., Hasan, G., Danin, M., Selinger, Z. and Hardie, R.C. (2000) Mol. Cell. Neurosci. **15**, 429–445.
301. Chyb, S., Raghu, P. and Hardie, R.C. (1999) Nature **397**, 255–259.
302. Yoshioka, T., Inoue, H., Kasama, T., Seyama, Y., Nakashima, S., Nozawa, Y. and Hotta, Y. (1986) J. Biochem. **98**, 657–662.
303. Chyb, S., Hevers, W., Forte, M, Wolfgang, W.J., Selinger, Z. and Hardie, R.C. (1999) J. Neurosci. (1999) **19**, 8799–8807.
304. Yagodin, S., Hardie, R.C., Lansdell, S.J., Millar, N.S., Mason, W.T. and Sattelle, D.B. (1998) Cell Calcium **23**, 219–228.
305. Xu, X.Z.S., Choudhury, A., Li, X.L. and Montell, C. (1998) J. Cell Biol. **142**, 545–555.
306. Adamski, F.M., Zhu, M.Y., Bahirae, F. and Shie, B.H.J. (1998) Biol. Chem. **273**, 17713–17719.
307. Oberwinkler, J. and Stavenga, D.G. (1998) J. Gen. Physiol. **112**, 113–124.
308. Oberwinkler, J. and Stavenga, D.G. (2000) J. Neurosci. **20**, 1701–1709.
309. Huber, A., Sander, P., Bahner, M. and Paulsen, R. (1998) FEBS Lett. **425**, 317–322.
310. Yoon, J., Cohen Ben-Ami, H., Hong, Y.S., Park, S., Strong, L.L.R., Bowman, J., Geng, C., Baek, K., Minke, B. and Pak, W.L. (2000) J. Neurosci. **20**, 649–659.
311. Aikawa, Y., Hara, H. and Watanabe, T. (1997) Biochem. Biophys. Res. Commun. **236**, 559–564.
312. Lu, C., Vihtellic, T.S., Hyde, D. and Li, T.J. (1999) J. Neurosci. **19**, 7317–7325.
313. Lev, S., Hernandez, J., Martinez, R., Chen, A., Plowman, G. and Schlessinger J. (1999) Mol. Cell Biol. **19**, 2278–2288.
314. Cook, B., Bar-Yaacov, M., Cohen Ben-Ami, H., Goldstein, R.E., Paroush, Z., Selinger, Z. and Minke, B. (2000) Nature Cell Biol. **2**, 296–301.
315. Haab, J.E., Vergara, C., Bacigalupo, I. and O'Day, P.M. (2000) J. Physiol. Lond. **523**, 6P.
316. Sullivan, K.M., Scott, K., Zuker, C.S. and Rubin, G.M. (2000) Proc. Natl. Acad. Sci., USA **97**, 5942–5947.
317. Xu, X.-Z.S., Chien, F., Butler, A., Salkoff, L. and Montell, C. (2000) Neuron **26**, 647–657.
318. Raghu, P., Usher, K., Jonas, S., Chyb, S., Polyanovsky, A. and Hardie, R.C. (2000) Neuron **26**, 169–179.

CHAPTER 10

Modeling Primary Visual Processes in Insect Photoreceptors

D.G. STAVENGA, J. OBERWINKLER and M. POSTMA

Department of Neurobiophysics, University of Groningen

Handbook of Biological Physics
Volume 3, edited by D.G. Stavenga, W.J. DeGrip and E.N. Pugh Jr

Contents

1. Introduction

Many insects are highly visual animals that heavily rely upon their light sensitive cells to gather optical information from their surroundings. From the optical signals, the light sensitive cells produce an electrical signal. The molecular components involved in generating this light response are gradually unraveled, and it is becoming apparent which of the different molecules interact, and how they do this (reviews [1,2] and see the previous chapters in this handbook [3–11]).

However, the knowledge gathered in this research, by molecular biology, optical and electrophysiological measurements, is rarely integrated to form a quantitative theory. The motivation for this chapter is the view that quantitative descriptions are essential for gaining insight into the mechanisms of biological cells; i.e., the functional possibilities and the physical limitations of a cell, especially of a neuron, can only be clearly seen by constructing quantitative models.

This review covers two rather distinct areas of photoreceptor research that have been treated with quantitative models. The first area investigates the visual pigments, specifically the wavelength dependence of their absorption characteristics and the cycling of the visual pigment molecules between the different states. The second area tries to quantitatively understand how changes in membrane permeability translate into changes of membrane currents and potentials, and cause changes of ion concentrations. These two topics will be discussed in the two major sections of this chapter. Most of the reviewed work has been done on fly photoreceptors, specifically *Drosophila*.

The first step in vision is the absorption of a photon by a visual pigment molecule, which then triggers the phototransduction process. In the following the spectral characteristics of insect photoreceptors will be emphasized. As is the case in vertebrate photoreceptors, the optical waveguide properties of the insect rhabdom(ere)s has allowed the development of spectral filters, presumably to improve colour discrimination. A quantitative treatment requires knowledge of the absorbance spectra of the visual pigments, which are treated first. Subsequently, the consequences of packing visual pigments in the optical waveguides are formally described, and then the consequences for the resulting sensitivity spectra of the photoreceptors are discussed.

In Section 2 we will review attempts to quantitatively understand how changes in membrane permeability translate into changes of membrane currents and potentials. The strong and non-inactivating currents that pass the membranes of light-stimulated insect photoreceptor cells cause changes of ion concentrations in insect photoreceptor cells. We will review the work that has been done to understand and predict these changes. As it will turn out, even spatial inhomogeneities in the distribution of the ion concentrations need to be taken into account.

2. Insect visual pigments

2.1. Absorbance spectra consist of bands

The principal material of a photoreceptor cell is the visual pigment. Insect visual pigments form, together with the visual pigments of all animals, a sub-class of the G-protein coupling proteins, which are integral membrane proteins. The chromophores of insect visual pigments known so far are retinal(dehyde) and 3-hydroxy-retinal [8,9,12,13], derivates of vitamin A1 and vitamin A3, respectively. Although Vogt [12] coined the generic name xanthopsins for vitamin A3-based visual pigments, we will here conform to the general practice by calling any visual pigment, irrespective of its chromophore, a rhodopsin. (To add to possible confusion, the photoactive yellow proteins, found in eubacteria, that have coumarine as chromophore, are now called xanthopsins [14].)

The prime characteristic of any rhodopsin is its absorbance spectrum. Rhodopsin absorbance spectra display several bands, named α, β, γ, etc., going from long to short wavelengths (Fig. 1). The α- and β-band are determined by the chromophore and its interaction with the protein, and the other bands by the amino acids of the protein moiety. Photons absorbed by the chromophore trigger photoconversion of the rhodopsin molecules [15–17] and thus the phototransduction process. Absorption in wavelength regions of the α- and β-bands hence is of principal interest for vision. Spectrophotometrical as well as physiological studies have shown that the range of the absorbance spectrum of an insect rhodopsin can be restricted to the ultraviolet, but there are also rhodopsins with absorbance spectra extending into the red [9]. Especially in insects [18], and in other arthropods [19], visual sensitivity in the ultraviolet is well developed. Indeed the absorption by the

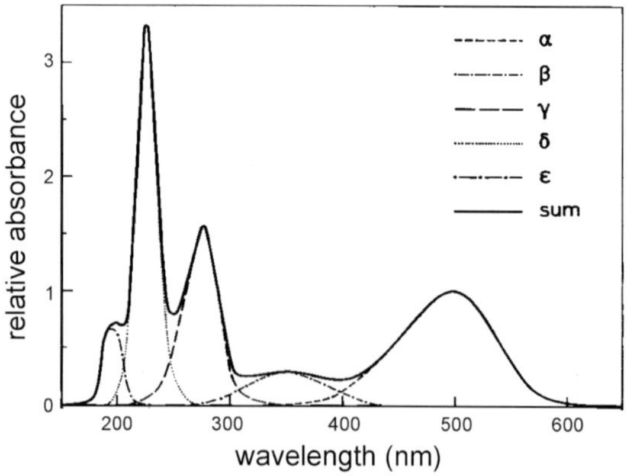

Fig. 1. Composite diagram of the absorbance spectrum of a visual pigment. The spectrum consists of several bands. The α- and β-band are determined by the chromophore and its interaction with the protein, and the other bands by the amino acids of the protein moiety (see [3,4,8,9]).

dioptric apparatus becomes substantial below 300 nm [17], contrary to the normal human eye, where absorption by the lens progressively increases below 400 nm [20].

2.2. *Templates*

For many studies into vision, notably those involving color, precise knowledge of the rhodopsin's absorbance spectrum is crucial. Although the molecular composition of many rhodopsins is now known into full detail (see Chapters 6 and 7 [8,9]), a theory predicting the spectral characteristics is still gravely lacking. Dartnall [21] was the first to tackle this problem with a phenomenological approach by proposing that the normalized α-band of any vitamin A1-based rhodopsin, when plotted on a frequency scale, has a standard shape, independent of the absorbance peak wavelength, λ_{max}. The accordingly devised Dartnall nomogram proved to be highly useful, but subsequent accurate measurements of rhodopsin absorbance spectra demonstrated clear inadequacies (e.g. [22]). Building on Dartnall's pioneering work, the evident research value of a reliable rhodopsin template has stimulated several investigators to extend and refine previous models [23,24].

For instance, Ebrey and Honig [25] constructed improved nomograms for the short-, middle-, and long-wavelength range, respectively, for both vitamin A1- and vitamin A2-based visual pigments. Subsequently, Metzler and Harris [26] provided an analytical expression derived from a lognormal function, and Dawis [27] approximated log absorbance curves with polynomial expressions for the three wavelength ranges. An invariant shape was regained by Barlow [28] by plotting the spectra as a function of $\lambda^{1/4} - \lambda_{max}^{1/4}$. Maximov [29] used these data to derive an analytical expression. Notably Mansfield [30] and MacNichol [31] showed that two unique invariant shapes, for vitamin A1- and A2-based visual pigments, respectively, emerge when plotting experimental spectra at a frequency scale relative to peak frequency, i.e., f/f_{max}. More recently, Partridge and DeGrip [32] very accurately measured the absorbance spectrum of purified bovine rhodopsin, and they derived a cubic polynomial fitting the long-wavelength limb as well as a Chebyshev polynomial fitting the complete measured spectrum. Stavenga et al. [33] extended this work by assuming that the absorbance spectrum of a visual pigment is an algebraic sum of absorbance bands and that each band conforms with a Gaussian-like, modified exponential function:

$$\varepsilon = A \exp[-a_0 x^2 (1 + a_1 x + a_2 x^2)] \tag{1}$$

where $x = -{}^{10}\log(f/f_{max}) = {}^{10}\log(\lambda/\lambda_{max})$, with λ_{max} the peak wavelength of the absorbance band. For instance, the α-band of a vitamin A1-based pigment is described by:

$$A = 1, \quad a_0 = 380, \quad a_1 = 6.09, \quad a_2 = 3a_1^2/8 = 13.9$$

and the β-band by:

$$A = 0.29, \quad a_0 = 247, \quad a_1 = 3.59, \quad a_2 = 3a_1^2/8 = 4.83$$

(for a complete set of parameter values, including the γ-band and the A2- and A4-visual pigments, see [33]). For most purposes the provided templates are quite adequate, but the exponential function appears to be inadequate at very long wavelengths. This problem was resolved by Lamb [34] who analyzed sensitivity spectra measured from human, monkey, bovine and squirrel and thus found that log normalized-spectral-sensitivity follows a unique curve in the long-wavelength range over more than 10 decades, plotted at a frequency scale relative to the peak frequency, $x = f/f_{max}$. The α-band thus is described by [34]:

$$S_\alpha = \{\exp[A(a-x)] + \exp[B(b-x)] + \exp[C(c-x)] + D\}^{-1} \tag{2}$$

where

$$A = 70, \quad a = 0.880, \quad B = 28.5, \quad b = 0.924,$$
$$C = -14.1, \quad c = 1.104, \quad D = 0.65$$

Most studies of the spectral absorbance by visual pigments and of the spectral sensitivity of photoreceptor cells have concentrated on the so-called visible wavelengths; not only due to the characteristics of the human set of rhodopsins, but also because the majority of visual pigments have α-bands peaking well above 400 nm. In the last decades extensive data for visual pigments absorbing mainly in the ultraviolet have accumulated. Partridge and DeGrip [32] and Hárosi [35] noticed that experimental spectra of ultraviolet visual pigments are no longer conforming to the spectral shapes of the longer-wavelength visual pigments. Consequently, the templates given by Eqs. (1) and (2) become aberrant when λ_{max} lies in the UV. This point was investigated in detail by Govardovskii et al. [36], by exploiting a wealth of new microspectrophotometrical measurements on A1- and A2-visual pigments of amphibians, reptiles and fishes. These authors capitalized on Lamb's formula for the α-band [34] and refined it by taking into account that the long-wavelength slope increases with decreasing λ_{max}, the peak wavelength of the α-band, when $\lambda_{max} < 500$ nm. They found that the optimal set of parameters describing the α-band of all A1-pigments, including those absorbing purely in the UV, is (with λ_{max} in nm) [36]:

$$A = 69.7, \quad a = 0.8795 + 0.0459 \exp[-(\lambda_{max} - 300)^2/11940],$$
$$B = 28, \quad b = 0.922, \quad C = -14.9, \quad c = 1.104, \quad D = 0.674$$

The α- and β-bands are only distinct when the α-peak wavelength is well above 400 nm. The β-band, located in the ultraviolet wavelength range, has remained ill-studied until recently, mainly due to technical reasons causing decreased spectrophotometrical sensitivity in the ultraviolet and increased errors by scattering. Stavenga et al. [33] assumed as a first approximation, due to paucity in reliable data, that the β-peak wavelength was identical for all visual pigments. However, Palacios et al. [37] demonstrated that the peak wavelengths of α- and β-band are more or less linearly related.

Govardovskii et al. [36] have analyzed the full spectral shape of their data set of visual pigments by subtracting the best fit for the α-band (Eq. (2)) from the complete spectrum. They thus obtained that the residue, representing the β-band, is Gaussian-shaped:

$$S_\beta = A_\beta \exp\left\{-\left[(\lambda - \lambda_{m\beta})/d\right]^2\right\} \tag{3}$$

where A_β is the amplitude of the β-band relative to that of the α-band, $\lambda_{m\beta}$ the position of the β-peak, and d a bandwidth parameter. The experimental spectra yielded as optimal parameter values (again with λ_{max} in nm):

$$A_\beta = 0.26, \quad \lambda_{m\beta} = 189 + 0.315\,\lambda_{max}, \quad d = -40.5 + 0.195\,\lambda_{max}$$

Govardovskii et al. [36] have also analyzed A2-based visual pigments and thus derived a complete set of parameters for the A2 absorbance spectra. Although up till now no insect A2 visual pigments have been identified, we add here the concluded set of parameters for the α-band of A2-pigments:

$$A = 62.7 + 1.834\ \exp[(\lambda_{max} - 625)/54.2],$$
$$a = 0.875 + 0.0268\ \exp[(\lambda_{max}665)/40.7],$$
$$B = 20.85, \quad b = 0.9101, \quad C = -10.37, \quad c = 1.1123, \quad D = 0.5343$$

and for the β-band:

$$A_\beta = 0.37, \quad \lambda_{m\beta} = 216.7 + 0.287\,\lambda_{max},$$
$$d = 317 - 1.149\,\lambda_{max} + 0.00124\,\lambda_{max}^2$$

The spectra of the A3-based visual pigments are presumably well-described by the A1-formulae and parameters, but a thorough analysis is lacking here.

3. Photochemistry of insect rhodopsins

3.1. Photochromism

When an invertebrate rhodopsin (R) molecule absorbs a photon, it converts to a thermostable metarhodopsin (M) via a few thermolabile intermediate states. Insect visual pigments hence are photochromic substances. The photochemical formalism is that of a two state molecular process:

$$\mathrm{R} \underset{k_M}{\overset{k_R}{\rightleftharpoons}} \mathrm{M}$$

The reaction equation is:

$$\frac{dC_R}{dt} = -k_R C_R + k_M C_M = -\frac{dC_M}{dt} \tag{4}$$

with C_R and C_M the concentrations of rhodopsin and metarhodopsin, respectively. It is often more convenient to consider the fractions, $f_R = C_R/C_0$ and

$f_M = C_M/C_0$, with $C_0 = C_R + C_M$ the total visual pigment concentration, i.e., $f_R + f_M = 1$:

$$\frac{df_R}{dt} = -k_R f_R + (1 - f_R)k_M = -(k_R + k_M)f_R + k_M \tag{5}$$

The rate constants represent the transition probabilities between the two states and are given by:

$$k_R = \int \beta_R(\lambda)I(\lambda)\,d\lambda \tag{6a}$$

and

$$k_M = \int \beta_M(\lambda)I(\lambda)\,d\lambda \tag{6b}$$

Here $\beta_R(\lambda)$ and $\beta_M(\lambda)$ are the photosensitivities of the two visual pigment states, and $I(\lambda)$ is the light intensity (or the photon flux per unit area and time). The photosensitivity is firstly determined by the molecular absorbance coefficient, α, i.e., the molecular cross-section for photon absorption, and secondly by the quantum efficiency, γ, i.e., the chance that an absorbed photon indeed successfully results in conversion. Or,

$$\beta_R(\lambda) = \alpha_R(\lambda)\gamma_R(\lambda) \tag{7a}$$

and

$$\beta_M(\lambda) = \alpha_M(\lambda)\gamma_M(\lambda) \tag{7b}$$

With a constant light intensity, i.e., k_R and k_M constant, the process of photoconversion is described by:

$$f_R(t) = (f_{R,0} - f_{R,\infty})e^{-(k_R+k_M)t} + f_{R,\infty} \tag{8}$$

where $f_{R,0}$ is the rhodopsin fraction at $t = 0$. $f_{R,\infty}$, the rhodopsin fraction in the steady state ($t = \infty$), is:

$$f_{R,\infty} = \frac{k_M}{k_R + k_M} \tag{9}$$

The conversion follows an exponential time course with a time constant $\tau = (k_R + k_M)^{-1}$.

The ratio of the quantum efficiencies γ_M and γ_R, $\varphi(\lambda) = \gamma_M(\lambda)/\gamma_R(\lambda)$, is believed to be independent of wavelength [18], or $\varphi(\lambda) = \varphi$. Irradiation with monochromatic light with wavelength λ yields a steady state where the rhodopsin fraction is:

$$f_{R,\infty}(\lambda) = \frac{\beta_M(\lambda)}{\beta_R(\lambda) + \beta_M(\lambda)} = \left[1 + \frac{\alpha_R(\lambda)}{\alpha_M(\lambda)}\varphi^{-1}\right]^{-1} \tag{10a}$$

and, the metarhodopsin fraction in the steady state then is:

$$f_{M,\infty}(\lambda) = \left[1 + \frac{\alpha_M(\lambda)}{\alpha_R(\lambda)}\varphi\right]^{-1} \tag{10b}$$

The isosbestic wavelength, λ_{iso}, is defined as the wavelength were $\alpha_R(\lambda) = \alpha_M(\lambda)$. Hence

$$f_{R,\infty}(\lambda_{iso}) = \varphi/(\varphi + 1) \tag{11}$$

or, if $\varphi = 1$, prolonged illumination with isosbestic-wavelength light results in 50% rhodopsin and 50% metarhodopsin.

When the illumination is monochromatic, the time constant becomes:

$$\tau(\lambda) = [k_R(\lambda) + k_M(\lambda)]^{-1} = [\{\beta_R(\lambda) + \beta_M(\lambda)\}I(\lambda)]^{-1} \tag{12}$$

The sum of the photosensitivities, called the relaxation spectrum [38,39], hence is:

$$\beta_{rel}(\lambda) = \beta_R(\lambda) + \beta_M(\lambda) = [\tau(\lambda)I(\lambda)]^{-1} \tag{13}$$

The relaxation spectrum can be determined by varying the wavelength of the monochromatic illumination and then measuring its intensity as well as the time constant of the induced photoconversion process.

As an example, the case of the *Drosophila* rhodopsin DRh1 is shown in Fig. 2. The peak wavelength of DRh1 is 486 nm and its metarhodopsin peaks at 566 nm (see [40] and also Fig. 5). Fig. 2 presents the normalized photosensitivity of the DRh1 rhodopsin, β_R, and that of the metarhodopsin, β_M, relative to the photosensitivity of rhodopsin (assuming a relative quantum efficiency $\varphi = 1$). The relaxation spectrum β_{rel} then follows from Eq. (13), and Eq. (9) yields the rhodopsin fraction resulting in the steady state at the given stimulus wavelength (Fig. 2). Fig. 2

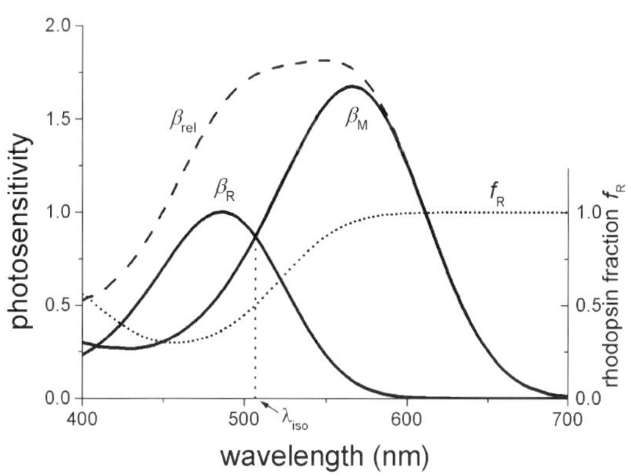

Fig. 2. Photosensitivity spectra of *Drosophila* DRh1 rhodopsin and its metarhodopsin, β_R and β_M, together with their sum, the relaxation spectrum, β_{rel}, and the rhodopsin fraction in the photosteady state, $f_R = f_{R,\infty}$. The photosensitivity spectra are normalized to the rhodopsin peak. It is assumed that the relative quantum efficiency $\varphi = 1$. The normalized photosensitivities then are identical to the normalized molecular absorbance coefficients.

shows that prolonged red light results in virtually 100% rhodopsin, whereas blue light reduces rhodopsin to ca. 30%.

The substantial spectral separation of the two absorbance bands hence allows extreme population of one of both thermostable states by appropriate monochromatic illumination. It is frequently thought that irradiation at the peak wavelengths is most effective to reach that goal (e.g. [40]). But in fact this needs careful consideration, as is exemplified in Fig. 3. Figs. 3(a) and (b) present the time course of the rhodopsin fraction during monochromatic illumination, calculated from Eq. (8), assuming the same light intensity at all stimulus wavelengths. Two cases are considered, where initially the rhodopsin fraction is maximal, $f_{R,0} = 1.0$, and minimal, $f_{R,0} = 0.3$, respectively. In the first case, illumination with 500 nm gives a more rapid decrease of rhodopsin than with 450 nm, but in the latter case the steady-state rhodopsin level is smaller. In the second case, illumination with 550 or 600 nm light causes a rapid increase of rhodopsin, but the final rhodopsin level is higher with 650 nm light. For the actual conversion to be achieved, one thus has to consider three factors: the relaxation spectrum, the steady state rhodopsin fraction, and the total number of the photons delivered by the light source, i.e., the product of stimulus intensity and duration.

The preceding treatment holds only for cases where the light intensity is spatially uniform. Usually visual pigments are densely packed within a light-guiding structure of the visual photoreceptor cell, and then light absorption causes gradients in the light intensity. This situation is discussed below.

3.2. Light absorption by a visual pigment in an optical waveguide

The principal eyes of insects are of the so-called compound eye type. A typical compound eye consists of a large number of (in principle) identical building blocks, the ommatidia. Each ommatidium consists of a small set of visual sense cells, the

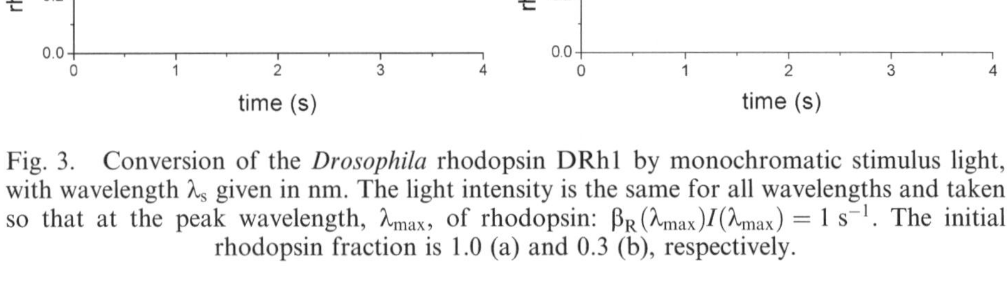

Fig. 3. Conversion of the *Drosophila* rhodopsin DRh1 by monochromatic stimulus light, with wavelength λ_s given in nm. The light intensity is the same for all wavelengths and taken so that at the peak wavelength, λ_{max}, of rhodopsin: $\beta_R(\lambda_{max})I(\lambda_{max}) = 1\ s^{-1}$. The initial rhodopsin fraction is 1.0 (a) and 0.3 (b), respectively.

photoreceptors, together with supporting cells, capped by a dioptrical apparatus, of which the facet lens is prominently recognizable in the intact eye. Each photoreceptor has a specialized organelle, the rhabdomere, constructed by strongly invaginating the membrane. There the visual pigment molecules are embedded in the rhabdomeric membrane, in microvilli, together with the molecules of the phototransduction machinery. The rhabdomere of a fly photoreceptor is a long cylindrical structure that acts as an independent, separate optical waveguide. In combination with facet lens, crystalline cone and surrounding pigment cells, it produces the good spatial vision that insects are endowed with. The precise shape of the angular sensitivity function of a photoreceptor is determined by the optics of the integral system, where the waveguide characteristics play a crucial role (reviews [41–44]).

Light absorption by the visual pigment molecules of a photoreceptor cell can be treated with Lambert–Beer's law. We first consider a fly rhabdomere of length L containing visual pigment molecules. When $J(z, \lambda)$ is the light flux at depth z, and $\kappa(z, \lambda)$ is the absorption coefficient of the rhabdomeric tissue (i.e., per unit length), then

$$dJ(z, \lambda) = -\kappa(z, \lambda)J(z, \lambda)\, dz \tag{14}$$

Light is propagated in the rhabdomere in waveguide modes, that extend beyond the boundary of the rhabdomere. The effective absorption coefficient of the rhabdomere then is determined by the local fractions, $f_R(z)$ and $f_M(z)$, of the two visual pigment states (assuming that the total visual pigment concentration, C_0, is constant throughout the rhabdomere) and the fraction of light propagated within the rhabdomere boundary, $\eta(z, \lambda)$:

$$\kappa(z, \lambda) = \eta(z, \lambda)[\alpha_R(\lambda)f_R(z) + \alpha_M(\lambda)f_M(z)]C_0 \tag{15}$$

When a total of $J_0(\lambda)$ photons enter the rhabdomere per second at wavelength λ, the number of photons transmitted per second, i.e., the transmission, then is

$$J_{tr}(\lambda) = J_0(\lambda) \exp\left\{-\int_0^L \kappa(z, \lambda)\, dz\right\} \tag{16}$$

It is often useful to consider the transmittance, the fraction of the incident light that is transmitted

$$T(\lambda) = \exp\left\{-\int_0^L \kappa(z, \lambda)\, dz\right\} \tag{17}$$

and the absorptance, the fraction of the incident light that is absorbed

$$A(\lambda) = 1 - T(\lambda) \tag{18}$$

The mean absorption coefficient of the rhabdomere is

$$\kappa_{av}(\lambda) = \frac{1}{L} \exp\left\{-\int_0^L \kappa(z, \lambda)\, dz\right\} \tag{19}$$

The absorbance, minus the decadic logarithm of the transmittance, hence is directly proportional to the mean absorption coefficient

$$E(\lambda) = -\log_{10}(T(\lambda)) = \log_{10} e\ \kappa_{av}(\lambda)L \tag{20}$$

Monochromatic stimulus light with wavelength λ_s causes the same visual pigment fractions in the steady state, $f_{R,\infty}(\lambda_s)$ and $f_{M,\infty}(\lambda_s)$, throughout the rhabdomere. The absorbance resulting at test wavelength λ_t then is

$$E(\lambda_t, \lambda_s) = \log_{10} e\ \eta(\lambda_t) C_0 L \left[\alpha_R(\lambda_t) f_{R,\infty}(\lambda_s) + \alpha_M(\lambda_t) f_{M,\infty}(\lambda_s)\right] \tag{21}$$

Calculations of the absorbance from transmission measurements requires knowledge of the incident light flux. When that cannot be reliably determined, valuable information often is obtained through the difference spectrum. The difference in absorbance with respect to the specific case that $f_{R,\infty}(\lambda_s^*) = 1$, or, $f_{M,\infty}(\lambda_s^*) = 0$, is

$$\begin{aligned} E(\lambda_t, \lambda_s) - E(\lambda_t, \lambda_s^*) &= \Delta E(\lambda_t, \lambda_s) \\ &= \log_{10} e\ \eta(\lambda_t) C_0 L [\alpha_M(\lambda_t) - \alpha_R(\lambda_t)] f_{M,\infty}(\lambda_s) \end{aligned} \tag{22}$$

The difference spectrum hence is proportional to the spectral difference in the molecular absorbance coefficients, $\Delta\alpha(\lambda) = \alpha_M(\lambda) - \alpha_R(\lambda)$, and to the fraction of metarhodopsin in the steady state, $f_{M,\infty}(\lambda)$, induced at stimulus wavelength λ_s. The difference spectrum for all stimulus wavelengths, λ_s, is zero at the isosbestic (test) wavelength: $\Delta E(\lambda_{iso}, \lambda_s) = 0$.

This is demonstrated in Fig. 4 for the case of the blowfly photoreceptor cell type R3, which contains a rhodopsin, rather similar to DRh1 of *Drosophila*, that absorbs maximally at 490 nm; its metarhodopsin absorbs maximally at 575 nm. Fig. 4(a) gives examples of absorbance difference spectra between steady states created by two stimulus wavelengths; the abscissa gives the value of the test wavelength, λ_t. The difference spectra between the photosteady states created by irradiation at two different stimulus wavelengths cross the zero-line at $\lambda_{iso} = 510$ nm. Fig. 4(b) gives an example of the absorbance difference between steady states created by a variety of stimulus wavelengths with respect to that where $f_M = 0$; the abscissa gives here the value of the stimulus wavelength, λ_s. The latter absorbance difference spectrum can be usefully simplified by normalization at $\lambda_s = \lambda_{iso}$. Equation (22) then yields:

$$\frac{\Delta E(\lambda_t, \lambda_s)}{\Delta E(\lambda_t, \lambda_{iso})} = \frac{f_M(\lambda_s)}{f_M(\lambda_{iso})} = \frac{1+\varphi}{1+\varphi(\alpha_M(\lambda_s)/\alpha_R(\lambda_s))} \tag{23}$$

When the relative quantum efficiency, φ, is approximately 1 (see [18]), the normalized spectrum of Eq. (23) is only a function of the ratio of the molecular absorbance coefficients. It immediately follows that long-wavelength irradiation of a rhodopsin with a strongly bathochromic shifted metarhodopsin, as DRh1 of *Drosophila*, yields a high-rhodopsin/low-metarhodopsin ratio, $f_R = 1$, in the steady-state; irradiation of DRh1 with blue light (at 450 nm) yields a low rhodopsin/high metarhodopsin photosteady state, $f_R = 0.3$.

Broad-band illumination causes intermediate values. For instance, intense white (0.3 s, Fig. 4(b)) light yields $f_R \approx 0.7$. The rhodopsin fraction is still higher under

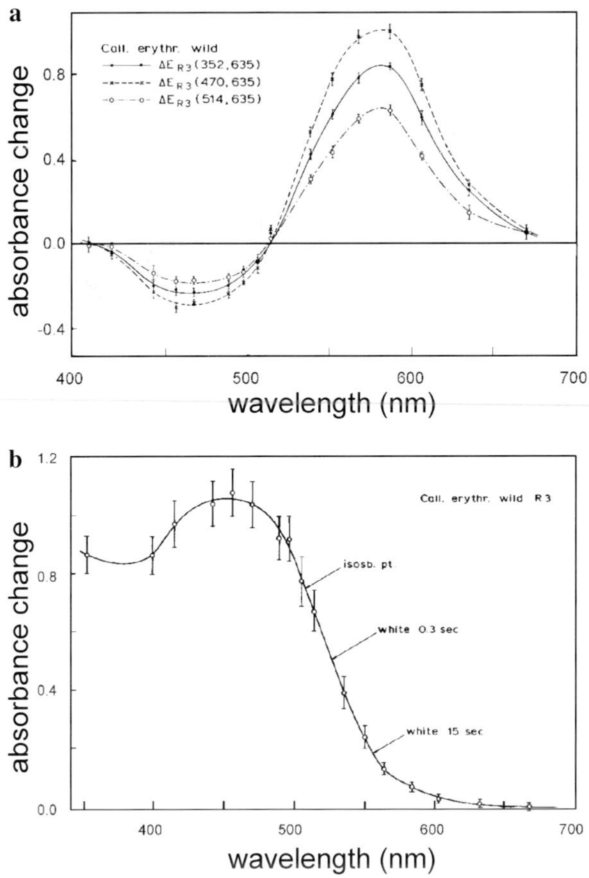

Fig. 4. Absorbance difference measurements calculated from *in vivo* transmission measurements on photoreceptor cell type R3 of the blowfly *Calliphora*. (a) Photosteady states were created by irradiation at 352, 470, 514 and 635 nm, and the absorbance difference was calculated from the transmission at each test wavelength with respect to that at 635 nm. (b) Photosteady states were created by irradiation at a variety of stimulus wavelengths, and the resulting transmission at 583 nm was subsequently measured. The absorbance difference then was calculated from the transmission values with respect to that after radiation with 665 nm light (after [45]).

normal light conditions, because then a blue-light absorbing, yellow-transmitting pupil mechanism is activated, resulting in $f_R \approx 0.9$ (see Section 6.5).

4. Spectral characteristics of insect visual pigments

4.1. Drosophila visual pigments

Compared to the extensive analyses of the shapes of the rhodopsin absorbance spectra, little is known of the spectral shapes of the metarhodopsin spectra. In their investigation of the complete set of visual pigments of the fruitfly *Drosophila*, Salcedo et al. [40] measured difference spectra from retinal extracts and analyzed them using the A1-formulae of Stavenga et al. [33]. Their results suggest that the A1-template, at least in first approximation, can be used for the metarhodopsins as well (see Fig. 5). The spectra of Fig. 5 show that the amplitude of the absorbance coefficient of metarhodopsin in the peak relative to that of rhodopsin varies between 1.5 and 1.7 (see also [18]). In addition to a change in absorbance often the peak wavelength shifts considerably. The peak wavelengths of the six *Drosophila* visual pigments are indicated in Fig. 5; e.g. the rhodopsin of Rh1, R486, absorbs maximally at 486 nm and its metarhodopsin, M566, at 566 nm.

The insect visual pigments investigated so far show that the spectral characteristics of both states are correlated. It appears that often the peak wavelengths are separated by a wide wavelength gap (see numbers in Fig. 5). Especially the UV-rhodopsins have metarhodopsins peaking near 480 nm, but there is also a group of rhodopsins peaking between 450 and 500 nm that have strongly bathochromic shifted metarhodopsins. It is a long-standing observation that rhodopsins peaking above 500 nm invariably have a hypsochromic-shifted metarhodopsin [46,47]. No interpretation in molecular terms has yet been forwarded, but this intriguing spectral characteristic has important consequences for the spectral properties of other, supporting pigments in the insect eye (see Section 6.4).

Although the rhodopsin and metarhodopsin spectra follow certain rules, there appears to be substantial play. Modification of parts of the molecule can have no effect on the spectrum of one state and distinctly change the spectrum of the other

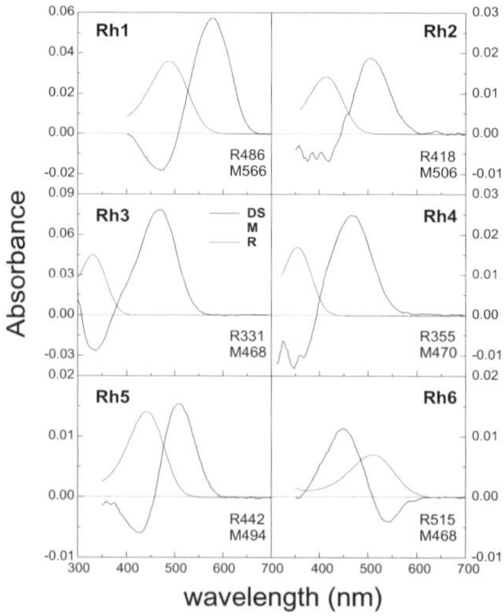

Fig. 5. Spectral characteristics of the visual pigments Rh1–Rh6 of the fruitfly *Drosophila*. Difference spectra (DS) where measured from eye extracts and fitted with absorbance spectra calculated from Eq. (1). The peak wavelengths (in nm) of the concluded rhodopsin (R) and metarhodopsin (M) spectra are indicated by the numbers (modified from [40]).

[48]. Assuming that the absorbance spectrum of an insect visual pigment is determined by the same rules as that of vertebrate visual pigments we can construct the absorbance spectra of the set of fruitfly visual pigments by using the formulae of Section 2.2. Together with the λ_{max} values ([40] and Fig. 5) Eqs. (2) and (3) then yields Fig. 6.

4.2. *The ultraviolet pigment of the owlfly* Ascalaphus

It is now well established that the photochemistry of insect visual pigments largely corresponds to that of vertebrate visual pigments, but there are slight, but important deviations. As in vertebrates, absorption of a photon by the rhodopsin molecule isomerizes the chromophore from its native 11-*cis* state to the all-*trans* configuration, causing the transformation of the whole pigment molecule, via a few thermally labile intermediate states, to the thermally stable metarhodopsin. The intermediates are thermostable below a certain critical temperature, and thus they can be studied sequentially by photoconversion of rhodopsin at low temperatures and then stepwise increasing the temperature. This is preferentially done on visual pigment extracts [49].

As an example, a few photochemical steps are shown for the UV-rhodopsin of the owlfly *Ascalaphus macaronius* (Neuroptera) [18,50], the first clear example of a purely UV-absorbing visual pigment (Fig. 7). Upon illumination the pigment can attain various states, as witnessed by their different absorbance spectra. The owlfly rhodopsin absorbs maximally in the UV at 345 nm. Absorption of a photon by rhodopsin (R) at −50°C yields lumirhodopsin (L). In this state the visual pigment absorbs much stronger than the native rhodopsin, indicative for the 11-*cis* to all-*trans* isomerization of the chromophore; the peak wavelength is slightly shifted to about 375 nm. The lumirhodopsin can be photoconverted back to rhodopsin at −50°C, but it is unstable at −15°C. Then the end photoproduct, metarhodopsin (M), arises (Fig. 7(a)). The strong bathochromic shift of the peak wavelength

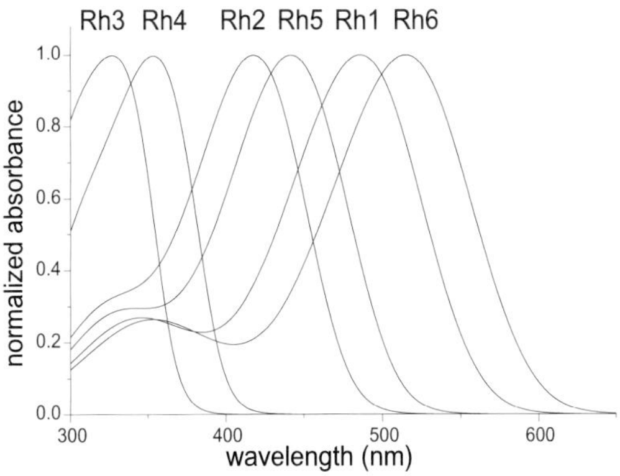

Fig. 6. Absorbance spectra of the complete set of *Drosophila* rhodopsins, calculated with Eqs. (2) and (3).

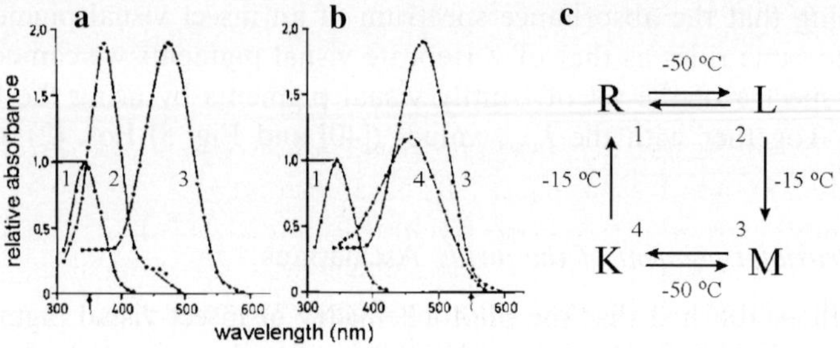

Fig. 7. Spectral characteristics of the UV-absorbing rhodopsin, and of its photoproducts, of the owlfly *Ascalaphus macaronius*. At low temperatures (−50°C) photon absorption results in conversion of rhodopsin, R (1 in a and b), to lumirhodopsin, L (2 in a), which can be photoconverted back into rhodopsin. Upon warming (−15°C) lumirhodopsin transforms to metarhodopsin, M (3 in a and b). At low temperatures this is photointerconvertible with an intermediate, K (4 in b), which above −15°C decays to rhodopsin. The chromophore in R and K takes the 11-*cis* configuration, and the chromophore in L and M is all-*trans*. The photochemical cycle is summarized in c (after [50]).

indicates distinct intramolecular conformation and/or charge changes. As in lumirhodopsin, the peak absorbance coefficient of metarhodopsin is about 1.8 times that of rhodopsin. Irradiation of metarhodopsin at −50°C yields an intermediate (K in Fig. 7(b)), that is photointerconvertible with metarhodopsin. The strongly reduced absorbance coefficient signifies the all-*trans* to 11-*cis* isomerization, but the peak wavelength shift to 460 nm suggests relatively minor changes in the conformation of the whole protein. At −15°C intermediate K decays thermally to rhodopsin.

Low temperature spectroscopy of visual pigments and their intermediates has been performed on several vertebrates as well as a few invertebrates (e.g. [49]). The decay scheme appears to follow a rather uniform temporal pattern, at least in the pathway of vertebrates, where rhodopsin transforms via bathorhodopsin, lumirhodopsin, and metarhodopsin to retinal and opsin (see e.g. Chapters 1 and 3 [3,5]). The low temperature studies demonstrate that intermediates themselves can also be photoconverted back to rhodopsin. This can be also achieved at physiological temperatures [51], but due to the brief lifetimes of the intermediates extreme irradiation intensities are required to cause noticeable amounts of photoconversions. With the light fluxes existing under normal, physiological conditions, the chances of photoconverting intermediates appear to be negligible. In practice, therefore, photoreconversion only occurs from the metarhodopsin state of invertebrate visual pigments. In other words, for all general spectral considerations it is virtually always fully adequate to consider insect visual pigments to exist either in the rhodopsin or the metarhodopsin configuration.

After having been created from rhodopsin, the thermostable metarhodopsins just sit and await photoconversion like the rhodopsin. Under normal daylight conditions, every visual pigment molecule of an insect eye therefore constantly shuttles back and forth from the rhodopsin to the metarhodopsin state; very roughly once in 1 min.

5. Dynamics of photoconversions

5.1. Laser spectroscopy

The time course of the photochemical processes following photon absorption can be monitored by laser spectroscopy [52–54]. As an example, Fig. 8 shows experiments performed on the virtually intact eye of a blowfly. In Fig. 8(a), the transmission was measured at 586 nm. Initially the visual pigment was in a photosteady state established by red (606 nm) light, causing a rhodopsin fraction, $f_{R,\infty}$, of ca. 100%. At $t = 0$ an intense wide-band, blue-green flash, lasting 10 ns, was delivered. This caused a sudden drop in transmission, followed by an initial increase and subsequent gradual decrease in transmission (Fig. 8(a)). This process can be described by a series of photochemical decay steps, where a rhodopsin (R) molecule photoconverts to a bathorhodopsin state (B), that subsequently further decays via lumirhodopsin (L) to the thermostable metarhodopsin (M):

$$R \rightarrow B \rightarrow L \rightarrow M$$

When initially $f_R = 1.0$ and the laser flash converts a fraction p of the R-molecules into the B-state, then the population of visual pigment molecules changes according to [54]:

$$f_R(t) = 1 - p \tag{24a}$$

$$f_B(t) = p \exp\left(-\frac{t}{\tau_B}\right) \tag{24b}$$

$$f_L(t) = p\frac{\tau_L}{\tau_L - \tau_B}\left[\exp\left(-\frac{t}{\tau_L}\right) - \exp\left(-\frac{t}{\tau_B}\right)\right] \tag{24c}$$

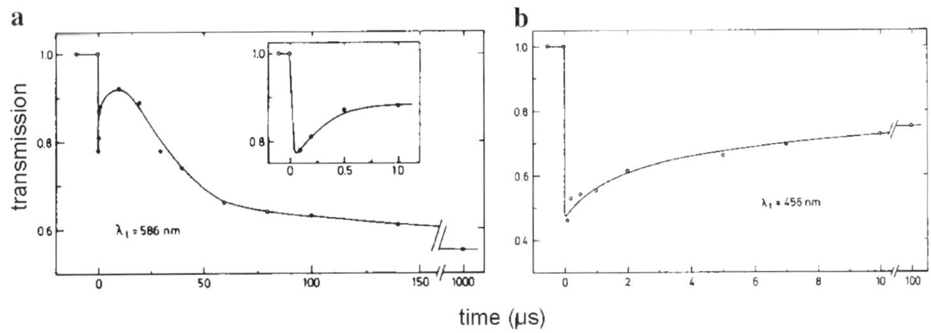

Fig. 8. Laser spectroscopical measurements of the (normalized) transmission of the eye of a blowfly at 586 (a) and 456 nm (b); the initial rhodopsin fractions were $f_R = 1.0$ (a) and $f_R = 0.3$ (b), respectively. At $t = 0$ an intense laser flash causes conversion of the visual pigment, showing intermediate states with μs lifetimes (from [54]).

and

$$f_{\mathrm{M}}(t) = p\frac{1}{\tau_{\mathrm{L}} - \tau_{\mathrm{B}}}\left[\tau_{\mathrm{L}}\left\{1 - \exp\left(-\frac{t}{\tau_{\mathrm{L}}}\right)\right\} - \tau_{\mathrm{B}}\left\{1 - \exp\left(-\frac{t}{\tau_{\mathrm{B}}}\right)\right\}\right] \tag{24d}$$

where τ_{B} and τ_{L} are the lifetimes of B and L, respectively. When the visual pigment population throughout the photoreceptor is identical, the absorption coefficient (Eq. (15)) is:

$$\kappa(\lambda, t) = \eta(\lambda)\left[\sum_i \alpha_i(\lambda) f_i(t)\right] C_0 \tag{25}$$

with i = R, B, L and M.

Normalization of the transmittance to the initial value (Fig. 8(a)) yields, with Eq. (17):

$$\frac{T(\lambda, t)}{T(\lambda, 0)} = \exp\left\{-\eta(\lambda) L C_0 \left[-\alpha_{\mathrm{R}}(\lambda) + \sum_i \alpha_i(\lambda) f_i(t)\right]\right\} \tag{26}$$

After correction for background due to straylight (see [54]), the data could be well fitted by this formula with a proper choice of the parameters, yielding the lifetimes of the B- and L-intermediate (Fig. 9(a)).

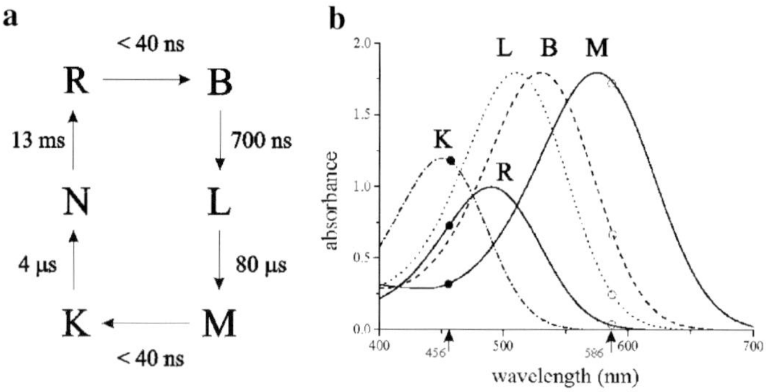

Fig. 9. (a) Photochemical cycle of blowfly rhodopsin as deduced from transmission measurements on blowfly eyes and from early receptor potential measurements (Section 5.2). The horizontal arrows represent the photoconversion plus thermal transitions to the earliest detected molecular states. The vertical arrows represent thermal decays, the time constants, or lifetimes, of which are given. (b) Absorbance spectra of thermostable states and intermediates of blowfly visual pigment. The absorbance spectra of rhodopsin (R) and metarhodopsin (M) were deduced from difference spectra, assuming applicability of the template formula Eq. (1), together with hypothetical spectra for the intermediate states, in order to interpret the data of Fig. 8. The open circles indicate the absorbance at 586 nm of the participants in the forward photoconversion pathway, from rhodopsin to metarhodopsin (Fig. 8(a)). The closed circles indicate the absorbances at 456 nm in the reverse pathway (Fig. 8(b)). As in the case of the UV rhodopsin of the owlfly (Fig. 7), it is assumed that bathorhodopsin (B) and lumirhodopsin (L) have a peak absorbance equivalent to that of metarhodopsin; intermediate K is assumed to absorb somewhat stronger than rhodopsin.

Figure 8(b) presents measurements on the reverse process, i.e., the reconversion of metarhodopsin to rhodopsin. Pre-illumination occurred with 467 nm, causing 70% metarhodopsin, or $f_R = 0.3$. The metarhodopsin was photoconverted by a broad-band orange laser flash, and the transmission was measured at test wavelength 456 nm. Analysis of the transmission time course with the same rationale as outlined above [54] demonstrated a fast decaying intermediate, K (Fig. 9(a)).

The spectra of B, L and K given in Fig. 9(b) are heuristic, and are added here to facilitate understanding of the results of Fig. 8. The spectra are based on the single-wavelength measurements of Fig. 8 and on the assumption, drawn from the case of the UV-rhodopsin of *Ascalaphus*, that intermediates with an all-*trans* chromophore have a similar peak absorbance as their metarhodopsin and that those with an 11-*cis* chromophore have a slightly higher peak absorbance than the rhodopsin.

The diagram of the photochemical cycle of blowfly rhodopsin (Fig. 9(a)) features an intermediate N, with lifetime 13 ms, in the reverse pathway, from M to R. This slowly decaying intermediate state was revealed by microspectrophotometrical measurements [55], stimulated by ERP experiments (see Section 5.2 [56]). It did not turn up in the laser experiments of Fig. 8(b), presumably because its absorbance at 456 nm is similar to that of rhodopsin.

The absorbance spectra of the intermediates can be obtained by performing the experiments of Figs. 8(a) and (b) at a variety of test wavelengths. This has not yet been attempted. So far no studies on the photochemistry of other insect visual pigments like those of Fig. 8 exist, but extensive work has been done on cephalopod visual pigments, extracted from retinal tissue, revealing several more intermediate states during photoconversion and reconversion of the rhodopsins [52,57]. The present insect data fit quite well to that work, indicating that the visual pigment photochemistry of invertebrates occurs along very similar lines.

5.2. *Early receptor potentials*

The light-induced transformations in a visual pigment molecule are accompanied by charge displacements, i.e., by currents. Since visual pigment molecules are integral membrane proteins, this current is measurable when the net direction is perpendicular to the membrane; for, the current then changes the charge of the membrane capacitor. This creates a change in membrane potential, called the early receptor potential (ERP) [20,58,59].

Pak and Lidington [60] discovered an ERP by electroretinographic recordings of *Drosophila* mutants. Delivering a bright orange flash to the eye of a visually defective mutant *norpA*, after blue preadaptation, yielded a positive waveform in the electroretinogram. This potential was coined the M-potential, because it was evidently coupled to conversion of metarhodopsin. Subsequent investigations [61,62] have shown that the M-potential is a complex summation of retina and lamina components, related to the ERP of the photoreceptors.

Figure 10 presents an intracellularly recorded ERP from a blowfly *Calliphora* photoreceptor (from [56]). The cell was preadapted to blue light causing 70% metarhodopsin. A subsequent bright, red flash reconverts this metarhodopsin to

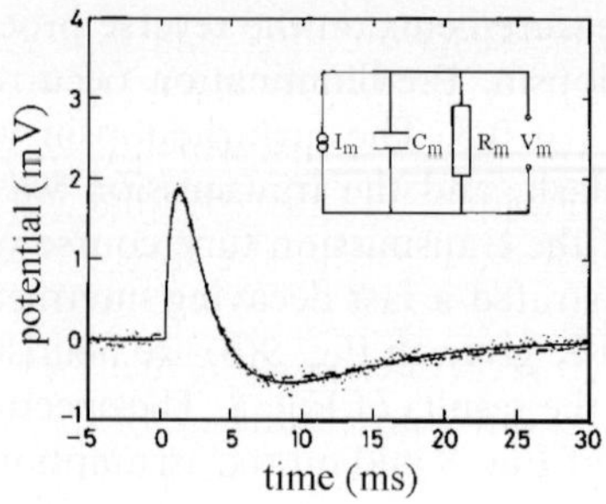

Fig. 10. Early receptor potential (ERP) measured by intracellular recording from a photoreceptor cell of a blowfly. The fit (continuous curve) to the data (dots) was obtained with the model of Eq. (27), where the current I_m is generated by converting metarhodopsin molecules to the rhodopsin state. The charge movement across the membrane's resistance, R_m, and capacitance, C_m, results in a change of the membrane potential, V_m (from [56]).

rhodopsin, which is accompanied by a clear biphasic potential change, the ERP. The inset of Fig. 10 shows the equivalent electrical circuit of the photoreceptor cell membrane. The time course of the membrane potential, V_m, occurring due to a membrane current, I_m, depends on the membrane resistance, R_m, and capacitance, C_m. Because an average blowfly photoreceptor has a membrane resistance in the dark adapted state of 25–35 MΩ [63] and its membrane surface is $1.7 \times 10^4\ \mu m^2$ [64], yielding a membrane capacity of 170 pF when taking the usual $1\ \mu F\,cm^{-2}$, its membrane time constant is about 5 ms.

The time course of the membrane potential can be modeled with the temporal characteristics of three components, of the flash, the membrane, and the thermal decay of intermediate N [56,59]:

$$V(t) = \frac{qN_M}{C_m} a_1 \left[\frac{a_3(1-B) - a_1}{(a_2 - a_1)(a_3 - a_1)} e^{-a_1 t} + \frac{a_3(1-B) - a_2}{(a_1 - a_2)(a_3 - a_2)} e^{-a_2 t} + \frac{a_3(1-B) - a_3}{(a_1 - a_3)(a_2 - a_3)} e^{-a_3 t} \right] \tag{27}$$

where q is the effective charge movement across the membrane due to a single isomerization, N_M the number of M-molecules converted to R, and B the charge transfer during N to R conversion divided by that during M to N conversion; $a_1 = 1/\tau_f$, where τ_f is the flash time constant; $a_2 = 1/\tau_c$, where $\tau_c = R_m C_m$ is the membrane time constant; $a_3 = 1/\tau_N$, where τ_N is the time constant of the thermal decay of N to R. The continuous curve in Fig. 10 represents a fit to the measured ERP using Eq. (27). The fit yielded $\tau_N = 13$ ms. Furthermore, it was found that $\tau_f = 0.7$ ms and $\tau_c = 7.0$ ms, which together with $R_m = 30$ MΩ gives $C_m = 230$ pF. With $N_M = 1.75 \times 10^8$, being 70% of the total number of visual pigment molecules per photoreceptor, the charge transferred outward during a single isomerization from M to N then appears to be $q = 5.0 \times 10^{-21}$ C, or 0.03 electron charge; the deduced charge transfer ratio is $B = 1.3$, or, the conversion of N to R drives 0.04

electron charge in the opposite direction. Flash-induced conversions of rhodopsin to metarhodopsin do not show up as changes in the membrane potential, because R to M conversion is complete within ca. 0.1 ms, which is much shorter than the membrane time constant, $R_m C_m$, and the net charge displacement then is small. But M to R conversion proceeds much slower, and the larger charge displacements are in opposite directions [55].

Evidently, the charge displacement across the membrane thickness that occurs in a single molecule is in fact minute, but a quite well measurable ERP occurs when a large fraction of the visual pigment molecules is rapidly converted. Comparable values were found for the ERP's of the visual pigment systems in other photoreceptor membranes [59,65,66].

A considerably improved situation is encountered when a photoreceptor cell is not studied with an intracellular electrode, but with a patch clamp electrode. The early receptor current (ERC) then can be measured with much better time resolution [67]. So far this technique has not yet been exploited for visual pigment studies.

6. Sensitivity spectrum

6.1. Self-screening

A rhodopsin molecule converted to metarhodopsin forms the trigger of the phototransduction process. The conversion of one rhodopsin molecule creates a so-called quantum bump, measurable as a short-lasting depolarization of the membrane, or as a brief inward current (Section 7). With more photons absorbed, the signal adds up, but the relationship between number of absorbed photons and receptor potential is complex, due to the numerous molecular components involved and their highly non-linear interactions. The last decade has seen considerable progress in delineation of the phototransduction process, but quantitative analyses have been offered for only a few cases. An exception is the estimation of a photoreceptor's spectral sensitivity.

The principle of univariance states that the spectral information of an absorbed photon is lost upon photoconversion of rhodopsin. The spectral sensitivity then can be derived by a constant criterion method, using a monochromatic light stimulus with intensity $I(\lambda)$ and duration Δt. The starting point then is the assumption that a certain criterion signal will be reached when the stimulus caused the conversion of a certain number of rhodopsin molecules, i.e., caused the creation of a certain number of metarhodopsin molecules. According to Eq. (4), the change in the concentration of rhodopsin molecules converting to metarhodopsin at $t = 0$ is:

$$\left(\frac{\mathrm{d}C_R}{\mathrm{d}t}\right)_{R\to M} = C_{R,0}\beta_R(\lambda)I(\lambda) = C_{R,0}\gamma_R\alpha_R(\lambda)I(\lambda) \tag{28}$$

or, the fraction of converted molecules during Δt is

$$\Delta f_M = \Delta f_{R\to M} = \Delta t f_{R,0}\gamma_R\alpha_R(\lambda)I(\lambda) \tag{29}$$

This is reached with an intensity

$$I(\lambda) = \frac{\Delta f_{\mathrm{M}}}{\Delta t f_{\mathrm{R},0} \gamma_{\mathrm{R}} \alpha_{\mathrm{R}}(\lambda)} \tag{30}$$

The spectral sensitivity, which is the inverse normalized criterion intensity, then is:

$$S(\lambda) = \left(\frac{I(\lambda)}{I(\lambda_{\max})}\right)^{-1} = \frac{\alpha_{\mathrm{R}}(\lambda)}{\alpha_{\mathrm{R}}(\lambda_{\max})} \tag{31}$$

In this case the spectral sensitivity equals the normalized absorbance spectrum of rhodopsin, which thus can be determined by electrophysiological measurements.

Of course, the situation has been simplified, because Eqs. (28)–(30) only hold when the light intensity everywhere in the photoreceptor waveguide is identical. In the case of a fly rhabdomere, the visual pigment is contained in a long optical waveguide, and then the intensity decreases exponentially. If the photon flux at depth z is $J(z, \lambda)$ and the cross-section of the rhabdomere is A_{r}, the number of converted molecules in a slice dz is (see Section 3):

$$\left(\frac{\mathrm{d}C_{\mathrm{R}}}{\mathrm{d}t}\right)_{\mathrm{R}\to\mathrm{M}} A_{\mathrm{r}}\,\mathrm{d}z = \eta(\lambda) J(z,\lambda) \beta_{\mathrm{R}}(\lambda) f_{\mathrm{R}} C_0\,\mathrm{d}z \tag{32}$$

The total number of created metarhodopsin molecules per unit time then is:

$$\begin{aligned}\int_0^L \left(\frac{\mathrm{d}C_{\mathrm{R}}}{\mathrm{d}t}\right)_{\mathrm{R}\to\mathrm{M}} A_{\mathrm{r}}\,\mathrm{d}z &= \int_0^L \eta(\lambda) J(z,\lambda) \beta_{\mathrm{R}}(\lambda) f_{\mathrm{R}} C_0\,\mathrm{d}z \\ &= J_0(\lambda)\gamma_{\mathrm{R}} \frac{\alpha_{\mathrm{R}}(\lambda)}{f_{\mathrm{R}}\alpha_{\mathrm{R}}(\lambda) + f_{\mathrm{M}}\alpha_{\mathrm{M}}(\lambda)} \\ &\quad \times (1 - \exp[-\eta(\lambda)\{f_{\mathrm{R}}\alpha_{\mathrm{R}}(\lambda) + f_{\mathrm{M}}\alpha_{\mathrm{M}}(\lambda)\}C_0 L])\end{aligned} \tag{33}$$

Taking the inverse and subsequent normalization yields the sensitivity spectrum. When $f_{\mathrm{M}} = 0$, i.e. $f_{\mathrm{R}} = 1$, this becomes:

$$S(\lambda) = \frac{1 - \exp[-\eta(\lambda)\alpha_{\mathrm{R}}(\lambda)C_0 L]}{1 - \exp[-\eta(\lambda_{\max})\alpha_{\mathrm{R}}(\lambda_{\max})C_0 L]} = \frac{A(\lambda)}{A(\lambda_{\max})} \tag{34}$$

where $A(\lambda)$ is the absorptance (Eq. (18)). In the pure rhodopsin case, the spectral sensitivity thus is the normalized absorptance. This result is in fact rather obvious, as the absorptance is the fraction of the incident light flux absorbed. With pure rhodopsin the conversion rate equals the quantum efficiency times the number of absorbed photons per unit time.

When the absorptance is moderate and the waveguide factor $\eta(\lambda)$ does not strongly vary in the spectral range of the visual pigment, Eq. (34) reduces to Eq. (31). But this is generally not the case, and then the sensitivity spectrum broadens due to so-called self-absorption or self-screening (Fig. 11; e.g. [18,39,68]).

The peak absorption coefficient of insect rhabdomeres is 0.004–0.009 μm^{-1}, taking a waveguide factor $\eta = 0.9$–1.0 [69]. In fruitfly photoreceptors, the length of

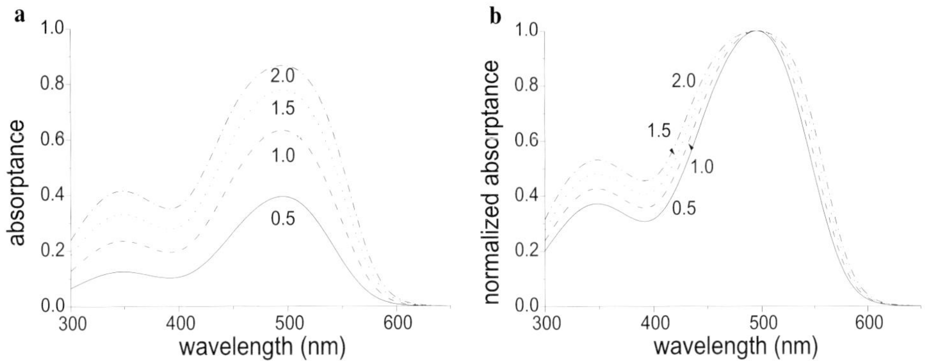

Fig. 11. Absorptance as a function of wavelength for a few values of the product of the peak absorption coefficient and the length of the photoreceptor: $\kappa_{av}(\lambda_{max})L$. The absorptance increases with this product (a), and the spectral sensitivity, which is equivalent to the normalized absorptance, broadens (b). (The waveguide factor η has been assumed to remain constant.)

a rhabdomere is ca. 80 μm [70] yielding $\kappa_{av}(\lambda_{max})L = 0.33–0.72$. Self-screening is then moderate. For a blowfly rhabdom of $L = 250$ μm, $\kappa_{av}(\lambda_{max})L = 1.1–2.0$, giving rise to noticeable broadening of the spectral sensitivity. Electrophysiological recordings of the main class of blowfly photreceptors R1–R6, that contain the main visual pigment, Rh1, indeed show substantial broadening (see Fig. 12). The spectra indicate effective absorption coefficients near the lower end of the range, which must be partly due to the fact that fly rhabdomeres are tapered [71], and hence the contribution to the absorptance accordingly diminishes, or, the effect of self-screening is reduced.

The expression for the sensitivity spectrum, Eq. (34), is slightly modified when $f_M > 0$. Measurements of sensitivity spectra in the presence and absence of metarhodopsin have been performed on droneflies to estimate the metarhodopsin absorbance spectrum by electrophysiological methods [72].

6.2. Sensitizing pigment

Flies reared on a vitamin A deprived diet have a low absolute light sensitivity, due to a low concentration of visual pigment [73]. Their sensitivity spectrum resembles the absorbance spectrum of a human rod rhodopsin, with a strong α-band and a small subsidiary β-band (Fig. 12(a)). However, the β-peak of the sensitivity spectrum of fly photoreceptors in normal conditions is much higher than the α-peak and has a fine structure [74]. The reason is that fly visual pigments, in addition to the chromophore 3-hydroxyretinal, bind 3-hydroxyretinol [12,75]. When this ultraviolet-absorbing compound is excited by a photon it transfers the absorbed energy to the chromophore, which then isomerizes as usual [76]. The 3-hydroxyretinol therefore acts as a sensitizing or antenna pigment (review [77]). Energy transfer can occur from the excited sensitizing pigment to both rhodopsin and metarhodopsin [78]. The

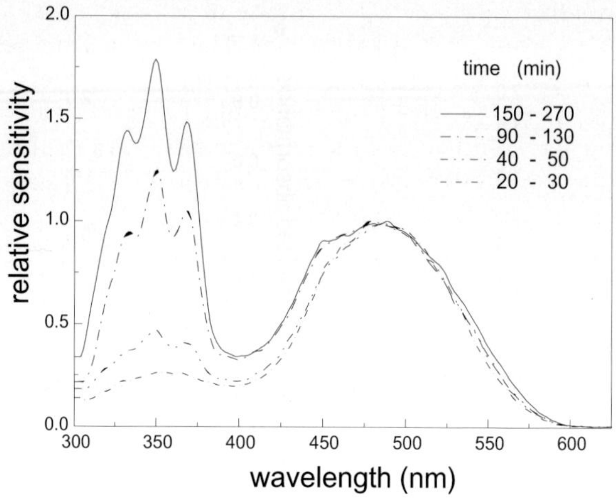

Fig. 12. Incorporation of a sensitizing pigment in the photoreceptors of vitamin A deprived blowflies. Retinoids administered to the eye are taken up by the retina, resulting in an enhanced sensitivity in the UV with respect to that in the blue-green. The enhancement is due to binding of a 3-hydroxyretinol to rhodopsin. UV-light absorbed by the 3-hydroxyretinol then results in transfer of energy to the chromophore of rhodopsin, 3-hydroxyretinal. The fine structure emerging in the UV is interpreted to be caused by a rigid binding of the 3-hydroxyretinol. The spectra were measured at the indicated time after supplying all-*trans* retinal to the eye (modified from [79]).

sensitization of rhodopsin by the UV-absorbing antenna pigment is the reason for the strongly enhanced spectral sensitivity in the ultraviolet, measured in physiological responses (review [74]).

The absolute sensitivity of vitamin A deprived flies is low, due to the necessity of vitamin A for producing rhodopsin. Supplying retinoids results firstly in enhancement of the sensitivity in the blue-green, but progressively also in enhancement of UV-sensitivity. Hamdorf et al. [79] have investigated this observation in an electrophysiological study on the speed of incorporation of various retinoids in the rhodopsin molecules of blowflies. The retinoids (e.g. retinal or retinol in either the all-*trans* or the 13-*cis* configuration) were applied with a most simple 'perfusion' technique. A drop containing dissolved retinoid was administered to the cornea of an immobilized blowfly *Calliphora*, white-eyed mutant *chalky*. The fly started with a very low visual pigment content, as it was reared on a vitamin A deprived diet. The spectral sensitivity then was measured at various times after application of the retinoid. As an example, Fig. 12 presents the sensitivity spectra, normalized to the sensitivity peak in the visible, measured after application of all-*trans*-retinal to the cornea. Initially, the sensitivity band in the UV is very low and smooth. Within a few hours this band has risen considerably and then features a prominent vibronic fine structure, with peaks at 333, 350 and 369 nm. These peaks prove the presence of 3-hydroxyretinol. Evidently, this derivative was enzymatically produced from the administered all-*trans*-retinal.

Figure 12 shows that with ample availability of retinoids, the UV-peak is much higher than the blue-green peak (the α-band). The initial hypothesis for the high UV-sensitivity of fly photoreceptors, namely that waveguide effects cause a relative enhancement of sensitivity in the UV [80], could be eliminated by a number of arguments. First, white-eyed flies were used in the experiments of Fig. 12, and then oblique, non-guided light contributes to light absorption [81]. Furthermore, the low sensitivity in the UV in flies with little rhodopsin would need an alternative explanation. The presence of sensitizing pigments in photosynthesis suggested an alternative interpretation that the characteristics of UV-sensitivity of flies are due to the action of a sensitizing pigment [76], and later experiments firmly established this new mechanism in the area of vision.

Vogt and Kirschfeld [82] have investigated the interaction of fly sensitizing pigment and rhodopsin in further detail. A sensitizing pigment can have a high energy transfer efficiency when donor and acceptor are at a sufficiently close range. According to the Förster dipole–dipole energy transfer mechanism the critical distance, where transfer efficiency $\gamma_S = 0.5$, is

$$R_0 = 310(Ik^2\gamma_F n^{-4})^{1/6} \tag{35}$$

where I is the spectral overlap integral (for the sensitizing pigment-rhodopsin complex calculated to be 1.39×10^{-10} cm^6 mol^{-1} [82]), k^2 the orientation factor (conservatively assumed to be 2/3, the value for fast isotropic motion), γ_F the fluorescence quantum yield (argued to be 0.033 for the complex), and n the refractive index of the medium (being 1.4). The resulting $R_0 = 3.0$ nm is smaller than the ca. 4 nm of the rhodopsin diameter, indicating the intimate connection of sensitizing pigment and rhodopsin. The distance between the two components of the energy transfer system, R, is related to the transfer efficiency by:

$$R = R_0(\gamma_S^{-1} - 1)^{1/6} \tag{36}$$

Ideally, the transfer efficiency is determined by fluorescence measurements, but this is not a feasible approach for the visual pigment *in situ*. Vogt and Kirschfeld [82] have ventured an alternative method, via electrophysiological measurements of the polarization sensitivity in the UV, at 359 nm, as a function of the ratio of the spectral sensitivities at 359 and at 490 nm (Fig. 13).

The spectral sensitivity of a photoreceptor with 100% rhodopsin having a sensitizing pigment absorbing only in the UV is:

$$S(\lambda) = \frac{\gamma_R E_R(\lambda) + \gamma_R\gamma_S E_S(\lambda)}{E_R(\lambda) + E_S(\lambda)} \frac{1 - 10^{-[E_R(\lambda)+E_S(\lambda)]}}{\gamma_R[1 - 10^{-E_R(\lambda_{max})}]} \tag{37}$$

where E_R is the absorbance of the photoreceptor due to rhodopsin and E_S that due to the sensitizing pigment. When $E_S = 0$ at $\lambda_{max} = 490$ nm, $S(490) = 1$.

Generally, insect photoreceptors exhibit strong polarization sensitivity, because the absorption of light depends on the direction of polarization (see e.g. [42]). The polarization sensitivity is defined as the ratio of the absorptances,

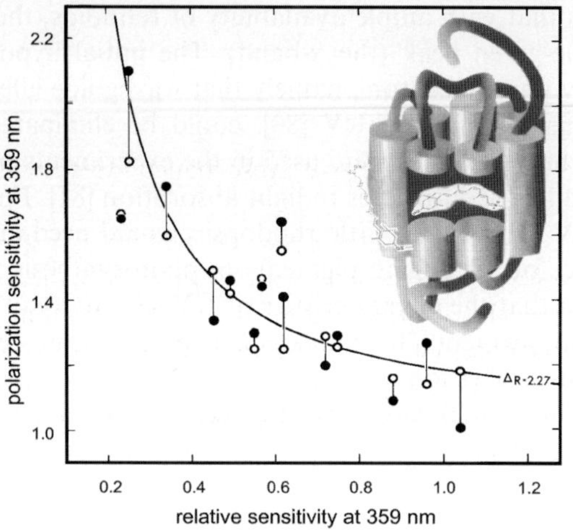

Fig. 13. Polarization sensitivity of housefly (*Musca*) photoreceptors in the UV vs the relative UV sensitivity for unpolarized light (from [82]). An increasing relative spectral sensitivity in the UV parallels a higher absolute sensitivity due to an increasing rhodopsin content of the photoreceptors. The UV polarization sensitivity decreases with increasing relative spectral sensitivity. This is explained by assuming that the rhodopsin has a constant, wavelength independent dichroism, $\Delta_R = 2.27$, and that no dichroism is due to the sensitizing pigment, 3-hydroxyretinol [82]. Inset: the sensitizing pigment is presumably bound at both ends to the rhodopsin molecule via hydrogen bonds (from [12]).

$P_S(\lambda) = A_{//}(\lambda)/A_{\perp}(\lambda)$, when $A_{//}(\lambda)$ is the absorptance for light polarized in the direction of maximal sensitivity and $A_{\perp}(\lambda)$ that for the perpendicular direction. The dichroism is the ratio of the absorbances; when only rhodopsin is present this is $\Delta_R = E_{R//}/E_{R\perp}$; and, when the absorbance is small, $P_S(\lambda) = \Delta_R(\lambda)$. In the presence of a sensitizing pigment, with transfer efficiency γ_S, the polarization sensitivity changes into

$$P_S(359) = \frac{E_{R\perp}(359)\Delta_R + \gamma_S E_S(359)}{E_{R\perp}(359)\gamma_S E_S(359)} \tag{38}$$

assuming that both the absorbance by the sensitizing pigment (E_S) and the absorbance by the visual pigment (E_R) are low, that the dichroism due to rhodopsin, $\Delta_R(\lambda)$, is constant, and that the sensitizing pigment has no dichroism, $\Delta_S = 1$.

The relative spectral sensitivity for unpolarized light then is given by

$$S(359) = \frac{E_{R\perp}(359)(1 + \Delta_R)/2 + \gamma_S E_S(359)}{E_{R\perp}(490)(1 + \Delta_R)/2} \tag{39}$$

Combining Eqs. (38) and (39) yields

$$P_S = \frac{S(359) + D}{S(359) - D} \tag{40}$$

with

$$D = a\frac{\Delta_R - 1}{\Delta_R + 1} \tag{41}$$

and $a = E_R(359)/E_R(490)$ the relative absorbance coefficient of rhodopsin at 359 nm.

The experimental results (Fig. 13) could be well-fitted with Eq. (40) by taking $a = 0.22$ and for the value of the dichroism of rhodopsin $\Delta_R = 2.27$. The transfer efficiency, γ_S, was calculated from Eq. (39), using measured sensitivity spectra. Because microspectrophotometry showed that always $E(359)/E(490) < 2$, it was concluded that $\gamma_S \geqslant 0.75$. Consequently, the distance between sensitizing pigment and chromophore must be $R \leqslant 2.5$ nm, i.e., about the radius of the rhodopsin molecule, suggesting an arrangement as pictured diagrammatically in the inset of Fig. 13 [12,75,77,82].

The high spectral sensitivity in the UV immediately suggests that more than one 3-hydroxyretinol per rhodopsin molecule is bound. Calculations of the energy transfer indicate that in retinoid-saturated eyes at least two sensitizing pigment molecules can be bound to one visual pigment molecule [79,82]. When ultimately 1000 rhodopsin molecules exist per microvillus [18], and 2000 sensitizers are supplied in a process with half time 100 min (Fig. 12), this implies that only once per 3 s a sensitizing molecule is bound to a visual pigment molecule in each microvillus.

Carotenes, like β-carotene, lutein or zeaxanthin cannot serve as precursors for the sensitizing pigment if applied to the cornea [79]. Interestingly, vitamin A2, 3-dehydroretinol, is taken up and can function as a sensitizing pigment as followed from the slightly different vibronic structure of the UV-band: the 350 and 369 nm peaks were shifted to 358 and 380 nm, respectively. However, after a few hours the normal UV fine structure emerged, indicating replacement of the 3-dehydroretinol by 3-hydroxyretinol [79].

6.3. Spectral filtering by photostable pigments

The sensitivity spectrum of a photoreceptor can substantially differ from its rhodopsin absorbance spectrum due to self-screening (Section 6.1). Much more important are spectral modulations due to filtering by other pigments. This occurs in insect eyes where the photoreceptor rhabdomeres are fused together, thus building the so-called fused rhabdom [83]. The mutual filtering of the different rhodopsins in one and the same waveguide counteracts the spectral broadening effect of self-screening. Photostable pigments can also act as effective filters, in a similar way as occurs by the colored oil droplets in bird and amphibian photoreceptors. This phenomenon is realized in several insect eyes, especially in flies and butterflies [70,84,85].

When the transmittance of the filtering pigment is $T_f(\lambda)$ and it is located distally of the photoreceptor, the incident light intensity $I(\lambda)$ then changes into $T_f(\lambda)I(\lambda)$. For cases where the absorbance is low and due to only one rhodopsin, the expression for the spectral sensitivity, Eq. (31), accordingly changes into (with wavelength of peak sensitivity λ_{max}, and neglecting the waveguide factor, η):

$$S(\lambda) = \left(\frac{I(\lambda)}{I(\lambda_{max})}\right)^{-1} = \frac{T_f(\lambda)\alpha(\lambda)}{T_f(\lambda_{max})\alpha(\lambda_{max})} \tag{42}$$

With suitable selective long-pass filtering the sensitivity spectrum can be substantially shifted with respect to the absorbance spectrum of the rhodopsin. The nearest equivalent to this (oil droplet) case actually is that of the eyes of tabanids. The corneal facet lenses of these flies are equipped with multilayer interference reflectors. Consequently, the transmission is reduced in a restricted wavelength band [86], possibly causing narrowed sensitivity spectra. Another case is the central photoreceptor, R7, of the housefly *Musca domestica*, that contains xantophyll pigments. These act as selective filters for the underlying R8 photoreceptor. In fact, these photostable pigments also affect the light absorption by the co-localized rhodopsin in the R7 photoreceptor [70,84].

Virtually all filtering combinations occur in the Japanese yellow swallowtail *Papilio xuthus*. This butterfly has at least five different receptor types, as follows from extensive measurements of their sensitivity spectra as well as of the polarization sensitivities [89,90]. The type of chromophore used by the *Papilio* rhodopsins is 3-hydroxyretinal [13,91]. Whereas some of the spectra indeed more or less conform to the general rhodopsin template shapes, some of the spectra, i.e., of the violet and single-peaked green receptors, are clearly at odds (Fig. 14). The deviations of the spectra from the expected spectral shapes suggested the presence of a specific UV-absorbing filter. A suitable candidate seemed to be a substance, emitting broad-band white light under ultraviolet excitation, that was discovered by fluorescence microscopy [87]. The fluorescing pigment appeared to be present only in a restricted number of randomly distributed ommatidia. Its excitation spectrum was limited to the UV and its emission spectrum strikingly corresponded to that of 3-hydroxyretinol [85].

To reveal the function of the fluorescing compound, electrophysiological recordings were performed with electrodes filled with the dye lucifer yellow. The photoreceptor cells were marked subsequent to measuring their sensitivity spectrum and were then observed with a fluorescence microscope. Invariably it appeared that the violet and single-peaked green cells were colocalized with the fluorescing substance.

Using the known absorbance spectrum of 3-hydroxyretinol (Figs. 14(a) and (b)) and varying the rhodopsin spectrum, the spectral sensitivity was calculated with the model of Section 3.2, extended as outlined below to the more complex situation of a fused rhabdom with several participating photoreceptor types, each with a different rhodopsin, and filtered by the photostable pigment, located within the rhabdom (see also [83]).

The light flux, $J(z, \lambda)$, along the length coordinate of the rhabdom, z, is described by (cf. Eq. (14)):

$$\mathrm{d}J(z,\lambda) = -\mathrm{d}z J(z,\lambda)\eta(\lambda)\left[\sum_j p_j \kappa_j(z)\varepsilon_j(\lambda) + \kappa_f(z)\varepsilon_f(\lambda)\right] \tag{43}$$

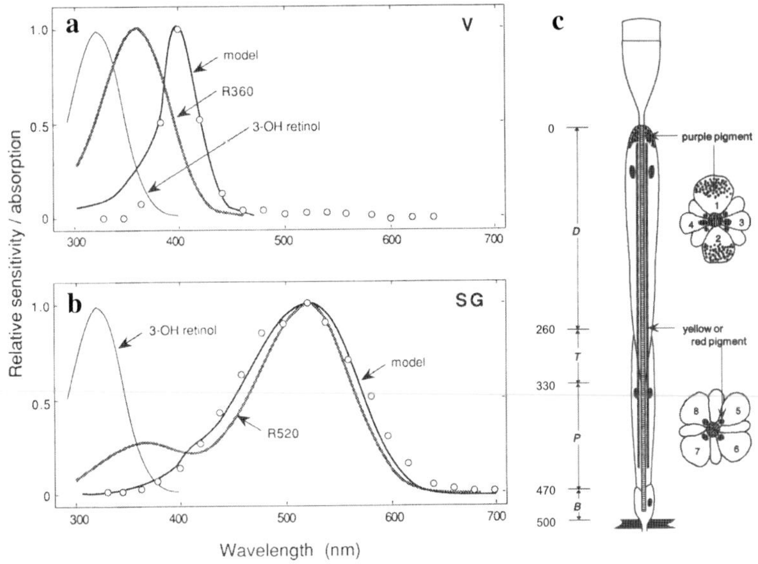

Fig. 14. Measured sensitivity spectra together with modeled spectra for the violet (V) and single-peaked green (SG) receptor of the Japanese yellow swallowtail *Papilio xuthus*. (a) The UV absorbing 3-hydroxyretinol is assumed to act as a UV filter for a rhodopsin with peak absorbance at 360 nm (R360), resulting in a narrow-band violet (V) receptor [85]. (b) When the ultraviolet filter acts on a photoreceptor with a rhodopsin absorbing maximally at 520 nm (R520) a single peaked green (SG) receptor results. The spectrum is somewhat broadened, because the receptor is about 300 μm long. (c) Anatomy of a *Papilio* ommatidium. The rhabdom consists of three tiers, a distal (D), proximal (P) and basal (B) tier. A transition layer (T) exists between distal and proximal tier. The depths are given in μm. Purple, yellow and red photostable pigments can act as optical filters. The 3-hydroxyretinol exists only in some of the ommatidia [87,88].

where p_j is the part of the rhabdom cross-section taken up by the rhabdomere of photoreceptor R_j; $\kappa_j = \alpha_{j,\max} C_j$ is the peak absorption coefficient of the tissue of photoreceptor R_j, with $\alpha_{j,\max}$ the molecular absorbance coefficient and C_j the concentration of rhodopsin j, respectively, assuming that all visual pigment molecules are in the rhodopsin state; $\varepsilon_j(\lambda)$ is the normalized spectral absorbance coefficient of the rhodopsin (cf. Eq. (1)); κ_f is the peak absorption coefficient due to the filtering pigment, and $\varepsilon_f(\lambda)$ its normalized spectral absorbance coefficient. The local change in the concentration of rhodopsin due to conversion to its metarhodopsin per unit time is for each of the photoreceptors given by Eq. (28). The total number of converted molecules per receptor then follows from the contribution of each photoreceptor to the rhabdom volume, which is known from anatomy (Fig. 14(c)). Of course, as in Eqs. (33) and (34), the conversion rate of visual pigment in a photoreceptor is proportional to its absorptance, i.e., the fraction of the incident photon flux absorbed. It hence is sufficient to compute the spectral absorptance of each photoreceptor. Normalization then yields the sensitivity spectrum.

By going through this procedure for a variety of rhodopsin combinations, and taking different peak absorption coefficients, satisfactory fits can be obtained for the measured sensitivity spectra. As shown in Figs. 14(a) and (b), a violet receptor results when a UV rhodopsin, with peak absorbance at 360 nm, is filtered by the ultraviolet-absorbing, fluorescing pigment, presumably 3-hydroxyretinol. Similarly, a single-peaked green receptor results when a green rhodopsin, with peak absorbance at 520 nm, is UV-filtered [85]. Clearly, the photoreceptor sensitivity spectra appear to become distinctly different from the rhodopsin spectra due to the spectral filtering effects.

6.4. Red screening pigments in insect eyes

A similar effect, but with quite different outcome is executed by red-coloured screening pigment granules that line the rhabdom in the *Papilio* eye (Fig. 14(c)). Their action is in fact rather similar to that of the oil droplets of bird eyes, acting as long-pass filters. The red pigment granules transmit long-wavelength light, but strongly absorb at the shorter wavelengths, and thus reduce the light flux there. The consequence is that the sensitivity spectra are narrowed and shifted towards the red. Red-colored pigments, concentrated in granules that are situated close to the rhabdom, in order to selectively enhance red sensitivity are widely used among insects [92–94].

Another usage of red screening pigments is encountered in fly eyes. There the red pigment is located in the screening pigment cells [70,81]. The high transmittance at the longer wavelengths helps to photoregenerate metarhodopsin molecules back to their rhodopsin (see Fig. 2; [39,45–47]).

6.5. Pigment granule migration; pupil mechanism

The spectral filters discussed so far are all non-mobile. Yet, the movement of colored pigment granules is a universal phenomenon among insect eyes, especially of the pigment granules inside the photoreceptor cells (review [46]). The pigment granules in fly photoreceptor cells are in the dark at a distance of the rhabdomere, but illumination causes pigment migration towards the optical waveguide. The yellow-coloured pigment granules there reduce the light flux, specifically in the blue, and hence enact the function of a light-controlling pupil mechanism [46,95]. Other effects are a shift of the sensitivity spectrum [96,97] and an increase of the rhodopsin content in the steady state when intense white light is applied (Fig. 4(b); [45]). Furthermore, the photoreceptor angular sensitivity is narrowed, which can be accurately modeled from the optical waveguide properties of the rhabdomere [98].

The mechanism driving the pigment migration is under control of the intracellular calcium concentration [99,100], which upon illumination of a dark-adapted photoreceptor increases enormously (Section 8; [101–104]). This then causes activation of the pupil mechanism, i.e., accumulation of the yellow pigment granules near the photoreceptor rhabdomere.

7. Quantitative descriptions of insect photoreceptor membrane currents

7.1. An early attempt to model the light response with a Hodgkin–Huxley type model

Before the early 1990s technical difficulties prevented the application of the patch- and voltage-clamp techniques to fly photoreceptor cells. Therefore, rather little was known about the currents that contribute to the electrical light response of these cells. A different situation was encountered in larger photoreceptor cells of arachnoid and crustacean animals, where voltage clamping with two electrodes could be employed [105]. In analogy with the model originally developed for *Limulus* photoreceptor cells [106], Muijser [107] argued that a minimal model (Fig. 15(a)) for the fly

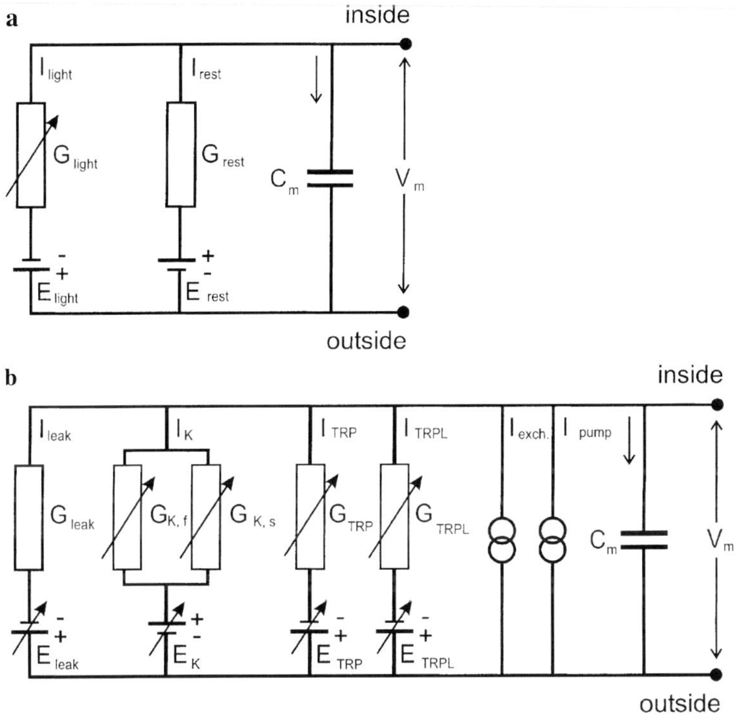

Fig. 15. The equivalent electrical circuits of fly photoreceptors used to model the light response. (a) The "minimal model" consisting only of 2 conductances [107], one of which was supposed to increase during illumination. (b) The model describing the electrical components of the blowfly photoreceptor cells [108,109], consisting of two types of voltage-gated K^+ conductances, two types of light-activated conductances (G_{trp} and G_{trpl}), a leak conductance, Na^+/Ca^{2+} exchangers and the Na^+/K^+ pump. The reversal potentials of the various conductances vary (indicated by the arrows), as the intra- and extracellular ion concentrations change considerably. Note that even this model, while describing the known electrogenic components of the *Calliphora* photoreceptor plasma membrane, is likely to be highly simplified. In *Drosophila* photoreceptor cells, additionally a Cl^- conductance [110] and a Ca^{2+} activated cation conductance [111] have been described.

photoreceptor potential had to contain at least two different types of conductances. One type, depolarizing the membrane, was assumed to increase during the light response. This conductance will here be called light-activated conductance, created by the light-activated channels. The second conductance was thought to be important for repolarizing the membrane after the light stimulation to the resting potential in darkness, and was initially assumed not to change during light stimulation.

This model was inspired by data obtained from *Limulus* ventral photoreceptor cells, that was – at the time – a much more accessible preparation, allowing two electrode voltage clamp, injection of drugs and good experimental control of the (extracellular) ion concentration [105]. With the minimum model of the light response (Fig. 15(a)), experimentally testable predictions could be formulated; e.g. the model predicts that the light response has a unique reversal potential and that the membrane resistance follows the membrane potential in a predictable manner. The subsequently performed experiments [107] to test these hypotheses showed that the minimum model was insufficient to explain the light response of fly photoreceptor cells. It hence was necessary to assume that the second, repolarizing membrane conductance also changed during light stimulation, albeit with a slower time constant than the first, light-activated conductance. An additional conclusion was that the reversal potential of the first, depolarizing conductance is close to 0 mV, indicating that this conductance is permeable to more than one ion-type [107].

From what will be detailed in the following, it will become clear that these conclusions still hold nowadays, although evidently the membrane of fly photoreceptor cells is much more complex than envisaged 20 years ago (Fig. 15(b)).

7.2. Modeling voltage gated K^+ channels

Using single-electrode voltage clamping and patch clamping Weckström et al. [111] described the presence of voltage-gated K^+ conductances in the photoreceptors of blowflies. It was subsequently established that the photoreceptor cells of all other dipteran species investigated possess these conductances [112–115]. The quantitative details of the voltage-gated K^+ conductances vary widely between the different species; e.g. *Drosophila* photoreceptor cells exhibit, in addition to a slowly and non-inactivating delayed rectifier current, fast and rapidly inactivating A-currents through channels formed by the Shaker gene products [112,113]. In the blowfly *Calliphora*, on the other side, only the non-inactivating delayed rectifier currents have been found [111]. In a comparative study, the inter-species differences in the voltage-gated K^+ currents were linked to the differing life-style of the various species [114]. The delayed rectifier K^+ currents observed in fly photoreceptor cells fit to a large extent the profile of the repolarizing conductance postulated by Muijser [107] (who, however, incorrectly assumed that this conductance was dependent on an increase of intracellular Ca^{2+}). The delayed rectifier currents repolarize the membrane after cessation of the light stimulus and have slower kinetics than the light-activated conductance.

Voltage-clamp measurements of the kinetics of the delayed rectifier K^+ [111] showed that in the photoreceptors of the blowfly *Calliphora* two components exist with distinct temporal kinetics, a fast and a slow component. These data were used by Gerster et al. [108] to provide a quantitative description of the delayed rectifier currents that allowed reproducing the voltage-clamp experiments (Fig. 16). The formalism employed was based on the Goldman–Hodgkin–Katz (GHK) equation, and therefore permeabilities rather than conductances were used. The equations that describe the voltage-gated K^+ current are:

$$I_K = S_m(P_{k,f} + P_{k,s})F\beta V_m \frac{[K^+]_i - [K^+]_o e^{-\beta V_m}}{1 - e^{-\beta V_m}}, \quad \beta = \frac{F}{RT} \tag{44}$$

where V_m indicates the membrane potential, S_m the membrane surface, F the Faraday constant, R the molar gas constant and T the temperature. $P_{k,f}$ and $P_{k,s}$ are the voltage dependent permeabilities, which can be derived with the following formulae:

$$\frac{dP_k}{dt} = P_{k,max}\frac{P_{k,\infty} - P_k}{\tau} \tag{45}$$

$$P_{k,\infty} = \left[1 + \exp\left(-\frac{V_m - V_s + a}{b}\right)\right]^{-1} \tag{46}$$

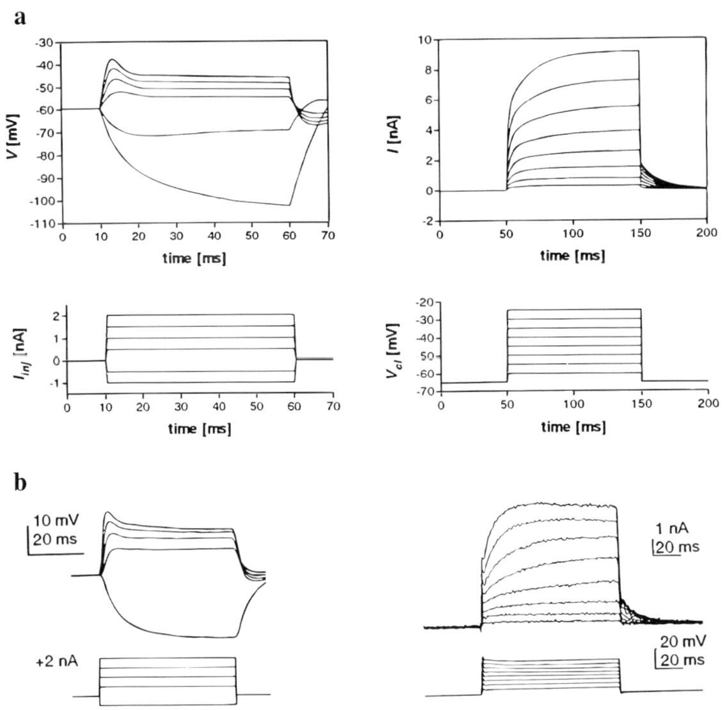

Fig. 16. (a) Simulations of current-clamp (left) and voltage-clamp (right) experiments compared to experimental data (from [108]).

and

$$\tau = \left[c\exp\left(\frac{V_m - V_s}{d}\right) + e\exp\left(\frac{V_m - V_s}{f}\right)\right]^{-1} \tag{47}$$

The deduced parameters (a to f and V_s) are listed in Table 1. With these equations it is possible to calculate the magnitude of the K^+ currents through the voltage-gated channels when the membrane potential and the intra- and extracellular concentrations of K^+ are known.

Gerster et al. [108] incorporated the description of the voltage dependent K^+ currents into a refined (electrical) model of the membrane of the photoreceptor cell (Fig. 15(b)). It proved necessary to assume a leak current, besides the light-activated channels and the voltage-dependent K^+ channels, in order to satisfactorily model the resting (dark) membrane potential (ca. −60 mV; e.g. [116]) and the input resistance (ca. 30 MΩ; [111,115]) of the photoreceptor cells. However, the molecular nature of this leak permeability is unknown. Gerster et al. [108] assumed it to be a selective Na^+ channel, and not to change its permeability during light stimulation.

During light stimulation, the rather large, sustained currents cause a considerable shift in the intra- and extracellular ion concentrations. In the extracellular space, these changes of the ion concentrations have been measured with ion-selective electrodes [117–119] as an increase of the extracellular K^+ concentration and a decrease of the extracellular Na^+ concentration. After the cessation of a light stimulus, in darkness, the concentrations recover to their initial values, due to the action of Na^+/K^+ pump molecules. These pump molecules, the presence of which has been shown by pharmacological [120] and immunohistochemical [121] methods, produce an outward (hyperpolarizing) current due to their electrogenic coupling ratio of 3 Na^+ against 2 K^+ ions. After a strong light stimulus has been turned off, the membrane potential shows a complex time course while returning to the resting potential, the so-called hyperpolarizing afterpotential, a pronounced undershoot, which lasts often for several tens of seconds. Such a hyperpolarizing afterpotential was linked to the action of the Na^+/K^+ pump current already in 1971 in barnacle photoreceptor cells [122] and subsequently also in fly photoreceptor cells [123]. During the hyperpolarizing afterpotential, the light-activated permeability can safely be assumed to be negligible, which allowed Gerster et al. [108] to reduce their model to the voltage-gated K^+ channels, the leak channel and the pump current. As the voltage-gated K^+ currents and the leak currents could be determined from the measured membrane potential, the magnitude of the Na^+/K^+ pump current can directly be calculated.

Table 1
Parameters for the fast and slow voltage-gated K^+ permeabilities in blowfly photoreceptors [108]

	a (mV)	b (mV)	c (s^{-1})	d (mV)	e (s^{-1})	f (mV)	V_s (mV)
Slow	51.6	9.4	900	13	3.7	−33.8	15
Fast	65	8.5	3.0×10^3	24.4	9.4×10^{-5}	−7.8	15

These calculations also allowed estimating the changes of the intracellular Na^+ concentration induced by the light stimulation. These changes were found to be surprisingly large, even exceeding 30 mM when using relatively weak light stimulation. Assuming that the activity of the Na^+/K^+ pump was solely dependent on the intracellular Na^+ concentration, Gerster et al. [108] found that the activity of the Na^+/K^+ pump could be described with a simple Hill function. The maximum current that the pump could produce was found to be quite large (ca. 1 nA), corroborating the findings of an earlier study [123].

7.3. Estimating the ionic composition of the currents through the light activated channels

The currents through the light activated channels can be directly measured by isolating the ommatidia from *Drosophila* and whole cell patch-clamping photoreceptor cells [124,125]. This technique, in conjunction with the powerful genetics available for this species, enabled rapid progress in the biophysical and molecular characterization of the light-activated channels [11]. The first measurements of the reversal potential of the light-activated current directly determined it to be at ca. 0 mV [124,125], confirming earlier suggestions [107]. These studies also confirmed that Ca^{2+} ions strongly permeated the light-activated channels, as had been previously shown by the influence of removing extracellular Ca^{2+} on the light response [107] and the pupil reaction [99,100].

It was subsequently found that the current through the light-activated channels has two components, one being sensitive to mutations in the trp gene, the other component being sensitive to mutations in the trpl gene from the same gene family [126,127]. The clear separability of these two components due to the availability of the mutants, allowed a detailed biophysical characterization of the two channel types. The trp-dependent channels are highly Ca^{2+} selective and have a small single channel conductance of ca. 8 pS under physiological conditions, while the trpl-dependent channels are relatively unselective cation channels and have a much larger single channel conductance of ca. 30 pS [128–130]; see also Chapter 9 [11]. This finding raises the – for modeling purposes highly important – question how much each of these channel types contributes to the light-activated current under physiological conditions. This question is complicated by the fact that the different channel types are subject to a variety of regulatory processes. The trp-dependent channels are blocked by Mg^{2+} ions in the physiological concentration range in a complex voltage-dependent fashion [110]. As Mg^{2+} itself permeates the light-activated channels, and hence might change its intra- and extracellular concentration, the possibility exists that the Mg^{2+} block is a physiologically important regulatory process that should be taken into account. Furthermore, the light-activated channels are strongly regulated by Ca^{2+}, which is not only dependent on the concentration of Ca^{2+}, but also on the time when Ca^{2+} has access to the channels [67]. Measurements under conditions that tried to mimic the physiological ionic conditions indicated that the trpl-dependent currents contribute very little to the light response, due to a rapid Ca^{2+}-dependent deactivation of the trpl-dependent channels [130]. Similarly, measurements of the membrane voltage in intact eyes of *trpl* mutants did not display discernible differences to light responses of wild-type flies [127].

Even when assuming that the contribution of the different channel types is equal and constant (e.g. only the trp-dependent channels contribute), the relative contributions of the different cation types that can permeate the light-activated channels can vary considerably, due to potentially large changes of the intra- and extracellular ion concentration and of the membrane potential. Gerster [109] attempted to calculate the magnitude of the currents for each ion type by assuming the previously established permeability ratios to be constant. As he used the measurements of the membrane potential of blowfly photoreceptor cells, he needed to restrict his analysis to the responses of short flashes, in order to minimize the unknown currents contributed by the Na^+/K^+ pump and the Na^+/Ca^{2+} exchangers (that he nevertheless tried to correct for). Using the membrane model including the voltage-gated K^+ channels (see Section 7.2), he could thus derive the total current through the light-activated channels. Employing the GHK-equation, this current then could be split up according to:

$$\frac{I_q}{I_{\text{tot}}} = \frac{z_q w_q f_q}{\sum_q z_q w_q f_q} \tag{48}$$

where w_q is the relative permeability for ion type q. The concentration- and voltage-dependent weighting factor f_q is given by:

$$f_q = \beta_q V_{\text{m}} \frac{C_{q,i} - C_{q,\text{o}} e^{-\beta_q V_{\text{m}}}}{1 - e^{-\beta_q V_{\text{m}}}} \quad \text{with} \quad \beta_q = \frac{z_q F}{RT} \tag{49}$$

$C_{q,i}$ and $C_{q,\text{o}}$ indicate the intra- and extracellular concentrations of ion type q.

By making discrete but small time-steps, the calculated currents could be used to update the intracellular ion concentrations that need to be fed back into the formula. Gerster [109] concluded that Ca^{2+} carries about 50% of the current through the light-sensitive channels. However, as Na^+ flows into the cell when Ca^{2+} is extruded again via the Na^+/Ca^{2+} exchanger at a ratio of at least 3 Na^+ ions against 1 Ca^{2+} ion, a large amount of Na^+ flows into the cells after the cessation of the light stimulus. The concomitant changes of the ion concentration were determined to be 4 mM for Na^+ and 1.5 μM for Ca^+, rather large values when considering the stimulus duration of 10 ms that triggered a light response lasting ca. 100 ms. The calculated changes of the ion concentrations, however, are averages over the entire cell volume. In the following we will discuss that such spatial averaging is – at least for the case of Ca^{2+} – inappropriate and massive inhomogeneities of the ion concentrations have to be taken into account.

8. Modeling the spatial distribution of Ca^{2+} in fly photoreceptors

8.1. Modeling the Ca^{2+} concentration in the microvilli

Employing Ca^{2+}-indicators and fluorescence imaging techniques, Ranganathan et al. [132] showed that the Ca^{2+} influx through the light-activated channels only occurred in the rhabdomeres. This was subsequently confirmed by immunohistological techniques that showed that the gene product of both the *trp* and the *trpl* gene are exclusively located in the rhabdomeric microvilli [127,133]. These findings indicated that it was impossible to neglect the spatial gradients of ions, which is especially true for the distribution of Ca^{2+}, as these ions act as important regulators of many proteins implicated in the phototransduction cascade, as in other cellular processes.

The light response is a superposition of many unitary events, so-called quantum bumps, that are triggered by the absorption of a single photon. By delivering weak light flashes to patch-clamped photoreceptor cells, a quantum bump can be measured as a short-lived, inward current, peaking on average at -9 pA [124,134,135]. Postma et al. [136] investigated how the Ca^{2+} concentration changes during a single quantum bump (Fig. 17(a)). The analysis was limited to the microvilli, since a single quantum bump is not expected to significantly raise the Ca^{2+} concentration in the cell body, due to the smallness of the current. The microvilli, however, have an exceptionally small volume (Fig. 18) indicating that the Ca^{2+} concentration there can actually rise substantially. The calculations for determining the Ca^{2+} concentration in the microvilli were done by splitting the light-activated current during an average quantum bump into the components carried by a single ion-type as described above. The flux of each ion species, being the movement of ions across the membrane of the microvilli, was combined with a one-dimensional diffusion equation that describes the movement of ions along the length of the microvilli. The ion concentrations in the cell body, as well as that on the extracellular side of the membrane were assumed to remain constant. Also the change in the permeability was taken to be the same along the length of the microvilli. However, due to diffusion, the intracellular ion concentrations exhibit strong spatial gradients within

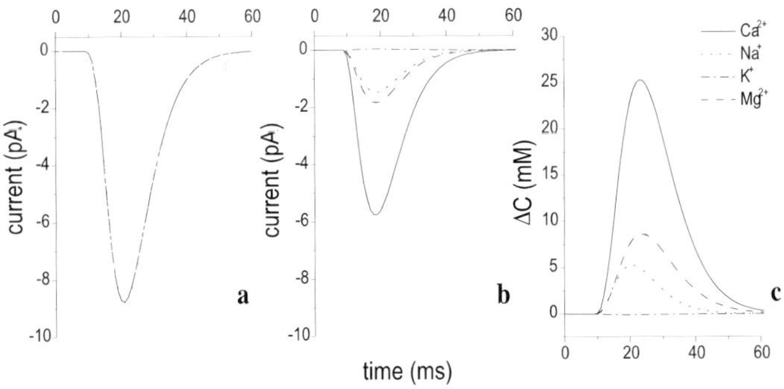

Fig. 17. Calculating the changes of the ion concentration inside a microvillus during a quantum bump. (a) The average of several quantum bumps (circles) was fitted by a gamma function (line) that was subsequently used as input for the model. (b) Using the GHK equation, the total current (a) was split into the components carried by one of the four physiologically relevant cations (b). (c) Combining the currents from b with the one-dimensional diffusion model for a single microvillus yields the changes of the ion concentrations, ΔC.

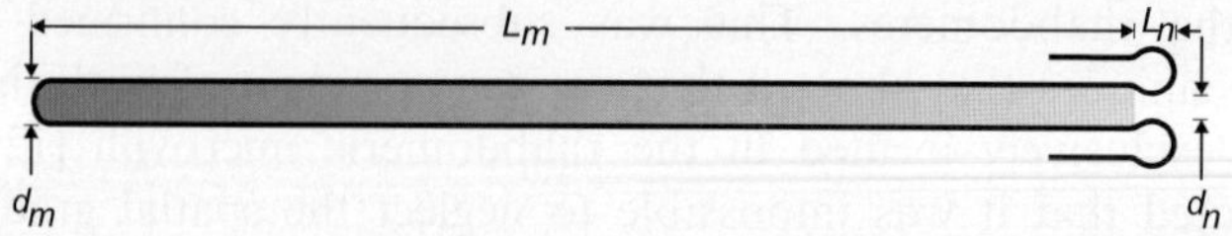

Fig. 18. Diagram of a single *Drosophila* microvillus. A microvillus is a tube-like protrusion from the photoreceptor cell body, with average length $L_{\mathrm{m}} = 1.5\ \mu\mathrm{m}$ and diameter $d_{\mathrm{m}} = 0.06\ \mu\mathrm{m}$, connected through a narrow neck to the cell soma. Upon light activation, cation ion channels open, giving rise, together with diffusion of ions through the neck, to large ion gradients along the length of the microvillus (from [136]).

the microvilli (Fig. 18). In the derivation of the partial currents for each ion type these gradients have to be taken into account. Equation (49) therefore needs to be modified into:

$$f_q = \frac{1}{S_{\mathrm{m}}} \int_{S_{\mathrm{m}}} \beta_q V_{\mathrm{m}} \frac{C_{q,i} - C_{q,\mathrm{o}} \mathrm{e}^{-\beta_q V_{\mathrm{m}}}}{1 - \mathrm{e}^{-\beta_q V_{\mathrm{m}}}}\, \mathrm{d}S \tag{50}$$

The current for each ion type can then again be calculated with Eq. (48). Fig. 17(b) shows the resulting partial currents for each ion type. These calculations indicate that more than 60% of the total current is carried by Ca^{2+} ions. From the currents, the changes of the ion concentrations inside the microvilli can be calculated for very small time steps. For longer periods of time, it is necessary to consider the diffusion of ions along the length of the microvillus and from the microvillus into the cell body. This can be implemented with a diffusion equation adapted to the geometry of a microvillus (Fig. 18):

$$\frac{\partial C_{q,i}(x)}{\partial t} = -\frac{2}{z_q F r_{\mathrm{m}}} i_q(x) + D_q \nabla^2 C_{q,i}(x) \tag{51}$$

where r_{m} is the radius of the microvillus, D_q is the diffusion coefficient of ion type q, and $i_q(x)$ is the current density for ion type q at the position x along the length of the microvillus. The resulting ion concentration changes are depicted in Fig. 17(c). Assuming that all the current flows into a single microvillus, enormously high values for the intracellular Ca^{2+} concentration were found, reaching up to 23 mM [136].

It is however known that large amounts of Ca^{2+} binding proteins are located in the microvilli; most significantly, 0.5 mM calmodulin has been found in the rhabdomeres [137]. Equally, the fatty acids in the microvillar membrane might bind significant amounts of Ca^{2+}, albeit with low affinity [138]. However, including these low-affinity buffers does not significantly change the peak of the calculated Ca^{2+} concentration (Fig. 19(a)). This can be understood by considering that the amount of inflowing Ca^{2+} is exceeding by far any reasonable concentration of the high-affinity Ca^{2+} buffers, resulting in a complete saturation of these buffers (Fig. 19(b)). The Ca^{2+} concentration then continuous to rise, until Ca^{2+} influx equals the diffusion of ions from the microvillus into the cell body (Fig. 19(c)).

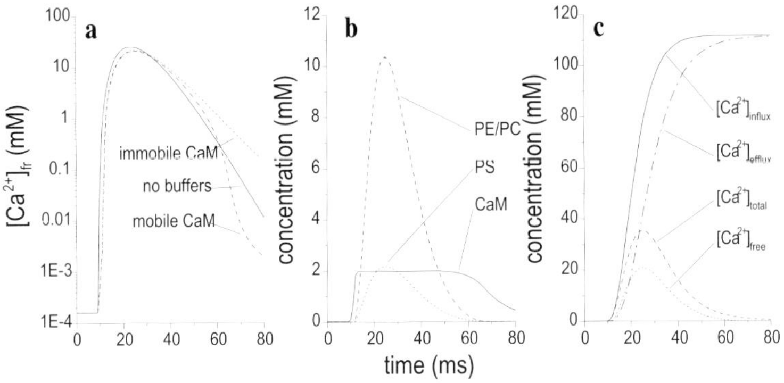

Fig. 19. The influence of Ca^{2+} buffers on the Ca^{2+} concentration in the microvillus. (a) Adding 0.5 mM calmodulin, either as immobile buffer or as mobile buffer does only marginally reduce the peak Ca^{2+} concentration. However, the time course with which the Ca^{2+} concentration diminishes changes strongly. Except for calmodulin, also the Ca^{2+} buffering effects of the phospholipids phosphatidylserine (PS), phosphatidylethanolamine (PE) and phosphatidylcholine (PC) were taken into account. (b) The amount of Ca^{2+} bound to the various buffers is shown during a quantum bump. While the high-affinity buffer calmodulin is immediately saturated, this does not happen with the low-affinity lipid Ca^{2+} buffers. (c) The Ca^{2+} balance during a quantum bump. The curves were obtained by integrating the amount of Ca^{2+} flowing in and the amount Ca^{2+} leaving the microvillus by diffusion into the cell body. The difference between the two curves is the time course of the total Ca^{2+} concentration. Subtracting the Ca^{2+} bound to buffers from the total Ca^{2+} concentration gives the free Ca^{2+} concentration.

Possibly, however, the assumption that all the current of a single bump flows across the membrane of a single microvillus, does not hold. When assuming that the current is distributed along several or even many microvilli, the peak Ca^{2+} concentration in the microvilli is reduced accordingly (Fig. 20). When the current flows into more than seven microvilli, the Ca^{2+} buffering molecules present there can efficiently reduce the Ca^{2+} concentration. It is, however, difficult to envisage how the signal originating from a single rhodopsin molecule can spread across several membranes to more than seven microvilli in the short time between absorption and the quantum bump, which can be as short as 10 ms [135].

Recent measurements of the Ca^{2+} concentration in the microvilli with fluorescent Ca^{2+} dyes indicate that the Ca^{2+} concentration rises indeed to very high values. Although the membrane potential was not clamped to the resting value (as in the quantum bump measurements), Ca^{2+} concentrations up to 0.6 mM have been measured [104].

In a recent study, Henderson et al. [135] found that the shape of the quantum bumps changed considerably when the extracellular Ca^{2+} concentration was reduced (i.e. from 1.5 to 0.05–0.2 mM), or when the intracellular Ca^{2+} buffer capacity was increased by adding high concentrations of fast Ca^{2+} chelators. Henderson et al. [135] concluded that during the early phase of Ca^{2+} influx, Ca^{2+} exerts a positive feedback on the phototransduction cascade and/or the light-activated

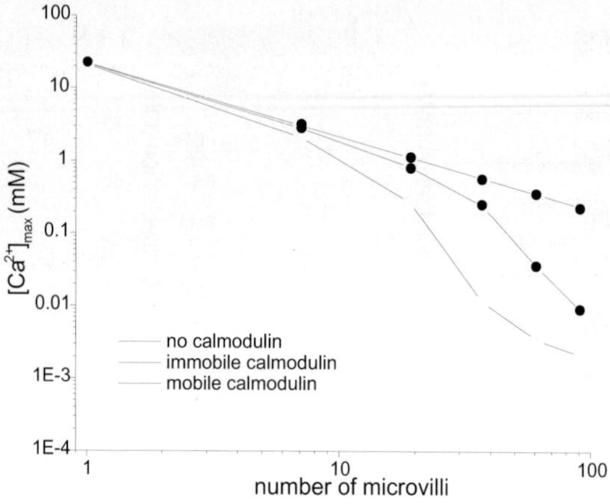

Fig. 20. The peak concentration of the Ca^{2+} concentration during a quantum bump reduces when the current is distributed over several microvilli. When assuming more than seven microvilli, high affinity Ca^{2+} buffers have a strong effect (see [136] for details).

channels, causing a rapid rise of the current. In the subsequent phase, the increased Ca^{2+} concentration down-regulates the light response. While similar results were already obtained in an earlier study, where the influx of rapidly uncaged Ca^{2+} on the light response was investigated [67], this study established that the high Ca^{2+} concentrations reached in the microvilli during even single quantum bumps have an important role in shaping the quantum bump.

8.2. *Modeling the Ca^{2+} diffusion in the cell body*

Ca^{2+} ions entering the rhabdomere through light-activated channels subsequently diffuse into the cell body. This process can be followed in cells injected with fluorescent Ca^{2+} indicators, in patch-clamped photoreceptor cells [132] as well as *in vivo* (Fig. 21(a); [139]). As expected, at the onset of light stimulation, the increase of the Ca^{2+} concentration in the cell body compared to that in the rhabdomere is much slower, and considerably lower values are reached (Figs. 21(c) and (d); [139]). Under sustained illumination (>0.5–1 s) however, the Ca^{2+} concentrations in the rhabdomere and the cell body have been found to be identical and to change only very slowly (Figs. 21(c) and (d); [139]).

Although the membrane potential is stable under this condition, the light-activated channels still are open and conduct a fairly strong inward current, that is still carried for ca. 50% by Ca^{2+} (according to the GHK equation). This inward current opposes the non-inactivating currents through the voltage-gated K^+ channels. As the Ca^{2+} concentration (in the rhabdomere and the cell body) changes only very slowly, this means that (most of) the Ca^{2+} flowing into the cell is also extruded, most probably by the strong currents produced by the Na^+/Ca^{2+} exchanger molecules [67,109,140]. Consequently, there must be a net diffusional flux of Ca^{2+} ions from the light-activated channels (located in the rhabdomere) to the place where the Ca^{2+}

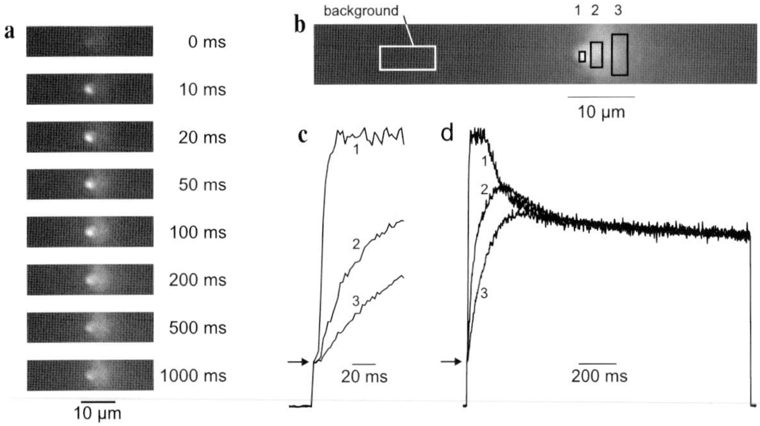

Fig. 21. Imaging the Ca^{2+} induced fluorescence of a Ca^{2+} indicator in a blowfly photoreceptor *in vivo*. (a) Already 10 ms after the light has been turned on, the rhabdomere shines up strongly, indicating that the Ca^{2+} concentration there has risen substantially. For longer illumination times, the Ca^{2+} spreads into the cell body and diminishes in the rhabdomere. (b) For quantitative analysis, three region of interests were defined, the first one being located in the rhabdomere. In these areas, the fluorescence intensity was averaged. Subsequently, the fluorescence intensity from the background was subtracted, and the resulting traces were normalized to the first measured fluorescence value after turning-on the light (arrows in (c) and d). (c) and (d) Time course of the normalized fluorescence in the three regions of interest. In the rhabdomere, the Ca^{2+} concentration displays a very rapid and large transient. After this transient has ceased, the normalized fluorescence – and hence the Ca^{2+} concentration – in all three regions is identical (from [139]).

extruding proteins are located. Thus, a gradient in the Ca^{2+} concentration will arise, the size of which depends on the location of the Ca^{2+} extrusion and the amount of mobile Ca^{2+} buffer present.

When the Ca^{2+} influx is homogeneous along the length of the rhabdomere, the size of the expected Ca^{2+} gradients can be calculated for the (two-dimensional) cross-section of the photoreceptor cell (that has for simplicity been assumed to be square [139]). When no mobile Ca^{2+} buffer is present and Ca^{2+} extrusion does not take place in the rhabdomere very large Ca^{2+} gradients are calculated (Fig. 22(a)–(c)). When the Ca^{2+} extrusion is located in the rhabdomere, a flat Ca^{2+} distribution is calculated (Fig. 22(d)).

However, the assumption of no mobile Ca^{2+} buffer is unrealistic, at least under the experimental conditions, because the added fluorescent Ca^{2+} indicator is a highly mobile Ca^{2+} buffer. These Ca^{2+} buffers are known to strongly reduce the Ca^{2+} gradients, as the net Ca^{2+} transport is then also accomplished by Ca^{2+} bound to the buffer [141,142]. Fig. 23 shows how different concentrations of Ca^{2+} buffers affect the expected gradients. The result from these calculations is that measurable Ca^{2+} gradients can be expected as long as the concentration of highly mobile Ca^{2+} buffers is lower than 700 μM. We have argued that this condition is met and therefore concluded that Ca^{2+} extrusion takes place close to the Ca^{2+} influx [139].

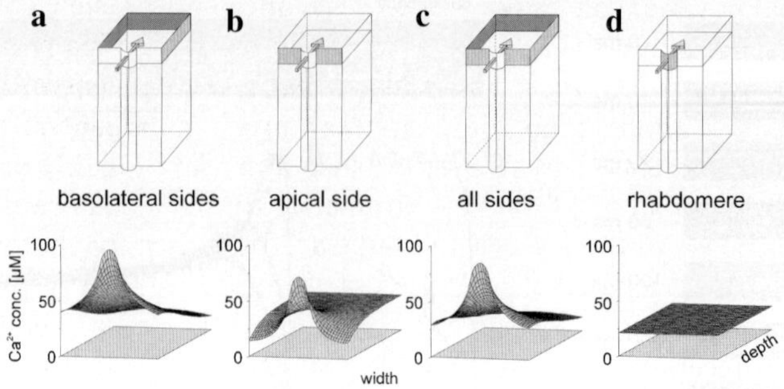

Fig. 22. Modeling the steady state distribution of Ca^{2+} for strongly light-adapted cells. A constant Ca^{2+} influx was assumed to be homogeneously distributed along the length of the rhabdomere. The Ca^{2+} extrusion was assumed to be located at: (a) the basolateral membranes, (b) the apical membranes (excluding the rhabdomere), (c) all sides (excluding the rhabdomere), and (d) only the rhabdomere. No mobile Ca^{2+} buffer was taken into account. Only in d there is no Ca^{2+} gradient in the steady state (from [139]).

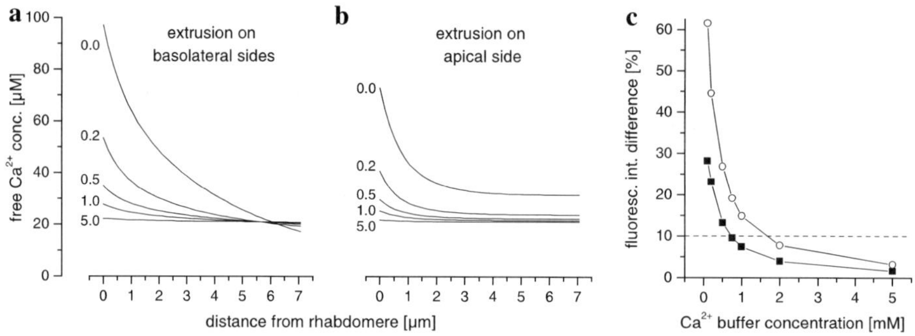

Fig. 23. The influence of mobile Ca^{2+} buffers on the steady state Ca^{2+} gradients. The calculations of Fig. 22 were repeated and the concentration of a mobile Ca^{2+} buffer (diffusion coefficient 220 $\mu m^2\ s^{-1}$) was varied. (a) and (b) show the Ca^{2+} concentration along a line from the rhabdomere to the opposite cell membrane. The numbers indicate the concentration of the Ca^{2+} buffer in mM. (c) From the curves in (a) and (b) the expected fluorescence intensity difference between the rhabdomere and a location close to the basolateral cell membrane (at 7 µm in (a) and (b) can be calculated. The expected fluorescence intensity is plotted as a function of the Ca^{2+} buffer concentration for the case where the extrusion was assumed on the basolateral sides (○) and on the apical sides (■). If the Ca^{2+} buffer concentration is lower than 700 µM, a measurable fluorescence intensity difference (i.e. larger than 10%) can be expected (from [139]).

8.3. Is an integrated model of the ion fluxes in photoreceptor cells already feasible?

So far, the described modeling efforts have all been deliberately limited to the study of specific questions and situations, e.g. only the microvilli and the light-response to a single photon [136] or the steady-state distribution of Ca^{2+} [139] were considered. However, under real-life conditions, the photoreceptor cells of flies are exposed to light stimuli that cover a wide range of intensities and are extremely dynamic. As the Ca^{2+} concentration plays a very important role in regulating many of the proteins involved in phototransduction, it seems likely that a comprehensive understanding of the light response and its regulation requires a much more general and integrated model than those discussed here. However, such a model probably requires a more detailed knowledge of some of the key-parameters than what is presently available. Most importantly, very little is known about Ca^{2+} buffers in fly photoreceptor cells; while Hardie [101] concluded that 99.8% of the inflowing Ca^{2+} was buffered in *Drosophila*, Oberwinkler and Stavenga [103] found much less Ca^{2+} buffer capacity in *Calliphora* photoreceptors. Furthermore, the Ca^{2+} buffering molecules might not be homogeneously distributed in the cell (as has been already shown for calmodulin [137,143]). For Ca^{2+} binding molecules other than calmodulin, the Ca^{2+}-affinities and the diffusion coefficients are not known. When modeling non-steady state conditions, these parameters can strongly influence the resulting Ca^{2+} distribution.

While it is known that the Na^+/Ca^{2+} exchanger can be directly activated by augmenting the intracellular Ca^{2+} concentration [67], its affinity for Ca^{2+} and its regulation by other ion concentrations or intracellular signaling processes are not yet characterized. Interestingly enough, two types of Na^+/Ca^{2+} exchangers have been found to be expressed by *Drosophila* photoreceptor cells [144,145]. From homology with the vertebrate cardiac Na^+/Ca^{2+} exchanger, the first type of exchanger is expected to transport 3 Na^+ ions for each Ca^{2+} ions, while the second type is dependent on K^+ [145]. As it is homologous to the vertebrate rod Na^+/Ca^{2+}–K^+ exchanger it might have the same coupling ratio, 4 Na^+ ions against 1 Ca^{2+} ion and 1 K^+ ion [146]. It will be a task for the future to determine the relative importance of these two exchanger types for the Ca^{2+} extrusion in fly photoreceptor cells.

9. Quantitative aspects of the early steps of the phototransduction cascade

When a rhodopsin is converted to metarhodopsin, the signal is transported to the light-activated ion channels by a molecular cascade that comprises a G_q-protein and a PLC (see Chapter 9 [11]). Only the first steps of this transduction cascadc arc presently amenable to a quantitative analysis, as the signal transduction from the PLC to the ion-channels is still controversially debated. Scott et al. [147] observed that severely reducing the amount of the α-subunit of the G_q-protein did not influence the shape and the size of the bump; equally, reducing the number of PLC molecules did not change the quantum bumps [148]. From these studies, it was concluded that a single activated G_q-protein activating a single PLC molecule is sufficient to produce a full sized quantum bump in *Drosophila* photoreceptor cells. This interpretation is strengthened by the fact that the shape and the size of quantum bumps are mainly determined by the various feedbacks that the inflowing Ca^{2+} exerts [135]. The mutants with severely reduced G_q-protein content, however, have a greatly reduced sensitivity, and the mutants with reduced PLC content have greatly increased latency times [148]. Scott and Zuker [148] integrated these data in a qualitative model in which the activated metarhodopsin becomes quickly inactivated. When the G_q-protein content is very low, this happens before a G_q-protein has had a chance of getting in contact with the rhodopsin. However, once a G_q-protein is activated, it stays active until it has reached a PLC molecule, which takes longer when the PLC molecules are rare.

Recently, strong evidence in favor of this model was produced by Cook et al. [149], who showed that in *Drosophila* G_q-proteins preferentially cleave the bound GTP (and thus become inactivated) when they are bound to a PLC molecule, similar to what has been shown for vertebrate PLC molecules [150]. In *Drosophila* mutants with severely reduced PLC content, the light response terminates only very slowly compared to wild-type flies when many active G_q-proteins are produced through strong light stimulation. In accordance with the proposed model this can be explained by assuming that the large number of activated G_q-proteins becomes inactivated at a rate that is proportional to the amount of available PLC molecules.

The possibility to manipulate the system through a large variety of mutants opens the way for a more quantitative analysis; e.g. it should be possible to determine the diffusional speed of the α-subunit of the G_q-protein and the time a single G_q-protein is bound to the PLC, which might be equivalent to the duration of quantum bump production that a single PLC molecule can accomplish. Equally, it might be possible to determine the average lifetime of an activated metarhodopsin molecule before it becomes inactivated by binding to an arrestin molecule [151] and the average time a G_q-protein needs to bind to an active metarhodopsin. The only presently available estimate for this value comes from the flash-induced responses of blowfly photoreceptors with a reduced rhodopsin content that were cooled to 10°C. Under these conditions, negligible activation of G_q-protein by metarhodopsin occurs within 3 to 5 ms [152].

References

1. Hardie, R.C. (1996) Curr. Biol. **6**, 1371–1373.
2. Montell, C. (1999) Ann. Rev. Cell Dev. Biol. **15**, 231–268.
3. DeGrip, W.J. and Rothschild, K.J. (2000) in: Hand. Biol. Phys. Vol. 3, eds D.G. Stavenga, W.J. DeGrip and E.N. Pugh, Jr. pp. 1–54, Elsevier, Amsterdam.
4. Mathies, R.A. and Lugtenburg, J. (2000) in: Hand. Biol. Phys. Vol. 3, eds D.G. Stavenga, W.J. DeGrip and E.N. Pugh, Jr. pp. 55–90, Elsevier, Amsterdam.
5. Hofmann, K.P. (2000) in: Hand. Biol. Phys. Vol. 3, eds D.G. Stavenga, W.J. DeGrip and E.N. Pugh, Jr. pp. 91–142, Elsevier, Amsterdam.
6. Molday, R.S. and Kaupp, U. (2000) in: Hand. Biol. Phys. Vol. 3, eds D.G. Stavenga, W.J. DeGrip and E.N. Pugh, Jr. pp. 143–181, Elsevier, Amsterdam.
7. Pugh, E.N. Jr, and Lamb, T.D. (2000) in: Hand. Biol. Phys. Vol. 3, eds D.G. Stavenga, W.J. DeGrip and E.N. Pugh, Jr. pp. 183–255, Elsevier, Amsterdam.
8. Yokoyama, S. and Yokoyama, R. (2000) in: Hand. Biol. Phys. Vol. 3, eds D.G. Stavenga, W.J. DeGrip and E.N. Pugh, Jr. pp. 257–296, Elsevier, Amsterdam.
9. Gaertner, W. (2000) in: Hand. Biol. Phys. Vol. 3, eds D.G. Stavenga, W.J. DeGrip and E.N. Pugh, Jr. pp. 297–388, Elsevier, Amsterdam.
10. Nasi, E., Pilar Gomez, M.del and Payne, R. (2000) in: Hand. Biol. Phys. Vol. 3, eds D.G. Stavenga, W.J. DeGrip and E.N. Pugh, Jr. pp. 389–448, Elsevier, Amsterdam.
11. Minke, B. and Hardie, R.C. (2000) in: Hand. Biol. Phys. Vol. 3, eds D.G. Stavenga, W.J. DeGrip and E.N. Pugh, Jr. pp. 449–525, Elsevier, Amsterdam.

12. Vogt, K. (1989) in: Facets of Vision, eds D.G. Stavenga and R.C. Hardie. pp. 134–151, Springer, Berlin.
13. Seki, T. and Vogt, K. (1998) Comp. Biochem. Physiol. **119B**, 53–64.
14. Perman, B., Srajer, V., Ren, Z., Teng, T., Pradervand, C., Ursby, T., Bourgeois, D., Schotte, F., Wulff, M., Kort, R., Hellingwerf, K. and Moffat, K. (1998) Science **279**, 1946–1950.
15. Goodeve, C.F., Lythgoe, R.J. and Schneider, E.E. (1942) Proc. R. Soc. London B **130**, 380–395.
16. Morton, R.A. (1972) in: Hand. Sens. Physiol. Vol. VII/1, ed H.J.A. Dartnall. pp. 33–68, Springer, Berlin.
17. Goldsmith, T.H. and Fernandez, H.R. (1968) J. Exp. Biol. **49**, 669–677.
18. Hamdorf, K. (1979) in: Hand. Sens. Physiol. Vol. VII/6A, ed H. Autrum. pp. 145–224, Springer, Berlin.
19. Marshall, J. and Oberwinkler, J. (1999) Nature **401**, 873–874.
20. Rodieck, R.W. (1973) The Vertebrate Retina. Principles of Structure and Function. Freeman, San Francisco.
21. Dartnall, H.J.A. (1953) Brit. Med Bull. **9**, 24–30.
22. Liebman, P. (1972) in: Hand. Sens. Physiol. Vol. VII/1, ed H.J.A. Dartnall. pp. 481–528, Springer, Berlin.
23. Bridges, C.D.B. (1972) in: Hand. Sens. Physiol. Vol. VII/1, ed H.J.A. Dartnall. pp. 417–480, Springer, Berlin.
24. Harósi, F. (1976) J. Gen. Physiol. **68**, 65–80.
25. Ebrey, T.G. and Honig, B. (1977) Vision Res. **17**, 147–151.
26. Metzler, D.E. and Harris, C.M. (1978) Vision Res. **18**, 1417–1420.
27. Dawis, S.M. (1981) Vision Res. **21**, 1427–1430.
28. Barlow, H.B. (1982) Vision Res. **22**, 635–643.
29. Maximov, V.V. (1988) Sensornye Sistemy **2**, 3–9.
30. Mansfield, R.J.W. (1985) in: The Visual System, eds A. Fein and J.S. Levine. pp. 89–106, Liss, New York.
31. MacNichol, E.F., Jr (1986) Vision Res. **26**, 1543–1556.
32. Partridge, J.C. and DeGrip, W.J. (1991) Vision Res. **31**, 619–630.
33. Stavenga, D.G., Smits, R.P. and Hoenders, B.J. (1993) Vision Res. **33**, 1011–1017.
34. Lamb, T.D. (1995) Vision Res. **35**, 3083–3091.
35. Harósi, F. (1994) Vision Res. **34**, 1359–1367.
36. Govardovskii, V.I., Fyhrquist, N., Reuter, T., Kuzmin, D.G. and Donner, K. (2000) Vis. Neurosci. in press.
37. Palacios, A.G., Goldsmith, T.H. and Bernard, G.D. (1996) Vis. Neurosci. **15**, 319–331.
38. Hochstein, S., Minke, B., Hillman, P. and Knight, B.W. (1978) Biol. Cybern. **30**, 23–32.
39. Stavenga, D.G. and Schwemer, J. (1984) in: Photoreception and Vision of Invertebrates, ed M.A. Ali. pp. 11–61, Plenum, New York.
40. Salcedo, E., Huber, A., Henrich, S., Chadwell, L.V., Chou, W.H., Paulsen, R. and Britt, S.G. (1999) J. Neurosci. **19**, 10716–10726.
41. Snyder, A.W., Menzel, R., eds (1975) Photoreceptor Optics. Springer, Berlin.
42. Snyder, A.W. (1979) in: Hand. Sens. Physiol. Vol. VII/6A, ed H. Autrum. pp. 225–313, Springer, Berlin.
43. Hateren, J.H. van (1989) in: Facets of Vision, eds D.G. Stavenga and R.C. Hardie. pp. 74–89, Springer, Berlin.
44. Nilsson, D.-E., Land, M. and Howard, J. (1988) J. Comp. Physiol. A **162**, 341–366.
45. Stavenga, D.G., Zantema, A. and Kuiper, J.W. (1973) in: Biochemistry and Physiology of Visual Pigments, ed H. Langer. pp. 175–180, Springer, Berlin.
46. Stavenga, D.G. (1989) in: Facets of Vision, eds D.G. Stavenga and R.C. Hardie. pp. 152–172, Springer, Berlin.
47. Stavenga, D.G. (1992) Trends Neurosci. **15**, 213–218.
48. Britt, S.G., Feiler, R., Kirschfeld, K. and Zuker, C.S. (1993) Neuron **11**, 29–39.

49. Yoshizawa, T. (1972) in: Hand. Sens. Physiol. Vol. VII/1, ed H.J.A. Dartnall. pp. 146–179, Springer, Berlin.
50. Hamdorf, K., Paulsen, R. and Schwemer, J. (1973) in: Biochemistry and Physiology of Visual Pigments, ed H. Langer. pp. 155–166, Springer, Berlin.
51. Williams, T.P. (1964) J. Gen. Physiol. **47**, 679–689.
52. Shichida, Y., Kobayashi, T., Ohtani, H. and Nagakura, S. (1978) Photochem. Photobiol. **27**, 335–341.
53. Kirschfeld, K., Feiler, R. and Minke, B. (1978) Z. Naturforsch. **33c**, 1009–1010.
54. Kruizinga, B., Kamman, R.L. and Stavenga, D.G. (1983) Biophys. Struct. Mech. **9**, 299–307.
55. Roebroek, J.G.H., Gagné, S. and Stavenga, D.G. (1989) J. Comp. Physiol. A **165**, 75–81.
56. Gagné, S., Roebroek, J.G.H. and Stavenga, D.G. (1989) Vision Res. **29**, 1670.
57. Doukas, A.G., Stefanic, V., Suzuki, T., Callender, R.H. and Alfano, R.R. (1980) Photobiochem. Photobiophys. **1**, 305–308.
58. Cone, R.A. and Pak, W.L. (1971) in: Hand. Sens. Physiol. Vol. I, ed W.R. Loewenstein. pp. 345–365, Springer, Berlin.
59. Hodgkin, A.L. and O'Bryan, P.M. (1977) J. Physiol. **267**, 737–766.
60. Pak, W.L. and Lidington, K.J. (1974) J. Gen. Physiol. **63**, 740–756.
61. Minke, B. and Kirschfeld, K. (1980) J. Gen. Physiol. **75**, 402.
62. Stephenson, R.S. and Pak, W.L. (1980) J. Gen. Physiol. **75**, 353–379.
63. Hateren, J.H. van (1986) J. Comp. Physiol. A **158**, 795–811.
64. Hardie, R.C., Franceschini, N., Ribi, W. and Kirschfeld, K. (1981) J. Comp. Physiol. A **145**, 139–152.
65. Rüppel, H. and Hagins, W.A. (1973) in: Biochemistry and Physiology of Visual Pigments, ed H. Langer. pp. 257–261, Springer, Berlin.
66. Lisman, J.E. and Bering, H. (1977) J. Gen. Physiol. **70**, 621–633.
67. Hardie, R.C. (1995) J. Neurosci. **15**, 889–902.
68. Gribakin, F. (1979) Int. Rev. Cytol. **57**, 127–184.
69. Warrant, E.J. and Nilsson, D.E. (1998) Vision Res. **38**, 195–207.
70. Hardie, R.C. (1985) in: Prog. Sens. Physiol. Vol. 5, ed D. Ottoson. pp. 1–79, Springer, Berlin.
71. Boschek, B. (1971) Z. Zellforsch. **118**, 369–409.
72. Tsukahara, Y. and Horridge, G.A. (1977) J. Comp. Physiol. A **114**, 233–251.
73. Stark, W.S., Ivanyshyn, A.M. and Greenberg, R.M. (1977) J. Comp. Physiol. A **121**, 289–305.
74. Hardie, R.C. (1986) Trends Neurosci. **9**, 419–423.
75. Vogt, K. and Kirschfeld, K. (1984) Naturwiss. **71**, 211–213.
76. Kirschfeld, K., Franceschini, N. and Minke, B. (1977) Nature **269**, 386–390.
77. Kirschfeld, K. (1986) in: The Molecular Mechanism of Photoreception, ed H. Stieve. pp. 31–49, Springer, Berlin.
78. Minke, B. and Kirschfeld, K. (1979) J. Gen. Physiol **73**, 517–540.
79. Hamdorf, K., Hochstrate, P., Höglund, G., Moser, M., Sperber, S. and Schlecht, P. (1992) J. Comp. Physiol. A **171**, 601–615.
80. Snyder, A.W. and Pask, C. (1973) J. Comp. Physiol. **84**, 59–76.
81. Goldsmith, T.H. (1965) J. Gen. Physiol. **49**, 265–287.
82. Vogt, K. and Kirschfeld, K. (1983) Biophys. Struct. Mech. **9**, 319–328.
83. Snyder, A.W., Menzel, R. and Laughlin, S.B. (1973) J. Comp. Physiol. **87**, 99–135.
84. Kirschfeld, K. (1982) Proc. R. Soc. London B Biol. Sci. **216**, 71–85.
85. Arikawa, K., Mizuno, S., Scholten, D.G., Kinoshita, M., Seki, T., Kitamoto, J. and Stavenga, D.G. (1999) Vision Res. **39**, 1–8.
86. Lunau, K. and Knüttel, H. (1995) Naturwiss. **82**, 432–434.
87. Arikawa, K. and Stavenga, D.G. (1997) J. Exp. Biol. **200**, 2501–2506.
88. Arikawa, K. (1999) in: Atlas of Arthropod Sensory Receptors, eds E. Eguchi and Y. Tominaga. pp. 23–32, Springer, Berlin.
89. Arikawa, K., Inokuma, K. and Eguchi, E. (1987) Naturwiss. **74**, 297–298.
90. Bandai, K., Arikawa, K. and Eguchi, E. (1992) J. Comp. Physiol. A **171**, 289–297.

91. Seki, T., Fujishita, S., Ito, M., Matsuoka, N. and Tsukida, K. (1987) Exp. Biol. **47**, 95–103.
92. Arikawa, K., Scholten, D.G., Kinoshita, M. and Stavenga, D.G. (1999) Zool. Sci. **16**, 17–24.
93. Ribi, W. (1978) Zool. Jb. Anat. **100**, 299–342.
94. Ribi, W. (1979) J. Comp. Physiol. A **132**, 1–9.
95. Franceschini, N. and Kirschfeld, K. (1976) Biol. Cybern. **21**, 181–203.
96. Hardie, R.C. (1979) J. Comp. Physiol. A **129**, 19–33.
97. Vogt, K., Kirschfeld, K. and Stavenga, D.G. (1982) J. Comp. Physiol. A **146**, 145–152.
98. Smakman, J.G., Hateren, J.H. van and Stavenga, D.G. (1986) Vision Res. **26**, 1019–1025.
99. Kirschfeld, K. and Vogt, K. (1980) Naturwiss. **67**, 516–517.
100. Howard, J. (1984) J. Exp. Biol. **113**, 471–475.
101. Hardie, R.C. (1996) J. Neurosci. **16**, 2924–2933.
102. Hardie, R.C. (1996) J. Photochem. Photobiol. B **35**, 83–89.
103. Oberwinkler, J. and Stavenga, D.G. (1998) J. Gen. Physiol. **112**, 113–124.
104. Oberwinkler, J. and Stavenga, D.G. (2000) J. Neurosci. **20**, 1701–1709.
105. Fain, G.L. and Lisman, J.E. (1981) Prog. Biophys. Molec. Biol. **37**, 91–147.
106. Fuortes, M.G.F. (1959) J. Physiol (Lond) **148**, 14–28.
107. Muijser, H. (1979) J. Comp. Physiol. **132**, 87–95.
108. Gerster, U., Stavenga, D.G. and Backhaus, W. (1997) J. Comp. Physiol. A **180**, 113–122.
109. Gerster, U. (1997) Vision Res. **37**, 2477–2485.
110. Hardie, R.C. and Mojet, M.H. (1995) J. Neurophysiol. **74**, 2590–2599.
111. Weckström, M., Hardie, R.C. and Laughlin, S.B. (1991) J. Physiol (Lond) **440**, 635–657.
112. Hardie, R.C., Voss, D., Pongs, O. and Laughlin, S.B. (1991) Neuron **6**, 477–486.
113. Hardie, R.C. (1991) J. Neurosci. **11**, 3079–3095.
114. Laughlin, S.B. and Weckström, M. (1993) J. Comp. Physiol. A **172**, 593–609.
115. Anderson, J. and Hardie, R.C. (1996) J. Comp. Physiol. A **178**, 513–522.
116. Laughlin, S.B. and Hardie, R.C. (1978) J. Comp. Physiol. **128**, 319–340.
117. Coles, J.A. (1986) Prog. Sens. Physiol. **6**, 105–138.
118. Ebbinghaus-Kintscher, U. (1994) Thesis, Ruhr University, Bochum.
119. Sandler, C. and Kirschfeld, K. (1992) J. Comp. Physiol. A **171**, 573–581.
120. Armon, E. and Minke, B. (1983) Biophys. Struct. Mech. **9**, 349–357.
121. Baumann, O., Lautenschläger, B. and Takeyasu, K. (1994) Cell Tissue Res. **275**, 225–234.
122. Koike, H., Brown, H.M. and Hagiwara, S. (1971) J. Gen. Physiol. **57**, 723–737.
123. Jansonius, N.M. (1990) J. Comp. Physiol. A **167**, 461–468.
124. Hardie, R.C. (1991) Proc. R. Soc. London B Biol. Sci. **245**, 203–210.
125. Ranganathan, R., Harris, G.L., Stevens, C.F. and Zuker, C.S. (1991) Nature **354**, 230–232.
126. Phillips, A.M., Bull, A. and Kelly, L.E. (1992) Neuron **8**, 631–642.
127. Niemeyer, B.A., Suzuki, E., Scott, K., Jalink, K. and Zuker, C.S. (1996) Cell **85**, 651–659.
128. Hardie, R.C. and Minke, B. (1992) Neuron **8**, 643–651.
129. Hardie, R.C., Reuss, H., Lansdell, S.J. and Millar, N.S. (1997) Cell Calcium **21**, 431–440.
130. Reuss, H., Mojet, M.H., Chyb, S. and Hardie, R.C. (1997) Neuron **19**, 1249–1259.
131. Hardie, R.C. (1995) J. Comp. Physiol. A **177**, 707–721.
132. Ranganathan, R., Bacskai, B.J., Tsien, R.Y. and Zuker, C.S. (1994) Neuron **13**, 837–848.
133. Huber, A., Sander, P., Gobert, A., Bahner, M., Hermann, R. and Paulsen, R. (1996) EMBO J. **15**, 7036–7045.
134. Hardie, R.C. and Minke, B. (1995) Cell Calcium **18**, 256–274.
135. Henderson, S.R., Reuss, H. and Hardie, R.C. (2000) J. Physiol (London) **524 Pt 1**, 179–194.
136. Postma, M., Oberwinkler, J. and Stavenga, D.G. (1999) Biophys. J. **77**, 1811–1823.
137. Porter, J.A., Yu, M., Doberstein, S.K., Pollard, T.D. and Montell, C. (1993) Science **262**, 1038–1042.
138. McLaughlin, S., Mulruine, T., Gresalfi, T., Vaio, G. and McLaughluin, A. (1981) J. Gen. Physiol. **77**, 475–487.
139. Oberwinkler, J. and Stavenga, D.G. (2000) Proc. Natl. Acad. Sci. USA **97**, 8578–8583.
140. Hochstrate, P. (1991) Z. Naturforsch. **44c**, 867–875.

141. Neher, E. and Augustine, G.J. (1992) J. Physiol. (Lond) **450**, 273–301.
142. Roberts, W.M. (1994) J. Neurosci. **14**, 3246–3262.
143. Porter, J.A., Hicks, J.L., Williams, D.S. and Montell, C. (1993) J. Cell Biol. **116**, 683–693.
144. Schwarz, E.M. and Benzer, S. (2000) Proc. Natl. Acad. Sci. USA **94**, 10249–10254.
145. Haug-Collet, K., Pearson, B., Webel, R., Szerencsei, R.T., Winkfein, R.J., Schnetkamp, P.P. and Colley, N.J. (1999) J. Cell Biol. **147**, 659–670.
146. Cervetto, L., Lagnado, L., Perry, R.J., Robinson, D.W. and McNaughton, P.A. (1989) Nature **337**, 740–743.
147. Scott, K., Sun, Y., Beckingham, K. and Zuker, C.S. (1997) Cell **91**, 375–383.
148. Scott, K. and Zuker, C.S. (1998) Nature **395**, 805–808.
149. Cook, B., Bar-Yaacov, M., Cohen Ben-Ami, H., Goldstein, R.E., Paroush, Z., Selinger, Z. and Minke, B. (2000) Nat. Cell Biol. **2**, 296–301.
150. Bernstein, G., Blank, J.L., Jhon, D.Y., Exton, J.H., Rhee, S.G. and Ross, E.M. (1992) Cell **70**, 411–418.
151. Alloway, P.G. and Dolph, P.J. (1999) Proc. Natl. Acad. Sci. USA **96**, 6072–6077.
152. Hamdorf, K. and Kirschfeld, K. (1980) Nature **283**, 859–860.

Subject Index